Advanced Color Image Processing and Analysis

Christine Fernandez-Maloigne
Editor

Advanced Color Image Processing and Analysis

Springer

Editor
Christine Fernandez-Maloigne
Xlim-SIC Laboratory
University of Poitiers
11 Bd Marie et Pierre Curie
Futuroscope
France

ISBN 978-1-4419-6189-1 ISBN 978-1-4419-6190-7 (eBook)
DOI 10.1007/978-1-4419-6190-7
Springer New York Heidelberg Dordrecht London

Library of Congress Control Number: 2012939723

Printed on acid-free paper

Springer is part of Springer Science+Business Media (www.springer.com)

Preface

Color is life and life is color!

We live our life in colors and the nature that surrounds us offers them all, in all their nuances, including the colors of the rainbow. Colors inspire us to express our feelings. We can be "red in the face" or "purple with rage." We can feel "blue with cold" in winter or "green with envy," looking at our neighbors' new car. Or, are we perhaps the black sheep of our family? ....

Color has accompanied us through the mists of time. The history of colors is indissociable, on the cultural as well as the economic level, from the discovery of new pigments and new dyes. From four or five at the dawn of humanity, the number of dyes has increased to a few thousands today.

Aristotle ascribed color and light to Antiquity. At the time, there was another notion of the constitution of colors: perhaps influenced by the importance of luminosity in the Mediterranean countries, clearness and darkness were dominating concepts compared to hues. Elsewhere, colors were only classified by their luminosity as white and black. Hues were largely secondary and their role little exploited. It should be said that it was rather difficult at that time to obtain dyes offering saturated colors. During the Middle Ages, the prevalence of the perception of luminosity continued to influence the comprehension of color, and this generally became more complicated with the theological connotations and with the dual nature of light declining in Lumen, the source of light of divine origin (for example, solar light) and Lux, which acquires a more sensory and perceptual aspect like the light of a very close wood fire, which one can handle. This duality is included in the modern photometric units where lumen is the unit that describes the flow of the source of light and Lux is the unit of illumination received by a material surface. This design based on clearness, the notion taken up by the painters of the Renaissance as well under the term of value, continues to play a major role, in particular for graphic designers who are very attached to the concept of the contrast of luminosity for the harmony of colors. In this philosophy, there are only two primary colors, white and black, and the other colors can only be quite precise mixtures of white and black. We can now measure the distance that separates our perception from that of the olden times.

Each color carries its own signature, its own vibration... its own universal language built over millennia! The Egyptians of Antiquity gave to the principal colors a symbolic value system resulting from the perception they had of natural phenomena in correlation with these colors: the yellow of the sun, the green of the vegetation, the black of the fertile ground, the blue of the sky, and the red of the desert. For religious paintings, the priests generally authorized only a limited number of colors: white, black, the three basic colors (red, yellow and blue), or their combinations (green, brown, pink and gray). Ever since, the language of color has made its way through time, and today therapeutic techniques use colors to convey this universal language to the unconscious, to open doors to facilitate the cure.

In the scientific world, although the fundamental laws of physics were discovered in the 1930s, colorimetrics had to await the rise of data processing to be able to use the many matrix algebra applications that it implies.

In the numerical world, color is of vital importance, as it is necessary to code and to model, while respecting the basic phenomena of the perception of its appearance, as we recall in Chaps. 1 and 2. Then color is measured numerically (Chap. 3), moves from one peripheral to another (Chap. 4), is handled (Chaps. 5–7), to extract automatically discriminating information from the images and the videos (Chaps. 8–11) to allow an automatic analysis. It is also necessary to specifically protect this information, as we show in Chap. 12, to evaluate its quality, with the metrics and standardized protocols described in Chap. 13. It is with the two applications in which color is central, the field of art and the field of medicine, that we conclude this work (Chaps. 14 and 15), which has brought together authors from all the continents.

Whether looked at as a symbol of joy or of sorrow, single or combined, color is indeed a symbol of union! Thanks to it, I met many impassioned researchers from around the world who became my friends, who are like the members of a big family, rich in colors of skin, hair, eyes, landscapes, and emotions. Each chapter of this will deliver to you a part of the enigma of digital color imaging and, within filigree, the stories of all these rainbow meetings. Good reading!

Contents

Chapter 1
Fundamentals of Color

M. James Shyu and Jussi Parkkinen

The color is the glory of the light

Jean Guitton

Abstract Color is an important feature in visual information reaching the human eye or an artificial visual system. The color information is based on the electromagnetic (EM) radiation reflected, transmitted, or irradiated by an object to be observed. Distribution of this radiation intensity is represented as a wavelength spectrum. In the standard approach, color is seen as human sensation to this spectrum on the wavelength range 380–780 nm. A more general approach is to manage color as color information carried by the EM radiation. This modern approach is not restricted to the limitations of human vision. The color can be managed, not only in a traditional three-dimensional space like RGB or $L^*a^*b^*$ but also in an n-dimensional spectral space. In this chapter, we describe the basis for both approaches and discuss some fundamental questions in color science.

Keywords Color fundamentals • Color theory • History of color theory • Colorimetry • Advanced colorimetry • Electromagnetic radiation • Reflectance spectrum • Metamerism • Standard observer • Color representation • Color space • Spectral color space • n-dimensional spectral space • Color signal • Human vision • Color detection system

---

M.J. Shyu (✉)
Department of Information Communications, Chinese Culture University, Taipei, Taiwan
e-mail: mjshyu@faculty.pccu.edu.tw

J. Parkkinen
School of Computing, University of Eastern Finland, Joensuu, Finland

School of Engineering, Monash University Sunway Campus, Selangor, Malaysia
e-mail: jussi@monash.edu

C. Fernandez-Maloigne (ed.), *Advanced Color Image Processing and Analysis*,
DOI 10.1007/978-1-4419-6190-7_1,

1.1 Everything Starts with Light

The ability of human beings to perceive color is fantastic. Not only does it make it possible for us to see the world in a more vibrant way, but it also creates the wonder that we can express our emotions by using various colors. In Fig. 1.1, the colors on the wooden window are painted with the meaning of bringing prosperity. In a way, we see the wonderful world through the colors as a window. There are endless ways to use, to interpret, and even to process color with the versatility that is in the nature of color. However, to better handle the vocabulary of color, we need to understand its attributes first. How to process as well as analyze color images for specific purposes under various conditions is another important subject which further extends the wonder of color.

In the communication between humans, color is a fundamental property of objects. We learn different colors in our early childhood and this seems to be obvious for us. However, when we start to analyze color more accurately and, for example, want to measure color accurately, it is not so obvious anymore. For accurate color measurement, understanding, and management, we need to answer the question: What is color?

Fig. 1.1 A colorful window with the theme of bringing prosperity (Photographed by M. James Shyu in Pingtong, Taiwan)

In a common use of the term and as an attribute of an object, color is treated in many ways in human communication. Color has importance in many different disciplines and there are a number of views to the color: in biology, color vision and colorization of plants and animals; in psychology, color vision; in medicine, eye diseases and human vision; in art, color as an emotional experience; in physics, the signal carrying the color information and light matter interaction; in chemistry, the molecular structure and causes of color; in technology, different color measuring and display systems; in cultural, studies color naming; and in philosophy, color as an abstract entity related to objects through language [2, 9, 28].

It is said that there is no color in the light—to quote Sir Isaac Newton, "For the Rays to speak properly are not coloured. In them there is nothing else than a certain Power and Disposition to stir up a Sensation of this or that Colour" [21, 26]. It is the perception of human vision that generates the feeling of color. It is the perceived color feeling of the human vision defining how we receive the physical property of light. Nevertheless, if color is only defined by human vision, it leaves all other animals "color blind." However, it is known that many animals see colors and have an even richer color world than human being [13, 19].

The new technological development in illumination and in camera and display technology requires new way of managing colors. RGB or other three-dimensional color representations are not enough anymore. The light-emitting diodes (LED) are coming into illumination and displays rapidly. There, the color radiation spectrum is so peaky that managing it requires a more accurate color representation than RGB. There exist also digital cameras and displays, where colors are represented by four or six colors. Also this technology requires new ways to express and compute color values.

Therefore, if we want to understand color thoroughly and be able to manage color in all purposes, where it is used today, we cannot restrict ourselves to the human vision. We have to look color through the signal, which causes color sensation by humans. This signal we call color signal or color spectrum.

1.2 Development of Color Theory

In color vocabulary, black and white are the first words to be used as color names [2]. After them when the language develops, come red and yellow. The vocabulary is naturally related to the understanding of nature. Therefore in ancient times, the color names were related to the four basic elements of the world, water, air, fire, and earth [9]. In ancient times, the color theory was developed by philosophers like Plato and Aristotle. For the later development of color theory, it is notable that white was seen as a basic color. Also the color mixtures were taken into theories, but each basic color was considered to be a single and separate entity [14].

Also from the point of view of the revolution of color theory by Newton [20], it is interesting to note that Aristotle had a seven basic color scale, where colors crimson,

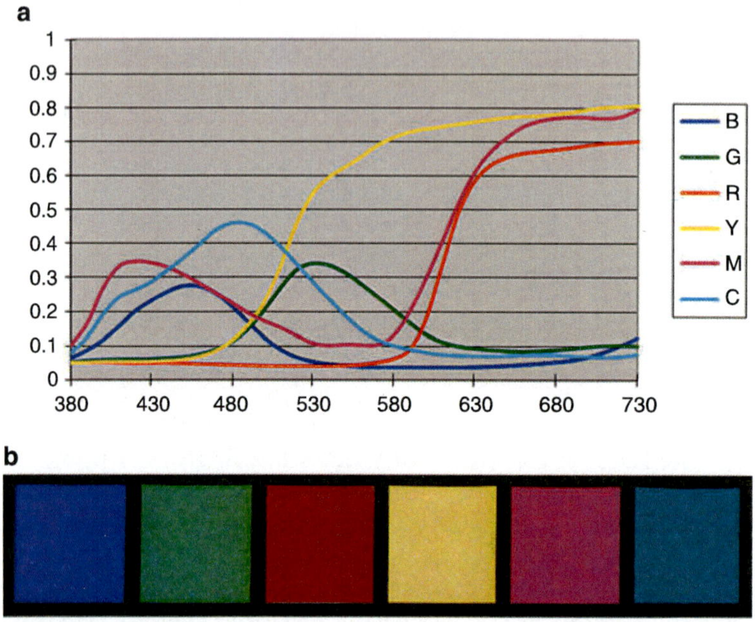

Fig. 1.2 (**a**) A set of color spectra (x-axis: wavelength from 380 to 730 nm, y-axis: reflectance factor) and (**b**) the corresponding colors

violet, leek-green, deep blue, and gray or yellow formed the color scale from black to white [9]. Aristotle also explains the color sensation so, that color sets the air in movement and that movement extends from object to the eye [24].

From these theories, one can see that already in ancient times, there exists the idea of some colors to be mixtures of primary colors and seven primary colors. Also, it is easy to understand the upcoming problems of Newton's description of colors, when the view was that each primary color is a single entity and the color sensation was seen as a kind of mechanical contact between light and the eye. The ancient way of thinking was strong until the Seventeenth century.

In the middle of the Seventeenth century, the collected information was enough to break the theory of ancient Greek about light and color. There were a number of experiments by prism and color in the early Seventeenth century. The credit for the discovery of the nature of light as a spectrum of wavelengths is given to Isaac Newton [20]. The idea that colors are formed as a combination of different component rays, which are immaterial by nature, was revolutionary at Newton's time. It broke the strong influence of ancient Greek thinking. This revolutionary idea was not easily accepted. A notable person was Johann Wolfgang von Goethe, who was still in the Nineteenth century opposing Newton's theory strongly [10].

Newton also presented colors in a color circle. In his idea, there were seven basic colors: violet, indigo, blue, green, yellow, orange, and red [14]. In the spectral approach to color as shown in Fig. 1.2, the wavelength scale is linear and continuing both ends, UV from short wavelengths and IR from long wavelengths. From the first look, the circle form is not natural for this physical signal. However, when the human perception of the different wavebands is considered, the circle form seems to be a good way to represent colors. The element, which connects the both ends of visible spectrum into a circle, is purple, which includes both red and violet part of spectrum.

The first to present colors in a circle form was the Finnish mathematician and astronomer Sigfrid Forsius in 1611 [14]. There are two different circle representation and both are based on idea to move from black to white through different color steps. Since Forsius and Newton, there are a number of presentations of colors on a circle. The circular form is used for the small number of basic colors. For continuous color tones, three-dimensional color coordinate systems form other shapes like cone (HSV) and cube (RGB).

Next important phase in the development of color science was the Nineteenth century. At that time, theories of human color vision were developed. In 1801, the English physicist and physician Thomas Young restated an earlier hypothesis by the English glassmaker George Palmer from the year 1777 [14]. According to these ideas, there are three different types of color-sensitive cells in the human retina. In their model, these cells are sensitive to red, green, and violet and to other colors which are mixtures of these principal pure colors. German physicist Hermann Helmholz studied this model further. He also provided the first estimates of spectral sensitivity curves for the retinal cells. This is known as the Young—Helmholz theory of color vision.

In the mid-Nineteenth century the Young—Helmholz theory was not fully accepted and in the mid-1870s German physician and physiologist Karl Hering presented his theory of human color vision [14]. His theory was based on four fundamental colors: red, yellow, green, and blue. This idea is the basis for the opponent color theory, where red—green and blue—yellow form opponent color pairs.

Both theories, the Young—Helmholz theory of color vision and the Hering opponent color theory, seemed to give a valid explanation to many observations about the human color vision. However, they were different even in the number of principal or fundamental colors. German physiologist Johannes von Kries proposed a solution to this confusion. He explained that the Young—Helmholz theory explained color vision on retinal color-sensitive cells level, and the Hering's opponent color theory was explaining color processes later in visual pathway [14]. This description was not accepted for some years, but currently it is seen as the basic view about the human color vision.

These ideas were bases for human color vision models, for the trichromatic color theories, and the standards of representing colors on a three-dimensional space. However, basing color representation and management on trichromatic theory of human color vision is very restrictive in many ways. Standard three-dimensional color coordinates are useful in many practical settings, where color is managed for humans to look at, especially under fixed illumination. However, there are also several drawbacks in the color representation based on human color vision.

Current level of measurement accuracy has led to a situation where in the equations for calculating color coordinates or color differences have become complicated. There are number of parameters, many without explaining the theory, but fitting the measurements to correspond the model. Furthermore, there are a number of issues, which cannot be managed by trichromatic color models. These include, e.g., fluorescence, metamerism, animal color vision, and transfer of accurate color information. To overcome these drawbacks, spectral color science has increased interest and is used more and more in color science.

As mentioned above, the basis of color is light, a physical signal of electromagnetic radiation. This radiation is detected by some detection system. If the system is human vision, then we consider traditional color. If we do not restrict the detection system, we consider the physical signal, color spectrum. Kuehni separates these approaches into color and spectral color.

1.3 Physical Attributes of Color

The color of an object can be defined as an object's physical attribute or as an object's attribute as humans see it. The first one is measurable attribute, but what humans see we cannot measure, since it happens in the human brain. In both definitions, the color information is carried to the color detector in the form of electromagnetic radiation. If the detector is human eye, seeing color of an object is based on how human eye senses the electromagnetic signal reaching the eye and how this sensory information is forwarded to the brain. In the artificial color vision systems, the signal reaches the detector and the detector response is related to the wavelength sensitivity of detector. This detected sensitivity information can then be managed the way the system requires.

The detector response D_i of ith detector to the color signal $l(\lambda)r(\lambda)$ is given as

$$D_i = \int l(\lambda)r(\lambda)s_i(\lambda)d\lambda, i = 1,n \tag{1.1}$$

where $l(\lambda)$ is the spectrum of illumination, $r(\lambda)$ is the reflectance spectrum of the object, $s_i(\lambda)$ is the sensitivity of the ith detector, and n is the number of detectors.

If the detector system has only one detector, it sees only intensity differences and not colors. Or we can also say that the detector sees only intensities of one color, i.e., color corresponding to the sensitivity $s(\lambda)$. For color sensation, at least two detectors w ith different wavelength sensitivities are needed ($n \geq 2$). The ratio of these different detector responses gives the color information. In the human eye, there are three types of wavelength-sensitive cone-cells ($n = 3$). These cells collect the color information from the incoming signal and human visual system converts it into color we see.

When we consider the color of an object, an essential part of color detection is the illumination. Since the color signal is originally light reflected (or radiated, or transmitted) from an object, the color of the illumination also affects to the detected object's color, term $l(\lambda)r(\lambda)$ in (1.1). A schematic drawing of detection of object color is shown in Fig. 1.3.

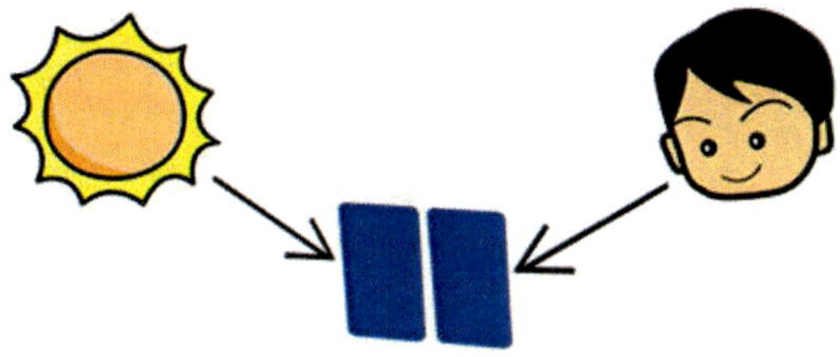

Fig. 1.3 The light source, color objects, and human visual system are needed to generate the perception of color

Here we have the approach that the color information is carried by the electromagnetic signal coming from the object and reaching the detector system. For this approach, we can set to the color signal certain assumptions.

Reflectance spectrum $r(\lambda)$ (or color spectrum $l(\lambda)r(\lambda)$) can be represented as a function r: $\Lambda \rightarrow$ R, which satisfies

$$\begin{aligned} &\text{(a)} && r(\lambda) \text{ is continuous on } \Lambda \\ &\text{(b)} && r(\lambda) \geq 0 \lambda \in \Lambda \qquad (1.2) \\ &\text{(c)} && \int |r(\lambda)|^2 d\lambda < \infty \end{aligned}$$

The proposition can be set due to the physical properties of the electromagnetic radiation. It means that reflectance (radiance or transmittance) spectra and color spectra can be thought as members of the square integrable function space, L^2. Since in practice the spectrum is formed as discrete measurements of the continuous signal, the spectra are represented as vectors in the space R^n. If spectra are represented in a low-dimensional space, they lose information, which causes problems like metamerism.

Using the vector space approach to the color, there are some questions to consider related to the color representation:

- What are the methods to manage color accurately?
- What is the actual dimensionality of color information?
- How to select the dimensions to represent color properly?

In the case of standard color coordinates, the dimensionality has been selected to be three. This is based on the models of the human color vision. Models are based on the assumption that there are three types of color sensitivity functions in the human retina.

In the spectral approach, originally the color signal is treated by using linear models [17, 18, 22, 23, 25]. The most popularly used and the standard method is the principal component analysis (PCA).

In this view, colors are represented as inner products between color spectrum and basis spectra of defined coordinate system. This approach unifies the ground of the different methods of the color representation and analysis. The basis spectra can be defined, e.g., by human response curves of three colors or by the interesting colors using some learning algorithm, depending on the needs and applications.

The use of inner products means seeing low-dimensional color representation as projection of original color signal onto a lower dimensional space. This leads to many theoretical approaches in the estimation of accurate color signal from the lower dimensional representation, like RGB. It is not possible to reconstruct the original color spectrum, from the RGB values of an object color. In theory, there is an infinite number of spectra, which produce the same RGB-value for spectra under fixed illumination conditions. However, if the original color spectra are from the certain limited region in the n-dimensional spectral space, a rather accurate reconstruction is possible to reach.

The considering of spectral color space as an n-dimensional vector space gives a basis for more general color theory to form. In this approach, human color vision and models based on that would be special cases. Theoretical frameworks, which have been studied as the basis for spectral color space, include, e.g., reproducing kernel Hilbert space [11, 23] and cylindrical spaces [16].

1.4 Standard Color Representation

In the case of human eye $n = 3$ in (1.1) and $s_i(\lambda)$'s are marked as $\bar{x}(\lambda), \bar{y}(\lambda), \bar{z}(\lambda)$ and called color matching functions [27]. This leads to the tristimulus values X, Y, and Z

$$
\begin{aligned}
\mathrm{X} &= k \int l(\lambda)\, r(\lambda) \bar{x}(\lambda)\, d\lambda \\
\mathrm{Y} &= k \int l(\lambda)\, r(\lambda) \bar{y}(\lambda)\, d\lambda \\
\mathrm{Z} &= k \int l(\lambda)\, r(\lambda) \bar{z}(\lambda)\, d\lambda \\
k &= 100 / \int l(\lambda) \bar{y}(\lambda)\, d\lambda
\end{aligned}
\tag{1.3}
$$

Moreover, three elements are involved for a human to perceive color on an object: light source, object, and observer. The physical property of the light source and the surface property of the object can be easily measured in their spectral power distribution with optical instruments. However, the observer's sensation of color cannot be measured directly by instruments since there is no place to gather a direct reading of perception. Equation (1.3) represents an implicit way to describe the human color perception in a numerical way which makes it possible to bring the human color perception into a quantitative form and to further compute or process it.

This implicit model to describe human color perception can be observed by the color-matching phenomena of two physically (spectrally) different objects which appear as the same color to the human eye, in the following equations:

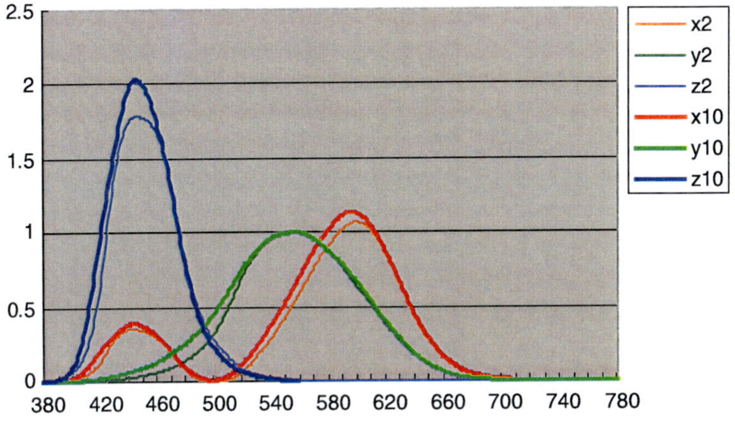

Fig. 1.4 Color matching functions for CIE standard observer in 2 and 10°-degree viewing angles

$$
\begin{aligned}
\int l(\lambda) r_1(\lambda) \bar{x}(\lambda) d\lambda &= \int l(\lambda) r_2(\lambda) \bar{x}(\lambda) d\lambda \\
\int l(\lambda) r_1(\lambda) \bar{y}(\lambda) d\lambda &= \int l(\lambda) r_2(\lambda) \bar{y}(\lambda) d\lambda \\
\int l(\lambda) r_1(\lambda) \bar{z}(\lambda) d\lambda &= \int l(\lambda) r_2(\lambda) \bar{z}(\lambda) d\lambda
\end{aligned}
\tag{1.4}
$$

Due to the integral operation in the equations, there can be two sets of different spectral reflectance of two objects that cause the equality to happen, i.e., make them appear as the same color. Furthermore, with the known (measurable) physical stimuli in the equations, if the unknown color-matching functions $(\bar{x}(\lambda), \bar{y}(\lambda), \bar{z}(\lambda))$ can be derived for the human visual system, it is possible to predict whether two objects of different spectral power distribution would appear as equal under this human visual color-matching model.

It was the Commission International de l'Eclairage (CIE) that in 1924 took the initiative to set up a Colorimetry Study Committee to coordinate the derivation of the color-matching functions [6]. Based on experimental color-mixture data and not on any particular theory of the color vision process, a set of color-matching functions for use in technical Colorimetry was first presented to the Colorimetry Committee at the 1931 CIE sessions [6]. This "1931 Standard Observer" as it was then called was based on observations made with colorimeters using field sizes subtending 2 degrees. In 1964, the CIE took a further step to standardizing a second set of color-matching functions as the "1964 Standard Observer" which used field sizes subtending 10 degrees. With these two sets of color-matching functions, shown in Fig. 1.4, it is possible to compute human color perception and subsequently open up promising research in the world of color science based on the model of human vision.

1.5 Metamerism

A property of color, which gives understanding about differences between human and spectral color vision approaches, is metamerism. Metamerism is a property, where two objects, which have different reflectance spectra, look the same under a certain illumination. As sensor responses, this is described in the form of (1.4). When the illumination changes, the object colors may look different.

The metamerism is a problem, e.g., in textile industry and in paper industry, if not taken care of. In paper industry, when a colored newspaper is used, a newspaper may be printed on papers produced on different days. If the required color is defined, e.g., by CIELAB coordinates and in the quality control only those values are monitored, color of different pages may look different under certain illumination although the pages appeared to have the same color under control illumination.

The metamerism is also used as a benefit. The most accurate way to reproduce the color of an object on the computer or TV screen would be the exact reconstruction of the original spectrum. This is not possible due to the limited number and shapes of the spectra of display primary colors. Therefore, a metameric spectrum of the original object is produced on the display and the object color looks to the human eye the same as the original color.

In the literature, metamerism is discussed mainly for the human visual system, but it can be generalized to any detection system with sufficient small number of detectors (Fig. 1.5). This means that

$$\int 1(\lambda)\mathrm{r}_1(\lambda)\mathrm{s}_\mathrm{i}(\lambda)\mathrm{d}\lambda = \int 1(\lambda)\mathrm{r}_2(\lambda)\mathrm{s}_\mathrm{i}(\lambda)\mathrm{d}\lambda \quad \text{for all} \quad \mathrm{i} \tag{1.5}$$

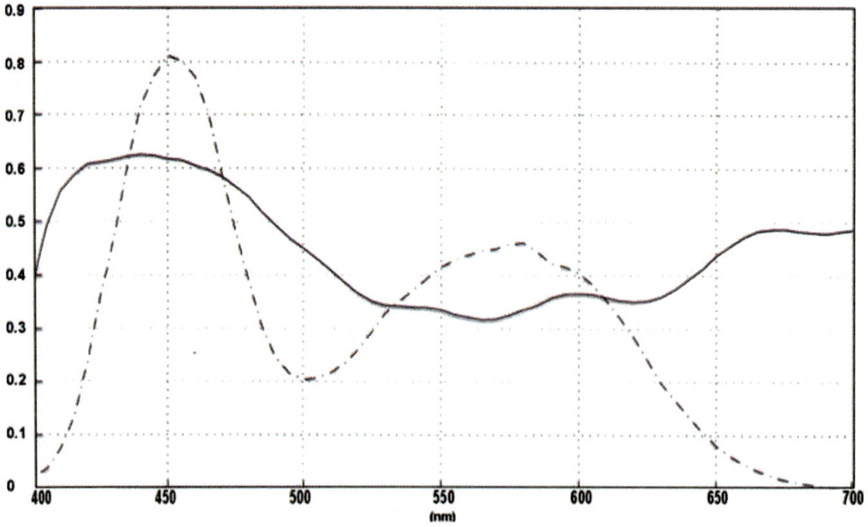

Fig. 1.5 Example of metamerism: two different reflectance curves from a metameric pair that could appear as the same color under specific illumination

Another aspect related to the color appearance under two different conditions is the color constancy. It is a phenomenon, where the observer considers the object color the same under different illuminations [3]. It means that the color is understood to be the same although the color signal reaching the eye is different under different illuminations. The color constancy can be seen as related to a color-naming problem [8]. The color constancy is considered in the Retinex theory, which is the basis, e.g., for the illumination change normalized color image analysis method [15].

In the color constancy, the background and the context, where the object is seen, are important for constant color appearance. If we look at a red paper under white illumination on a black background, it looks the same as that of a white paper under red illumination on a black background [8]. This implicit model to describe human color perception can be observed by the color-matching phenomena of two physically different objects that appear to be the same color to human eyes.

1.6 Measuring Physical Property or Perceptual Attribute of Color

The measurement of color can be done in various ways. In printing and publishing, the reflection densitometer has been used historically in prepress and pressroom operations for color quality control. ISO standard 5/3 for *Density Measurement—Spectral Conditions* defines a set of weightings indicating the standard spectral response for Status A, Status M, and Status T filters [1]. Reflectance density (D_R) is calculated from spectral reflectance according to the following equation:

$$D_R = -\log_{10}[\Sigma r(\lambda)\Pi(\lambda)/\Sigma\Pi(\lambda)] \tag{1.6}$$

where

$r(\lambda)$ is the reflectance value at wavelength λ of the object measured
$\Pi(\lambda)$ is the spectral product at wavelength λ for the appropriate density response

It is well known that densitometers can be used to evaluate print characteristics such as consistency of color from sheet to sheet, color uniformity across the sheet, and color matching of the proof. According to (1.5), one can find that for two prints of the same ink, if the reflectance values $r(\lambda)$ are the same, it is certain that the density measures will be the same, i.e., the color of the prints will appear to be the same. However, it is also known that for two inks whose narrow-band density values have been measured as identical could appear as different colors to the human eye if their spectral characteristics are different in the insensitive dead zone of the filter [5]. It must be pointed out that due to the spectral product at each wavelength, prints even with the same density values but not with the same ink have not necessarily the same spectral reflectance values, i.e., they can appear as different colors to the human eye. Since the spectral product in densitometry is not directly related to

Fig. 1.6 (**a**) Gray patches with the same color setting appear as the same color. (**b**) The same patches in the center appear as different levels of gray due to the "simultaneous contrast effect" where the background influence makes the central color patches appear different

human visual response, the density measure can only guarantee the equality of the physical property of the same material, not the perceptual attribute of the color that appears.

There are similarities and differences between Densitometry and Colorimetry. Both involve integration with certain spectral weightings, but only the spectral weighting in the color-matching functions in Colorimetry is directly linked to the responsivity of human color vision. The measurement of color in the colorimetric way defined in (1.3) is therefore precisely related to the perceptual attribute of human color vision.

On the other hand, the resulting values of Colorimetry are more into the perceptual measurements of human color response. By definition in (1.4), if the spectral reflectance of $r_1(\lambda)$ and $r_2(\lambda)$ are exactly the same, this "spectral matching" method can of course create the sensation of two objects of the same color. However, it is not necessary to constrain the reflectance of the two objects to be exactly the same, as long as the integration results are the same, the sensation of color equality would occur, which is referred as "*colorimetric matching.*" These two types of matching are based on the same physical properties of the light source and the same adaptation status of the visual system, which is usually referred as "*fundamental Colorimetry*" (or simple CIE XYZ tristimulus system).

Advanced Colorimetry usually refers to the color processing that goes beyond the matching between simple solid color patches or pixels, where spatial influence, various light sources, different luminance levels, different visual adaptation, and various appearance phenomena are involved in a cross media environment. These are the areas on which active research into Color Imaging focuses and the topics covered in the subsequent chapters. One example is shown in Fig. 1.6a where all the gray patches are painted with the same R, G, and B numbers and appear as the same color in such circumstances. However, the same gray patches with different background color patches now appear as different levels of gray as shown in Fig. 1.6b. This so-called "simultaneous contrast effect" gives a good example of how "advanced Colorimetry" has to deal with subjects beyond the matching among simple color patches where spatial influence and background factors, etc. are taken into consideration.

1.7 Color Spaces: Linear and NonLinear Scales

Measurement of physical property is a very common activity in modern life. Conveying the measured value by a scale number enables the quantitative description of certain property, such as length and mass. A uniform scale ensures that the fundamental operations of algebra (addition, subtraction, equality, less than, greater than, etc.) are applicable. It is therefore possible to apply mathematical manipulation within such a scale system. In the meantime, establishing a perceptual color-matching system is the first step toward color processing. Deriving a color scale system (or color space) is the second step, which makes color image processing and analysis a valid operation.

Establishing a color scale is complex because physical property is much easier to be accessed than the sensation of human color perception. For example, a gray scale with equal increment in a physical property like the reflectance factor (in 0.05 difference) is shown in Fig. 1.7a. It is obvious that in this scale the reflectance factor does not yield an even increment in visual sensation. As stated by Fechner's law—the sensation increases linearly as a function of the logarithm of stimulus intensity [4,7]—it is known that a certain nonlinear transformation is required to turn physical stimulus intensity into the perceived magnitude of a stimulus. Based on this concept the CIE in 1976 recommended two uniform color spaces, CIELAB and CIELUV. The following is a brief description in computing the CIELAB values from the reflectance value of an object.

Take the tristimulus values X, Y, Z from (1.3),

$$\begin{aligned} L^* &= 116(Y/Y_n)^{1/3} - 16 \\ a^* &= 500\left[(X/X_n)^{1/3} - (Y/Y_n)^{1/3}\right] \\ b^* &= 200\left[(Y/Y_n)^{1/3} - (Z/Z_n)^{1/3}\right] \end{aligned} \tag{1.7}$$

where Y/Y_n, X/X_n, and $Z/Z_n > 0.008856$; more details in CIE 15.2[4]

X, Y, and Z are tristimulus values of the object measured
X_n, Y_n, and Z_n are the tristimulus values of a reference white object

Fig. 1.7 Gray scales in physical and perceptual linear space: (**a**) a gray scale with a linear increment of the reflectance factor (0.05) and (**b**) a gray scale with a visually linear increment of the L* (Lightness) value in the CIELAB coordinate

L^* is the visual lightness coordinate
a^* is the chromatic coordinate ranged approximately from red to green
b^* is the chromatic coordinate ranged approximately from yellow to blue

Important criteria in designing the CIELAB color space are making the coordinates visually uniform and maintaining the opponent hue relationship according to the human color sensation. This equal CIELAB L^* increment is used to generate the gray scale in Fig. 1.7b, where it turns out to appear as a much smoother gradation than another gray scale with an equal reflectance factor increment shown in Fig. 1.7a.

It is important to note that the linear scale of a physical property, like the equal increment of the reflectance factor, does not yield linear visual perception. It is necessary to perform a certain nonlinear transformation from the physical domain to the perceptual domain which is perceived as a linear scale by the human visual system. As more and more research is dedicated to color science and engineering, it has been discovered that the human visual system can adjust automatically to a different environment by various adaptation processes. What kind of nonlinear processing is needed to predict human color image perception from measured physical property under various conditions therefore definitely deserves intense analysis and study, which is covered also in the following chapters. There are many more color spaces and models, like S-CIELAB, CIECAM02, iCAM, and spectral process models for various color imaging processing and analysis for specific conditions.

As shown above, color can be treated in two ways: as a perceived property by humans or as physical signal causing color detection in a detection system. Color is very common in our daily life yet not directly accessible. Scientists have derived mathematical models to define color properties. Engineers control devices to generate different colors. Artists know how to express their emotions by various colors. In a way, the study of color image processing and analysis is to bring more use of color into our lives. As shown in Fig. 1.8, the various colors can be interpreted as completeness in accumulating wisdom. No matter how complicated is our practice of color imaging science and technology, making life interesting and colorful is an ultimate joy.

1.8 Concluding Remarks

At the end of this chapter, we have a short philosophical discussion about color. In general texts and discussions the term "color" is not used rigorously. It can mean the human sensation, it can mean the color signal reflected from an object, and it can be a property of an object itself. In the traditional color approach, it is connected to the model of human color vision. Yet the same vocabulary is used when considering the animal vision, although the animal (color) vision systems may vary very much from that of human. In order to analyze and manage the color, we need to define

Fig. 1.8 Colorful banners are used in Japanese traditional buildings (Photographed by M. James Shyu in Kyoto, Japan)

color well. In this chapter, and in the book, the spectral approach is described in addition to the traditional color representation. In the spectral approach, color means the color signal originated from the object and reaching the color detection system. Both approaches are used in this book depending on the topic of the chapter.

In the traditional color science, black and white, and gray levels in between, are called achromatic light. This means that they differ from each other only by radiant intensity, or luminous intensity in photometrical terms. Other light is chromatic. Hence, one may say that black, white, and gray are not colors. This is a meaningful description only, if we have a fixed detection system, which is well defined. In the traditional color approach, the human color vision is considered to be based on a fixed detection system. Studies about human cone sensitivities and cone distribution show that this is not the case [12].

In the spectral approach, the "achromaticity" of black, white, and gray levels is not so obvious. In the spectral sense, the ultimate white is the equal energy white, for which the spectrum intensity is a constant maximum value over the whole wavelength range. When we start to decrease the intensity for some wavelength, the spectrum changes and at a certain point the spectrum represents a color in traditional means. If we consider white not to be a color, we have to define "epsilon" for each wavelength by which the change from the white spectrum makes it a color. Also, white spectrum can be seen as the limit of sequence of color spectra. This means that in the traditional color approach, the limit of sequence of colors is not color.

Blackness, whiteness, and grayness are also dependent on detection system. Detected signal looks white, when all the wavelength-sensitive sensors give the

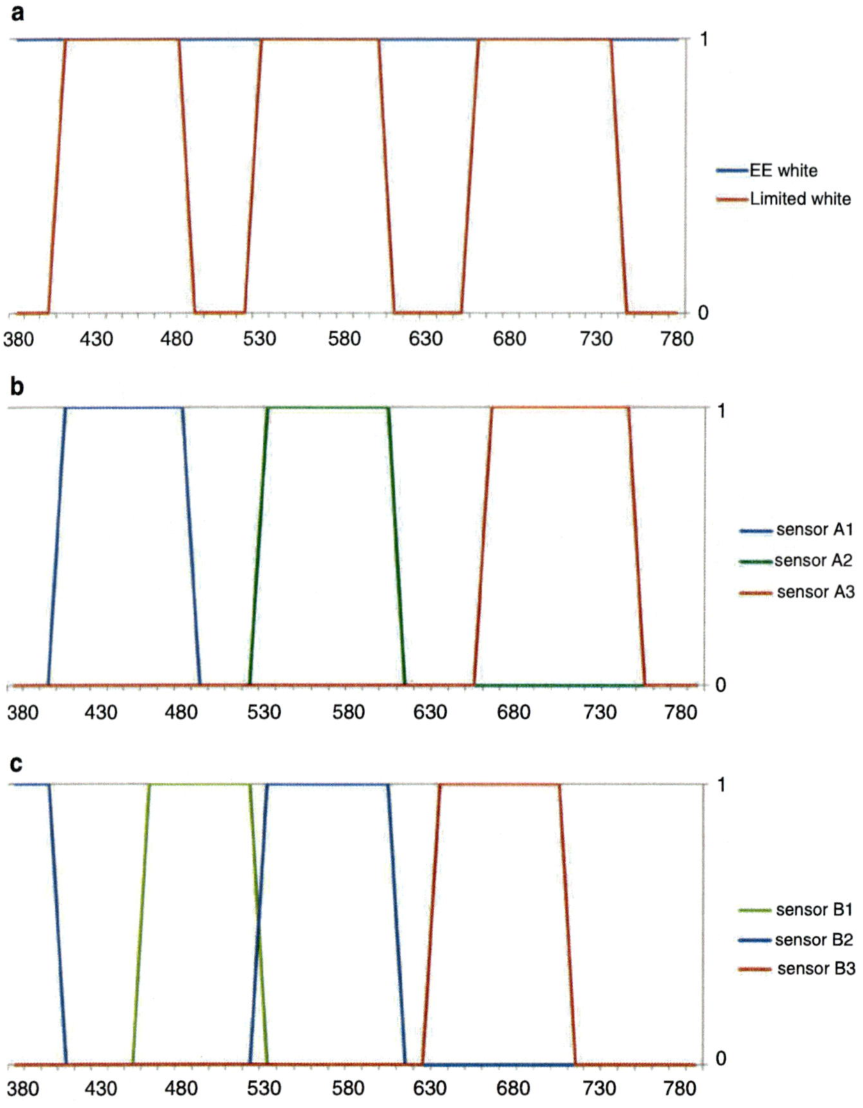

Fig. 1.9 White is a relative attribute. (**a**) Two spectra, equal energy white (*blue line*) and a spectrum which look white for sensors A, but colored to sensors B (*red line*). (**b**) Sensors A, sensitivity functions have the same shape as the "limited white" (*red line*) spectrum on (a). (**c**) Sensors B, sensitivity functions does not match with spectrum "limited white" (*red line*) in (**a**)

maximum response. In Fig. 1.9a there are two color signals, which both "looks white" for the theoretical color detection system given in Fig. 1.9b. But if we change the detector system to one shown in Fig. 1.9c, the other white is a colored signal, since not all the sensors have maximum input.

With this small discussion, we want to show that in the color science there is a need and development into direction of generalized color. In this approach, color is not restricted to the human visual system, but its basis is in a measurable and well-defined color signal. Signal, which originates from the object, reaches the color detection system, and carries the full color information of the object. The traditional color approach is shown to be powerful tool to manage color for human vision. The well defined models are useful tools also in the future, but the main restriction, uncertainty in understanding of detection system, needs much research also in the future.

References

1. ANSI CGATS.3–1993 (1993) Graphic technology—Spectral measurement and colorimetric computation for graphic arts images. NPES
2. Berlin B, Kay P (1969) Basic color terms: their universality and evolution University of California Press, Berkeley, CA
3. Berns R (2000) Billmeyer and Salzman's principles of color technology 3rd edn. Wiley, New York
4. Boynton RM (1984) Psychophysics, in optical radiation measurement. In: Bartleson CJ, Franc Grum (eds) Visual measurement Vol 5 Academic Press p 342
5. Brehm PV (1992) Introduction to densitometry. Graphic Communications Association 3rd Revision
6. Publication CIE No 15.2 (1986) Colorimetry. Second edition
7. Fechner G (1966) Elements of psychophysics. In: Adler HE, Howes DH, Boring EG (editors and translators), Vol I Holt, New York
8. Foster DH (2003) Does colour constancy exist? Trends in Cognitive Sciences 7(10)439–443
9. John Gage (1995) Colour and culture: practice and meaning from antiquity to abstraction. Thames and Hudson, Singapore
10. Goethe JW, Farbenlehre Z, Cotta T (1810) In: Goethe JWv, English version: Theory of Colors. MIT, Cambridge, MA, USA, 1982
11. Heikkinen V (2011) Kernel methods for estimation and classification of data from spectral imaging. PhD thesis, University of Eastern Finland, Finland
12. Hofer H, Carroll J Neitz J, Neitz M, Williams DR (2005) Organization of the human trichromatic cone mosaic J Neurosci 25(42):9669–9679
13. Jacobs GH (1996) Primate photopigments and primate color vision. Proc Natl Acad Sci 93(2):577–581
14. Kuehni RG (2003) Color space and its divisions: color order from antiquity to the present Wiley, Hoboken, NJ, USA
15. Land E (1977) The Retinex theory of color vision. Sci Am 237(6):108–128
16. Lenz R (2001) Estimation of illumination characteristics IEEE Trans Image Process 10(7):1031–1038
17. Maloney LT (1986) Evaluation of linear models of surface spectral reflectance with small numbers of parameters J Opt Soc Am A 3:1673–1683
18. Maloney LT, Wandell B (1986) Color constancy: a method for recovering surface spectral reflectance. J Opt Soc Am A 3:29–33
19. Menzel R, Backhaus W (1989) Color vision in honey bees: Phenomena and physiological mechanisms. In: Stavenga D, Hardie N (eds) Facets of vision. Berlin 281–297
20. Sir Isaac Newton (1730) Opticks: or, a treatise of the reflections, refractions, inflections and colours of light, 4th edn. Innys, London

21. George Palmer, Theory of Colors and Vision", from Selected papers on Colorimetry—Fundamentals. In D.L. MacAdam (ed.), SPIE Milestone Series Volume MS77, pp. 5–8, SPIE Optical Engineering Press, 1993 (Originally printed by Leacroft 1777 reprinted from Sources of Color Science, pp. 40–47, MIT Press, 1970)
22. Parkkinen JPS, Jaaskelainen T, Oja E (1985) Pattern recognition approach to color measurement and discrimination Acta Polytechnica Scandinavica: Appl Phys 1(149):171–174
23. Parkkinen JPS, Hallikainen J, Jaaskelainen T (1989) Characteristic spectra of Munsell colors. J Opt Soc Am A 6:318–322
24. Wade NJ (1999) A natural history of vision MIT Press, Cambridge, MA, USA 2nd printing
25. Wandell B (1985) The synthesis and analysis of color images, NASA Technical Memorandum 86844. Ames Research Center, California, USA, pp 1–34
26. William David Wright, The CIE Contribution to Colour Technology, 1931 to 1987, pp. 2–5, in Inter-Society Color Council News, Number 368, July/August, 1997
27. Wyszecki G, Stiles W (1982) Color science: concepts and methods, quantitative data and formulae 2nd edition Wiley, New York
28. Zollinger H (1999) Color: a multidisciplinary approach. Wiley, Weinheim

Chapter 2
CIECAM02 and Its Recent Developments

Ming Ronnier Luo and Changjun Li

The reflection is for the colors what the echo is for the sounds

Joseph Joubert

Abstract The development of colorimetry can be divided into three stages: colour specification, colour difference evaluation and colour appearance modelling. Stage 1 considers the communication of colour information by numbers. The second stage is colour difference evaluation. While the CIE system has been successfully applied for over 80 years, it can only be used under quite limited viewing conditions, e.g., daylight illuminant, high luminance level, and some standardised viewing/illuminating geometries. However, with recent demands on crossmedia colour reproduction, e.g., to match the appearance of a colour or an image on a display to that on hard copy paper, conventional colorimetry is becoming insufficient. It requires a colour appearance model capable of predicting colour appearance across a wide range of viewing conditions so that colour appearance modelling becomes the third stage of colorimetry. Some call this as advanced colorimetry. This chapter will focused on the recent developments based on CIECAM02.

Keywords Color appearance model • CAM • CIECAM02 • Chromatic adaptation transforms • CAT • Colour appearance attributes • Visual phenomena • Uniform colour spaces

M.R. Luo (✉)
Zheijiang University, Hangzhou, China
University of Leeds, Leeds, UK
e-mail: M.R.Luo@Leeds.ac.uk

C. Li
Liaoning University of Science and Technology, Anshan, China

C. Fernandez-Maloigne (ed.), *Advanced Color Image Processing and Analysis*,
DOI 10.1007/978-1-4419-6190-7_2,

2.1 Introduction

The development of colorimetry [1] can be divided into three stages: colour specification, colour difference evaluation and colour appearance modelling. Stage 1 considers the communication of colour information by numbers. The Commission Internationale de l'Eclairage (CIE) recommended a colour specification system in 1931 and later, it was further extended in 1964 [2]. The major components include standard colorimetric observers, or colour matching functions, standard illuminants and standard viewing and illuminating geometry. The typical colorimetric measures are the tristimulus value (X,Y,Z), chromaticity coordinates (x, y), dominant wavelength, and excitation purity.

The second stage is colour difference evaluation. After the recommendation of the CIE specification system in 1931, it was quickly realised that the colour space based on chromaticity coordinates was far from a uniform space, i.e., two pairs of stimuli having similar perceived colour difference would show large difference of the two distances from the chromaticity diagram. Hence, various uniform colour spaces and colour difference formulae were developed. In 1976, the CIE recommended CIELAB and CIELUV colour spaces [2] for presenting colour relationships and calculating colour differences, More recently, the CIE recommended the CIEDE2000 [3] for evaluating colour differences.

While the CIE system has been successfully applied for over 80 years, it can only be used under quite limited viewing conditions, for example, daylight illuminant, high luminance level, and some standardised viewing/illuminating geometries. However, with recent demands on cross-media colour reproduction, for example, to match the appearance of a colour or an image on a display to that on hard copy paper, conventional colorimetry is becoming insufficient. It requires a colour appearance model capable of predicting colour appearance across a wide range of viewing conditions so that colour appearance modelling becomes the third stage of colorimetry. Some call this as advanced colorimetry.

A great deal of research has been carried out to understand colour appearance phenomena and to model colour appearance. In 1997, the CIE recommended a colour appearance model designated CIECAM97s [4,5], in which the "s" represents a simple version and the "97" means the model was considered as an interim model with the expectation that it would be revised as more data and better theoretical understanding became available. Since then, the model has been extensively evaluated by not only academic researchers but also industrial engineers in the imaging and graphic arts industries. Some shortcomings were identified and the original model was revised. In 2002, a new model: CIECAM02 [6, 7] was recommended, which is simpler and has a better accuracy than CIECAM97s.

The authors previously wrote an article to describe the developments of CIECAM97s and CIECAM02 [8]. The present article will be more focused on the recent developments based on CIECAM02. There are six sections in this chapter. Section 2.2 defines the viewing conditions and colour appearance terms used in CIECAM02. Section 2.3 introduces some important colour appearance data sets which were used for deriving CIECAM02. In Sect. 2.4, a brief introduction of different chromatic adaptation transforms (CAT) leading to the CAT02 [8], embedded in CIECAM02, will be given. Section 2.5 gives various visual phenomena predicted by CIECAM02. Section 2.6 summarises some recent developments of the CIECAM02. For example, the new uniform colour spaces based on CIECAM02 by Luo et al. (CAM02-UCS, CAM02-SCD and CAM02-LCD) [9] will be covered. Xiao et al. [10–12] extended CIECAM02 to predict the change in size of viewing field on colour appearance, known as size effect. Fu et al. [13] has extended the CIECAM02 for predicting colour appearances of unrelated colours presented in mesopic region. Finally, efforts were paid to modify the CIECAM02 in connection with international color consortium (ICC) profile connection space for the colour management [14]. In the final section, the authors point out a concept of the universal model based on CIECAM02.

2.2 Viewing Conditions and Colour Appearance Attributes

The step-by-step calculation of CIECAM02 is given in Appendix. In order to use CIECAM02 correctly, it is important to understand the input and output parameters of the model. Figure 2.1 shows the viewing parameters, which define the viewing conditions, and colour appearance terms, which are predicted by the model. Each of them will be explained in this section. X_W, Y_W, Z_W are the tristimulus values of the reference white under the test illuminant; L_A specifies the luminance of the adapting field; Y_b defines the luminance factor of background; the definition of surround will be introduced in later this section.

The output parameters from the model include Lightness (J), Brightness (Q), Redness–Greenness (a), Yellowness–Blueness (b), Colourfulness (M), Chroma (C), Saturation (s), Hue composition (H), and Hue angle (h). These attributes will also be defined in this section.

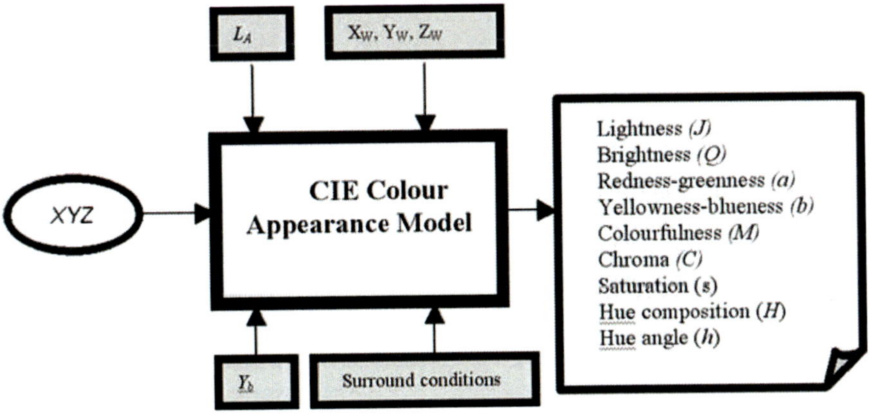

Fig. 2.1 A schematic diagram of a CIE colour appearance model

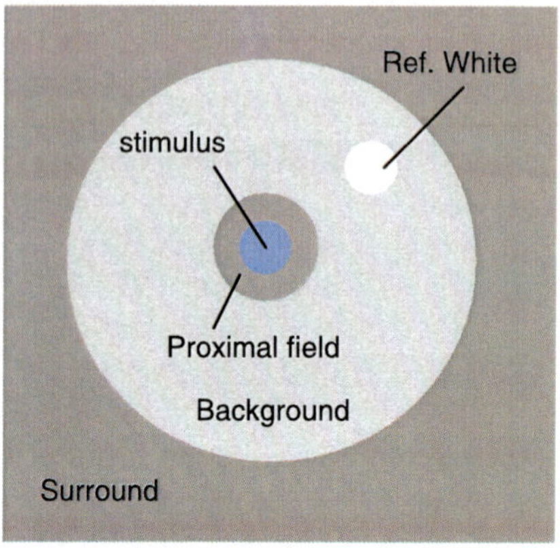

Fig. 2.2 Configuration for viewing colour patches of related colours

2.2.1 Viewing Conditions

The aim of the colour appearance model is to predict the colour appearance under different viewing conditions. Various components in a viewing field have an impact on the colour appearance of a stimulus. Hence, the accurate definition of each component of the viewing field is important. Figures 2.2–2.4 are three configurations considered in this chapter: colour patches for related colours, images for related colours, and patches for unrelated colours. The components in each configuration will be described below. Note that in the real world, objects are normally viewed in a complex context of many stimuli; they are known as "related" colours. An "unrelated colour" is perceived by itself, and is isolated, either completely or partially, from any other colours. Typical examples of unrelated colours are signal lights, traffic lights, and street lights, viewed in a dark night.

2.2.1.1 Stimulus

In Figs. 2.2 and 2.4 configurations, the stimulus is a colour element for which a measure of colour appearance is required. Typically, the stimulus is taken to be a uniform patch of about 2° angular subtense. A stimulus is first defined by the tristimulus values (X,Y,Z) measured by a tele-spectroradiometer (TSR) and then normalised against those of reference white so that Y is the percentage reflection factor.

In Fig. 2.3 configuration, the stimulus becomes an image. The pixel of each image is defined by device independent coordinates such as CIE XYZ or CIELAB values.

Fig. 2.3 Configuration for viewing images

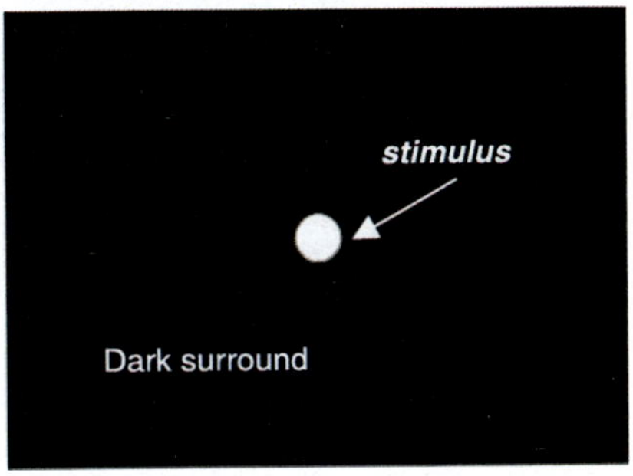

Fig. 2.4 Configuration for viewing unrelated colours

2.2.1.2 Proximal Field

In Fig. 2.2 configuration, proximal field is the immediate environment of the colour element considered, extending typically for about 2° from the edge of that colour element in all or most directions. Currently, proximal field is not used in CIECAM02. It will be applied when simultaneous contrast effect to be introduced in the future.

This element is not considered in Figs. 2.3 and 2.4 configurations.

2.2.1.3 Reference White

In Fig. 2.2 configuration, the reference white is used for scaling lightness (see later) of the test stimulus. It is assigned to have a lightness of 100. It is measured by a TSR again to define the tristimulus values of the light source (X_W, Y_W, Z_W) in cd/m^2 unit. The parameter of L_W (equal to Y_W) in the model defines the luminance of the light source. When viewing unrelated colours, there is no such element. For viewing images, the reference white will be the white border (about 10 mm) surrounding the image.

The reference white in this context can be considered as the "adopted white" i.e., the measurement of "a stimulus that an observer who is adapted to the viewing environment would judge to be perfectly achromatic and to have a reflectance factor of unity (i.e., have absolute colorimetric coordinates that an observer would consider to be the perfect white diffuser)" [ISO 12231] For viewing an image, there could be some bright areas such as a light source or specularly reflecting white objects, possibly illuminated by different sources. In the latter case, the "adapted white" (the actual stimulus which an observer adapted to the scene judges to be equivalent to a perfect white diffuser) may be different from the adopted white measured as above.

2.2.1.4 Background

In Fig. 2.2 configuration, background is defined as the environment of the colour element considered, extending typically for about 10° from the edge of the proximal field in all, or most directions. When the proximal field is the same colour as the background, the latter is regarded as extending from the edge of the colour element considered. Background is measured by a TSR to define background luminance, L_b. In CIECAM02, background is defined by the luminous factor, $Y_b = 100 \times L_b/L_W$.

There is no such element for Fig. 2.4 configuration, normally in complete darkness. For viewing images (Fig. 2.3), this element can be the average Y value for the pixels in the entire image, or frequently, a Y value of 20, approximate an L* of 50 is used.

2.2.1.5 Surround

A surround is a field outside the background in Fig. 2.2 configuration, and outside the white border (reference white) in Fig. 2.3. Surround includes the entire room or the environment. Figure 2.4 configuration has a surround in complete darkness.

Surround is not measured directly, rather the surround ratio is determined and used to assign a surround. The surround ratio, S_R, can be computed:

$$S_R = L_{SW}/L_{DW}, \tag{2.1}$$

where L_{SW} is the luminance of the surround white and L_{DW} is the luminance of the device white. L_{SW} is a measurement of a reference white in the surround field while L_{DW} is a measurement of the device white point for a given device, paper or peak white. If SR is 0, then a dark surround is appropriate. If S_R is less than 0.2, then a dim surround should be used while an S_R of greater than or equal to 0.2 corresponds to an average surround. Different surround "average," "dim," "dark" leads to different

Table 2.1 Parameter settings for some typical applications

Example	Ambient illumination in lux (or cd/m^2)	Scene or device white luminance	L_A *in cd/m^2*	Adopted white point	S_R	Surround
Surface colour evaluation in a light booth	1,000 (318.3)	318.30 cd/m^2	60	Light booth	1	Average
Viewing self-luminous display at home	38 (12)	80 cd/m^2	20	Display and ambient	0.15	Dim
Viewing slides in dark room	0 (0)	150 cd/m^2	30	Projector	0	Dark
Viewing self-luminous display under office illumination	500 (159.2)	80 cd/m^2	15	Display	2	Average

parameters (F: incomplete adaptation factor; Nc: chromatic induction factor and c: impact of surround) used in CIECAM02. Table 2.1 define S_R values in some typical examples in real applications.

2.2.1.6 Adapting Field

For Fig. 2.2 configuration, adapting field is the total environment of the colour element considered, including the proximal field, the background and the surround, and extending to the limit of vision in all directions. For Fig. 2.3 image configuration, it can be approximated the same as background, i.e., approximate an L* of 50.

The luminance of adapting field is expressed as L_A, which can be approximated by $L_W \times Y_b/100$, or by L_b.

Photopic, Mesopic and Scotopic Vision

Another parameter is also very important concerning the range of illumination from the source. It is well known that rods and cones in our eyes are not uniformly distributed on the retina. Inside the foveola (the central 1° field of the eye), there are only cones; outside, there are both cones and rods; in the area beyond about 40° from the visual axis, there are nearly all rods and very few cones. The rods provide monochromatic vision under low luminance levels; this scotopic vision is in operation when only rods are active, and this occurs when the luminance level is less than about 0.1 cd/m^2. Between this level and about 10 cd/m^2, vision involves a mixture of rod and cone activities, which is referred to as mesopic vision. It requires luminance of about 10 cd/m^2 for photopic vision in which only cones are active.

2.2.2 *Colour Appearance Attributes*

CIECAM02 predicts a range of colour appearance attributes. For each attribute, it will be accurately defined mainly following the definitions of CIE International Lighting Vocabulary [15]. Examples will be given to apply them in the real-world situation, and finally the relationship between different attributes will be introduced.

2.2.2.1 Brightness (Q)

This is a visual perception according to which an area appears to exhibit more or less light. This is an openended scale with a zero origin defining the black.

The brightness of a sample is affected by the luminance of the light source used. A surface colour illuminated by a higher luminance would appear brighter than the same surface illuminated by a lower luminance. This is known as "Steven Effect" (see later).

Brightness is an absolute quantity, for example, a colour appears much brighter when it is viewed under bright outdoor sunlight than under moonlight. Hence, their Q values could be largely different.

2.2.2.2 Lightness (J)

This is the brightness of an area judged relative to the brightness of a similarly illuminated reference white.

It is a relative quantity, for example, thinking a saturated red colour printed onto a paper. The paper is defined as reference white having a lightness of 100. By comparing the light reflected from both surfaces in the bright sunlight, the red has a lightness of about 40% of the reference white (J value of 40). When assessing the lightness of the same red colour under the moonlight against the same reference white paper, the lightness remains more or less the same with a J of 40.

It can be expressed by $J = Q_S/Q_W$, where Q_S and Q_W are the brightness values for the sample and reference white, respectively.

2.2.2.3 Colourfulness (M)

Colourfulness is that attribute of a visual sensation according to which an area appears to exhibit more or less chromatic content.

This is an open-ended scale with a zero origin defining the neutral colours. Similar to the brightness attribute, the colourfulness of a sample is also affected by luminance. An object illuminated under bright sunlight would appear more colourful than when viewed under moonlight, such as M value changes from 2000 to 1 with a ratio of 2000.

Fig. 2.5 An image to illustrate saturation

2.2.2.4 Chroma (C)

This is the colourfulness of an area judged as a proportion of the brightness of a similarly illuminated reference white. This is an open-ended scale with a zero origin representing neutral colours. It can be expressed by $C = M/Q_W$.

The same example is given here, a saturated red printed on a white paper. It has a colourfulness of 50 against the white paper having a brightness of 250 when viewed under sunlight. When viewed under dim light, colourfulness reduces to 25 and brightness of paper also reduces to half. Hence, the C value remains unchanged.

2.2.2.5 Saturation (S)

This is the colourfulness of an area judged in proportion to its brightness as expressed by $s = M/Q$, or $s = C/J$. This scale runs from zero, representing neutral colours, with an open end.

Taking Figs. 2.3–2.5 as an example, the green grass under sunlight is bright and colourful. In contrast, those under the tree appear dark and less colourful. Because they are the same grass in the field, we know that they have the same colour, but their brightness and colourfulness values are largely different. However, their saturation values will be very close because it is the ratio between brightness and colourfulness. Similar example can also be found in the image on the brick wall. Hence, saturation could be a good measure for detecting the number and size of objects in an image.

2.2.2.6 Hue (H and H)

Hue is the attribute of a visual sensation according to which an area appears to be similar to one, or to proportions of two, of the perceived colours red, yellow, green and blue.

CIECAM02 predicts hue with two measures: hue angle (h) ranging from 0° to 360°, and hue composition (H) ranging from 0, through 100, 200, 300, to 400 corresponding to the psychological hues of red, yellow, green, blue and back to red. These four hues are the psychological hues, which cannot be described in terms of any combinations of the other colour names. All other hues can be described as a mixture of them. For example, an orange colour should be described as mixtures of red and yellow, such as 60% of red and 40% of yellow.

2.3 Colour Appearance Data Sets

Colour appearance models based on colour vision theories have been developed to fit various experimental data sets, which were carefully generated to study particular colour appearance phenomena. Over the years, a number of experimental data sets were accumulated to test and develop various colour appearance models. Data sets investigated by CIE TC 1-52 CAT include: Mori et al. [16] from the Color Science Association of Japan, McCann et al. [17] and Breneman [18] using a haploscopic matching technique; Helson et al. [19], Lam and Rigg [20] and Braun and Fairchild [21] using the memory matching technique; and Luo et al. [22, 23] and Kuo and Luo [24] using the magnitude estimation method. These data sets, however, do not include visual saturation correlates. Hence, Juan and Luo [25, 26] investigated a data set of saturation correlates using the magnitude estimation method. The data accumulated played an important role in the evaluation of the performance of different colour appearance models and the development of the CIECAM97s and CIECAM02.

2.4 Chromatic Adaptation Transforms

Arguably, the most important function of a colour appearance model is chromatic adaptation transform. CAT02 is the chromatic adaptation transformation imbedded in CIEAM02. This section covers the developments towards this transform.

Chromatic adaptation has long been extensively studied. A CAT is capable of predicting corresponding colours, which are defined as pairs of colours that look alike when one is viewed under one illuminant (e.g., D65[1]) and the other is under

[1]In this chapter we will use for simplified terms "D65" and "A" instead of the complete official CIE terms: "CIE standard illuminant D65" and "CIE standard illuminant A".

a different illuminant (e.g., A). The following is divided into two parts: light and chromatic adaptation, and the historical developments of Bradford transform [20], CMCCAT2000 [27] and CAT02.

2.4.1 *Light and Chromatic Adaptation*

Adaptation can be divided into two: light and chromatic. The former is the adaptation due to the change of light levels. It can be further divided into two: light adaptation and dark adaptation. Light adaptation is the decrease in visual sensitivity upon an increase in the overall level of illumination. An example occurs when entering a bright room from a dark cinema. Dark adaptation is opposite to light adaptation and occurs, for example, when entering a dark cinema from a well-lit room.

2.4.2 *Physiological Mechanisms*

The physiology associated with adaptation mainly includes rod–cone transition, pupil size (dilation and constriction), receptor gain and offset. As mentioned earlier, the two receptors (cones and rods) functioning entirely for photopic (above approximately 10 cd/m^2) and for scotopic (below approximately 0.01 cd/m^2), respectively. Also, both are functioning in mesopic range between the two (approximately from 0.01 $cd/m^2$to 10 cd/m^2).

The pupil size plays an important role in adjusting the amount of light that enters the eye by dilating or constricting the pupil: it is able to adjust the light by a maximum factor of 5. During dark viewing conditions, the pupil size is the largest. Each of the three cones responds to light in a nonlinear manner and is controlled by the gain and inhibitory mechanisms.

Light and dark adaptations only consider the change of light level, not the difference of colour between two light sources (up to the question of Purkinje shift due to the difference in the spectral sensitivity of the rods and cones). Under photopic adaptation conditions, the difference between the colours of two light sources produces chromatic adaptation. This is responsible for the colour appearance of objects, and leads to the effect known as colour constancy (see also Chap. 2: Chromatic constancy). The effect can also be divided into two stages: a "chromatic shift" and an "adaptive shift". Consider, for example, what happens when entering a room lit by tungsten light from outdoor daylight. We experience that all colours in the room instantly become reddish reflecting the relative hue of the tungsten source. This is known as the "colorimetric shift" and it is due to the operation of the sensory mechanisms of colour vision, which occur because of the changes in the spectral power distribution of the light sources in question. After a certain short adaptation period, the colour appearances of the objects become more normal. This is caused by the fact that most of coloured objects in the real world are more or less colour constant (they do not change their colour appearance under different illuminants). The most obvious example is white paper always appears white regardless of which illuminant it is viewed under. The second stage is called the "adaptive shift" and it is caused by physiological changes and by a cognitive mechanism, which is based upon an observer's knowledge of the colours in the scene content in the viewing field. Judd [28] stated that "the processes by means of which an observer adapts to the illuminant or discounts most of the effect of non-daylight illumination are complicated; they are known to be partly retinal and partly cortical".

2.4.3 *Von Kries Chromatic Adaptation*

The von Kries coefficient law is the oldest and widely used to quantify chromatic adaptation. In 1902, von Kries [29] assumed that, although the responses of the three cone types (RGB)[2] are affected differently by chromatic adaptation, the spectral sensitivities of each of the three cone mechanisms remain unchanged. Hence, chromatic adaptation can be considered as a reduction of sensitivity by a constant factor for each of the three cone mechanisms. The magnitude of each factor depends upon the colour of the stimulus to which the observer is adapted. The relationship, given in (2.2), is known as the von Kries coefficient law.

$$\begin{aligned} R_c &= \alpha \cdot R, \\ G_c &= \beta \cdot G, \\ B_c &= \gamma \cdot B, \end{aligned} \tag{2.2}$$

where R_c, G_c, B_c and R, G, B are the cone responses of the same observer, but viewed under test and reference illuminants, respectively. α, β and γ are the von Kries coefficients corresponding to the reduction in sensitivity of the three cone mechanisms due to chromatic adaptation. These can be calculated using (2.3).

$$\alpha = \left(\frac{R_{wr}}{R_w}\right); \quad \beta = \left(\frac{G_{wr}}{G_w}\right); \quad \gamma = \left(\frac{B_{wr}}{B_w}\right), \tag{2.3}$$

where

$$\frac{R}{R_w} = \frac{R_c}{R_{wr}}, \quad \frac{G}{G_w} = \frac{G_c}{G_{wr}}, \quad \frac{B}{B_w} = \frac{B_c}{B_{wr}}, \tag{2.4}$$

[2]In this chapter the RGB symbols will be used for the cone fundamentals, in other chapters the reader will find the LMS symbols. The use of RGB here should not be confused with the *RGB* primaries used in visual colour matching.

Here R_{wr}, G_{wr}, B_{wr}, and R_{w}, G_{w}, B_{w} are the cone responses under the reference and test illuminants, respectively. Over the years, various CATs have been developed but most are based on the von Kries coefficient law.

2.4.4 *Advanced Cats: Bradford, CMCCAT20000 and CAT02*

In 1985, Lam and Rigg accumulated a set of corresponding colour pairs. They used 58 wool samples that had been assessed twice by a panel of five observers under D65 and A illuminants. The memory-matching technique was used to establish pairs of corresponding colours. In their experiment, a subgroup of colours was first arranged in terms of chroma and hue, and each was then described using Munsell H V/C coordinates. The data in H V/C terms, were then adjusted and converted to CIE 1931 XYZ values under illuminant C. Subsequently, the data under illuminant C were transformed to those under illuminant D65 using the von Kries transform. They used this set of data to derive a chromatic transform known as BFD transform now. The BFD transform can be formulated as the following:

2.4.4.1 Bfd Transform [20]

Step 1:

$$\begin{pmatrix} R \\ G \\ B \end{pmatrix} = \frac{1}{Y} M_{\mathrm{BFD}} \begin{pmatrix} X \\ Y \\ Z \end{pmatrix} \quad \text{with } M_{\mathrm{BFD}} = \begin{pmatrix} 0.8951 & 0.2664 & 0.1614 \\ -0.7502 & 1.7135 & 0.0367 \\ 0.0389 & -0.0685 & 1.0296 \end{pmatrix}.$$

Step 2:

$$\begin{pmatrix} R_{\mathrm{c}} \\ G_{\mathrm{c}} \\ B_{\mathrm{c}} \end{pmatrix} = \begin{pmatrix} R_{\mathrm{wr}}/R_{\mathrm{w}} & & \\ & G_{\mathrm{wr}}/G_{\mathrm{w}} & \\ & & B_{\mathrm{wr}}/B_{\mathrm{w}}^{p} \end{pmatrix} \begin{pmatrix} R \\ G \\ \mathrm{sign}(B)|B|^{p} \end{pmatrix} \quad \text{with}$$

$$p = (B_{\mathrm{w}}/B_{\mathrm{wr}})^{0.0834}.$$

Step 3:

$$\begin{pmatrix} X_{\mathrm{c}} \\ Y_{\mathrm{c}} \\ Z_{\mathrm{c}} \end{pmatrix} = M_{\mathrm{BFD}}^{-1} \begin{pmatrix} YR_{\mathrm{c}} \\ YG_{\mathrm{c}} \\ YB_{\mathrm{c}} \end{pmatrix}.$$

Note that the BFD transform is a nonlinear transform. The exponent p in step 2 for calculating the blue corresponding spectral response can be considered as a modification of the von Kries type of transform. The BFD transform performs much better than the von Kries transform. In 1997, Luo and Hunt [30] in 1997 modified the step 2 in the above BFD transform by introducing an adaptation factor D. The new step 2 becomes,

Step 2'

$$\begin{pmatrix} R_{\mathrm{c}} \\ G_{\mathrm{c}} \\ B_{\mathrm{c}} \end{pmatrix} = \begin{pmatrix} [D(R_{\mathrm{wr}}/R_{\mathrm{w}}) + 1 - D]R \\ [D(G_{\mathrm{wr}}/G_{\mathrm{w}}) + 1 - D]G \\ [D(B_{\mathrm{wr}}/B_{\mathrm{w}}^{p}) + 1 - D]\mathrm{sign}(B)|B|^{p} \end{pmatrix},$$

where

$$D = F - F/[1 + 2L_A^{1/4} + L_A^2/300].$$

The transform consisting of Step 1, Step 2' and Step 3 was then recommended by the colour measurement committee (CMC) of the society of dyers and colourists (SDC) and, hence, was named as the CMCCAT97. This transform is included in the CIECAM97s for describing colour appearance under different viewing conditions. The BFD transform was originally derived by fitting only one data set, Lam and Rigg. Although it gave a reasonably good fit to many other data sets, it predicted badly the McCann data set. In addition, the BFD and CMCCAT97 include an exponent p for calculating the blue corresponding spectral response. This causes uncertainty in reversibility and complexity in the reverse mode. Li et al. [31] addressed this problem and provided a solution by including an iterative approximation using the Newton method. However, this is unsatisfactory in imaging applications where the calculations need to be repeated for each pixel. Li et al. [27] gave a linearisation version by optimising the transform to fit all the available data sets, rather than just the Lam and Rigg set. The new transform, named CMCCAT2000, is given below.

2.4.4.2 Cmccat2000

Step 1:

$$\begin{pmatrix} R \\ G \\ B \end{pmatrix} = M_{00} \begin{pmatrix} X \\ Y \\ Z \end{pmatrix} \text{ with } M_{00} = \begin{pmatrix} 0.7982 & 0.3389 & -0.1371 \\ -0.5918 & 1.5512 & 0.0406 \\ 0.0008 & 0.0239 & 0.9753 \end{pmatrix}.$$

Step 2:

$$\begin{pmatrix} R_{\mathrm{c}} \\ G_{\mathrm{c}} \\ B_{\mathrm{c}} \end{pmatrix} = \begin{pmatrix} [D(Y_{\mathrm{w}}/Y_{\mathrm{wr}})(R_{\mathrm{wr}}/R_{\mathrm{w}}) + 1 - D]R \\ [D(Y_{\mathrm{w}}/Y_{\mathrm{wr}})(G_{\mathrm{wr}}/G_{\mathrm{w}}) + 1 - D]G \\ [D(Y_{\mathrm{w}}/Y_{\mathrm{wr}})(B_{\mathrm{wr}}/B_{\mathrm{w}}) + 1 - D]B \end{pmatrix}$$

with

$$D = F\{0.08\log_{10}[0.5(L_{A1} + L_{A2})] + 0.76 - 0.45(L_{A1} - L_{A2})/(L_{A1} + L_{A2})\}.$$

Step 3:

$$\begin{pmatrix} X_c \\ Y_c \\ Z_c \end{pmatrix} = M_{00}^{-1} \begin{pmatrix} R_c \\ G_c \\ B_c \end{pmatrix}.$$

The CMCCAT2000 not only overcomes all the problems with respect to reversibility discussed above, but also gives a more accurate prediction than other transforms of almost all the available data sets.

During and after the development of the CMCCAT2000, scientists decided to drop the McCann et al. data set because the experiment was carried out under a very chromatic adapting illuminant. Its viewing condition is much different from all the other corresponding data sets. Hence, it would be better to optimising the linear chromatic adaptation transform via fitting all the corresponding data sets without the McCann et al. data set. The new matrix obtained by the authors, now named the CAT02 matrix, is given by

$$M_{02} = \begin{pmatrix} 0.7328 & 0.4296 & -0.1624 \\ -0.7036 & 1.6975 & 0.0061 \\ 0.0030 & 0.0136 & 0.9834 \end{pmatrix},$$

which was first included in the appendix of our paper [32] in 2002. At the same time, Nathan Moroney (Chair of CIETC8-01 at that time) proposed a new formula for D function:

$$D = F\left[1 - \frac{1}{3.6}\mathrm{e}^{\frac{-L_A-42}{92}}\right]. \tag{2.5}$$

The CMCCAT2000 with the new matrix and D formula given by (2.5) becomes the CAT02.

At a later stage, CIE TC 8-01 *Colour Appearance Modelling for Colour Management Systems* had to choose a linear chromatic transform for CIECAM02. Multiple candidates such as CMCCAT2000 [27], the sharp chromatic transform [33] developed by Finlayson et al., and CAT02 [6–8] were proposed for use as a von Kries type transform. All had similar levels of performance with respect to the accuracy of predicting various combinations of previously derived sets of corresponding colours. In addition to the sharpening of the spectral sensitivity functions, considerations used to select the CIE transform included the degree of backward compatibility with CIECAM97s and error propagation properties by combining the forward and inverse linear CAT, and the data sets which were used during the optimisation process. Finally, CAT02 was selected because it is compatible with CMCCAT97 and was optimised using all available data sets except the McCann et al. set, which includes a very chromatic adapting illuminant.

Figure 2.6 illustrates 52 pairs of corresponding colours predicted by CIECAM02 (or its chromatic adaptation transform, CAT02) from illuminant A (open circles of vectors) to S_E (open ends of vectors) plotted in the CIE $u'v'$ chromaticity diagram for the 2° observer. The open circle colours have a value of L^* equal

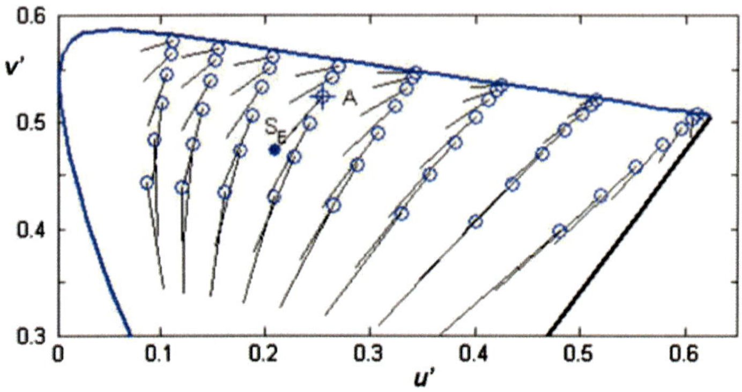

Fig. 2.6 The corresponding colours predicted by the CIECAM02 from illuminant A (*open circles* of vectors) to illuminant S_E (*open ends* of vectors) plotted in CIE $u'v'$ chromaticity diagram for the CIE 1931 standard colorimetric observer. The plus (+) and the dot (•) represent illuminants A and S_E, respectively

to 50 according to CIELAB under illuminant A. These were then transformed by the model to the corresponding colours under illuminant S_E(the equi-energy illuminant). Thus, the ends of each vector represent a pair of corresponding colours under the two illuminants. The input parameters are (the luminance of adapting field) $L_A = 63.7\,\mathrm{cd/m^2}$ and average surround. The parameters are defined in the Appendix.

The results show that there is a systematic pattern, i.e., for colours below v' equal to 0.48 under illuminant A the vectors are predicted towards the blue direction under the illuminant S_E. For colours outside the above region, the appearance change is in a counterclockwise direction, i.e., red colours shift to yellow, yellow to green and green to cyan as the illuminant changes from A to S_E.

2.5 Colour Appearance Phenomena

This section describes a number of colour appearance phenomena studied by various researchers in addition to the chromatic adaptation as described in the earlier section. The following effects are also well understood.

2.5.1 Hunt Effect

Hunt [34] studied the effect of light and dark adaptation on colour perception and collected data for corresponding colours via a visual colorimeter using the haploscopic matching technique, in which each eye was adapted to different viewing conditions and matches were made between stimuli presented in each eye.

The results revealed a visual phenomena known as Hunt effect [34]. It refers to the fact that the colourfulness of a colour stimulus increases due to the increase of luminance. This effect highlights the importance of considering the absolute luminance level in colour appearance models, which is not considered in traditional colorimetry.

2.5.2 Stevens Effect

Stevens and Stevens [35] asked observers to make magnitude estimations of the brightness of stimuli across various adaptation conditions. The results showed that the perceived brightness contrast increased with an increase in the adapting luminance level according to a power relationship.

2.5.2.1 Surround Effect

Bartleson and Breneman [36] found that the perceived contrast in colourfulness and brightness increased with increasing illuminance level from dark surround, dim surround to average surround. This is an important colour appearance phenomenon to be modelled, especially for the imaging and graphic arts industries where, on many occasions, it is required to reproduce images on different media under quite distinct viewing conditions.

2.5.3 Lightness Contrast Effect

The lightness contrast effect [37] reflects that the perceived lightness increases when colours are viewed against a darker background and vice versa. It is a type of simultaneous contrast effect considering the change of colour appearance due to different coloured backgrounds. This effect has been widely studied and it is well known that a change in the background colour has a large impact on the perception of lightness and hue. There is some effect on colourfulness, but this is much smaller than the effect on lightness and hue [37].

2.5.4 Helmholtz–Kohlrausch Effect

The Helmholtz–Kohlrausch [38] effect refers to a change in the brightness of colour produced by increasing the purity of a colour stimulus while keeping its luminance constant within the range of photopic vision. This effect is quite small compared with others and is not modelled by CIECAM02.

2.5.5 Helson–Judd Effect

When a grey scale is illuminated by a light source, the lighter neutral stimuli will exhibit a certain amount of the hue of the light source and the darker stimuli will show its complementary hue, which is known as the Helson–Judd effect [39]. Thus for tungsten light, which is much yellower than daylight, the lighter stimuli will appear yellowish, and the darker stimuli bluish. This effect is not modelled by CIECAM02.

2.6 Recent Developments of CIECAM02

Recently, several extensions to the CIECAM02 have been made, which have widened the applications of the CIECAM02. In this section, the extensions for predicting colour discrimination data sets, size effects and unrelated colour appearance in the mesopic region. Besides, recent developments from CIETC8-11 will be reported as well.

2.6.1 CIECAM02-Based Colour Spaces

CIECAM02 [6, 7] includes three attributes in relation to the chromatic content: chroma (C), colourfulness (M) and saturation (s). These attributes together with lightness (J) and hue angle (h) can form three colour spaces: J, a_C, b_C, J, a_M, b_M and J, a_s, b_s where

$$\begin{array}{l} a_{\mathrm{C}} = C \cdot \cos(h) \\ b_{\mathrm{C}} = C \cdot \sin(h) \end{array}, \quad \begin{array}{l} a_{\mathrm{M}} = M \cdot \cos(h) \\ b_{\mathrm{M}} = M \cdot \sin(h) \end{array}, \quad \begin{array}{l} a_{\mathrm{s}} = s \cdot \cos(h) \\ b_{\mathrm{s}} = s \cdot \sin(h). \end{array}$$

It was also found [40] that the CIECAM02 space is more uniform than the CIELAB space. Thus, the CIECAM02 space is used as a connection space for the gamut mapping in the colour management linked with the ICC profile [41, 42]. Further attempts have been also made by the authors to extend CIECAM02 for predicting available colour discrimination data sets, which include two types, for Large and Small magnitude Colour Differences, designated by LCD and SCD, respectively. The former includes six data sets with a total 2,954 pairs, having an average 10 ΔE^*_{ab} units over all the sets. The SCD data with a total of 3,657 pairs having an average 2.5 ΔE^*_{ab} units, are a combined data set used to develop the CIE 2000 colour difference formula: CIEDE2000[3].

Li et al. [43] found that a colour space derived using $J, a_{\mathrm{M}}, b_{\mathrm{M}}$ gave the most uniform result when analysed using the large and small colour difference data sets. Hence, various attempts [9, 43] were made to modify this version of CIECAM02 to fit all available data sets. Finally, a simple, generic form, (2.6) was found that

Table 2.2 The coefficients for CAM02-LCD, CAM02-SCD and CAM02-UCS

Versions	CAM02 -LCD	CAM02-SCD	CAM02-UCS
K_L	0.77	1.24	1.00
c_1	0.007	0.007	0.007
c_2	0.0053	0.0363	0.0228

adequately fitted all available data.

$$J' = \frac{(1+100\cdot c_1)\cdot J}{1+c_1\cdot J},$$
$$M' = (1/c_2)\cdot \ln(1+c_2\cdot M), \tag{2.6}$$

where c_1 and c_2 are constants given in Table 2.2.

The corresponding colour space is J', a'_M, b'_M where $a'_M = M'\cdot\cos(h)$, and $b'_M = M'\cdot\sin(h)$. The colour difference between two samples can be calculated in J', a'_M, b'_M space using (2.7).

$$\Delta E' = \sqrt{(\Delta J'/K_L)^2 + \Delta a'^2_M + \Delta b'^2_M}, \tag{2.7}$$

where $\Delta J', \Delta a'_M$ and $\Delta b'_M$ are the differences of J', a'_M and b'_M between the "standard" and "sample" in a pair. Here, K_L is a lightness parameter and is given in Table 2.2.

Three colour spaces named CAM02-LCD, CAM02-SCD and CAM02-UCS were developed for large, small and combined large and small differences, respectively. The corresponding parameters in (2.6) and (2.7) are listed in Table 2.2.

The three new CIECAM02 based colour spaces, together with the other spaces and formulae were also tested by Luo et al. [9]. The results confirmed that CAM02-SCD and CAM02-LCD performed the best for small and large colour difference data sets. When selecting one UCS to evaluate colour differences across a wide range, CAM02-UCS performed the second best across all data sets. The authors have been recommending using CAM02-UCS for all applications.

Figure 2.7 shows the relationship between *CIECAM02 J* and *CAM02-UCS J'* and Fig. 2.8 shows the relationship between *CIECAM02 M* and *CAM02-UCS M'*. It can be seen that CIECAM02 J is less than CAM02-UCS J' except at the two ends, while CIECAM02 M is greater than CAM02-UCS M' except when $M = 0$. Thus in order to have a more uniform space, CIECAM02 J should be increased and CIECAM02 M should be decreased.

The experimental colour discrimination ellipses used in the previous studies [44, 45] were also used for comparing different colour spaces. Figures 2.9 and 2.10 show the ellipses plotted in CIELAB and CAM02-UCS spaces, respectively. The size of the ellipse was adjusted by a single factor in each space to ease visual comparison. For perfect agreement between the experimental results and a uniform colour space, all ellipses should be constant radius circles. Overall, it can be seen that the ellipses in CIELAB (Fig. 2.9) are smaller in the neutral region and gradually increase in size as chroma increases. In addition, the ellipses are

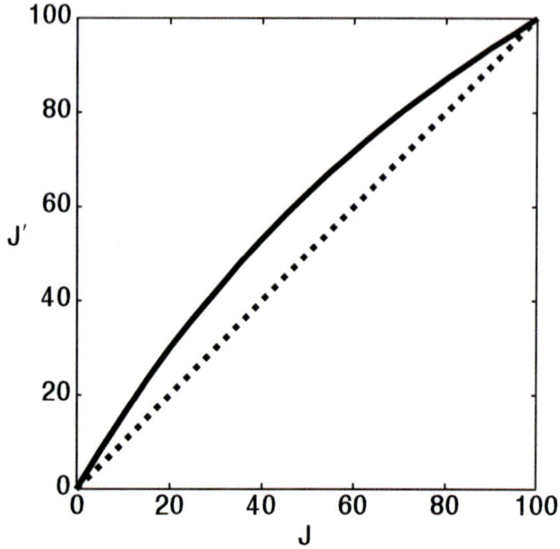

Fig. 2.7 The *full line* shows the relationship between J and J' and the *dotted line* is the 45° line

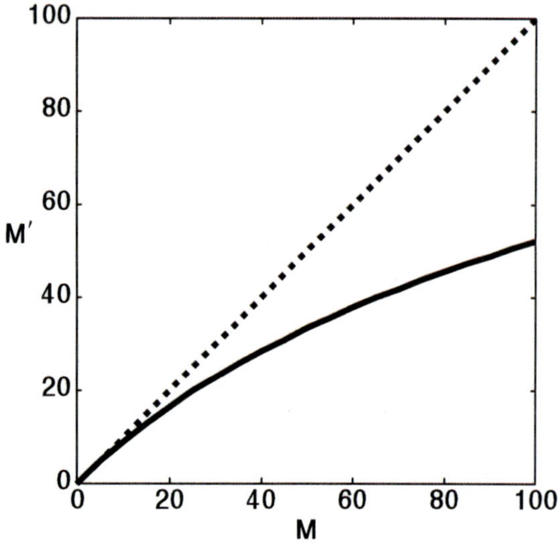

Fig. 2.8 The *full line* shows the relationship between M and M′ and the *dotted line* is the 45° line

orientated approximately towards the origin except for those in the blue region in CIELAB space. All ellipses in CAM02-UCS (Fig. 2.10) are approximately equal-sized circles. In other words, the newly developed CAM02-UCS is much more uniform than CIELAB.

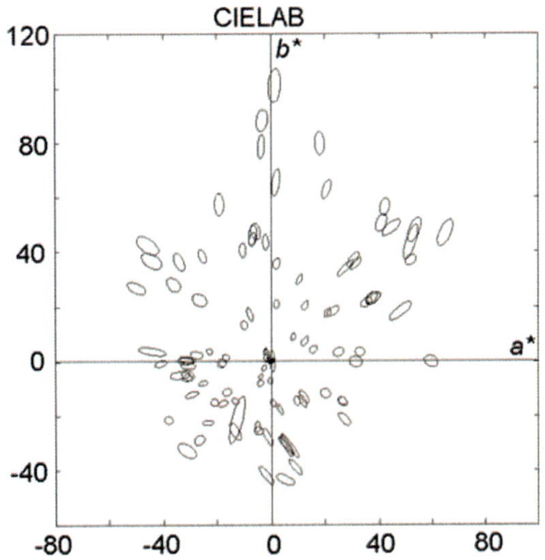

Fig. 2.9 Experimental chromatic discrimination ellipses plotted in CIELAB

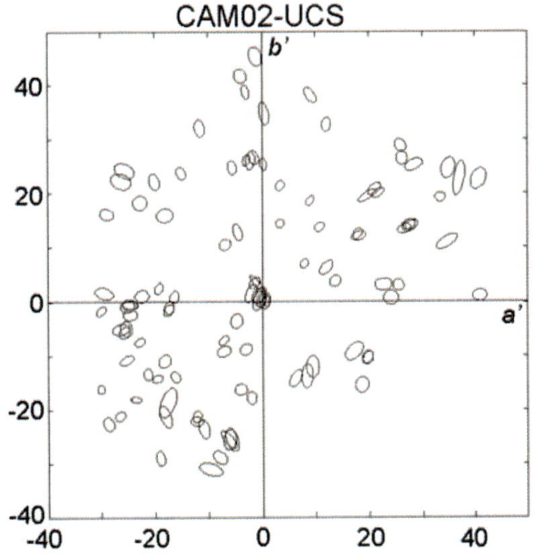

Fig. 2.10 Experimental chromatic discrimination ellipses plotted in CAM02-UCS

2.6.2 *Size Effect Predictions Based on CIECAM02*

The colour size effect is a colour appearance phenomenon [10–12], in which the colour appearance changes according to different sizes of the same colour stimulus. The CIE 1931 (2°) and CIE 1964 (10°) standard colorimetric observers were recommended by the CIE to represent human vision in smaller and larger than 4° viewing fields, respectively [2]. However, for a colour with a large size, such as over 20° viewing field, no standard observer can be used. The current

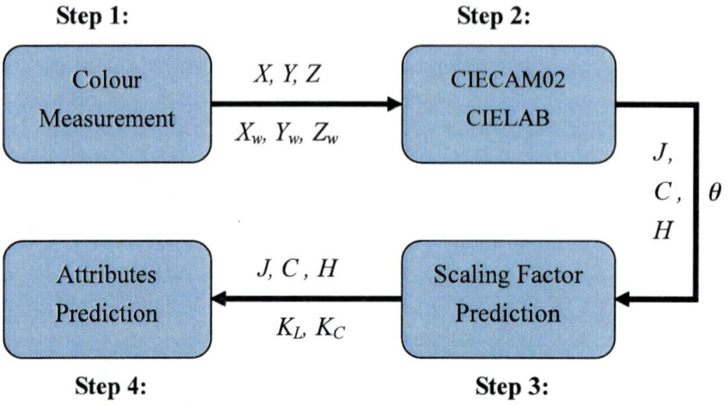

Fig. 2.11 The flow chart of size effect correction model based on CIECAM02

CIECAM02 is capable of predicting human perceptual attributes under various viewing conditions. However, it cannot predict the colour size effect. The size effect has been interested in many applications. For example, in the paint industry, the paints purchased in stores usually do not appear the same comparing between those shown in the packaging and painted onto the walls in a real room. This also causes great difficulties for homeowners, interior designers and architects when they select colour ranges. Furthermore, the display size tends to become larger. Colour size effect has also been greatly interested by display manufacturers in order to precisely reproduce or to enhance the source images on different sizes of colour displays.

With the above problems in mind, the CIE established a technical committee, TC1-75, *A comprehensive model for colour appearance* with one of aims to take colour size effect into account in the CIECAM02 colour appearance model [7]. In the recent work of Xiao et al. [10–12], six different sizes from 2° to 50° of same colours were assessed by a panel of observers using colour-matching method to match surface colours using a CRT display. The colour appearance data were accumulated in terms of CIE tristimulus values. A consistent pattern of colour appearance shifts was found according to different sizes for each stimulus. The experimental results showed that attributes of lightness and chroma increase with the increase of the physical size of colour stimulus. But the hue (composition) is not affected by the change of physical size of colour stimulus. Hence, a model based on CIECAM02 for predicting the size effect was derived. The model has the general structure shown in Fig. 2.11.

Step 1 calculates or measures tristimulus values X, Y, Z of a 2° stimulus size under a test illuminant X_W, Y_W, Z_W, and provides a target stimulus size θ; next, Step 2 predicts the appearance attributes J, C and H using CIECAM02 for colours with 2° stimulus size; and Step 3 computes the scaling factors K_J and K_C via the following formulae:

$$K_J = -0.007\theta + 1.1014,$$

$$K_C = 0.008\theta + 0.94.$$

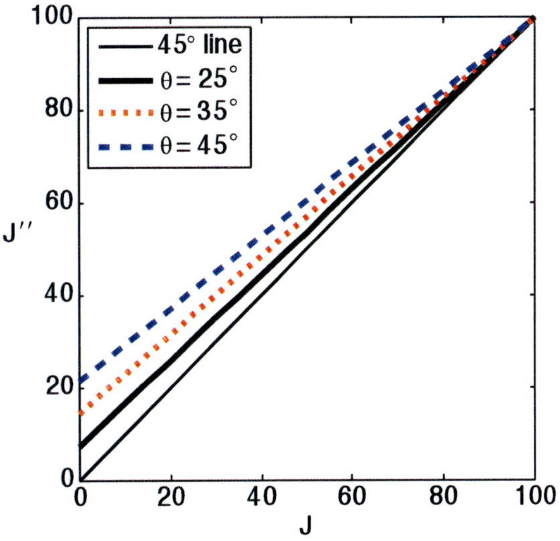

Fig. 2.12 The size effect corrected attributes J'' vs CIECAM02 J under viewing angles being 25°, (*thick solid line*), 35° (*dotted line*) and 45° (*dashed line*), respectively. The *thin solid line* is the 45° line where $J = J''$

Finally in Step 4, the colour appearance attributes J'', C'' and H'' for the target stimulus size θ are predicted using the formulae:

$$J'' = 100 + K_{\mathrm{J}} \times (J - 100), \tag{2.8}$$

$$C'' = K_{\mathrm{C}} \times C, \tag{2.9}$$

$$H'' = H. \tag{2.10}$$

The earlier experimental results [10] were used to derive the above model.

Figure 2.12 shows the corrected attributes J'' of 25°, 35° and 45°, respectively, plotted against J at 2° viewing field. The thick solid line is the corrected J'' when viewing field is 25°; the dotted line corresponds to the J'' with viewing angle being 35°. The dashed line is the J'' with viewing angle of 45°. The thin solid line is the 45° line where $J = J''$. The trend is quite clear as shown in Fig. 2.12, i.e., an increase of lightness for a larger viewing field. For example, when $J = 60$ with a size of 2°, J'' values are 62.9, 65.7 and 68.5 for sizes of 25°, 35° and 45°, respectively. However, when $J = 10$ with a size of 2°, J''s become 16.6, 22.9 and 29.2 for 25°, 35° and 45°, respectively. This implies that the large effect is mainly occurred for the dark colour region.

Figure 2.13 shows the corrected attributes C'' of 25°, 35° and 45°, respectively plotted against C at 2° viewing field. Vertical axis is the size effect corrected C''. The thick solid line is the corrected C'' when viewing angle is 25°; the dotted line corresponds to the C'' with viewing angle being 35°. The dashed line is the C'' with viewing angle of 45°. The thin solid line is the 45° line where $C = C''$. Again, a clear trend in Fig. 2.13 is shown that an increase of chroma for a larger viewing field. For example, when C is 60 with a size of 2°, C'' values are 68.4, 73.2 and 78.0 for sizes

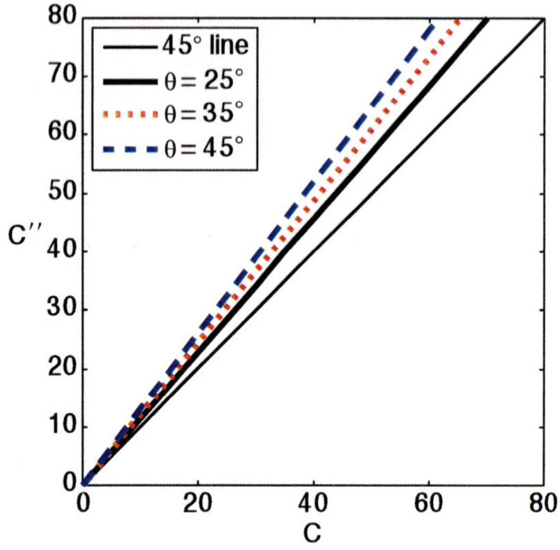

Fig. 2.13 The size effect corrected attributes C'' vs CIECAM02 C under viewing being 25°, (*thick solid line*), 35° (*dotted line*) and 45° (*dashed line*), respectively. The *thin solid line* is the 45° line where $C = C''$

of 25°, 35° and 45°, respectively. However, when C is 10 with a size of 2°, C''s become 11.4, 12.2 and 13.0 for 25°, 35° and 45°, respectively. This implies that the large effect in mainly occurrs in the high chroma region.

2.6.3 Unrelated Colour Appearance Prediction Based on CIECAM02

As mentioned at the beginning of this chapter, unrelated colours are important in relation to safety issues (such as night driving). It includes signal lights, traffic lights and street lights, viewed on a dark night. These colours are important in connection with safety issues. The CIECAM02 was derived for predicting colour appearance for related colours and it cannot be used for predicting unrelated colour appearance. The CAM97u derived by Hunt [46] can be used for predicting unrelated colour appearance. However, the model was not tested since there was no available visual data for unrelated colours. Fu et al. [13] carried out the research work recently. They accumulated a set of visual data using the configuration in Fig. 2.4. The data were accumulated for the colour appearance of unrelated colours under photopic and mesopic conditions. The effects of changes in luminance level and stimulus size on appearance were investigated. The method used was magnitude estimation of brightness, colourfulness and hue. Four luminance levels (60, 5, 1 and 0.1, cd/m^2) were used. For each of the first three luminance levels, two stimulus sizes (10°, 2°, 1° and 0.5°) were used. Ten observers judged 50 unrelated colours. A total of 17,820 estimations were made. The observations were carried out in a completely darkened room, after about 20 min adaptation; each test colour was presented on its own. Brightness and colourfulness were found to decrease with decreases of both luminance level and stimulus size. The results were used to further extend CIECAM02 for predicting unrelated colours under both photopic and mesopic conditions. The model includes parameters to reflect the effects of luminance level and stimulus size. The model is described below:

Inputs:

Measure or calculate the luminance L and chromaticity x,y of the test colour stimulus corresponding to CIE colour-matching functions (2° or 10°). The parameters are the same as CIECAM02 except that the test illuminant is equal energy illuminant (S_E, i.e., $X_W = Y_W = Z_W = 100$), and $L_A = 1/5$ of the adapting luminance, and the surround parameters are set as those under the dark viewing condition. As reported by Fu et al. [13], when there is no reference illuminant to compare with (such as assessing unrelated colours), S_E illuminant can be used by assuming no adaptation takes place for unrelated viewing condition.

Step 1: Using the CIECAM02 (Steps 0–8, Step 10, ignore the calculation of Q and s) to predict the (cone) achromatic signal A, colourfulness (M) and hue (H_C).

Step 2: Modify the achromatic signal A since there is a contribution from rod response using the formula:

$$A_{\text{new}} = A + k_A A_S \text{ with } A_S = (2.26L)^{0.42}.$$

Here, k_A depends on luminance level and viewing angle size of the colour stimulus.

Step 3: Modify the colourfulness M predicted from CIECAM02 using the following formula:

$$M_{\text{new}} = k_M M.$$

Here, k_M depends on luminance level and viewing angle size of the colour stimulus.

Step 4: Predict the new brightness using the formula:

$$Q_{\text{new}} = A_{\text{new}} + M_{\text{new}}/100.$$

Outputs: Brightness Q_{new}, colourfulness M_{new} and hue composition H_C.

Note that the hue composition H_C is the same as predicted by CIECAM02. The above model was tested using the visual data [13].

Figure 2.14 shows the brightness and colourfulness changes for a red colour of medium saturation (relative to S_E, $h_{uv} = 355°$, and $s_{uv} = 1.252$) as predicted by the new model under different luminance levels. The luminance levels were varied from

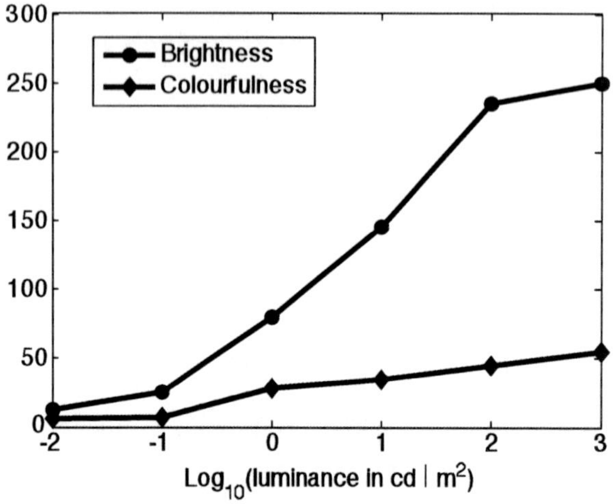

Fig. 2.14 The brightness and colourfulness predicted by the new model for a sample varying in luminance level with 2° stimulus size

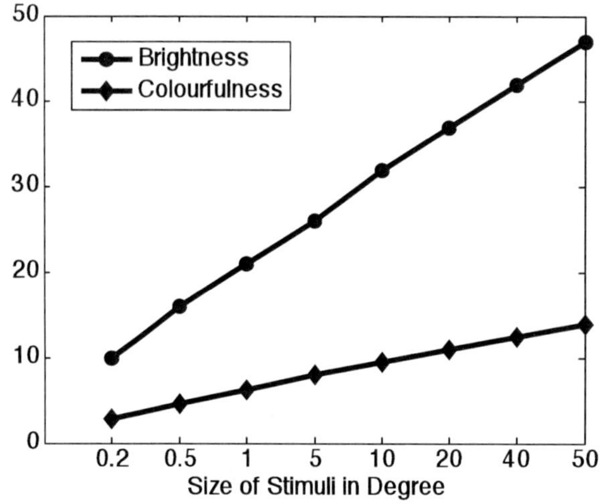

Fig. 2.15 The brightness and colourfulness predicted by the new model for a sample varying in stimulus size at 0.1 cd/m^2 luminance level

0.01 to 1000 cd/m^2, and L_A was set at one fifth of these values. The ratio Y_b/Y_w set at 0.2. Figure 2.15 shows the brightness and colourfulness changes, for the same red colour, predicted by the new model for different stimulus sizes ranging from 0.2° to 40°. The luminance level (L) was set at 0.1 cd/m^2. It can be seen that brightness and colourfulness increase when luminance increases up to around 100 cd/m^2, and they also increase when stimulus size increases. These trends reflect the phenomena found in Fu et al.'s study, i.e. when luminance level increases, colours become brighter and more colourful, and larger colours appear brighter and more colourful than smaller sized colours; however, below a luminance of 0.1 cd/m^2 and above a luminance of 60 cd/m^2, and below a stimulus size of 0.5° and above a stimulus size of 10^0, these results are extrapolations, and must be treated with caution.

2.6.4 *Problems with CIECAM02*

Since the recommendation of the CIECAM02 colour appearance model [6, 7] by CIE TC8-01 Colour appearance modelling for colour management systems, it has been used to predict colour appearance under a wide range of viewing conditions, to specify colour appearance in terms of perceptual attributes, to quantify colour differences, to provide a uniform colour space and to provide a profile connection space for colour management. However, some problems have been identified and various approaches have been proposed to repair the model to enable it to be used in practical applications. During the 26th session of the CIE, held in Beijing in July 2007, a Technical Committee, TC8-11 CIECAM02 Mathematics, was formed to modify or extend the CIECAM02 model in order to satisfy the requirements of a wide range of industrial applications. The main problems that have been identified can be summarised as follows:

1. Mathematical failure for certain colours
2. The CIECAM02 colour domain is smaller than that of ICC profile connection space
3. The HPE matrix
4. The brightness function

Each problem will be reviewed in turn and then a possible solution that either repairs the problem or extends the model will be given as well. Note that all notations used in this paper have the same meaning as those in CIE Publication 159 [7].

2.6.4.1 Mathematical Failure

It has been found that the Lightness function:

$$J = 100(A/A_{\mathrm{w}})^{cz}$$

gives a problem for some colours. In fact Li and Luo [47] have shown that $A_{\mathrm{w}} > 0$, but for some colours, the achromatic signal

$$A = \left[2R'_a + G'_a + (1/20)B'_a - 0.305\right] \mathrm{N_{bb}}$$

can be negative; thus, the ratio in the bracket for the J function is negative which gives problem when computing J. At the beginning, it has been suggested that the source of the problem is the CAT02 transform which, for certain colours, predicts negative tristimulus values. Several approaches have been made on modifying the CAT02 matrix. Brill and Süsstrunk [48–50] found that the red and green CAT02 primaries lie outside the HPE triangle and called this as the "Yellow-Blue" problem. They suggested that the last row of the CAT02 matrix can be changed to 0, 0, 1. The changed matrix is denoted by M_{BS}. It has been found that for certain colours, using matrix M_{BS} works well, but using matrix M_{02} does not. However, this repair seems to correct neither the prediction of negative tristimulus values for the CAT02 nor the failure of CIECAM02.

Another suggestion is equivalent to set $R'_a \geq 0.1$, i.e., if $R'_a < 0.1$, then set $R'_a = 0.1$, if $R'_a \geq 0.1$, then R'_a does not change. Similar considerations are applied to G'_a and B'_a. Thus, under this modification, the achromatic signal A is non-negative. However, this change causes new problem with the inverse model.

Li et al. [51] gave a mathematical approach for obtaining CAT02 matrix. The approach has two constraints. The first one is to ensure the CAT02 predict corresponding colours with non-negative tristimulus values under all the illuminants considered for all colours located on or inside the CIE chromaticity locus. The second one is to fit all the corresponding colour data sets. This approach indeed ensures the CAT02 with the new matrix predicts corresponding colours with non-negative tristimulus values which is important in many applications. However, this approach does not solve the mathematical failure problem for the CIECAM02.

Recently, Li et al. [14] proposed a mathematical approach for ensuring the achromatic signal A being non-negative, at the same time the CIECAM02 should fit all the colour appearance data sets. Finally the problem is formulated as a constrained non-linear optimisation problem. By solving the optimization problem, a new CAT02 matrix was derived. With this new matrix, it was found that the mathematical failure problem of the CIECAM02 is overcome for all the illuminants considered. Besides, they also found that if the CAT02 with the HPE matrix, the mathematical failure problem is also overcome for any illuminant. More important, the HPE matrix makes the CIECAM02 simpler. All the new matrices are under the evaluation of the CIE TC8-11.

2.6.4.2 CIECAM02 Domain is Smaller than that of ICC Profile Connection Space

The ICC has developed and refined a comprehensive and rigorous system for colour management [52]. In an ICC colour management work flow, an input colour is mapped from a device colour space into a colorimetric description for specific viewing conditions (called the profile connection space—PCS). The PCS is selected as either CIE XYZ or Lab space under illuminant D50 and the 2° observer. Generally speaking, the input and output devices have different gamuts and, hence, a gamut mapping is involved. Gamut mapping in XYZ space can cause problems because of the perceptual non-uniformity of that colour space. Lab space is not a good space for gamut mapping since lines of constant hue are not generally straight lines, especially in the blue region [53]. CIECAM02 has been shown to have a superior perceptual uniformity as well as better hue constancy [40]. Thus, the CIECAM02 space has been selected as the gamut mapping space.

However, the ICC PCS can contain non-physical colours, which cause problems when transforming to CIECAM02 space, for example, in the Lightness function J defined above and the calculation of the parameter defined by

$$t = \frac{(50000/13)N_c N_{cb} e_t (a^2+b^2)^{1/2}}{R'_a + G'_a + (21/20)B'_a}.$$

When computing J, the value of A can be negative and when computing t, $R'_a + G'_a + (21/20)B'_a$ can be or near zero. One approach [41, 42] to solving these problems is to find the domain of CIECAM02 and to pre-clip or map colour values outside of this domain to fall inside or on this domain boundary, and then the CIECAM02 model can be applied without any problems. The drawbacks of this approach are that a two step transformation is not easily reversible to form a round trip solution and clipping in some other colour space would seem to defeat much of the purpose of choosing CIECAM02 as the gamut mapping space. Another approach [54] is to extend CIECAM02 so that it will not affect colours within its normal domain but it will still work, in the sense of being mathematically well defined, for colours outside its normal domain. To investigate this, the J function and the non-linear post-adaptation functions in the CIECAM02 were extended. Furthermore, scaling factors were introduced to avoid the difficulty in calculating the t value. Simulation results showed this extension of CIECAM02 works very well and full details can be found in the reference [54]. This approach is also under the evaluation of the CIE TC8-11.

2.6.4.3 The HPE Matrix;

Kuo et al. [55] found that the sum of the first row of the HPE matrix (eq. (12)) is different from unity, which causes a non-zero value of a and b when transforming the test light source to the reference (equal-energy) light source under full adaptation. Hence, a slight change to the matrix should be made. For example, the top right element -0.07868 could be changed to -0.07869. In fact, Kuo et al. [55] suggested changing each element in the first row slightly.

2.6.4.4 The Brightness Function

The brightness function of CIECAM02 is different from the brightness function of the older CIECAM97s model. The major reason for the change [56] was because of the correction to the saturation function (s). However, it has been reported that the brightness prediction of CIECAM02 does not correlate well with the appropriate visual data [57]. More visual brightness data is needed to clarify the brightness function.

2.7 Conclusion

This chapter describes the CIECAM02 in great details. Furthermore, more recent works have been introduced to extend its functions. Efforts were made to reduce the problems such as mathematical failure for the computation of the lightness attribute.

Overall, the CIECAM02 is capable of accurately predicting colour appearance under a wide range of viewing conditions. It has been proved to achieve successfully cross-media colour reproduction (e.g., the reproduction of an image on a display, on a projection screen or as hard copy) and is adopted by the Microsoft Company in their latest colour management system, window color system (WCS). With the addition of CAM02-UCS uniform colour space, size effect and unrelated colours, it will become a comprehensive colour appearance models to serve most of the applications.

Appendix: CIE Colour Appearance Model: CIECAM02

Part 1: The Forward Mode

Input: X, Y, Z (under test illuminant $X_{\mathrm{w}}, Y_{\mathrm{w}}, Z_{\mathrm{w}}$)

Output: Correlates of lightness J, chroma C, hue composition H, hue angle h, colourfulness M, saturation s and brightness Q

Illuminants, viewing surrounds set up and background parameters
(See the note at the end of this Appendix for determining all parameters)

Adopted white in test illuminant: $X_{\mathrm{w}}, Y_{\mathrm{w}}, Z_{\mathrm{w}}$
Background in test conditions: Y_{b}
(Reference white in reference illuminant: $X_{\mathrm{wr}} = Y_{\mathrm{wr}} = Z_{\mathrm{wr}} = 100$, which are fixed in the model)
Luminance of test-adapting field (cd/m^2) : L_{A}
All surround parameters are given in Table 2.3 below
Note that for determining the surround conditions, see the note at the end of this Appendix. N_c and F are modelled as a function of c, and can be linearly interpolated as shown in the Fig. 2.16 below, using the above points

Step 0: Calculate all values/parameters which are independent of input samples

$$\begin{pmatrix} R_{\mathrm{w}} \\ G_{\mathrm{w}} \\ B_{\mathrm{w}} \end{pmatrix} = M_{\mathrm{CAT02}} \cdot \begin{pmatrix} X_{\mathrm{w}} \\ Y_{\mathrm{w}} \\ Z_{\mathrm{w}} \end{pmatrix}, \quad D = F \cdot \left[1 - \left(\frac{1}{3.6}\right) \cdot \mathrm{e}^{\left(\frac{-L_{\mathrm{A}}-42}{92}\right)}\right].$$

Note if D is greater than one or less than zero, set it to one or zero, respectively.

$$D_{\mathrm{R}} = D \cdot \frac{Y_{\mathrm{w}}}{R_{\mathrm{w}}} + 1 - D, \quad D_{\mathrm{G}} = D \cdot \frac{Y_{\mathrm{w}}}{G_{\mathrm{w}}} + 1 - D, \quad D_{\mathrm{B}} = D \cdot \frac{Y_{\mathrm{w}}}{B_{\mathrm{w}}} + 1 - D,$$

$$F_L = 0.2\, k^4 \cdot (5L_A) + 0.1(1 - k^4)^2 \cdot (5L_A)^{1/3},$$

Table 2.3 Surround parameters

	F	c	N_c
Average	1.0	0.69	1.0
Dim	0.9	0.59	0.9
Dark	0.8	0.535	0.8

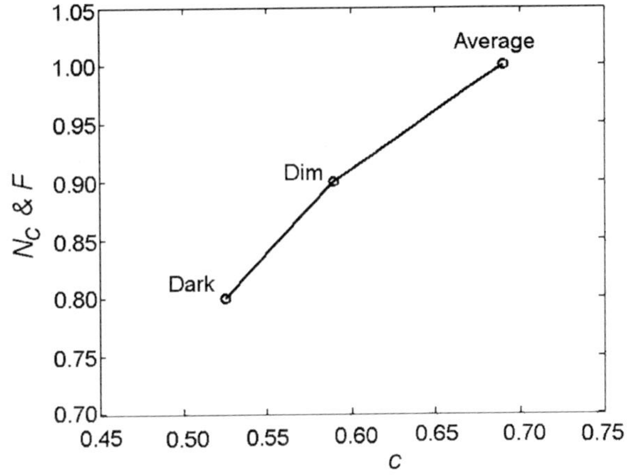

Fig. 2.16 N_c and F varies with c

where $k = \frac{1}{5 \cdot L_A + 1}$.

$$n = \frac{Y_b}{Y_w}, \quad z = 1.48 + \sqrt{n}, \quad N_{bb} = 0.725 \cdot \left(\frac{1}{n}\right)^{0.2}, \quad N_{cb} = N_{bb},$$

$$\begin{pmatrix} R_{wc} \\ G_{wc} \\ B_{wc} \end{pmatrix} = \begin{pmatrix} D_R \cdot R_w \\ D_G \cdot G_w \\ D_B \cdot B_w \end{pmatrix}, \quad \begin{pmatrix} R'_w \\ G'_w \\ B'_w \end{pmatrix} = M_{HPE} \cdot M_{CAT02}^{-1} \cdot \begin{pmatrix} R_{wc} \\ G_{wc} \\ B_{wc} \end{pmatrix},$$

$$M_{CAT02} = \begin{pmatrix} 0.7328 & 0.4296 & -0.1624 \\ -0.7036 & 1.6975 & 0.0061 \\ 0.0030 & 0.0136 & 0.9834 \end{pmatrix},$$

$$M_{HPE} = \begin{pmatrix} 0.38971 & 0.68898 & -0.07868 \\ -0.22981 & 1.18340 & 0.04641 \\ 0.00000 & 0.00000 & 1.00000 \end{pmatrix},$$

$$R'_{aw} = 400 \cdot \left(\frac{\left(\frac{F_L \cdot R'_w}{100}\right)^{0.42}}{\left(\frac{F_L \cdot R'_w}{100}\right)^{0.42} + 27.13} \right) + 0.1,$$

$$G'_{\mathrm{aw}} = 400 \cdot \left(\frac{\left(\frac{F_{\mathrm{L}} \cdot G'_{\mathrm{w}}}{100}\right)^{0.42}}{\left(\frac{F_{\mathrm{L}} \cdot G'_{\mathrm{w}}}{100}\right)^{0.42} + 27.13} \right) + 0.1,$$

$$B'_{\mathrm{aw}} = 400 \cdot \left(\frac{\left(\frac{F_{\mathrm{L}} \cdot B'_{\mathrm{w}}}{100}\right)^{0.42}}{\left(\frac{F_{\mathrm{L}} \cdot B'_{\mathrm{w}}}{100}\right)^{0.42} + 27.13} \right) + 0.1,$$

$$A_{\mathrm{w}} = \left[2 \cdot R'_{\mathrm{aw}} + G'_{\mathrm{aw}} + \frac{B'_{\mathrm{aw}}}{20} - 0.305 \right] \cdot N_{\mathrm{bb}}.$$

Note that all parameters computed in this step are needed for the following calculations. However, they depend only on surround and viewing conditions; hence, when processing pixels of image, they are computed once for all. The following computing steps are sample dependent.

Step 1: Calculate (sharpened) cone responses (transfer colour-matching functions to sharper sensors)

$$\begin{pmatrix} R \\ G \\ B \end{pmatrix} = M_{\mathrm{CAT02}} \cdot \begin{pmatrix} X \\ Y \\ Z \end{pmatrix},$$

Step 2: Calculate the corresponding (sharpened) cone response (considering various luminance level and surround conditions included in D; hence, in D_{R}, D_{G} and D_{B})

$$\begin{pmatrix} R_{\mathrm{c}} \\ G_{\mathrm{c}} \\ B_{\mathrm{c}} \end{pmatrix} = \begin{pmatrix} D_{\mathrm{R}} \cdot R \\ D_{\mathrm{G}} \cdot G \\ D_{\mathrm{B}} \cdot B \end{pmatrix},$$

Step 3: Calculate the Hunt-Pointer-Estevez response

$$\begin{pmatrix} R' \\ G' \\ B' \end{pmatrix} = M_{\mathrm{HPE}} \cdot M_{\mathrm{CAT02}}^{-1} \cdot \begin{pmatrix} R_{\mathrm{c}} \\ G_{\mathrm{c}} \\ B_{\mathrm{c}} \end{pmatrix},$$

Step 4: Calculate the post-adaptation cone response (resulting in dynamic range compression)

$$R'_{\mathrm{a}} = 400 \cdot \left(\frac{\left(\frac{F_{L} \cdot R'}{100}\right)^{0.42}}{\left(\frac{F_{L} \cdot R'}{100}\right)^{0.42} + 27.13} \right) + 0.1.$$

If R' is negative, then

Table 2.4 Unique hue data for calculation of hue quadrature

	Red	Yellow	Green	Blue	Red
i	1	2	3	4	5
h_i	20.14	90.00	164.25	237.53	380.14
e_i	0.8	0.7	1.0	1.2	0.8
H_i	0.0	100.0	200.0	300.0	400.0

$$R'_{\mathrm{a}} = -400 \cdot \left(\frac{\left(\frac{-F_L \cdot R'}{100}\right)^{0.42}}{\left(\frac{-F_L \cdot R'}{100}\right)^{0.42} + 27.13} \right) + 0.1$$

and similarly for the computations of G'_{a}, and B'_{a}, respectively.

Step 5: Calculate Redness–Greenness (a) , Yellowness–Blueness (b) components and hue angle (h):

$$a = R'_{\mathrm{a}} - \frac{12 \cdot G'_{\mathrm{a}}}{11} + \frac{B'_{\mathrm{a}}}{11},$$

$$b = \frac{(R'_{\mathrm{a}} + G'_{\mathrm{a}} - 2 \cdot B'_{\mathrm{a}})}{9},$$

$$h = \tan^{-1}\left(\frac{b}{a}\right)$$

make sure h between 0 and 360°.

Step 6: Calculate eccentricity (e_t) and hue composition (H), using the unique hue data given in Table 2.4; set $h' = h + 360$ if $h < h_1$, otherwise $h' = h$. Choose a proper $i (i = 1,2,3$ or $4)$ so that $h_i \le h' < h_{i+1}$. Calculate

$$e_t = \frac{1}{4} \cdot \left[\cos\left(\frac{h' \cdot \pi}{180} + 2 \right) + 3.8 \right],$$

which is close to, but not exactly the same as, the eccentricity factor given in Table 2.4.

$$H = H_i + \frac{100 \cdot \frac{h' - h_i}{e_i}}{\frac{h' - h_i}{e_i} + \frac{h_{i+1} - h'}{e_{i+1}}}.$$

Step 7: Calculate achromatic response A

$$A = \left[2 \cdot R'_{\mathrm{a}} + G'_{\mathrm{a}} + \frac{B'_{\mathrm{a}}}{20} - 0.305 \right] \cdot N_{\mathrm{bb}}.$$

Step 8: Calculate the correlate of lightness

$$J = 100 \cdot \left(\frac{A}{A_{\mathrm{w}}} \right)^{c \cdot z}.$$

Step 9: Calculate the correlate of brightness

$$Q = \left(\frac{4}{c}\right) \cdot \left(\frac{J}{100}\right)^{0.5} \cdot (A_{\mathrm{w}} + 4) \cdot F_{\mathrm{L}}^{0.25}.$$

Step 10: Calculate the correlates of chroma (C), colourfulness (M) and saturation (s)

$$t = \frac{\left(\frac{50000}{13} \cdot N_{\mathrm{c}} \cdot N_{\mathrm{cb}}\right) \cdot e_t \cdot \left(a^2 + b^2\right)^{1/2}}{R'_{\mathrm{a}} + G'_{\mathrm{a}} + \left(\frac{21}{20}\right) \cdot B'_{\mathrm{a}}},$$

$$C = t^{0.9} \cdot \left(\frac{J}{100}\right)^{0.5} \cdot (1.64 - 0.29^{n})^{0.73},$$

$$M = C \cdot F_{\mathrm{L}}^{0.25},$$

$$s = 100 \cdot \left(\frac{M}{Q}\right)^{0.5}.$$

Part 2: The Reverse Mode

Input: J or Q; C, M or s; H or h
Output: X,Y,Z (under test illuminant $X_{\mathrm{w}},Y_{\mathrm{w}},Z_{\mathrm{w}}$)

Illuminants, viewing surrounds and background parameters are the same as those given in the forward mode. See notes at the end of this Appendix calculating/defining the luminance of the adapting field and surround conditions.

Step 0: Calculate viewing parameters
Compute all $F_{\mathrm{L}}, n, z, N_{\mathrm{bb}} = N_{\mathrm{bc}}, R_{\mathrm{w}}, G_{\mathrm{w}}, B_{\mathrm{w}}$, D, $D_{\mathrm{R}}, D_{\mathrm{G}}, D_{\mathrm{B}}$, $R_{\mathrm{wc}}, G_{\mathrm{wc}}, B_{\mathrm{wc}}$, $R'_{\mathrm{w}}, G'_{\mathrm{w}}, B'_{\mathrm{w}}$ $R'_{\mathrm{aw}}, G'_{\mathrm{aw}}, B'_{\mathrm{aw}}$ and A_{w} using the same formulae as in Step 0 of the Forward model. They are needed in the following steps. Note that all data computed in this step can be used for all samples (e.g., all pixels for an image) under the viewing conditions. Hence, they are computed once for all. The following computing steps are sample dependent.

Step 1: Obtain J, C and h from H, Q, M, s
The entering data can be in different combination of perceived correlates, i.e., J or Q; C, M, or s; and H or h. Hence, the followings are needed to convert the others to J, C, and h.

Step 1–1: Compute J from Q (if start from Q)

$$J = 6.25 \cdot \left[\frac{c \cdot Q}{(A_{\mathrm{w}} + 4) \cdot F_{\mathrm{L}}^{0.25}}\right]^2.$$

Step 1–2: Calculate C from M or s

$$C = \frac{M}{F_{\mathrm{L}}^{0.25}} \text{ (if start from } M\text{)}$$

$$\mathrm{Q} = \left(\frac{4}{c}\right) \cdot \left(\frac{\mathrm{J}}{100}\right)^{0.5} \cdot (\mathrm{A_w}+4.0) \cdot F_L^{0.25}$$

and $C = \left(\frac{s}{100}\right)^2 \cdot \left(\frac{Q}{F_{\mathrm{L}}^{0.25}}\right)$ (if start from s)

Step 1–3: Calculate h from H (if start from H)
The correlate of hue (h) can be computed by using data in Table 2.4 in the Forward mode.
Choose a proper i ($i = 1,2,3$ or 4) so that $H_i \le H < H_{i+1}$.

$$h' = \frac{(H - H_i) \cdot (e_{i+1}h_i - e_i \cdot h_{i+1}) - 100 \cdot h_i \cdot e_{i+1}}{(H - H_i) \cdot (e_{i+1} - e_i) - 100 \cdot e_{i+1}}.$$

Set $h = h' - 360$ if $h' > 360$, otherwise $h = h'$.

Step 2: Calculate t, e_t, p_1, p_2 and p_3

$$t = \left[\frac{C}{\sqrt{\frac{J}{100}} \cdot (1.64 - 0.29^n)^{0.73}}\right]^{\frac{1}{0.9}},$$

$$e_t = \frac{1}{4} \cdot \left[\cos\left(h \cdot \frac{\pi}{180} + 2\right) + 3.8\right],$$

$$A = A_{\mathrm{w}} \cdot \left(\frac{J}{100}\right)^{\frac{1}{c \cdot z}},$$

$$p_1 = \left(\frac{50000}{13} \cdot N_c \cdot N_{\mathrm{cb}}\right) \cdot e_t \cdot \left(\frac{1}{t}\right), \text{ if t} \neq 0,$$

$$p_2 = \frac{A}{N_{\mathrm{bb}}} + 0.305,$$

$$p_3 = \frac{21}{20},$$

Step 3: Calculate a and b
If $t = 0$, then $a = b = 0$ and go to Step 4
(be sure transferring h from degree to radian before calculating $\sin(h)$ and $\cos(h)$)
If $|\sin(h)| \ge |\cos(h)|$, then

$$p_4 = \frac{p_1}{\sin(h)},$$

$$b = \frac{p_2 \cdot (2+p_3) \cdot \left(\frac{460}{1403}\right)}{p_4 + (2+p_3) \cdot \left(\frac{220}{1403}\right) \cdot \left(\frac{\cos(h)}{\sin(h)}\right) - \left(\frac{27}{1403}\right) + p_3 \cdot \left(\frac{6300}{1403}\right)},$$

$$a = b \cdot \left(\frac{\cos(h)}{\sin(h)}\right).$$

If $|\cos(h)| > |\sin(h)|$, then

$$p_5 = \frac{p_1}{\cos(h)},$$

$$a = \frac{p_2 \cdot (2+p_3) \cdot \left(\frac{460}{1403}\right)}{p_5 + (2+p_3) \cdot \left(\frac{220}{1403}\right) - \left[\left(\frac{27}{1403}\right) - p_3 \cdot \left(\frac{6300}{1403}\right)\right] \cdot \left(\frac{\sin(h)}{\cos(h)}\right)},$$

$$b = a \cdot \left(\frac{\sin(h)}{\cos(h)}\right).$$

Step 4: Calculate R'_a, G'_a and B'_a

$$R'_\mathrm{a} = \frac{460}{1403} \cdot p_2 + \frac{451}{1403} \cdot a + \frac{288}{1403} \cdot b,$$

$$G'_\mathrm{a} = \frac{460}{1403} \cdot p_2 - \frac{891}{1403} \cdot a - \frac{261}{1403} \cdot b,$$

$$B'_\mathrm{a} = \frac{460}{1403} \cdot p_2 - \frac{220}{1403} \cdot a - \frac{6300}{1403} \cdot b.$$

Step 5: Calculate R', G' and B'

$$R' = \mathrm{sign}(R'_\mathrm{a} - 0.1) \cdot \frac{100}{F_L} \cdot \left[\frac{27.13 \cdot |R'_\mathrm{a} - 0.1|}{400 - |R'_\mathrm{a} - 0.1|}\right]^{\frac{1}{0.42}}.$$

Here, $\mathrm{sign}(x) = \begin{cases} 1 & \text{if } x > 0 \\ 0 & \text{if } x = 0 \\ -1 & \text{if } x < 0 \end{cases}$, and similarly computing G', and B' from G'_a, and B'_a.

Step 6: Calculate R_C, G_C and B_C (for the inverse matrix, see the note at the end of the Appendix)

$$\begin{pmatrix} R_\mathrm{c} \\ G_\mathrm{c} \\ B_\mathrm{c} \end{pmatrix} = M_{\mathrm{CAT02}} \cdot M_{\mathrm{HPE}}^{-1} \cdot \begin{pmatrix} R' \\ G' \\ B' \end{pmatrix}.$$

Step 7: Calculate R, G and B

$$\begin{pmatrix} R \\ G \\ B \end{pmatrix} = \begin{pmatrix} \frac{R_c}{D_R} \\ \frac{G_c}{D_G} \\ \frac{B_c}{D_B} \end{pmatrix}.$$

Step 8: Calculate X, Y and Z (for the coefficients of the inverse matrix, see the note at the end of the Appendix)

$$\begin{pmatrix} X \\ Y \\ Z \end{pmatrix} = M_{\mathrm{CAT02}}^{-1} \cdot \begin{pmatrix} R \\ G \\ B \end{pmatrix}.$$

Notes to Appendix

1. It is recommended to use the matrix coefficients given below for the inverse matrix M_{CAT02}^{-1} and M_{HPE}^{-1}:

$$M_{\mathrm{CAT02}}^{-1} = \begin{pmatrix} 1.096124 & -0.278869 & 0.182745 \\ 0.454369 & 0.473533 & 0.072098 \\ -0.009628 & -0.005698 & 1.015326 \end{pmatrix},$$

$$M_{\mathrm{HPE}}^{-1} = \begin{pmatrix} 1.910197 & -1.112124 & 0.201908 \\ 0.370950 & 0.629054 & -0.000008 \\ 0.000000 & 0.000000 & 1.000000 \end{pmatrix}$$

2. For implementing the CIECAM02, the testing data and the corresponding results from the forward and reverse modes can be found from reference 7.
3. The L_A is computed using (2.11)

$$L_A = \left(\frac{E_W}{\pi}\right) \cdot \left(\frac{Y_b}{Y_W}\right) = \frac{L_W \cdot Y_b}{Y_W}, \tag{2.11}$$

where $E_w = \pi \cdot L_w$ is the illuminance of reference white in lux unit; L_w the luminance of reference white in cd/m^2 unit, Y_b the luminance factor of the background and Y_w the luminance factor of the reference white.

References

1. Luo MR (1999) Colour science: past, present and future. In: MacDonald LW and Luo MR (Eds) Colour imaging: vision and technology. Wiley, New York, 384–404
2. CIE Technical Report (2004) Colorimetry, 3rd ed. Publication 15:2004, CIE Central Bureau, Vienna.

3. Luo MR, Cui GH, Rigg B (2001) The development of the CIE 2000 colour difference formula. Color Res Appl 26:340-350.
4. Luo MR, Hunt RWG (1998) The structure of the CIE 1997 colour appearance model (CIECAM97s). Color Res Appl 23:138–146
5. CIE (1998) The CIE 1997 interim colour appearance model (simple version), CIECAM97s. CIE Publication 131, CIE Central Bureau, Vienna, Austria.
6. Moroney N, Fairchild MD, Hunt RWG, Li C, Luo MR, Newman T (2002) The CIECAM02 color appearance model, Proceedings of the 10th color imaging conference, IS&T and SID, Scottsdale, Arizona, 23–27
7. CIE (2004) A colour appearance model for colour management systems: CIECAM02, CIE Publication 159 CIE Central Bureau, Vienna, Austria
8. Luo MR and Li CJ (2007) CIE colour appearance models and associated colour spaces, Chapter 11 of the book: colorimetry-understanding the CIE System. In: Schanda J (ed) Wiley, New York
9. Luo MR, Cui GH, Li CJ and Rigg B (2006) Uniform colour spaces based on CIECAM02 colour appearance model. Color Res Appl 31:320–330
10. Xiao K, Luo MR, Li C, Hong G (2010) Colour appearance prediction for room colours, Color Res Appl 35:284–293
11. Xiao K, Luo MR, Li CJ, Cui G, Park D (2011) Investigation of colour size effect for colour appearance assessment, Color Res Appl 36:201–209
12. Xiao K, Luo MR, Li CJ (2012) Color size effect modelling, Color Res Appl 37:4–12
13. Fu CY, Li CJ, Luo MR, Hunt RWG, Pointer MR (2007) Quantifying colour appearance for unrelated colour under photopic and mesopic vision, Proceedings of the 15th color imaging conference, IS&T and SID, Albuquerque, New Mexico, 319–324
14. Li CJ, Chorro-Calderon E, Luo MR, Pointer MR (2009) Recent progress with extensions to CIECAM02, Proceedings of the 17th color imaging conference, IS&T and SID, Albuquerque, New Mexico 69–74
15. CIE Publ. 17.4:1987, International lighting vocabulary, the 4th edition
16. Mori L, Sobagaki H, Komatsubara H, Ikeda K (1991) Field trials on CIE chromatic adaptation formula. Proceedings of the CIE 22nd session, 55–58
17. McCann JJ, McKee SP, Taylor TH (1976) Quantitative studies in Retinex theory: a comparison between theoretical predictions and observer responses to the 'color mondrian ' experiments. Vision Res 16:445–458
18. Breneman EJ (1987) Corresponding chromaticities for different states of adaptation to complex visual fields. J Opt Soc Am A 4:1115–1129
19. Helson H, Judd DB, Warren MH (1952) Object-color changes from daylight to incandescent filament illumination. Illum Eng 47:221–233
20. Lam KM (1985) Metamerism and colour constancy. Ph.D. thesis, University of Bradford, UK
21. Braun KM, Fairchild MD (1996) Psychophysical generation of matching images for cross-media colour reproduction. Proceedings of 4th color imaging conference, IS&T, Springfield, Va., 214–220
22. Luo MR, Clarke AA, Rhodes PA, Schappo A, Scrivener SAR, Tait C (1991) Quantifying colour appearance. Part I. LUTCHI colour appearance data. Color Res Appl 16:166–180
23. Luo MR, Gao XW, Rhodes PA, Xin HJ, Clarke AA, Scrivener SAR (1993) Quantifying colour appearance, Part IV: transmissive media. Color Res Appl 18:191–209
24. Kuo WG, Luo MR, Bez HE (1995) Various chromatic adaptation transforms tested using new colour appearance data in textiles. Color Res Appl 20:313–327
25. Juan LY, Luo MR (2000) New magnitude estimation data for evaluating colour appearance models. Colour and Visual Scales 2000, NPL, 3-5 April, UK
26. Juan LY, Luo MR (2002) Magnitude estimation for scaling saturation. Proceedings of 9th session of the association internationale de la couleur (AIC Color 2001), Rochester, USA, (June 2001), Proceedings of SPIE 4421, 575–578
27. Li CJ, Luo MR, Rigg B, Hunt RWG (2002) CMC 2000 chromatic adaptation transform: CMCCAT2000. Color Res Appl 27:49–58

28. Judd DB (1940), Hue, saturation, and lightness of surface colors with chromatic illumination. J Opt Soc Am 30:2–32
29. Kries V (1902), Chromatic adaptation, Festschrift der Albrecht-Ludwig-Universitat (Fribourg), [Translation: MacAdam DL, Sources of Color Science, MIT Press, Cambridge, Mass. (1970)]
30. Luo MR, Hunt RWG (1998) A chromatic adaptation transform and a colour inconstancy index. Color Res Appl 23:154–158
31. Li CJ, Luo MR, Hunt RWG (2000) A revision of the CIECAM97s Model. Color Res Appl 25:260–266
32. Hunt RWG, Li CJ, Juan LY, Luo MR (2002), Further improvements to CIECAM97s. Color Res Appl 27:164–170
33. Finlayson GD, Süsstrunk S (2000) Performance of a chromatic adaptation transform based on spectral sharpening. Proceedings of IS&T/SID 8th color imaging conference, 49–55
34. Hunt RWG (1952) Light and dark adaptation and perception of color. J Opt Soc Am 42:190–199
35. Stevens JC, Stevens SS (1963) Brightness functions: effects of adaptation. J. Opt Soc Am 53:375–385
36. Bartleson CJ, Breneman EJ (1967) Brightness perception in complex fields. J. Opt Soc Am 57:953–957
37. Luo MR, Gao XW, Sciviner SAR (1995) Quantifying colour appearance, Part V, Simultaneous contrast. Color Res Appl 20:18–28
38. Wyszecki G, Stiles WS (1982) Color Science: concepts and methods, Quantitative data and formulae. Wiley, New York
39. Helson H (1938) Fundamental problems in color vision. I. The principle governing changes in hue, saturation, and lightness of non-selective samples in chromatic illumination. J Exp Psych 23:439–477
40. CIE Publ. 152:2003, Moroney N, Han Z (2003) Field trials of the CIECAM02 colour appearance, Proceedings of the 25th session of the CIE, San Diego D8-2–D8-5.
41. Tastl I, Bhachech M, Moroney N, Holm J (2005) ICC colour management and CIECAM02, Proceedings of the 13th of CIC, p 318
42. Gury R, Shaw M (2005) Dealing with imaginary color encodings in CIECAM02 in an ICC workflow. Proceedings of the 13th of CIC, pp 217–223
43. Li CJ, Luo MR, Cui GH (2003) Colour-difference evaluation using colour appearance models. The 11th Color Imaging Conference, IS&T and SID, Scottsdale, Arizona, November, 127–131
44. Luo MR, Rigg B (1986) Chromaticity–discrimination ellipses for surface colours. Color Res Appl 11:25–42
45. Berns RS, Alman DH, Reniff L, Snyder GD, Balonon-Rosen MR (1991) Visual determination of suprathreshold color-difference tolerances using probit analysis. Color Res Appl 16:297–316
46. Hunt RWG (1952) Measuring colour, 3rd edition, Fountain Press, Kingston-upon-Thames, 1998
47. Li CJ, Chorro-Calderon E, Luo MR, Pointer MR (2009) Recent progress with extensión to CIECAM02, Seventeenth Colour Imaging Conference, Final Program and Proceedings, 69–74
48. Brill MH (2006) Irregularity in CIECAM02 and its avoidance. Color Res Appl 31(2):142–145
49. Brill MH, Susstrunk S (2008) Repairing gamut problems in CIECAM02: a progress report. Color Res Appl 33(5):424–426
50. Süsstrunk S, Brill M (2006) The nesting instinct: repairing non nested gamuts in CIECAM02. 14th SID/IS&T color imaging conference
51. Li CJ, Perales E, Luo MR, Martínez-Verdú F, A Mathematical approach for predicting non-negative tristimulus values using the CAT02 chromatic adaptation transform, Color Res Appl (in press)
52. ISO 15076-1 (2005) Image technology, colour management-Architecture, profile format and data structure-Part I: based on ICC.1:2004-10, http://www.color.org
53. Moroney N (2003) A hypothesis regarding the poor blue constancy of CIELAB. Color Res Appl 28(5):371–378

54. W. Gill GW (2008) A solution to CIECAM02 numerical and range issues, Proceedings of the 16th of color imaging conference, IS&T and SID, Portland, Oregan 322–327
55. Kuo CH, Zeise E, Lai D (2006) Robust CIECAM02 implementation and numerical experiment within an ICC workflow. Proceedings of the 14th of CIC, pp 215–219
56. Hunt RWG, Li CJ, Luo MR (2002) Dynamic cone response functions for modes of colour appearance. Color Res Appl 28:82–88
57. Paula J, Alessi P (2008) Private communication pursuit of scales corresponding to equal perceptual brightness, personal correspondence

Chapter 3
Colour Difference Evaluation

Manuel Melgosa, Alain Trémeau, and Guihua Cui

In the black, all the colors agree

Francis Bacon

Abstract For a pair of homogeneous colour samples or two complex images viewed under specific conditions, colour-difference formulas try to predict the visually perceived (subjective) colour difference starting from instrumental (objective) colour measurements. The history related to the five up-to-date CIE-recommended colour-difference formulas is reviewed, with special emphasis on the structure and performance of the last one, CIEDE2000. Advanced colour-difference formulas with an associated colour space (e.g., DIN99d, CAM02, Euclidean OSA-UCS, etc.) are also discussed. Different indices proposed to measure the performance of a given colour-difference formula (e.g., PF/3, STRESS, etc.) are reviewed. Among current trends on colour-difference evaluation, it can be mentioned the research activities carried out by different CIE Technical Committees (e.g., CIE TC's 1-55, 1-57, 1-63, 1-81 and 8-02), the need of new reliable experimental datasets, the development of colour-difference formulas based on IPT and colour-appearance models, and the concept of "total differences," which considers the interactions between colour properties and other object attributes like texture, translucency, and gloss.

M. Melgosa (✉)
Departamento de Optica, Facultad de Ciencias, Universidad de Granada, Spain
e-mail: mmelgosa@ugr.es

A. Trémeau
Laboratory Hubert Curien, UMR CNRS 5516, Jean Monnet University, Saint-Etienne, France
e-mail: alain.tremeau@univ-st-etienne.fr

G. Cui
VeriVide Limited, Leicester, LE19 4SG, United Kingdom
e-mail: guihua.cui@gmail.com

C. Fernandez-Maloigne (ed.), *Advanced Color Image Processing and Analysis*,
DOI 10.1007/978-1-4419-6190-7_3,

Keywords Colour-difference formula • Uniform colour space • CIELUV • CIELAB • CIE94 • CIEDE2000 • DIN99 • CAM02-SCD • CAM02-LCD • CAM02-UCS • S-CIELAB • IPT • PF/3 • STRESS

3.1 Introduction

From two homogeneous colour stimuli, we can ask ourselves what is the magnitude of the perceived colour difference between them. Of course, this question may also be asked in the case of more complex stimuli like two colour images. In fact, to achieve a consistent answer to the previous question, we must first specify the experimental observation conditions: for example, size of the stimuli, background behind them, illuminance level, etc. It is well known that experimental illuminating and viewing conditions (the so-called "parametric effects") play an important role on the magnitude of perceived colour differences, as reported by the International Commission on Illumination (CIE) [1]. Specifically, to avoid the spread of experimental results under many different observation conditions, in 1995 the CIE proposed [2] to analyze just 17 "colour centers" well distributed in colour space (Table 3.1), under a given set of visual conditions similar to those usually found in industrial practice, which are designated as "reference conditions," and are as follows:

Illumination: D65 source
Illuminance: 1000 lx
Observer: Normal colour vision
Background field: Uniform, neutral grey with $L^* = 50$

Table 3.1 The 17 colour centers proposed by CIE for further coordinated research on colour-difference evaluation [2]. Bold letters indicate five colour centers used as experimental controls, which were earlier proposed with the same goal by CIE (A.R. Robertson, Color Res. Appl. 3, 149–151, 1978)

Name	L^*_{10}	a^*_{10}	b^*_{10}
1. Grey	**62**	**0**	**0**
2. Red	**44**	**37**	**23**
3. Red, high chroma	44	58	36
4. Orange	63	13	21
5. Orange, high chroma	63	36	63
6. Yellow	**87**	**−7**	**47**
7. Yellow, high chroma	87	−11	76
8. Yellow-green	65	−10	13
9. Yellow-green, high chroma	65	−30	39
10. Green	**56**	**−32**	**0**
11. Green, high chroma	56	−45	0
12. Blue-green	50	−16	−11
13. Blue-green, high chroma	50	−32	−22
14. Blue	**36**	**5**	**−31**
15. Blue, high chroma	34	7	−44
16. Purple	46	12	−13
17. Purple, high chroma	46	26	−26

Viewing mode: Object
Sample size: Greater than four degrees
Sample separation: Direct edge contact
Sample colour-difference magnitude: Lower than 5.0 ΔE^*_{ab}
Sample structure: Homogeneous (without texture)

The perceived visual difference between two colour stimuli is often designated as ΔV, and it is just the subjective answer provided by our visual system. It must be mentioned that large inter- and intra-observer variability (sometimes designated as accuracy and repeatability, respectively) can be found determining visual colour differences, even in carefully designed experiments on colour difference evaluation [3]. Intra and inter-observer variability was rarely considered in old experiments (e.g., the pioneer MacAdam's experiment [4] producing x,y chromaticity discrimination ellipses involved just one observer), but it is essential in modern experiments on colour differences [5], because individuals give results that rarely correlate with those of population. Although different methods have been proposed to obtain the visual difference ΔV in a colour pair, the two most popular ones are the "anchor pair" and "grey scale" methods. In "anchor pair" experiments [6], the observer just reports whether the colour difference in the colour pair is smaller or greater than the one shown in a fixed neutral colour pair. In "grey scale" experiments [7], the observer compares the colour difference in the test pair with a given set of neutral colour pairs with increasing colour-difference magnitudes, choosing the one with the closest colour difference to the test pair (Fig. 3.1). Commercial grey-scales with colour-difference pairs in geometrical progression are currently available, [8–10] although the most appropriate grey scale to be used in experiments is now questioned [11]. It has been reported [12] that "anchor pair" and "grey scale" experiments conduct only to qualitative analogous results.

In many industrial applications, it is highly desirable to predict the subjective visual colour difference ΔV from objective numerical colour specifications; specifically, the tristimulus values measurements of the two samples in a colour pair. This is just the main goal of the so-called "colour-difference formulas". A colour-difference formula can be defined as a computation providing a non-negative value ΔE from the tristimulus values of the two samples in a colour pair. It is worth to mention that in modern colour-difference formulas, additional information on parameters related to observation conditions is also considered to compute ΔE:

$$\Delta E = f\,(X_1, Y_1, Z_1, X_2, Y_2, Z_2, \text{Observation Conditions Parameters}) \qquad (3.1)$$

While ΔV is the result of a *subjective* measurement like the average of the visual assessments performed by a panel of observers, using a specific method and working under fixed observation conditions, ΔE is an *objective* measurement which can be currently performed using colorimetric instrumentation. Obviously, the main goal is to achieve a ΔE analogous to ΔV for any colour pair in colour space and under any visual set of observational conditions. In this way, complex tasks like visual pass/fail decisions in a production chain could be done in a completely automatic way

Fig. 3.1 A yellow colour pair of textile samples, together with a grey scale for visual assessment of the colour difference in such a pair. A colour mask may be employed to choose a colour pair in the grey scale, or to have a test pair with the same size than those in the grey scale. Photo from Dr. Michal Vik, Technical University of Liberec, Czech Republic

Fig. 3.2 Visual versus instrumental color-difference evaluation: example of quality control using a colorimeter. Photo from "Precise Color Communication", Konica-Minolta Sensing, Inc., 1998

(Fig. 3.2). However, it must be recognized that this is a very ambitious goal, because in fact it is intended to predict the final answer of our visual system, currently unknown in many aspects. Anyway, important advances have been produced in colour-difference measurement, as will be described in the next section.

Three different steps can be distinguished in the history of modern colorimetry: *colour matching*, *colour differences*, and *colour appearance*. *Colour matching* culminates with the definition of the tristimulus values X, Y, Z: Two stimuli, viewed under identical conditions, match for a specific standard observer when their tristimulus values are equal. This defines "basic colorimetry" and is the basis for numerical colour specification [13]. However, when tristimulus values are unequal, the match may not persist, depending on the magnitude of dissimilarity (i.e., larger or not than a threshold difference). Tristimulus values X, Y, Z (or x, y, Y coordinates) should never be used as direct estimates of *colour differences*. Colour-difference formulas can be used to measure the dissimilarity between two colour stimuli of the same size and shape, which are observed under the same visual conditions. Relating numerical differences to perceived colour differences is one of the challenges of so-called "advanced colorimetry". Finally, *colour appearance* is concerned with the description of what colour stimuli look like under a variety of visual conditions. More specifically, a colour appearance model provides a viewing condition-specific method for transforming tristimulus values to and/or from perceptual attribute correlates [14]. Application of such models open up a world of possibilities for the accurate specification, control and reproduction of colour, and may eventually include in the future the field of colour differences [15].

3.2 The CIE Recommended Colour-Difference Formulas

As described by Luo [16], first colour-difference formulas were based on the Munsell system, followed by formulas based on MacAdam's data, and linear (or non-linear) transformations of tristimulus values X, Y, Z. The interested reader can find useful information in the literature [17–19] about many colour-difference formulas proposed in the past. Colour-difference formulas have been considered in CIE programs since the 1950s, and in this section we will focus on the five up-to-date CIE-recommended colour-difference formulas.

The first CIE-recommended colour-difference formula was proposed in 1964 as the Euclidean distance in the CIE U^*, V^*, W^* colour space. This space was actually based on MacAdam's 1960 uniform colour scales (CIE 1960 UCS), which intended to improve the uniformity of the CIE 1931 x, y chromaticity diagram. In 1963, Wyszecki added the third dimension to this space. The currently proposed CIE Colour Rendering Index [20] is based on the CIE 1964 U^*, V^*, W^* colour-difference formula.

A landmark was achieved in 1976 with the joint CIE recommendation of the CIELUV and CIELAB colour spaces and colour-difference formulas [21]. As described by Robertson [22] in 1976 the CIE recommended the use of two approximately uniform colour spaces and colour-difference formulas, which were chosen from among several of similar merit to promote uniformity of practice, pending the development of a space and formula giving substantially better correlation with visual judgments. While the CIELAB colour-difference formula ΔE^*_{ab} had the advantage that it was very similar to the Adams–Nickerson (ANLAB40) formula, already adopted by several national industrial groups, the CIELUV colour-difference formula ΔE^*_{uv} had the advantage of a linear chromaticity diagram, particularly useful in lighting applications. It can be said that the CIELAB colour-difference formula was soon accepted by industry: while in 1977 more than 20 different colour-difference formulas were employed in the USA industry, 92% of these industries had adopted CIELAB in 1992 [23]. Because there are no fixed scale factors between the results provided by two different colour-difference formulas, the uniformity of practice (standardization) achieved by CIELAB was an important achievement for industrial practice. It should be said that a colour difference between 0.4 and 0.7 CIELAB units is approximately a just noticeable or threshold difference, although even lower values of colour differences are sometimes managed by specific industries. Colour differences between contiguous samples in colour atlases (e.g., Munsell Book of Color) are usually greater than 5.0 CIELAB units, being designated as large colour differences.

After the proposal of CIELAB, many CIELAB-based colour-difference formulas were proposed with considerable satisfactory results [24]. Among these CIELAB-based formulas, it is worth mentioning the CMC [25] and BFD [26] colour-difference formulas. The CMC formula was recommended by the Colour Measurement Committee of the Society of Dyers and Colourists (UK), and integrated into some ISO standards. CIELAB lightness, chroma, and hue differences are properly weighted in the CMC formula, which also includes parametric factors dependent on visual conditions (e.g., the CMC lightness differences have half value for textile samples). In 1995 the CIE proposed the CIE94 colour-difference formula [27], which may be considered a simplified version of CMC. CIE94 was based on most robust trends in three reliable experimental datasets, proposing simple corrections to CIELAB (linear weighting functions of the average chroma for the CIELAB chroma and hue differences), as well as parametric factors equal to 1.0 under the so-called "reference conditions" (see Introduction). It can be said that CIE94 adopted a versatile but too conservative approach adopting only the most well-known CIELAB corrections, like the old chroma-difference correction already suggested by McDonald for the ANLAB formula in 1974 [28].

In 2001 the CIE recommended its last colour-difference formula, CIEDE2000 [29]. From a combined dataset of reliable experimental data containing 3,657 colour pairs from four different laboratories, the CIEDE2000 formula was developed [30]. The CIEDE2000 formula has the same final structure as the BFD [26] formula. Five corrections to CIELAB were included in CIEDE2000: A weighting function for lightness accounting for the "crispening effect" produced by an achromatic background with lightness $L^* = 50$; a weighting function for chroma identical to the one adopted by the previous CIE94 formula; a weighting function for hue which is dependent on both hue and chroma; a correction of the a^* coordinate for neutral colours; and a rotation term which takes account of the experimental chroma and hue interaction in the blue region. The most important correction to CIELAB in CIEDE2000 is the chroma correction [31]. CIEDE2000 also includes parametric factors with values $k_L = k_C = k_H = 1$ under the "reference conditions" adopted by CIE94 and mentioned in the previous section. Starting from CIELAB, the mathematical equations defining the CIEDE2000 [29] colour-difference formula, noted ΔE_{00}, are as follows:

$$\Delta E_{00} = \sqrt{\left(\frac{\Delta L'}{k_L S_L}\right)^2 + \left(\frac{\Delta C'}{k_C S_C}\right)^2 + \left(\frac{\Delta H'}{k_H S_H}\right)^2 + R_T \left(\frac{\Delta C'}{k_C S_C}\right)\left(\frac{\Delta H'}{k_H S_H}\right)} \quad (3.2)$$

First, for each one of the two colour samples, designated as "b" ("batch") and "s" ("standard"), a localized modification of the CIELAB coordinate a^* is made:

$$L' = L^* \quad (3.3)$$

$$a' = (1+G)\, a^* \quad (3.4)$$

$$b' = b^* \quad (3.5)$$

$$G = 0.5\left(1 - \sqrt{\frac{\overline{C^{*7}_{ab}}}{\overline{C^{*7}_{ab}} + 25^7}}\right) \quad (3.6)$$

where the upper bar means arithmetical mean of standard and batch. Transformed a', b' are used in calculations of transformed chroma and hue angle, in the usual way [21]:

$$C' = \sqrt{a'^2 + b'^2} \quad (3.7)$$

$$h' = \arctan\left(\frac{b'}{a'}\right) \quad (3.8)$$

Lightness-, chroma-, and hue-differences employed in (3.2) are computed as follows:

$$\Delta L' = L'_b - L'_s \quad (3.9)$$

$$\Delta C' = C'_b - C'_s \quad (3.10)$$

$$\Delta h' = h'_b - h'_s \quad (3.11)$$

$$\Delta H' = 2\sqrt{C'_b C'_s}\sin\left(\frac{\Delta h'}{2}\right) \quad (3.12)$$

The "weighting functions" for lightness, chroma, and hue, where once again the upper bars means arithmetical mean of standard and batch, are as follows:

$$S_L = 1 + \frac{0.015\left(\overline{L'} - 50\right)^2}{\sqrt{20 + \left(\overline{L'} - 50\right)^2}} \quad (3.13)$$

$$S_C = 1 + 0.045\overline{C'} \quad (3.14)$$

$$S_H = 1 + 0.015\,\overline{C'}T \tag{3.15}$$

$$T = 1 - 0.17\cos\left(\overline{h'} - 30^\circ\right) + 0.24\cos\left(2\overline{h'}\right) + 0.32\cos\left(3\overline{h'} + 6^\circ\right) - 0.20\cos\left(4\overline{h'} - 63^\circ\right) \tag{3.16}$$

Finally, the rotation term R_T is defined by the next equations:

$$R_T = -\sin\left(2\Delta\theta\right)R_C \tag{3.17}$$

$$\Delta\theta = 30\exp\left\{-\left[\left(\overline{h'} - 275^\circ\right)/25\right]^2\right\} \tag{3.18}$$

$$R_C = 2\sqrt{\frac{\overline{C'}^7}{\overline{C'}^7 + 25^7}} \tag{3.19}$$

Statistical analyses confirmed that CIEDE2000 significantly improved both CIE94 and CMC colour-difference formulas, for the experimental combined dataset employed at its development [30], and therefore it was proposed to the scientific community. Figure 3.3 shows that experimental colour discrimination ellipses [30] in CIELAB a^*b^* plane are in very good agreement with predictions made by the CIEDE2000 colour-difference formula.

Sharma et al. [32] have pointed out different problems in the computation of CIEDE2000 colour differences, which were not detected at its development. Specifically, these problems come from $\Delta h'$ and $\overline{h}'$ values when samples are placed in different angular sectors, which leads to discontinuities in T (3.16) and $\Delta\theta$ (3.18) values. In the worst case, these discontinuities produced a deviation of 0.27 CIEDE2000 units for colour differences up to 5.0 CIELAB units, and were around 1% for threshold ($\Delta E^*_{ab} < 1.0$) colour differences, which can be considered negligible in most cases.

Currently CIEDE2000 is the CIE-recommended colour-difference formula, and CIE TC 1–57 "Standards in Colorimetry" is now in the way to propose this formula as a CIE standard. Anyway, CIEDE2000 cannot be considered a final answer to the problem of colour difference evaluation [33]. At this point, it is very interesting to note that CIEDE2000 (and also CIE94) are CIELAB-based colour-difference formulas which have not an associated colour space, as it should be desirable and discussed in the next section.

3.3 Advanced Colour-Difference Formulas

Under this epigraph, we are going to mention some recent colour-difference formulas with an associated colour space; that is, alternative colour spaces to CIELAB where the simple Euclidean distance between two points provides the corresponding colour difference.

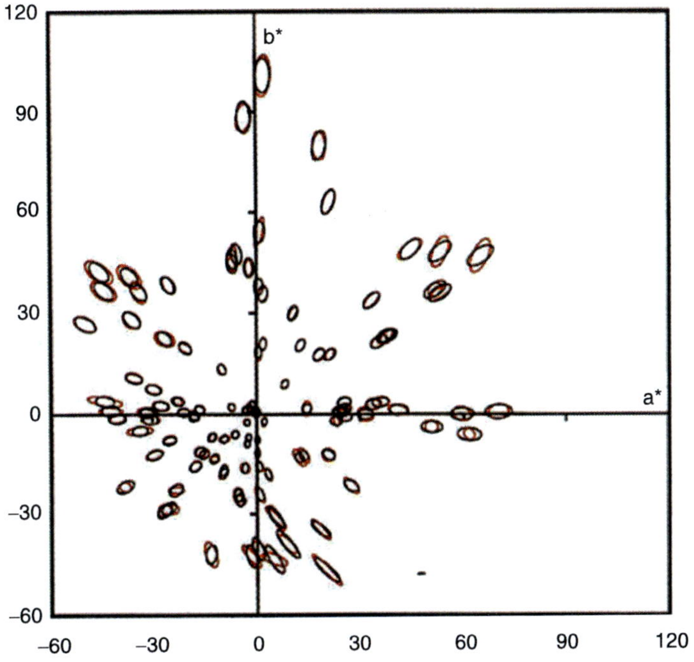

Fig. 3.3 Experimental colour discrimination ellipses in CIELAB a*b* for the BFD and RIT-DuPont datasets *(red)*, compared with predictions made by the CIEDE2000 colour-difference formula *(black)* [30]

In 1999, K. Witt proposed the DIN99 colour-difference formula (Witt, 1999, DIN99 colour-difference formula, a Euclidean model, Private communication), later adopted as the German standard DIN6176 [34]. The DIN99 colour space applies a logarithmic transformation on the CIELAB lightness L^*, and a rotation and stretch on the chroma plane a^*b^*, followed by a chroma compression inspired in the CIE94 weighting function for chroma. The DIN99 colour-difference formula is just the Euclidean distance in the DIN99 colour space. In 2002, Cui et al. [35] proposed different uniform colour spaces based on DIN99, the DIN99d being the one with the best performance. In DIN99d space the tristimulus value X was modified by subtracting a portion of Z to improve the performance in the blue region, as suggested by Kuehni [36]. Equations defining the DIN99d colour-difference formula, noted as ΔE_{99d}, are as follows:

$$\Delta E_{99d} = \sqrt{\Delta L_{99d}^2 + \Delta a_{99d}^2 + \Delta b_{99d}^2} \tag{3.20}$$

where the symbol "Δ" indicates differences between batch and standard samples in the colour pair. For each one of the two samples (and also the reference white), the next equations based on CIELAB L^*, a^*, b^* coordinates are applied:

$$X' = 1.12X - 0.12Z \tag{3.21}$$

$$L_{99d} = 325.22\ln(1 + 0.0036L^*) \tag{3.22}$$

$$e = a^*\cos(50°) + b^*\sin(50°) \tag{3.23}$$

$$f = 1.14[-a^*\sin(50°) + b^*\cos(50°)] \tag{3.24}$$

where the new e and f coordinates are the result of a rotation and re-scaling of the CIELAB a^*b^* coordinates, and L_{99d} is not too different to CIELAB lightness L^*.

$$G = \sqrt{e^2 + f^2} \tag{3.25}$$

$$C_{99d} = 22.5\ln(1 + 0.06G) \tag{3.26}$$

where this new chroma C_{99d} is a compression of CIELAB chroma C^*_{ab}. Finally:

$$h_{99d} = \arctan(f/e) + 50° \tag{3.27}$$

$$a_{99d} = C_{99d}\cos(h_{99d}) \tag{3.28}$$

$$b_{99d} = C_{99d}\sin(h_{99d}) \tag{3.29}$$

In 2006, on the basis of the CIECAM02 colour appearance model [14], three new Euclidean colour-difference formulas were proposed [37] for small (CAM02-SCD), large (CAM02-LCD), and all colour differences (CAM02-UCS). In these CAM02 formulas a non-linear transformation to CIECAM02 lightness J, and a logarithmic compression to the CIECAM02 colourfulness M were applied. The corresponding equations are as follows:

$$\Delta E_{CAM02} = \sqrt{(\Delta J'/K_L)^2 + (\Delta a')^2 + (\Delta b')^2} \tag{3.30}$$

$$J' = \frac{(1 + 100c_1)J}{1 + c_1 J} \tag{3.31}$$

$$M' = (1/c_2)\ln(1 + c_2 M) \tag{3.32}$$

$$a' = M'\cos(h) \tag{3.33}$$

$$b' = M'\sin(h) \tag{3.34}$$

where J, M, and h are the CIECAM02 lightness, colourfulness, and hue angle values, respectively. In addition, the $\Delta J'$, $\Delta a'$, and $\Delta b'$ are the J', a', and b' differences between the standard and batch in a colour pair. Finally, the parameter K_L has values 0.77, 1.24, and 1.00 for the CAM02-LCD, CAM02-SCD, and CAM02-UCS formulas, respectively, while $c_1 = 0.007$ for all these formulas, and c_2 has values 0.0053, 0.0363, and 0.0228, for the CAM02-LCD, CAM02-SCD, and CAM02-UCS formulas, respectively [37]. The results achieved by these CAM02 formulas are very encouraging: embedded uniform colour space in the CIECAM02 colour appearance model can be useful to make successful predictions of colour differences; that is, colour difference may be a specific aspect of colour appearance. Berns and Xue have also proposed colour-difference formulas based on the CIECAM02 colour appearance model [38].

OSA-UCS is a noteworthy empirical colour system for large colour differences developed in 1974 by the Optical Society of America's committee on Uniform Color Scales [39]. In this system, the straight lines radiating from any colour sample are geodesic lines with uniform colour scales. Thus, OSA-UCS was adopted to develop a CIE94-type colour-difference formula, valid under D65 illuminant and CIE 1964 colorimetric observer [40]. This formula was latter refined with chroma and lightness compressions, achieving an Euclidean colour-difference formula based also on the OSA-UCS space [41]. The equations conducting to this Euclidean formula, noted as ΔE_E, are as follows:

$$\Delta E_E = \sqrt{(\Delta L_E)^2 + (\Delta G_E)^2 + (\Delta J_E)^2} \tag{3.35}$$

$$L_E = \left(\tfrac{1}{b_L}\right) \ln\left[1 + \tfrac{b_L}{a_L}(10 L_{OSA})\right] \text{ with } a_L = 2.890,\ b_L = 0.015 \tag{3.36}$$

where OSA-UCS lightness, L_{OSA}, which takes account of the Helmholtz–Kohlrausch effect, is computed from the CIE 1964 chromaticity coordinates x_{10}, y_{10}, Y_{10} using the equations:

$$L_{OSA} = \left\{5.9\left[\left(Y_0^{1/3} - \frac{2}{3}\right) + 0.042(Y_0 - 30)^{1/3}\right] - 14.4\right\}\frac{1}{\sqrt{2}} \tag{3.37}$$

$$Y_0 = Y_{10}\left(4.4934 x_{10}^2 + 4.3034 y_{10}^2 - 4.2760 x_{10} y_{10} - 1.3744 x_{10} - 2.5643 y_{10} + 1.8103\right) \tag{3.38}$$

and the coordinates G_E and J_E are defined from:

$$G_E = -C_E \cos(h) \tag{3.39}$$

$$J_E = C_E \sin(h) \tag{3.40}$$

$$C_E = \left(\tfrac{1}{b_C}\right) \ln\left[1 + \tfrac{b_C}{a_C}(10 C_{OSA})\right] \text{ with } a_C = 1.256,\ b_C = 0.050 \tag{3.41}$$

$$C_{OSA} = \sqrt{G^2 + J^2} \tag{3.42}$$

$$h = \arctan\left(-\tfrac{J}{G}\right) \tag{3.43}$$

with J and G coordinates defined, for the D65 illuminant, from the transformations:

$$\begin{pmatrix} J \\ G \end{pmatrix} = \begin{bmatrix} 2(0.5735 L_{OSA} + 7.0892) & 0 \\ 0 & -2(0.7640 L_{OSA} + 9.2521) \end{bmatrix}$$

$$\begin{bmatrix} 0.1792 & 0.9837 \\ 0.9482 & -0.3175 \end{bmatrix} \begin{pmatrix} \ln\left(\frac{A/B}{0.9366}\right) \\ \ln\left(\frac{B/C}{0.9807}\right) \end{pmatrix} \tag{3.44}$$

$$\begin{pmatrix} A \\ B \\ C \end{pmatrix} = \begin{bmatrix} 0.6597 & 0.4492 & -0.1089 \\ -0.3053 & 1.2126 & 0.0927 \\ -0.0374 & 0.4795 & 0.5579 \end{bmatrix} \begin{pmatrix} X_{10} \\ Y_{10} \\ Z_{10} \end{pmatrix} \tag{3.45}$$

In 2008, Berns [13] proposed a series of colour-difference spaces based on multi-stage colour vision theory and line integration. These colour spaces have a similar transformation from tristimulus values to IPT space [42] to model multi-stage colour vision theory. First, a CIECAM02's chromatic adaptation transformation ensures the colour appearance property in the first step of the model:

$$\begin{pmatrix} X_{\text{Illuminant E}} \\ Y_{\text{Illuminant E}} \\ Z_{\text{Illuminant E}} \end{pmatrix} = M_{CAT02}^{-1} M_{VK} M_{CAT02} \begin{pmatrix} X \\ Y \\ Z \end{pmatrix} \tag{3.46}$$

where tristimulus values ranged between 0 and 1 following the transformation. M_{CAT02} is the matrix employed in CIECAM02 [14] to transform XYZ to pseudo-cone fundamentals, RGB. M_{VK} is the von Kries diagonal matrix in RGB. Illuminant E was selected because for either CIE standard observer, $X = Y = Z = 1$. Second, a constrained linear transformation from tristimulus values to pseudo-cone fundamentals is performed to simulate the linear processing at the cones of the human visual system:

$$\begin{pmatrix} L \\ M \\ S \end{pmatrix} = \begin{pmatrix} e_1 & e_2 & e_3 \\ e_4 & e_5 & e_6 \\ e_7 & e_8 & e_9 \end{pmatrix}_{\text{cones}} \begin{pmatrix} X_{\text{Illuminant E}} \\ Y_{\text{Illuminant E}} \\ Z_{\text{Illuminant E}} \end{pmatrix} \tag{3.47}$$

where $(e_1 + e_2 + e_3) = (e_4 + e_5 + e_6) = (e_7 + e_8 + e_9) = 1$. These row sums were optimization constraints and were required to maintain illuminant E as the reference illuminant. Third, an exponential function was used for the nonlinear stage, where γ defined the exponent (the same for all three cone fundamentals):

$$\begin{pmatrix} L' \\ M' \\ S' \end{pmatrix} = \begin{pmatrix} L^{1/\gamma} \\ M^{1/\gamma} \\ S^{1/\gamma} \end{pmatrix} \tag{3.48}$$

Fourth, the compressed cone responses were transformed to opponent signals:

$$\begin{pmatrix} W' \Leftrightarrow K' \\ R' \Leftrightarrow G' \\ Y' \Leftrightarrow B' \end{pmatrix} = \begin{pmatrix} 100 & 0 & 0 \\ 0 & 100 & 0 \\ 0 & 0 & 100 \end{pmatrix} \begin{pmatrix} o_1 & o_2 & o_3 \\ o_4 & o_5 & o_6 \\ o_7 & o_8 & o_9 \end{pmatrix}_{\text{opponency}} \begin{pmatrix} L' \\ M' \\ S' \end{pmatrix} \tag{3.49}$$

where $(o_1+o_2+o_3)=1$; $(o_4+o_5+o_6)=(o_7+o_8+o_9)=0$. These rows constraints generated an opponent-type system and were also used as optimization constraints. The fifth step was to compress the chromaticness dimensions to compensate for the chroma dependency (3.52):

$$\begin{pmatrix} L^E \\ a^E \\ b^E \end{pmatrix} = \begin{pmatrix} W' \Leftrightarrow K' \\ (R' \Leftrightarrow G')\,f(C) \\ (Y' \Leftrightarrow B')\,f(C) \end{pmatrix} \tag{3.50}$$

$$C = \sqrt{(R' \Leftrightarrow G')^2 + (Y' \Leftrightarrow B')^2} \tag{3.51}$$

$$f(C) = \frac{\ln\left(1+\beta_C\overline{C}\right)}{\beta_C\overline{C}} \tag{3.52}$$

where $\overline{C}$ indicates the arithmetical average of the chroma of the two samples in the colour pair.

Finally, Berns' models adopted the Euclidean distance as the measure of colour differences, and all the previous parameters were optimized [13] to achieve a minimum deviation between visual and computed colour differences for the RIT-DuPont dataset [6]:

$$\Delta E^E = \sqrt{(\Delta L^E)^2 + (\Delta a^E)^2 + (\Delta b^E)^2} \tag{3.53}$$

Recently, Shen and Berns [43] have developed an Euclidean colour space IPT-EUC, claiming it as a potential candidate for a unique colour model for both describing colour and measuring colour differences.

Euclidean colour spaces can be also developed by either analytical or computational methods to map the non-linear, non-uniform colour spaces to linear and uniform colour spaces based on the clues provided by different colour difference formulas optimized for reliable experimental datasets [44, 45].

3.4 Relationship Between Visual and Computed Colour Differences

As stated at the beginning of this chapter, the main goal of a colour-difference formula is to achieve a good relationship between what we see (ΔV) and what we measure (ΔE), for any two samples in colour space, viewed under any visual conditions. Therefore, how to measure this relationship between subjective (ΔV) and objective (ΔE) data is an important matter in this field. Although just a simple plot of ΔE_i against ΔV_i (where $i=1,\ldots,N$ indicates the number of colour pairs), may be an useful tool, different mathematical indexes have been employed to measure the strength of the relationship between $\Delta V$and ΔE, as described in this section.

PF/3 is a combined index which was proposed by Guan and Luo [7] from previous metrics suggested by Luo and Rigg [26], which in turn employed the γ and *CV* metrics proposed by Alder et al. [46] and the V_{AB} metric proposed by Shultze [47]. The corresponding defining equations are as follows:

$$\log_{10}(\gamma) = \sqrt{\frac{1}{N}\sum_{i=1}^{N}\left[\log_{10}\left(\frac{\Delta E_i}{\Delta V_i}\right) - \overline{\log_{10}\left(\frac{\Delta E_i}{\Delta V_i}\right)}\right]^2} \tag{3.54}$$

$$V_{AB} = \sqrt{\frac{1}{N}\sum_{i=1}^{N}\frac{(\Delta E_i - F\Delta V_i)^2}{\Delta E_i F \Delta V_i}} \text{ with } F = \sqrt{\frac{\sum_{i=1}^{N}\Delta E_i/\Delta V_i}{\sum_{i=1}^{N}\Delta V_i/\Delta E_i}} \tag{3.55}$$

$$CV = 100\sqrt{\frac{1}{N}\sum_{i=1}^{N}\frac{(\Delta E_i - f\Delta V_i)^2}{\left(\overline{\Delta E_i}\right)^2}} \text{ with } f = \frac{\sum_{i=1}^{N}\Delta E_i \Delta V_i}{\sum_{i=1}^{N}\Delta V_i^2} \tag{3.56}$$

$$PF/3 = \frac{100\left[(\gamma - 1) + V_{AB} + \frac{CV}{100}\right]}{3} \tag{3.57}$$

where N indicates the number of colour pairs (with visual and computed differences ΔV_i and ΔE_i, respectively), F and f are factors adjusting the ΔE_i and ΔV_i values to the same scale, and the upper bar in a variable indicates the arithmetical mean. For perfect agreement between ΔE_i and ΔV_i, *CV* and V_{AB} should equal zero and γ should equal one, in such a way that *PF*/3 should equal zero. A higher *PF*/3 value indicates worse agreement. Guan and Luo [7] state that *PF*/3 gives roughly the typical error in the predictions of ΔV_i as a percentage: for example, a 30% error in all pairs corresponds approximately to γ of 1.3, V_{AB} of 0.3, and *CV* of 30 leading to *PF*/3 of 30.

The decimal logarithm of γ is the standard deviation of the $\log_{10}(\Delta E_i/\Delta V_i)$. This metric was adopted because ΔE_i values should be directly proportional to ΔV_i, and the ratio $\Delta E_i/\Delta V_i$ should be constant. The standard deviation of the values of this ratio could be used as a measure of agreement, but this would give rise to anomalies which can be avoided by considering the logarithms of the $\Delta E_i/\Delta V_i$ values [25]. Natural logarithms have sometimes been employed to define the γ index, but the standard version of *PF*/3 uses decimal logarithms. The V_{AB} and *CV* values express the mean square root of the ΔE_i values with respect to the ΔV_i values (scaled by the F or f coefficients), normalized to appropriate quantities. Therefore, the V_{AB} and *CV* indices could be interpreted as two coefficients of variations, and the F and f factors as slopes of the plot of ΔE_i against ΔV_i (although they are not exactly the slope of the linear-regression fit). In earlier papers [25, 26], the product-moment correlation coefficient r was also employed as another useful measure to test the relationship between ΔE_i and ΔV_i. However, the r coefficient was not included in final *PF*/3 definition because it was found to be quite inconsistent with the other three indices for different experimental and theoretical datasets [7,16,47]. The main reason to propose the *PF*/3 index was that sometimes different measures led to different conclusions; for example, one formula performed the best according to *CV* while using V_{AB} other different formula provided the most accurate prediction. Thus, it was considered useful to avoid making a decision as to which of the metrics was the best, and provide a single value to evaluate the strength of the relationship between ΔE_i and ΔV_i [16]. Anyway, although *PF*/3 was widely employed in recent colour-difference literature, other indices have been also employed in this field. For example, the "wrong decision" percentage [48] is employed in acceptability experiments, the coefficient of variation of tolerances was employed by Alman et al. [49] and the linear correlation coefficients also continue being used by some researchers [50,51].

Any flaw in γ, *CV*, or V_{AB} is immediately transferred to *PF*/3, which is by definition an eclectic index. In addition, *PF*/3 cannot be used to indicate the significance of the difference between two colour-difference formulas with respect to a given set of visual data, because the statistical distribution followed by *PF*/3 is unknown. This last point is an important shortcoming for the *PF*/3 index, because the key question is not just to know that a colour-difference formula has a lower *PF*/3 than other for a given set of reliable visual data, but to know whether these two colour-difference formulas are or not statistically significant different for these visual data. From the scientific point of view, it is not reasonable to propose a new colour-difference formula if it is not significantly better than previous formulas, for different reliable visual datasets. In addition, industry is reluctant to change colour-difference formulas they are familiar to, in such a way that these changes must be based on the achievement of statistically significant improvements.

In a recent paper [52] the *STRESS* index has been suggested as a good alternative to *PF*/3 for colour-difference evaluation. *STRESS* comes from multidimensional scaling [53], and is defined as follows:

$$STRESS = 100\left(\frac{\sum(\Delta E_i - F_1\Delta V_i)^2}{\sum F_1^2\Delta V_i^2}\right)^{1/2} \quad \text{with}\, F_1 = \frac{\sum \Delta E_i^2}{\sum \Delta E_i \Delta V_i} \tag{3.58}$$

It can be proved that $STRESS^2$ is equal to $(1 - r^2)$, where r is equal to the correlation coefficient when imposing the restriction that regression line should pass through the origin (i.e., restricted regression). *STRESS* is always in the range [0, 100], low *STRESS* values indicating good performance of a colour-difference formula. But the key advantage of *STRESS* with respect to *PF*/3 is that F-test can be performed to know on the statistical significance of the differences between two colour-difference formulas A and B for a given set of visual data. Thus, the next conclusions can be achieved from values of next parameter F:

$$F = \left(\frac{STRESS_A}{STRESS_B}\right)^2 \tag{3.59}$$

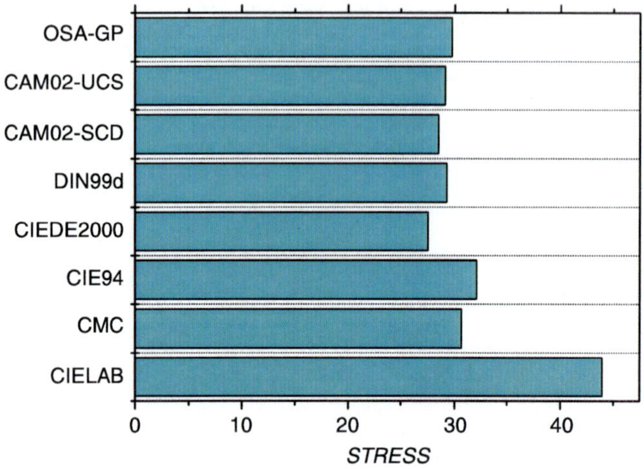

Fig. 3.4 Computed STRESS values using different colour-difference formulas for the combined dataset employed at CIEDE2000 development [30]

- The colour-difference formula A is significantly better than B when $F < F_C$
- The colour-difference formula A is significantly poorer than B when $F > 1/F_C$
- The colour-difference formula A is insignificantly better than B when $F_C \leq F < 1$
- The colour-difference formula A is insignificantly poorer than B when $1 < F \leq 1/F_C$
- The colour-difference formula A is equal to B when $F = 1$

where F_C is the critical value of the two-tailed F distribution with 95% confidence level and $(N-1, N-1)$ degrees of freedom.

STRESS can be also employed to measure inter- and intra-observer variability [54]. *STRESS* values from reliable experimental datasets using different advanced colour-difference formulas have been reported in the literature [55]. Thus, Fig. 3.4 shows *STRESS* values found for the combined dataset employed at CIEDE2000 development [30] (11,273 colour pairs), using the following colour-difference formulas: CIELAB, [21] CMC, [25] CIE94, [27] CIEDE2000, [29] DIN99d, [35] CAM02-SCD, [37] CAM02-UCS, [37] and OSA-GP [41]. For this combined dataset, the worst colour-difference formula (highest *STRESS*) was CIELAB, and the best (lowest *STRESS*) CIEDE2000. It can be added that, from F-test results, for this specific dataset CIELAB performed significantly poorer than any of the remaining colour-difference formulas, while CIEDE2000 was significantly better than any of the remaining colour-difference formulas. Of course, different results can be found for other experimental datasets [55], but best advanced colour-difference formulas hardly produce *STRESS* values lower than 20, which should be in part attributable to internal inconsistencies in the experimental datasets employed. Methods for mathematical estimation of such kind of inconsistencies have been suggested [56] concluding that, for the experimental dataset employed at CIEDE2000 development [29, 30], only a few colour pairs with very small colour differences have a low degree of consistency.

3.5 Colour Differences in Complex Images

Most complex images are not made up of large uniform fields. Therefore, discrimination and appearance of fine patterned colour images differ from similar measurements made using large homogeneous fields [57]. Direct application of previously mentioned colour-difference formulas to predict complex image difference (e.g., using a simple pixel by pixel comparison method) does not give satisfactory results. Colour discrimination and appearance is a function of spatial pattern. In general, as the spatial frequency of the target goes up (finer variations in space), colour differences become harder to see, especially differences along the blue-yellow direction. So, if we want to apply a colour-difference formula to colour images, the patterns of the image have to be taken into account.

Different spatial colour-difference metrics have been suggested, the most famous one being the one proposed by Zhang and Wandell [58] in 1996, known as S-CIELAB. S-CIELAB is a "perceptual colour fidelity" metric. It measures how accurate the reproduction of a colour image is to the original when viewed by a human observer. S-CIELAB is a spatial extension of CIELAB, where two input images are processed like made in the human visual system, before conventional CIELAB colour differences are applied pixel by pixel. Specifically, the steps followed by S-CIELAB are as follows: (1) each pixel (X,Y,Z) in the input images is translated to an opponent colour space, consisting of one luminance and two chrominance components; (2) each one of these three components is passed through a spatial filter that is selected according to the spatial sensitivity of the human visual system to this component, taken into account visual conditions; (3) the filtered images are transformed back into the CIE X,Y,Z format; (4) finally, the colour differences can be computed using the conventional CIELAB colour-difference formula, and the average of these colour differences for all pixels could then be used to represent the difference between two complex images. In fact, this idea can be applied using at the end any colour-difference formula; for example, in 2003 Johnson and Fairchild [59] applied the S-CIELAB framework replacing CIELAB by the CIEDE2000 colour-difference formula. Recently, Johnson et al. [60] have also pointed out that, for image difference calculations, the ideal opponent colour space would be both linear and orthogonal, such that the linear filtering is correct and any spatial processing on one channel does not affect the others, proposing a new opponent colour space and corresponding spatial filters specifically designed for image colour-difference calculations.

The evaluation of colour differences in complex images requires the corresponding images be carefully selected, as suggested by standardization organisms, avoiding potential bias from some kind of images [61]. Experimental methods employed to compare image quality must also be carefully considered [62]. While some results indicate a clear advantage of S-CIELAB with respect to CIELAB analyzing colour differences in complex images [63], other results [64] suggest no clear improvements using spatial colour-difference models, and results dependent on image content. Recent CIE Publication 199–2011 [65] provides useful information related to methods for evaluating colour differences in images.

3.6 Future Directions

Colour differences have been an active field of research since the 1950s trying to respond to industrial requirements in important topics like colour control, colour reproduction, etc. CIE-proposed colour-difference formulas have played an important positive role in the communication between buyers and sellers, as well as among different industries. The CIE recommendations of CIE94 and CIEDE2000 colour-difference formulas in 1995 and 2001, respectively, are eloquent examples of significant work and advances in this scientific area. Currently, research on colour differences continues, in particular, within some CIE Technical Committees in Divisions 1 and 8, as shown by the following examples: CIE TC1–55 (chairman: M. Melgosa) is working on the potential proposal of a uniform colour space for industrial colour-difference evaluation; CIE TC1–57 (chairman: A. Robertson) "Standards in colorimetry" has proposed the CIEDE2000 colour-difference formula as a CIE standard; CIE TC1–63 (chairman: K. Richter) has studied the range of validity of the CIEDE2000 colour-difference formula, concluding with the proposal of the new CIE TC1–81 (chairman: K. Richter) to analyze the performance of colour-difference formulas for very small colour differences (visual thresholds); CIE TC8–02 (chairman: M.R. Luo) studied colour differences in complex images [65].

Another important aspect in colour-difference research is the need of new reliable experimental datasets which can be used to develop better colour-difference formulas. New careful determinations of visual colour differences under well-defined visual conditions, together with their corresponding uncertainties, are highly desirable [66]. At the same time it is also very convenient to avoid an indiscriminate use of new colour-difference formulas, which should affect negatively industrial colour communication. New colour-difference formulas are only interesting if they can prove a statistically significant improvement with respect to previous ones, for several reliable experimental datasets.

There is an increasing activity aimed at incorporating colour-appearance models into practical colour-difference specification. For example, a colour appearance model could incorporate the effects of the background and luminance level on colour-difference perception, in such a way that the associated colour-difference formula could be applied to a wide set of visual conditions, in place of just a given set of "reference conditions". A colour appearance model would also make it possible to directly compare colour differences measured for different viewing conditions or different observers. Colour appearance models would also make it possible to calculate colour differences between a sample viewed in one condition and a second sample viewed in another different condition. As stated by Fairchild [15], "it is reasonable to expect that a colour difference equation could be optimized in a colour appearance space, like CIECAM02, with performance equal to, or better than equations like CIE94 and CIEDE2000."

In many situations colour is the most important attribute of objects' visual appearance, but certainly it is not the only one. At least, gloss, translucency, and texture may interact with colour and contribute to the so-called "total difference". Total difference models including colour differences plus coarseness or glint differences have been proposed in recent literature [67, 68].

Acknowledgments To our CIMET Erasmus-Mundus Master students (http://www.master-erasmusmundus-color.eu/) enrolled in the "Advanced Colorimetry" course during the academic years 2008–2009 and 2009–2010, who contributed with their questions and comments to improve our knowledge in the field of colour-difference evaluation. This work was partly supported by research project FIS2010–19839, Ministerio de Educación y Ciencia (Spain), with European Regional Development Fund (ERDF).

References

1. CIE Publication 101 (1993) Parametric effects in colour-difference evaluation. CIE Central Bureau, Vienna
2. Witt K (1995) CIE guidelines for coordinated future work on industrial colour-difference evaluation. Color Res Appl 20:399–403
3. Kuehni RG (2009) Variability in estimation of suprathreshold small color differences. Color Res Appl 34:367–374
4. MacAdam DL (1942) Visual sensitivities to color differences in daylight. J Opt Soc Am 32:247–274
5. Shen S, Berns RS (2011) Color-difference formula performance for several datasets of small color differences based on visual uncertainty. Color Res Appl 36:15–26
6. Berns RS, Alman DH, Reniff L, Snyder GD, Balonon-Rosen MR (1991) Visual determination of suprathreshold color-difference tolerances using probit analysis. Color Res Appl 16:297–316
7. Guan S, Luo MR (1999) Investigation of parametric effects using small colour-differences. Color Res Appl 24:331–343
8. ISO 105-A02:1993 Tests for Colour Fastness-Part A02: Gray Scale for Assessing Change in Colour, International Organization for Standardization Geneva, Switzerland. http://www.iso.org
9. AATCC Committee RA36, AATCC Evaluation Procedure 1 (2007) Gray scale for color change. AATCC, NC, Research Triangle Park. http://www.aatcc.org
10. Fastness Tests Co-ordinating Committee (F.T.C.C.) Publication XI (1953) The development of the geometric grey scales for fastness assessment. J Soc Dyers Colour 69:404–409
11. Cárdenas LM, Shamey R, Hinks D (2009) Development of a novel linear gray scale for visual assessment of small color differences. AATCC Review 9:42–47
12. Montag ED, Wilber DC (2003) A comparison of color stimuli and gray-scale methods of color difference scaling. Color Res Appl 28:36–44
13. Berns RS (2008) Generalized industrial color-difference based on multi-stage color vision and line-element integration. Óptica Pur Appl 41:301–311

14. CIE Publication 159:2004 (2004) A colour appearance model for colour management systems: CIECAM02. CIE Central Bureau, Vienna
15. Fairchild MD (2005) Colour Appearance Models, 2nd edn. Wiley, New York
16. Luo MR (2002) Development of colour-difference formulae. Rev Prog Color 32:28–39
17. McDonald R (1982) A review of the relationship between visual and instrumental assessment of colour difference, part 1. J Oil Colour Chem Assoc 65:43–53
18. McDonald R (1982) A review of the relationship between visual and instrumental assessment of colour difference, part 2. J Oil Colour Chem Assoc 65:93–106
19. Wit K (2007) CIE color difference metrics. In: Schanda J (ed) Chapter 4 in Colorimetry-Understanding the CIE System, Wiley, New York
20. CIE Publication 13.3 (1995) Method of measuring and specifying colour rendering properties of light sources. CIE Central Bureau, Vienna
21. CIE 15:2004 (2004) Colorimetry, 3rd edn. CIE Central Bureau, Vienna
22. Robertson AR (1990) Historical development of CIE recommended color difference equations. Color Res Appl 15:167–170
23. Kuehni RG (1990) Industrial color-difference: progress and problems. Color Res Appl 15:261–265
24. Melgosa M (2000) Testing CIELAB-based color-difference formulas. Color Res Appl 25:49–55
25. Clarke FJJ, McDonald R, Rigg B (1984) Modification to the JPC79 colour-difference formula. J Soc Dyers Colour 100:128–132
26. Luo MR, Rigg B (1987) BFD(l:c) colour-difference formula. Part 1 – Development of the formula. J Soc Dyers Colour 103:86–94
27. CIE Publication 116 (1995) Industrial colour-difference evaluation. CIE Central Bureau, Vienna
28. McDonald R (1974) The effect of non-uniformity in the ANLAB color space on the interpretation of visual colour differences. J Soc Dyers Colour 90:189–198
29. CIE Publication 142 (2001) Improvement to industrial colour-difference evaluation. CIE Central Bureau, Vienna
30. Luo MR, Cui G, Rigg B (2001) The development of the CIE 2000 colour-difference formula: CIEDE2000. Color Res Appl 26:340–350
31. Melgosa M, Huertas R, Berns RS (2004) Relative significance of the terms in the CIEDE2000 and CIE94 color-difference formulas. J Opt Soc Am A 21:2269–2275
32. Sharma G, Wu W, Dalal EN (2005) The CIEDE2000 color-difference formula: implementation notes, supplementary test data, and mathematical observations. Color Res Appl 30:21–30
33. Kuehni RG (2002) CIEDE2000, milestone or final answer? Color Res Appl 27:126–128
34. 6176 DIN (2000) Farbmetrische Bestimmung von Farbabständen bei Köroerfarben nach der DIN-99-Formel. DIN Deutsche Institut für Normung e.V, Berlin
35. Cui G, Luo MR, Rigg B, Roesler G, Witt K (2002) Uniform colour spaces based on the DIN99 colour-difference formula. Color Res Appl 27:282–290
36. Kuehni RG (1999) Towards an improved uniform color space. Color Res Appl 24:253–265
37. Luo MR, Cui G, Li C (2006) Uniform colour spaces based on CIECAM02 colour appearance model. Color Res Appl 31:320–330
38. Xue Y (2008) Uniform color spaces based on CIECAM02 and IPT color difference equations. MD Thesis, Rochester Institute of Technology, Rochester, NY
39. MacAdam DL (1974) Uniform color scales. J Opt Soc Am 64:1691–1702
40. Huertas R, Melgosa M, Oleari C (2006) Performance of a color-difference formula based on OSA-UCS space using small-medium color differences. J Opt Soc Am A 23:2077–2084
41. Oleari C, Melgosa M, Huertas R (2009) Euclidean color-difference formula for small-medium color differences in log-compressed OSA-UCS space. J Opt Soc Am A 26:121–134
42. Ebner F, Fairchild MD (1998) Development and testing of a color space (IPT) with improved hue uniformity. In: Proceedings of 6th Color Imaging Conference, 8–13, IS&T, Scottsdale, AZ
43. Shen S (2008) Color difference formula and uniform color space modeling and evaluation. MD Thesis, Rochester Institute of Technology, Rochester, NY

44. Thomsen K (2000) A Euclidean color space in high agreement with the CIE94 color difference formula. Color Res Appl 25:64–65
45. Urban P, Rosen MR, Berns RS, Schleicher D (2007) Embedding non-euclidean color spaces into Euclidean color spaces with minimal isometric disagreement. J Opt Soc Am A 24:1516–1528
46. Alder C, Chaing KP, Chong TF, Coates E, Khalili AA, Rigg B (1982) Uniform chromaticity scales – New experimental data. J Soc Dyers Colour 98:14–20
47. Schultze W The usefulness of colour-difference formulae for fixing colour tolerances. In: Proceedings of AIC/Holland (Soesterberg 1972) 254–265
48. McLaren K (1970) Colour passing—Visual or instrumental? J Soc Dyers Colour 86:389–392
49. Alman DH, Berns RS, Snyder GD, Larsen WA (1989) Performance testing of color-difference metrics using a color tolerance dataset. Color Res Appl 14:139–151
50. Gibert JM, Dagà JM, Gilabert EJ, Valldeperas J and the Colorimetry Group (2005) Evaluation of colour difference formulae. Color Technol 121:147–152
51. Attridge GG, Pointer MR (2000) Some aspects of the visual scaling of large colour differences-II. Color Res. Appl 25:116–122
52. García PA, Huertas R, Melgosa M, Cui G (2007) Measurement of the relationship between perceived and computed color differences. J Opt Soc Am A 24:1823–1829
53. Coxon APM (1982) The user's guide to multidimensional scaling. London, Heinemann
54. Melgosa M, García PA, Gómez-Robledo L, Shamey R, Hinks D, Cui G, Luo MR (2011) Notes on the application of the standardized residual sum of squares index for the assessment of intra- and inter-observer variability in color-difference experiments. J. Opt. Soc. Am. A. 28:949–953
55. Melgosa M, Huertas R, Berns RS (2008) Performance of recent advanced color-difference formulas using the standardized residual sum of squares index. J. Opt. Soc. Am. A 25:1828–1834
56. Morillas S, Gómez-Robledo L, Huertas R, Melgosa M (2009) Fuzzy analysis for detection of inconsistent data in experimental datasets employed at the development of the CIEDE2000 colour difference formula. J Mod Optic 56:1447–1456
57. Wandell BA (1996) Photoreceptor sensitivity changes explain color appearance shifts induced by large uniform background in dichoptic matching. Vis Res 35:239–254
58. Zhang XM, Wandell BA (1996) A spatial extension to CIELAB for digital color image reproduction. Proc Soc Information Display 27:731–734
59. Johnson GM, Fairchild MD (2003) A top down description of S-CIELAB and CIEDE2000. Color Res Appl 28:425–435
60. Johnson GM, Song X, Montag E, Fairchild MD (2010) Derivation of a color space for image color difference measurements. Color Res Appl 35:387–400
61. International Standardization Organization (ISO) Graphic technology—Prepress digital data exchange. Part 1, ISO 12640–1 (1997), Part 2, ISO 12640–2 (2004), Part 3 ISO 12640–3 (2007)
62. International Standardization Organization (ISO) (2005) Photography—Psychophysical experimental method to estimate image quality. Parts 1, 2 and 3, ISO 20462
63. Aldaba MA, Linhares JM, Pinto PD, Nascimento SM, Amano K, Foster DH (2006) Visual sensitivity to color errors in images of natural scenes. Vis Neurosci 23:555–559
64. Lee DG (2008) A colour-difference model for complex images on displays. Ph.D. Thesis, University of Leeds, UK
65. CIE Publication 199:2011 (2011) Methods for evaluating colour differences in images. CIE Central Bureau, Vienna
66. Melgosa M (2007) Request for existing experimental datasets on color differences. Color Res Appl 32:159
67. Huang Z, Xu H, Luo MR, Cui G, Feng H (2010) Assessing total differences for effective samples having variations in color, coarseness, and glint. Chinese Optics Letters 8:717–720
68. Dekker N, Kirchner EJJ, Supèr R, van den Kieboom GJ, Gottenbos R (2011) Total appearance differences for metallic and pearlescent materials: Contributions from color and texture. Color Res Appl 36:4–14

Chapter 4
Cross-Media Color Reproduction and Display Characterization

Jean-Baptiste Thomas, Jon Y. Hardeberg, and Alain Trémeau

The purest and most thoughtful minds are those which love color the most

John Ruskin

Abstract In this chapter, we present the problem of cross-media color reproduction, that is, how to achieve consistent reproduction of images in different media with different technologies. Of particular relevance for the color image processing community is displays, whose color properties have not been extensively covered in previous literature. Therefore, we go more in depth concerning how to model displays in order to achieve colorimetric consistency.

The structure of this chapter is as follows: After a short introduction, we introduce the field of cross-media color reproduction, including a brief description of current standards for color management, the concept of colorimetric characterization of imaging devices, and color gamut mapping. Then, we focus on state of the art and recent research in the colorimetric characterization of displays. We continue by considering methods for inverting display characterization models; this is an essential step in cross-media color reproduction, before discussing briefly quality factors, based on colorimetric indicators. Finally, we draw some conclusions and outline some directions for further research.

J.-B. Thomas (✉)
Laboratoire Electronique, Informatique et Image, Université de Bourgogne, Dijon, France
e-mail: jib.thomas@gmail.com

J.Y. Hardeberg
The Norwegian Color Research Laboratory, Gjøvik University College, Gjøvik, Norway
e-mail: jon.hardeberg@hig.no

A. Trémeau
Laboratory Hubert Curien, UMR CNRS 5516, Jean Monnet University, Saint-Etienne, France
e-mail: alain.tremeau@univ-st-etienne.fr

C. Fernandez-Maloigne (ed.), *Advanced Color Image Processing and Analysis*,
DOI 10.1007/978-1-4419-6190-7_4,

Keywords Color management • Cross-media color reproduction • Colorimetric device characterization • Gamut mapping • Displays • Inverse model

4.1 Introduction

Digital images today are captured and reproduced using a plethora of different imaging technologies (e.g., digital still cameras based on CMOS or CCD sensors, Plasma or Liquid Crystal Displays, inkjet, or laser printers). Even within the same type of imaging technology, there are many parameters which influence the processes, resulting in a large variation in the *color behavior* of these devices.

It is therefore a challenge to achieve color consistency throughout an image reproduction workflow, even more so since such image reproduction workflows tend to be highly distributed and generally uncontrolled. This challenge is relevant for a wide range of users, from amateurs of photography to professionals of the printing industry. And as we try to advocate in this chapter, it is also highly relevant to researchers within the field of image processing and analysis.

In the next section we introduce the field of cross-media color reproduction, including a brief description of current standards for color management, the concept of colorimetric characterization of imaging devices, and color gamut mapping. Then, in Sect. 4.3 we focus on state of the art and recent research in the characterization of displays. In Sect. 4.4, we consider methods for inverting display characterization models; this is an essential step in cross-media color reproduction, before discussing quality factors, based on colorimetric indicators, briefly in Sect. 4.5. Finally, in Sect. 12.5 we draw some conclusions and outline some directions for further research.

4.2 Cross-Media Color Reproduction

When using computers and digital media technology to acquire, store, process, and reproduce images of colored objects or scenes, a digital color space is used, typically *RGB*, describing each color as a combination of variable amounts of the primaries red, green, and blue. Since most imaging devices *speak RGB* one may think that there is no problem with this. However, every individual device has its own definition of *RGB*, i.e., for instance for output devices such as displays, for the same input *RGB* values, different devices will produce significantly different colors. It usually suffices to enter the TV section of an home electronics store to be reminded of this fact.

So, therefore, the *RGB* color space is usually not standardized, and every individual imaging device has its own definition of it, i.e., its very own relationship between the displayed or acquired *real-world* color and the corresponding *RGB* digital color space. Achieving color consistency throughout a complex and distributed color reproduction workflow with several input and output devices is therefore a serious challenge; achieving such consistency defines the research field of cross-media color reproduction.

The main problem is thus to determine the relationships between the different devices's color *languages*, analogously to color *dictionaries*. As we will see in the next sections, a standard framework has been defined (color management system), in which dictionaries (profiles) are defined for all devices; between their native color language and a common, device-independent language. Defining these dictionaries by characterizing the device's behavior is described in Sect. 4.2.2, while Sect. 4.2.3 addresses the problem of when a device simply does not have rich enough *vocabulary* to reproduce the colors of a certain image.

4.2.1 Color Management Systems

By calibrating color peripherals to a common standard, color management system (CMS) software and architecture makes it easier to match colors that are scanned to those that appear on the monitor and printer, and also to match colors designed on the monitor, using, e.g., CAD software, to a printed document. Color management is highly relevant to persons using computers for working with art, architecture, desktop publishing or photography, but also to non-professionals, as, e.g., when displaying and printing images downloaded from the Internet or from a Photo CD.

To obtain faithful color reproduction, a CMS has two main tasks. First, colorimetric characterization of the peripherals is needed, so that the *device-dependent* color representations of the scanner, the printer, and the monitor can be linked to a *device-independent* color space, the profile connection space (PCS). This is the process of *profiling*. Furthermore, efficient means for processing and converting images between different representations are needed. This task is undertaken by the color management module (CMM).

The industry adoption of new technologies such as CMS depends strongly on standardization. The international color consortium (ICC, http://www.color.org) plays a very important role in this concern. The ICC was established in 1993 by eight industry vendors for the purpose of creating, promoting, and encouraging the standardization and evolution of an open, vendor-neutral, cross-platform CMS architecture and components.

For further information about color management system architecture, as well as theory and practice of successful color management, refer to the ICC specification [47] or any recent textbooks on the subject [40].

Today there is wide acceptance of the ICC standards, and different studies such as one by [71] have concluded that color management solutions offered by different vendors are approximately equal, and that color management has passed the breakthrough phase and can be considered a valid and useful tool in color image reproduction.

However, there is still a long way to go, when it comes to software development (integration of CMS in operating systems, user-friendliness, simplicity, etc.), research in cross-media color reproduction (better color consistency, gamut mapping, color appearance models, etc.), and standardization. Color management is a very active area of research and development, though limited by our knowledge on the human perception process. Thus in the next sections, we will briefly review different approaches to the colorimetric characterization of image acquisition and reproduction devices.

4.2.2 Device Colorimetric Characterization

Successful cross-media color reproduction needs the calibration and the characterization of each color device. It further needs a color conversion algorithm, which permits to convert color values from one device to another.

In the literature, the distinction between calibration and characterization can vary substantially, but the main idea usually remains the same. For instance, some authors will consider a tone response curve establishment as a part of the calibration, others as a part of the characterization. These difference does not mean too much in practice and is just a matter of terminology. Let us consider the following definition:

The *calibration* process put a device in a fixed state, which will not change with time. For a color device, it consists in setting up the device. Settings can be position, brightness, contrast, and sometimes primaries and gamma, etc.

The *characterization* process can be defined as understanding and modeling the relationship between the input and the output, in order to control a device for a given calibration set-up. For a digital color device, this means either to understand the relationship between a digital value input and a produced color for an output color device (printer, display) or, in the case of an input color device (camera, scanner), to understand the relationship between the acquired color and the digital output value. Usually, a characterization model is mostly static, and is relying on the capability of the device to remain in a fixed state, thus on the calibration step.

As stated above, the characterization of a color device is a modeling step, which permits to relate the digital value that characterizes the device and the actual color defined in a standard color space, such as *CIEXYZ*. There are different approaches to modeling a device.

One can consider a *physical* approach, which will aim to determine a set of physical parameters of a device, and uses these in a physical model based on the technology definition. Such an approach has been extensively used for CRT displays, and also it is quite common for cameras. In this case, the resulting accuracy will be constrained by how well the device fits the model hypothesis and how accurate the related measurements were taken. Commonly a physical device model consists in a two steps process. First, a linearization of the intensity response curves of the individual channels, i.e., the relation between the digital value and the corresponding intensity of light. The second step is typically colorimetric linear transform (i.e., a 3x3 matrix multiplication). The characteristics of the colorimetric transform is based on the chromaticity of the device primaries.

Another approach consists in fitting a data set with any *numerical* model. In this case, the accuracy will depend on the number of data, on their distribution and on the interpolation method used. Typically a numerical model would require more measurement, but would make no assumption on the device behavior. We can note that the success of such a model will depend also on the capacity of the model to fit with the technology anyway.

For a numerical method, depending on the interpolation method used, one have to provide different sets of measures in order to optimize the model determination. This implies to first define which color space is used to make all the measures. The *CIEXYZ* color space seems at first to be the best choice considering that some numerical method would use its vectorial space properties successfully, particularly additivity, in opposition with *CIELAB*. An advantage is that it is *absolute* and can be used as an intermediary color space to a uniform color space, *CIELAB*, which is recommended by the *CIE* for measuring the color difference when we will evaluate the model accuracy (the ΔE in *CIELAB* color space). However, since we define the error of the model, and often the cost function of the optimization process as an Euclidean distance in *CIELAB*, this color space can be a better choice.

These sets of measures can be provided using a specific (optimal) color chart, or a first approach can be to use a generic color chart, which allows to define a first model of characterization.

However, it has been shown that it is of major importance to have a good distribution of the data everywhere in the gamut of the device and more particularly on the faces and the edges of the gamut, which is roughly fitting with the edges and faces of the *RGB*-associated cube. These faces and edges define the color gamut of the color device. The problem with acquisition device such as cameras is that the lighting conditions are changing, and it is difficult to have a dedicated data set of patches to measure for every possible condition. Thus, optimized color charts have been designed, for which the spectral characteristics of the color patches are designed carefully.

Another possibility is that, based on a first rough or draft model, one can provide an optimal data set to measure, which takes into account the nonlinearity of the input device. There are several methods to minimize errors due to the nonlinear response of devices. By increasing the number of patches, we can tighten the mesh's sampling. This method can be used to reach a lower error. Unfortunately, it might not improve much the maximum error. To reduce it, one can decide to over-sample some particular area of the color space. The maximum error is on the boundaries of the gamut, since there are fewer points to interpolate, and in the low luminosity areas, as our eyes can easily see small color differences in dark colors. Finally, one can solve this nonlinearity problem by using a nonlinear data set distribution, which provides a quite regular sampling in the *CIELAB* color space.

4.2.2.1 Characterization of Input Devices

An input device has the ability to transform color information of a scene or an original object into digital values. A list of such devices would include digital still cameras, scanners, camcorders, etc. The way it transforms the color information is usually based on (three) spectral filters with their highest transmission or resulting color around a Red, Green, and Blue part of the spectrum. The intensity for each filter will be related to the *RGB* values. A common physical model of such a device is given as

$$\rho = f(\nu) = \int L(\lambda)R(\lambda)S(\lambda)\mathrm{d}\lambda, \tag{4.1}$$

where ρ is the actual digital value output, ν is the nonlinearized value, $L(\lambda)$, $R(\lambda)$, $S(\lambda)$ are the spectral power distribution of the illuminant, the spectral reflectance of the object, and the spectral sensitivity of the sensor, including a color filter.

The input device calibration includes the setup of the time exposure, the illumination (for a scanner), the contrast setup, the color filters, etc.

In the case of input devices, let us call the forward transform the transform which relates the acquired color with the digital value, e.g., conversion from *CIEXYZ* to *RGB*. Meanwhile the inverse transform will estimate the acquired color given digital value caught by the device, e.g., converts from *RGB* to *CIEXYZ*.

The input device characterization can be done using a physical modeling or a combination of numerical methods. In the case of a physical modeling, the tone response curves will have to be retrieved; the spectral transmission of the color filters may have to be retrieved too, in order to determine their chromaticities, thus establishing the linear transform between intensity linearized values and the digital values. This last part requires usually a lot of measurements, and may require the use of a monochromator or an equivalent expensive tool. In order to reduce this set of measurements, one needs to make some assumptions and to set some constraints to solve the related inverse problem. Such constraints can be the modality of the spectral response of a sensor or that the sensor response curve can be fitted with just a few of the first Fourier coefficients, see, e.g., [10,36,76]. Such models would mostly use the *CIEXYZ* color space or another space which has the additivity property.

Johnson [51] gives good advice for achieving a reliable color transformation for both scanners and digital cameras. In his paper, one can find diverse characterization procedures, based on the camera colorimetric evaluation using a set of test images. The best is to find a linear relationship to map the output values to the input target (each color patch). The characterization matrix, once more, provides the transformation applied to the color in the image. In many cases, the regression analysis shows that the first order linear relationship is not satisfactory and a higher order relationship or even nonlinear processing is required (log data, gamma correction, or S-shape, e.g.). Lastly, if a matrix cannot provide the transformation, then a look-up table (LUT) will be used. Unfortunately, the forward transform can be complicated and quite often produces artifacts [51]. Possible solutions to the problems of linear transformations encountered by Johnson are least-squares fitting, nonlinear transformations, or look-up tables with interpolation. In the last case, any scanned pixel can be converted into tristimulus values via the look-up table(s) and interpolation is used for intermediate points which do not fall in the table itself. This method is convenient for applying a color transformation when a first order solution is not relevant. It can have a very high accuracy level if the colors are properly selected.

The colorimetric characterization of a digital camera was analyzed by [45]. An investigation was done to determine the influence of the polynomial used for interpolation and the possible correlation between the RGB channels. The channel independence allows us to separate the contribution of spectral radiance from the three channels. Hong et al. [45] also checked the precision of the model with respect to the training samples' data size provided and the importance of the color precision being either 8 or 12 bits. According to the authors, there are two categories of color characterization methods: either spectral sensitivity based (linking the spectral sensitivity to the CIE color-matching functions) or color target based (linking color patches to the CIE color-matching functions). These two solutions lead to the same results, but the methods and devices used are different. Spectral sensitivity analysis requires special equipment like a radiance meter and a monochromator; while a spectrophotometer is the only device needed for the color target-based solution. Typical methods like 3D look-up tables with interpolation and extrapolation, least square polynomials modeling and neural networks can be used for the transformation between *RGB* and *CIEXYZ* values, but in this article, polynomial regression is used. As for each experiment only one parameter (like polynomial order, number of quantization levels, or size of the training sample) changes, the ΔE^*_{ab} difference is directly linked to the parameter.

Articles published on this topic are rare, but characterization of other input devices with a digital output operates the same way. Noriega et al. [67] and [37] further propose different transformation techniques. These articles discuss the colorimetric characterization of a scanner and a negative film. In the first article [67], the authors decided to use least squares fitting, LUTs and distance-weighted interpolation. The originality comes from the use of the Mahalanobis distance used to perform the interpolation. The second article [37] deals with the negative film characterization. Distance-weighted interpolation, Gaussian interpolation neural networks, and nonlinear models have been compared using Principal Component Analysis. In these respective studies, the models were trained with the Mahalanobis distance (still using the color difference as a cost function) and neural networks.

4.2.2.2 Characterization of Output Devices

An output device in this context is any device that will reproduce a color, such as printers, projection systems, or monitors. In this case, the input to the device is a digital value, and we will call the forward transform the transform that predicts the color displayed for a given input, e.g., *RGB* to *CIEXYZ*. The inverse or backward transform will then define which digital value we have to input to the device to reproduce a wanted color, e.g., *CIEXYZ* to *RGB*.

The characterization approach for output devices and media is similar to that of input devices. One has to determine a model based on more or less knowledge of the physical behavior of the device, and more or less measurement of color patches and mathematical approximation/interpolation. Since displays are covered in depth in Sect. 4.3, we will here briefly discuss printer characterization.

We can distinguish between two kinds of printer characterization models, the computational and the physical ones. Typically, for a 4-colorant CMYK printer the computational approach consists in building a grid in four dimensions, a multidimensional look-up table (mLUT). The estimation of the resulting color for a given colorant combination will then be calculated by multidimensional interpolation in the mLUT. An important design trade-off for such modeling is between the size of the mLUT and the accuracy of the interpolation.

The physical models attempt to imitate the physics involved in the printing device. Here also these models can be classified into two subtypes with regard to the assumptions they make and their complexity [90]: regression-based and first-principal models. Regression-based models are rather simple and works with a few parameters to predict a printer output while first-principal model will closely imitate the physics of the printing process by taking into account multiple light interactions between the paper and the ink layers, for instance. Regression-based models are commonly used to model the behavior of digital printing devices.

During the last century, printing technology has evolved and the printer models as well. Starting from a single-colorant printing device the Murray-Davies model predicts the output spectral reflectances of a single-colorant coverage value knowing the spectral reflectance of the paper and maximum colorant coverage value. This model was extended to color by [64]. The prediction of a colorant combination is the summation of all the colorants involved in the printing process weighted by their coverage on the paper. All the colorants are referring to all the primaries (cyan, magenta, and yellow in case of a CMY printer) plus all the combination between them plus the paper, these colors are called the Neugebauer primaries (NP). Later the interaction of light penetrating and scattering into the paper was added to these models by [95], as form of an exponent known as the n factor. For more information about printer characterization, refer, e.g., to [38].

4.2.3 Color Gamut Considerations

A color gamut is the set of all colors that can be produced by a given device or that are present in a given image. Although these sets are in principle discrete, gamuts are most often represented as volumes or blobs in a 3D color space using a gamut boundary descriptor [7]. When images are to be reproduced between different devices, the problem of gamut mismatch has to be addressed. This is usually referred to as color gamut mapping. There is a vast amount of literature about the gamut-mapping problem, see, for instance, a recent book by [63].

To keep the image appearance, some constraints are usually considered while doing a gamut mapping:

- Preserve the gray axis of the image and aim for maximum luminance contrast.
- Reduce the number of out-of-gamut colors.
- Minimize hue shifts.
- Increase the saturation.

CIELAB is one of the most often used color spaces for gamut mapping, but there are deficiencies in the uniformity of hue angles in the blue region. To prevent this shift, one can use Hung and Berns' data to correct the *CIELAB* color space [21]. To map a larger source gamut into a smaller destination gamut of a device with a reduced lightness dynamic range, often a linear lightness remapping process is applied. It suffers from a global reduction in the perceived lightness contrast and an increase in the average lightness of the remapped image. It is of utmost importance to preserve the lightness contrast. An adaptive lightness rescaling process has been developed by [22]. The lightness contrast of the original scene is increased before the dynamic range compression is applied to fit the input lightness range into the destination gamut. This process is known as a sigmoidal mapping function, the shape of this function aids in the dynamic range mapping process by increasing the image contrast and by reducing the low-end textural defects of hard clipping.

We can categorize different types of pointwise gamut-mapping technics (See Fig. 4.1); *gamut clipping* only changes the colors outside the reproduction gamut while *gamut compression* changes all colors from the original gamut. The knee function rescaling preserves the chromatic signal through the central portion of the gamut, while compressing the chromatic signal near the edges of the gamut. The sigmoid-like chroma mapping function has three linear segments; the first segment preserves the contrast and colorimetry, the second segment is a mid-chroma boost (increasing chroma), and the last segment compresses the out-of-gamut chroma values into the destination gamut.

Spatial gamut mapping has become an active field of research in the recent years [35, 56]. In contrast to the conventional color gamut-mapping algorithms, where the mapping can be performed once and for all and stored as a look-up table, e.g., in an ICC profile, the spatial algorithms are image dependent by nature. Thus, the algorithms have to be applied for every single image to be reproduced, and make direct use of the gamut boundary descriptors many times during the mapping process.

Quality assessment is also required for the evaluation of gamut-mapping algorithms, and extensive work has been carried out on subjective assessment [32]. This evaluation is long, tiresome, and even expensive. Therefore, objective assessment methods are preferable. Existing work on this involves image quality metrics, e.g., by [17, 44]. However, these objective methods can still not replace subjective assessment, but can be used as a supplement to provide a more thorough evaluation.

Recently, [4] presented a novel, computationally efficient, iterative, spatial gamut-mapping algorithm. The proposed algorithm offers a compromise between

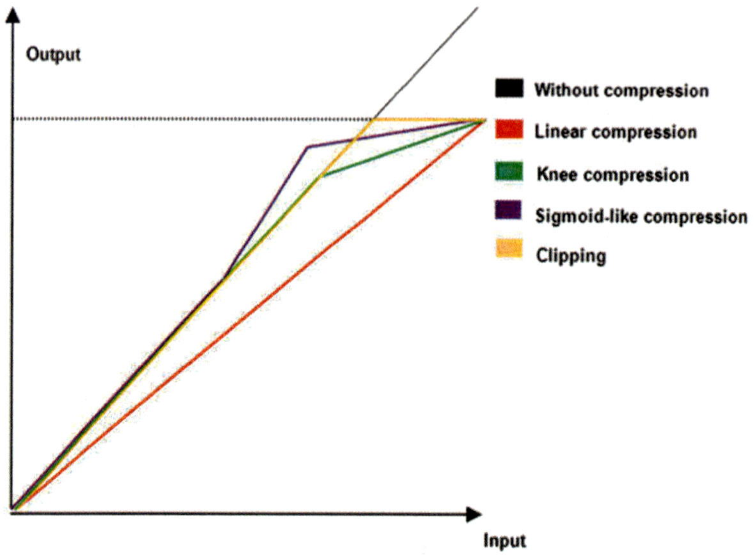

Fig. 4.1 Scheme of typical gamut-mapping techniques

the colorimetrically optimal gamut clipping and the most successful spatial methods. This is achieved by the iterative nature of the method. At iteration level zero, the result is identical to gamut clipping. The more we iterate, the more we approach an optimal, spatial, gamut-mapping result. Optimal is defined as a gamut-mapping algorithm that preserves the hue of the image colors as well as the spatial ratios at all scales. The results show that as few as five iterations are sufficient to produce an output that is as good or better than that achieved in previous, computationally more expensive, methods. Unfortunately, the method also shares some of the minor disadvantages of other spatial gamut-mapping algorithms: halos and desaturation of flat regions for particularly difficult images. There is therefore much work left to be done in this direction, and one promising idea is to incorporate knowledge of the strength of the edges.

4.3 Display Color Characterization

This section will study in depth display colorimetric characterization. Although many books investigate color device characterization, they mostly focus on printers or cameras, which have been far more difficult to characterize than displays during the CRT era; thus, mostly a simple linear model and a gamma correction were addressed in books when considering displays. With the emergence of new technologies used to create newer displays in the last 15 years, a lot of work has been done concerning this topic, and a new bibliography and new methods have

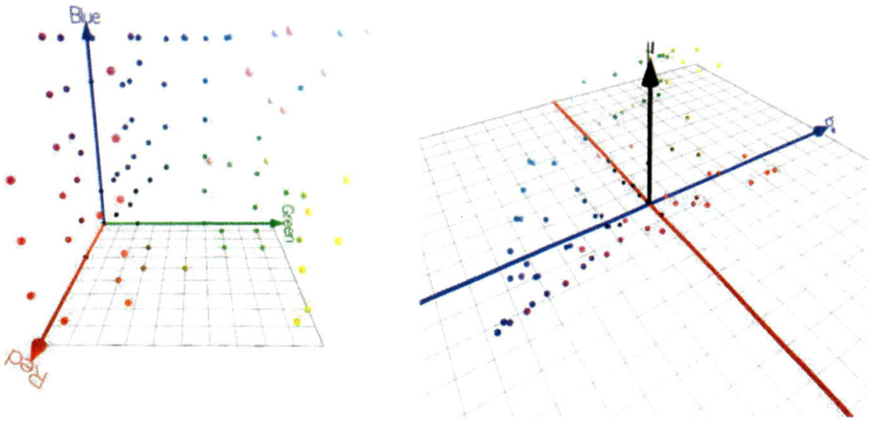

Fig. 4.2 3D look-up table for a characterization process from *RGB* to *CIELAB*

appeared. Many methods have been borrowed from printers or camera though, but the way to reproduce colors and the assumptions one can do are different when talking about displays, so the results or the explanation of why a model is good or not are slightly different. We propose to discuss the state of the art and the major trends about display colorimetric characterization in this section.

4.3.1 State of the Art

Many color characterization methods or models exist; we can classify them in three groups. In a first one, we find the models, which tend to model physically the color response of the device. They are often based on the assumption of independence between channels and of chromaticity constancy of primaries. Then, a combination of the primary tristimulus at the full intensity weighted by the luminance response of the display relatively to a digital input can be used to perform the colorimetric transform. The second group can be called numerical models. They are based on a training data set, which permits optimization of the parameters of a polynomial function to establish the transform. The last category consists of 3D LUT-based models. Some other methods can be considered as hybrid. They can be based on a data set and assume some physical properties of the display, such as in the work of [16].

4.3.1.1 3D LUT Models

The models in the 3D LUT group are based on the measurement of a defined number of color patches, i.e., we know the transformation between the input values (i.e., *RGB* input values to a display device) and output values (i.e., *CIEXYZ* or *CIELAB* values) measured on the screen by a colorimeter or spectrometer in a small number of color space locations (see Fig. 4.2). Then this transformation is generalized to the whole space by interpolation. Studies assess that these methods can achieve accurate results [11, 80], depending on the combination of the interpolation method used [2,5,18,53,66], the number of patches measured, and on their distribution [80] (note that some of the interpolation methods cited above cannot be used with a non-regular distribution). However, to be precise enough, a lot of measurements are typically required, i.e., a $10 \times 10 \times 10$ grid of patches measured in [11]. Note that such a model is technology independent since no assumptions are made about the device but that the display will always have the same response at the measurement location. Such a model needs high storage capacity and computational power to handle the 3D data. The computational power is usually not a problem since Graphic Processor Units can perform this kind of task easily today [26]. The high number of measurements needed is a greater challenge.

4.3.1.2 Numerical Models

The numerical models suppose that the transform can be approximated by a set of equations, usually an n-order polynomial function. The parameters are retrieved using an n-order polynomial regression process based on measurements. The number of parameters required involves a significant number of measurements, depending on the order of the polynomial function. The advantage of these models is that they take into account channel interdependence by applying cross components factors in the establishment of the function [54,55,83]. More recently, an alternative method has been proposed by [89] who removed the three-channel crosstalk from the model, considering that the inter-channel dependence is only due to two-channel crosstalk, thus reducing the required number of measurements. They obtained results as accurate as when considering the three-channel crosstalk.

Radial basis function (RBF) permits to use a sum of low-order polynomials instead of one high-order polynomial and has been used successfully in different works [26, 27, 79, 80]. Mostly polyharmonic splines are used, which include thin plate splines (TPS) that [75] used for printers too. TPS are a subset of polyharmonic splines (bi-harmonic splines). Sharma and Shaw [75] recalled the mathematical framework and presented some applications and results for printer characterization. They showed that using TPS, they achieved a better result than in using local polynomial regression. They showed that by using a smoothing factor, error in measurement impact can be avoided at the expense of the computational cost that optimize this parameter, similar results were observed by [26]. However, [75] did study neither data distribution influence (but they stated that the data distribution can improve the accuracy in their conclusion) nor the use of other kernels for interpolation. This aspect has been studied by [26], in which main improvements were in the optimization of the selection of the data used to build the model in an iterative way.

4.3.1.3 Physical Models

Physical models are historically widely used for displays, since the CRT technology follows well the assumptions cited above [13, 19, 29]. Such a model typically first aims to linearize the intensity response of the device. This can be done by establishing a model that assumes the response curve to follow a mathematical function, such as a gamma law for CRT [13, 14, 28, 74], or an S-shaped curve for LCD [58, 59, 94]. Another way to linearize the intensity response curve is to generalize measurements by interpolation along the luminance for each primary [68]. The measurement of the luminance can be done using a photometer. Some approaches propose as well a visual response curve estimation, where the 50% luminance point for each channel is determined by the user to estimate the gamma value [28]. This method can be generalized to the retrieval of more luminance levels in using half-toned patches [62, 65]. Recently a method to retrieve the response curve of a projection device using an uncalibrated camera has been proposed by [8] and extended by [62]. Note that it has been assumed that the normalized response curve is equivalent for all the channels, and that only the gray level response curve can be retrieved. In the case of a doubt about this assumption, it is useful to retrieve the three response curves independently. Since visual luminance matching for the blue channel is a harder task, it is of use to perform an intensity matching for the red and green channel, and a chromaticity matching or gray balancing for the blue one [57]. This method should not be used with projectors though, since they show a large chromaticity shift with the variation of input for the pure primaries.

A model has been defined by [91, 92] for DLP projectors using a white segment in the color wheel. In their model, the characteristics of the luminance of the white channel is retrieved with regard to additive property of the display, given the four-tuplet (R, G, B, W) from an input (d_r, d_g, d_b).

The second step of these models is commonly the use of a 3×3 matrix containing primary tristimulus values at full intensity to build the colorimetric transform from luminance to an additive independent color space. The primaries can be estimated by measurement of the device channels at full intensity, using a colorimeter or a spectroradiometer, assuming their chromaticity constancy. In practice this assumption does not hold perfectly, and the model accuracy suffers from that. The major part of the non-constancy of primaries can be corrected by applying a black offset correction [50]. Some authors tried to minimize the chromaticity non-constancy in finding the best chromaticity values of primaries (optimizing the components of the 3×3 matrix) [30]. Depending on the accuracy required, it is also possible to use generic primaries such as *sRGB* for some applications [8], or data supplied by the manufacturer [28]. However, the use of a simple 3×3 matrix for the colorimetric transform leads to inaccuracy due to the lack of channel independence and of chromaticity constancy of primaries. An alternative approach has been derived in the masking model and modified masking model, which takes into account the cross-talk between channels [83]. Furthermore, the lack of chromaticity constancy can be critical, particularly for LCD technology, which has been shown to fail this assumption [20, 58]. The piecewise linear model assuming variation in chromaticity (PLVC) [34] is not subject to this effect, but has not been widely used since [68] demonstrated that among the models they tested in their article, the PLVC and the piecewise linear-assuming chromaticity constancy (PLCC) models were of equivalent accuracy for the CRT monitors they tested. With the last one requiring less computation, it has been more used than the former one. These results have been confirmed in studies on CRT technology [68, 69], especially with a flare correction [50, 86]. On DLP technology when there is a flare correction, results can be equivalent; however, PLVC can give better results on LCDs [86].

Other models exist, such as the two-steps parametric model proposed by [16]. This model assumes separation between chromaticity and intensity, and is shown to be accurate, with average ΔE^*_{ab}'s around 1 or below for one DLP projector and a CRT monitor. The luminance curve is retrieved, as for other physical models, but the colorimetric transform is based on 2D interpolation in the chromaticity plane based on a set of saturated measured colors.

4.3.1.4 The Case of Subtractive Displays

An analog film-projection system in a movie theater was studied by [3]. A Minolta CS1000 spectrophotometer was used to find the link between the RGB colors of the image and the displayed colors. For each device, red, green, blue, cyan, magenta, yellow, and gray levels were measured. The low luminosity levels didn't allow a precise color measurement with the spectrophotometer at their disposal. For the 35 mm projector, it was found that the color synthesis is not additive, since the projection is based on a subtractive method. It is difficult to model the transfer function of this device; the measures cannot be reproduced as both measure and projection angles change. Moreover, the luminance is not the same all over the projected area. The subtractive synthesis, by removing components from the white source, cannot provide the same color sensation as a cinema screen or a computer screen, which is based on additive synthesis of red, green, and blue components. Subtractive cinema projectors are not easy to characterize as the usual models are for additive synthesis. The multiple format transformations and data compression led to data lost and artifacts.

Ishii [49] shows the gamut differences between CRT monitors (*RGB* additive method) and printed films (*CMY* dyes subtractive method). The main problem for a physical modeling is the tone shift. In a matching process from a CRT to a film, both gamut difference and mapping algorithm are important. During the production step, the minor emulsion changes and chemical processes can vary and then make small shifts on the prints, leading to a shift on the whole production. An implementation of a 3D LUT was successfully applied to convert color appearance from CRT to film display.

4.3.2 Physical Models

4.3.2.1 Display Color Characterization Models

Physical models are easily invertible, do not require a lot of measurements, require a little computer memory, and do not require high computing power. So, they can be used in real time. Moreover, the assumptions of channel independence and chromaticity constancy are appropriate for the CRT technology. However, these assumptions (and others such as spatial uniformity, both in luminance and in chromaticity, view angle independence, etc.) do not fit so well with some of today's display technologies. For instance, the colorimetric characteristic of a part of an image in a Plasma Display is strongly dependent of what is happening in the surrounding [25] for energy economy reasons. In LC technology, which has become the leader for displays market, these common assumptions are not valid. Making such assumptions can reduce drastically the accuracy of the characterization. For instance, a review of problems faced in LC displays has been done by [94]. Within projection systems, the large amount of flare induces a critical chromaticity shift of primaries.

In the same time, the computing power has become less and less a problem. Some models not used in practice because of their complexity can now be highly beneficial for display color characterization. This section provides definitions, analysis, and discussion about display color characterization models. We do not detail hybrid methods or numerical methods in this section because they show less interest for modeling purpose, and we do prefer to refer the reader to the papers cited above. 3D LUT-based method are more considered in the part concerning model's inversion.

In 1983, [28] wrote what is considered to be the pioneer article in the area of physical models for display characterization. In this work, the author stated that a power function can be used, but is not the best to fit with the luminance response curve of a CRT device. Nevertheless, the well-known "gamma" model that considers a power function to approximate the luminance response curve of a CRT display is still currently widely used.

Whichever shape the model takes, the principle remains the same. First, it estimates the luminance response of the device for each channel, using a set of functions monotonically increasing such as (4.2). Note that the results of these functions can also be estimated with any interpolation method, since the problem of monotonicity that can arise during the inversion process is taken into account. This step is followed by a colorimetric transform.

4.3.2.2 Response Curve Retrieval

We review here two types of models. The models of the first type are based on functions, the second type is the PLCC model. This model is based on linear interpolation of the luminance response curve and its accuracy has been demonstrated by [68] who found it the best among the models they tested (except in front of the PLVC model for chromatic accuracy).

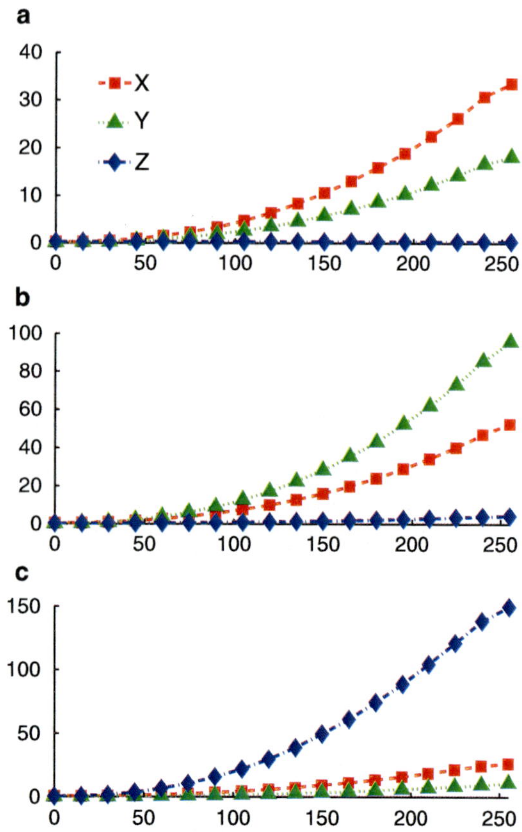

Fig. 4.3 Response curve in X, Y and Z for an LCD display in function of the digital input for, respectively, the red (**a**), green (**b**) and blue (**c**) channel

For function-based model, the function used is the power function for CRT devices, which is still the most used, even if it has been shown that it does not fit well LC technology [33]. It has been shown that for other technologies, there is no reason to try to fit the device response with a gamma curve, especially for an LCD technology that shows an S-shaped response curve in most cases (Fig. 4.3) and an S-curve model can be defined [58, 59, 94]. However, the gamma function is still often used, mainly because it is easy to estimate the response curve with a few number of measurements, or using estimations with a visual matching pattern.

The response in luminance for a set of digital values input to the device can be expressed as follows:

$$
\begin{aligned}
Y_R &= f_r(D_r) \\
Y_G &= f_g(D_g) \\
Y_B &= f_b(D_b),
\end{aligned}
\tag{4.2}
$$

where f_r, f_g, and f_b are functions that give the Y_R, Y_G, and Y_B contribution in luminance of each primary independently for a digital input D_r, D_g, D_b. Note that for CRT devices, after normalization of the luminance and digital value, the function can be the same for each channel. This assumption is not valid for LCD technology [73], and is only a rough approximation for DLP-based projection systems, as seen, for instance, in the work of [72].

For a CRT, for the channel $h \in \{r, g, b\}$, this function can be expressed as

$$Y_H = (a_h d_h + b_h)^{\gamma_h}, \tag{4.3}$$

where $H \in \{R, G, B\}$ is the equivalent luminance from a channel $h \in \{r, g, b\}$ for a normalized digital input d_h, with $d_h = \frac{D_h}{2^n - 1}$. D_h is the digital value input to a channel h and n is the number of bits used to encode the information for this channel. a_h is the gain and b_h is the internal offset for this channel. These parameters are estimated empirically using a regression process.

This model is called gain-offset-gamma (GOG) [12, 48, 55]. If we make the assumption that there is no internal offset and no gain, $a = 1$ and $b = 0$, it becomes the simple "gamma" model.

Note that for luminance transforms, polynomials can be fitted better in the logarithmic domain or to cube root function than in the linear domain because the eye response to signal intensity is logarithmic (Weber's law). For gamma-based models, it has been shown that a second order function with two parameters such as $Log(Y_H) = b_h \times Log(d_h) + c_h \times (Log(d_h))^2$ [1] gives better results[28] and that two gamma curves should be combined for a better accuracy in low luminance[6].

For an LCD, it has been shown by [58, 59] that an S-shaped curve based on four coefficients per channel can fit well the intensity response of the display.

$$Y_H = A_h \times g_h(d_h) = A_h \times \frac{d_h^{\alpha_h}}{d_h^{\beta_h} + C_h}, \tag{4.4}$$

with the same notation as above, and with A_h, α_h, β_h, and C_h parameters obtained using the least-squares method. This model is called S-curve I.

The model S-curve II considers the interaction between channels. It has been shown in [58, 59, 94] that the gradient of the original S-curve function fits the importance of the interaction between channels. Then this component can be included in the model in order to take this effect into account.

$$\begin{aligned} Y_R &= A_{rr} \times g_{Y_R Y_R}(d_r) + A_{rg} \times g'_{Y_R Y_G}(d_g) + A_{rb} \times g'_{Y_R Y_B}(d_b), \\ Y_G &= A_{gr} \times g'_{Y_G Y_R}(d_r) + A_{gg} \times g_{Y_G Y_G}(d_g) + A_{gb} \times g'_{Y_G Y_B}(d_b), \\ Y_B &= A_{br} \times g'_{Y_B Y_R}(d_r) + A_{bg} \times g'_{Y_B Y_G}(d_g) + A_{bb} \times g_{Y_B Y_B}(d_b), \end{aligned} \tag{4.5}$$

[1]Note that [68] added a term to this equation, which became $Log(Y_H) = a + b_h \times Log(d_h) + c_h.(Log(d_h))^2$.

where $g(d)$ and its first-order derivative $g'(d)$ are

$$g(d) = \frac{d^\alpha}{d^\beta + C}, \quad g'(d) = \frac{(\alpha - \beta)x^{\alpha+\beta-1} + \alpha C x^{\alpha-1}}{(x^\beta + C)^2}. \tag{4.6}$$

To ensure the monotonicity of the functions for the S-curve models I and II, some constraints on the parameters have to be applied. We let the reader refer to the discussion in the original article [59] for that matter.

For the PLCC model, the function f is approximated by a piecewise linear interpolation between the measurements. The approximation is valid for a large enough amount of measurements (16 measurements per channel in [68]). This model is particularly useful when no information is available about the shape of the display luminance response curve.

4.3.2.3 Colorimetric Transform

A colorimetric transform is then performed from the (Y_R, Y_G, Y_B) "linearized" luminance to the *CIEXYZ* color tristimulus.

$$\begin{bmatrix} X \\ Y \\ Z \end{bmatrix} = \begin{bmatrix} X_{r,\max} & X_{g,\max} & X_{b,\max} \\ Y_{r,\max} & Y_{g,\max} & Y_{b,\max} \\ Z_{r,\max} & Z_{g,\max} & Z_{b,\max} \end{bmatrix} \times \begin{bmatrix} Y_R \\ Y_G \\ Y_B \end{bmatrix}, \tag{4.7}$$

where the matrix components are the tristimulus colorimetric values of each primary, measured at their maximum intensity.

Using such a matrix for the colorimetric transform supposes perfect additivity and chromaticity constancy of primaries. These assumptions have been shown to be acceptable for CRT technology [19, 29].

The channel interdependence observed in CRT technology is mainly due to an insufficient power supply and an inaccuracy of the electron beams, which meet inaccurately the phosphors [54]. In LC technology, it comes from the overlapping of the spectral distribution of primaries (the color filters), and from the interferences between the capacities of two neighboring subpixels [72, 94]. In DLP-DMD projection devices, there is still some overlapping between primaries and inaccuracy at the level of the DMD mirrors.

Considering the assumption of chromaticity constancy, it appears that when there is a flare [54], either a black offset (internal flare) or an ambient flare (external flare), added to the signal, the assumption of chromaticity constancy is not valid anymore. Indeed, the flare is added to the output signal and the lower the luminance level of the primaries, the more the flare is a significant fraction of the resulting stimulus. This leads to a hue shift toward the black offset chromaticity. Often the flare has a "gray" (nearly achromatic) chromaticity; thus, the chromaticities of the primaries shift to a "gray" chromaticity (Fig. 4.4, left part). Note that the flare "gray" chromaticity does not necessarily correspond to the achromatic point of the device (Fig. 4.4). In fact,

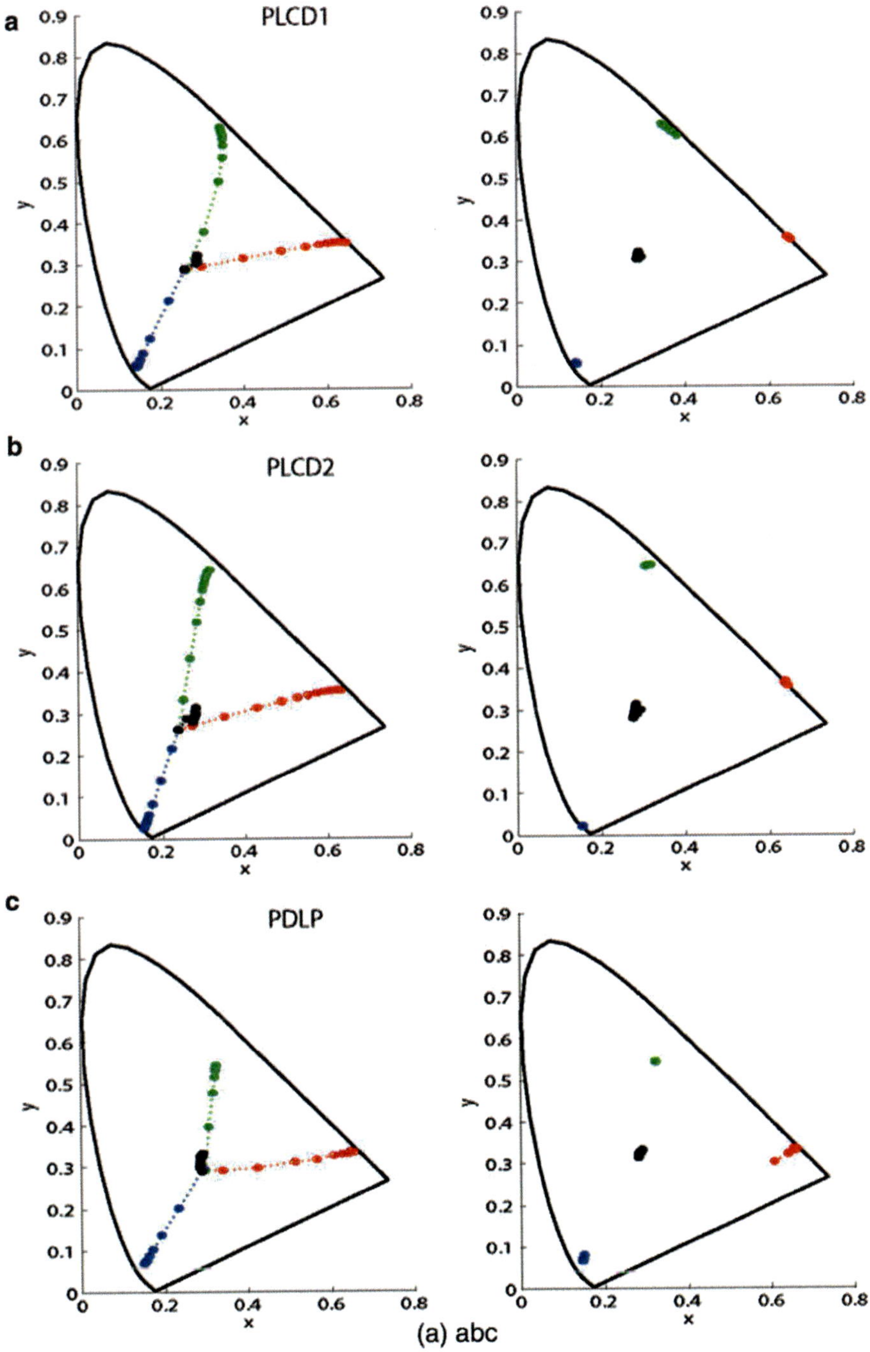

Fig. 4.4 Chromaticity tracking of primaries with variation of intensity. The left part of the figure shows it without black correction. On the right, one can see the result with a black correction performed. All devices tested in our PLVC model study are shown, a-PLCD1, b-PLCD2, c-PDLP, d-MCRT, e-MLCD1, f-MLCD2. Figures from [86]

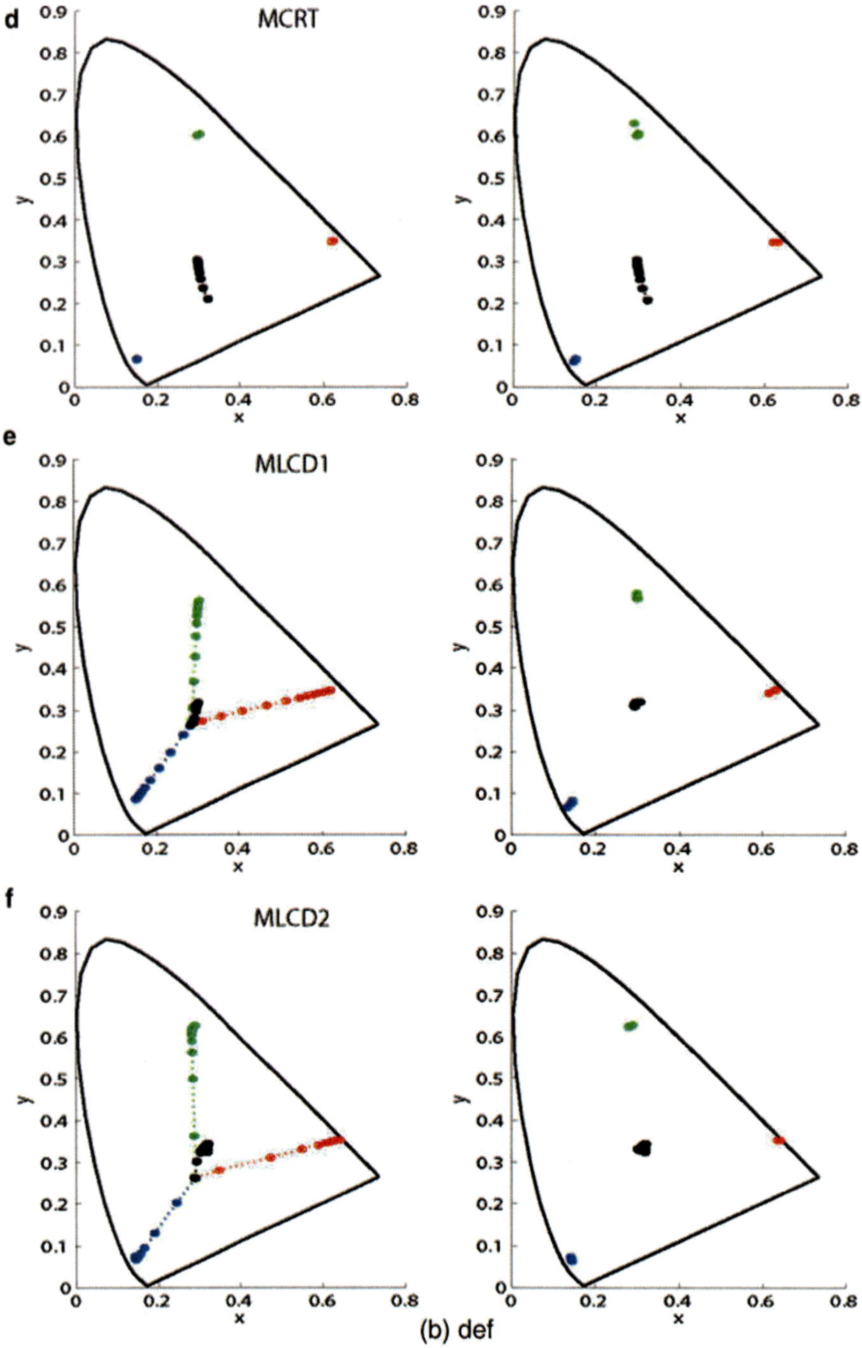

Fig. 4.4 (continued)

in the tested LCD devices (Fig. 4.4a, b, e, f), we can notice the same effect as in the work of [61]: the black level chromaticity is bluish because of the poor filtering power of the blue filter in the low wavelength.

The flare can be taken all at once as the measured light for an input $(d_{r,k}, d_{g,k}, d_{b,k}) = (0,0,0)$ to the device. Then it includes ambient and internal flare.

The ambient flare comes from any light source reflecting on the display screen. If the viewing conditions do not change it remains constant, can be measured and taken into account, or can be simply removed in setting up a dark environment (note that for a projection device, there is always an amount of light that lights the room, coming from the bulb through the ventilation hole).

The internal flare, which is the major part of chromaticity inconstancy at least in CRT technology [54], is coming from the black level. In CRT technology, it has been shown that in setting the brightness to a high level, the black level increases to a non-negligible value [54]. In LC technology, the panel let an amount of light passing through due to a leakage of the crystal to stop all the light. In DLP technology, an amount of light can be not absorbed by the "black absorption box," and is focused on the screen via the lens.

On Fig. 4.4, one can see the chromaticity shift to the flare chromaticity with the decreasing of the input level. We have performed these measurements in a dark room, then the ambient flare is minimized, and only the black level remains. After black level subtraction, the chromaticity is more constant (Fig. 4.4), and a new model can be set up in taking that into account [43, 50, 54, 55].

The gamma models reviewed above have been extended in adding an offset term. Then the GOG can become a gain-offset-gamma-offset (GOGO) model [46, 54, 55].

The previous equation (4.2) becomes:

$$Y_H = (a_h d_h + b_h)^{\gamma_h} + c, \tag{4.8}$$

where c is a term containing all the different flares in presence. If we consider the internal offset b_h as null, the model becomes gain-gamma-offset (GGO) [46].

A similar approach can be used for the PLCC model. When the black correction [50] is performed, we name it PLCC* in the following. The colorimetric transform used then is (4.9) that permits to take the flare into account during the colorimetric transformation. For the S-curve models, the black offset is taken into account in the matrix formulation in the original papers.

If we consider that mathematically, the linear transform from the linearized *RGB* to *CIEXYZ* needs to associate the origin of *RGB* to the origin of *CIEXYZ* in order to respect the vectorial space property of additivity and homogeneity. Thus, the original transform of the origin of *RGB* to *CIEXYZ* needs to be translated of $[-X_k - Y_k - Z_k]$. However, in doing that we modify the physical reality and we need to translate the result of the transformation of $[X_k Y_k Z_k]$. We can formulate these transforms such as in (4.9).

$$\begin{bmatrix} X \\ Y \\ Z \end{bmatrix} = \begin{bmatrix} X_{r,\max} - X_k & X_{g,\max} - X_k & X_{b,\max} - X_k & X_k \\ Y_{r,\max} - Y_k & Y_{g,\max} - Y_k & Y_{b,\max} - Y_k & Y_k \\ Z_{r,\max} - Z_k & Z_{g,\max} - Z_k & Z_{b,\max} - Z_k & Z_k \end{bmatrix} \times \begin{bmatrix} Y_R \\ Y_G \\ Y_B \\ 1 \end{bmatrix}. \tag{4.9}$$

The A_k's, $A \in \{X,Y,Z\}$, come from a black level estimation.

Such a correction permits the achievement of better results. However, on the right part of Fig. 4.4, one can see that even with the black subtraction, the primary chomaticities do not remain perfectly constant. On Fig. 4.4, right-a, it remains a critical shift especially for the green channel.

Several explanations are involved. First, there is a technology contribution. For LC technology, the transmittance of the cells of the panel changes within the input voltage [20, 93]. This leads to a chromaticity shift when changing the input digital value. For different LC displays, we notice a different shift in chromaticity; this is due to the combination backlight/LC with the color filters. Since the filters transmittances are optimized taking into account the transmittance shift of the LC cells, the display can achieve good chromaticity constancy. For CRT, there are less problems due to the same phosphors properties, as well for DLP as the light and the filters remain the same.

However, even with the best device, there is still a small amount of non-constancy. This leads to a discussion about the accuracy of the measured black offset. Indeed, the measurement devices are less accurate in the low luminance. Berns et al. [15] proposed a way to estimate the best black offset value. A way to overcome the problems linked with remaining inaccuracy for LCD devices has been presented by [30]. It consists in the replacement of the full intensity measurement of primary chromaticities colorimetric values by the optimum values in the colorimetric transformation matrix. It appears that the chromaticity shift is a major issue for LCD. Sharma [73] stated that for LCD devices, the assumption of chromaticity constancy was weaker than the channel interdependence.

More models that linearize the transform exist. In this section, we presented the ones that appeared to us as the more interesting or the more known.

4.3.2.4 Piecewise Linear Model Assuming Variation in Chromaticity

Defining the piecewise linear model assuming variation in chromaticity (PLVC) in this section has many motivations. First, it is the first display color characterization model introduced in the literature as far as we know. Secondly, it is a hybrid method, considering that it is based on data measurement and assumes a small amount of hypothesis on the behavior of the display. Finally, there is a section in next chapter devoted to the study of this model.

According to [68], the first persons who have introduced the PLVC were [34] in 1980. Note that it preceded the well-known article from [28]. Further studies have been performed afterward on CRT [50, 68, 69], and recently on more recent technologies [86]. This model does not consider the channel interdependence, but does model the chromaticity shift of the primaries. In this section, we recall the principles of this model, and some features that characterize it.

Knowing the tristimulus values of X, Y, and Z for each primary as a function of the digital input, assuming additivity, the resulting color tristimulus values can be expressed as the sum of tristimulus values for each component (i.e., primary) at the given input level. Note that in order not to add several times the black level, it is removed from all measurements used to define the model. Then, it is added to the result, to return to a correct standard observer color space [50, 69]. The model is summarized and generalized in (9.3) for N primaries, and illustrated in (4.11) for a three primaries *RGB* device, following an equivalent formulation as the one given by [50].

For an N primary device, we consider the digital input to the ith primary, $d_i(m_i)$, with i an integer $\in [0,N]$, and m_i an integer limited by the resolution of the device (i.e., $m_i \in [0,255]$ for a channel coded on 8 bits). Then, a color $CIEXYZ(\ldots,d_i(m_i),\ldots)$ can be expressed by:

$$X(\ldots,d_i(m_i),\ldots) = \sum_{i=0,j=m_i}^{i=N-1} [X(d_i(j)) - X_k] + X_k,$$

$$Y(\ldots,d_i(m_i),\ldots) = \sum_{i=0,j=m_i}^{i=N-1} [Y(d_i(j)) - Y_k] + Y_k,$$

$$Z(\ldots,d_i(m_i),\ldots) = \sum_{i=0,j=m_i}^{i=N-1} [Z(d_i(j)) - Z_k] + Z_k \tag{4.10}$$

with X_k, Y_k, Z_k the color tristimulus coming out from a $(0,\ldots,0)$ input.

We illustrate this for a three primaries *RGB* device, with each channel coded on 8 bits. The digital input are $d_r(i)$, $d_g(j)$, $d_b(l)$, with i,j,l integers $\in [0,255]$. In this case, a $CIEXYZ(d_r(i),d_g(j),d_b(l))$ can be expressed by:

$$\begin{aligned} X(d_r(i),d_g(j),d_b(l)) &= [X(d_r(i)) - X_k] + [X(d_g(j)) - X_k] + [X(d_b(l)) - X_k] + X_k, \\ Y(d_r(i),d_g(j),d_b(l)) &= [Y(d_r(i)) - Y_k] + [Y(d_g(j)) - Y_k] + [Y(d_b(l)) - Y_k] + Y_k, \\ Z(d_r(i),d_g(j),d_b(l)) &= [Z(d_r(i)) - Z_k] + [Z(d_g(j)) - Z_k] + [Z(d_b(l)) - Z_k] + Z_k. \end{aligned} \tag{4.11}$$

If the considered device is a *RGB* primaries device, thus the transformation between digital *RGB* values and *RGB* device's primaries is as direct as possible. The A_k, $A \in \{X,Y,Z\}$ are obtained by accurate measurement of the black level. The $[A(d_i(j)) - A_k]$, are obtained by one dimensional linear interpolation with the measurement of a ramp along each primary. Note that any 1-D interpolation method can be used. In the literature, the piecewise linear interpolation is mostly used.

Studies of this model have shown good results, especially on dark and mid-luminance colors. When the colors reach higher luminance, the additivity assumption is less true for CRT technology. Then the accuracy decreases (depending on the device properties). More precisely, [68, 69] stated that chromaticity error is lower for the PLVC than for the PLCC in low luminance. This is due to the setting of primaries colorimetric values at maximum intensity in the PLCC. Both models show inaccuracy for high luminance colors due to channel interdependence. Jimenez Del Barco et al. [50] found that for CRT technology, the higher level of brightness in the settings leads to a non-negligible amount of light for a (0,0,0) input. This light should not be added three times, and they proposed a correction for that.[2] They found that the PLVC model was more accurate in medium to high luminance colors. Inaccuracy is more important in low luminance, due to inaccuracy of measurements, and in high luminance, due to channel dependencies. Thomas et al. [86] demonstrated that this model is more accurate than usual linear models (PLCC, GOGO) for LCD technology, since it takes into account the chromaticity shift of primaries that is a key features for characterizing this type of display. More results for this model are presented in the next chapter.

4.4 Model Inversion

4.4.1 State of the Art

The inversion of a display color characterization model is of major importance for color reproduction since it provides the set of digital values to input to the device in order to display a desired color.

Among the models or methods used to achieve color characterization, we can distinguish two categories. The first one contains models that are practically invertible (either analytically, or in using simple 1D LUT) [13, 14, 29, 50, 54, 55, 68], such as the PLCC, the black-corrected PLCC*, the GOG, or GOGO models. The second category contains the models or methods, which are not practically invertible directly. and that show difficulties to be applied. Models of this second category require other methods to be inverted in practice. We can list some typical problems and methods used to invert these models:

- Some conditions have to be verified, such as in the masking model [83].
- A new matrix might have to be defined by regression in numerical models [54, 55, 89].
- A full optimization process has to be set up for each color, such as in S-curve model II [58, 59] in the modified masking model, [83] or in the PLVC model [50, 68, 84].
- The optimization process can appear only for one step of the inversion process, as in the PLVC [68] or in the S-curve I [58, 59] models.

[2]Equations (4.10) and (4.11) are based on the equation proposed by [50], and take that into account.

- Empirical methods based on 3-D LUT (look-up table) can be inverted directly [11], using the same geometrical structure. In order to have a better accuracy, however, it is common to build another geometrical structure to yield the inverse model. For instance, it is possible to build a draft model to define a new set of color patches to be measured [80].

The computational complexity required to invert these models makes them seldom used in practice, except the full 3-D LUT, whose major drawback is that it requires a lot of measurements. However, these models do have the possibility to take into account more precisely the device color-reproduction features, such as interaction between channels or chromaticity inconstancy of the primaries. Thus, they are often more accurate than the models of the first category.

4.4.2 *Practical Inversion*

Models such as the PLCC, the black-corrected PLCC*, the GOG, or GOGO models [13, 14, 29, 50, 54, 55, 68] are easily inverted since they are based on linear algebra and on simple functions. For these models, it is sufficient to invert the matrix of (4.7). Then we have:

$$\begin{bmatrix} Y_R \\ Y_G \\ Y_B \end{bmatrix} = \begin{bmatrix} X_{r,\max} & X_{g,\max} & X_{b,\max} \\ Y_{r,\max} & Y_{g,\max} & Y_{b,\max} \\ Z_{r,\max} & Z_{g,\max} & Z_{b,\max} \end{bmatrix}^{-1} \times \begin{bmatrix} X \\ Y \\ Z \end{bmatrix}. \tag{4.12}$$

Once the linearized $\{Y_R, Y_G, Y_B\}$ have been retrieved, the intensity response curve function is inverted as well to retrieve the $\{d_r, d_g, d_b\}$ digital values. This task is easy for a gamma-based model or for an interpolation-based one. However, for some models such as the S-curve I, an optimization process can be required (note that this response curve can be used to create a 1D LUT).

4.4.3 *Indirect Inversion*

When the inversion becomes more difficult, it is of use to set an optimization process using the combination of the forward transform and the color difference (often the euclidean distance) in a perceptually uniform color space, such as *CIELAB*, as cost function. This generally leads to better results than usual linear models, depending on the forward model, but is computationally expensive, and cannot be implemented in real time. It is then of use to set a 3-D LUT based on the forward model. Note that it does not mean that an optimization process is useless, since it can help to design a good LUT.

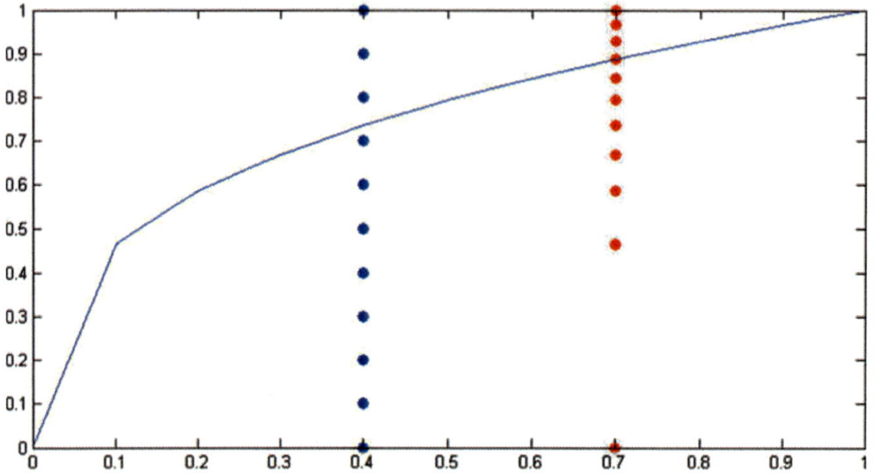

Fig. 4.5 The transform between *RGB* and *CIELAB* is not linear. Thus while using a linear interpolation based on data regularly distributed in *RGB*, the accuracy is not the same everywhere in the colorspace. This figure shows a plot of regularly distributed data in a linear space (blue dot, left) and the resulting distribution after a cubic root transform (that mimics *CIELAB* transform)(red dots, right)

Such a model is defined by the number and the distribution of the color patches used in the LUT, and by the interpolation method used to generalize the model to the entire space. In this subsection, we review some basic tools and methods. We distinguish works on displays from more general works, which have been performed in this way either in a general purpose or especially for printers. One of the major challenges for printers is the problem of measurement, which is really restrictive, and many works have been carried out in using a 3-D LUT for the color characterization of these devices. Moreover, since printer devices are highly nonlinear, their colorimetric models are complex. So it has been customary in the last decade to use a 3-D complex LUT for the forward model, created by using an analytical forward model, both to reduce the amount of measurements and to perform the color space transform in a reasonable time. The first work we know about creating a LUT based on the forward model is a patent from [81]. In this work, the LUT is built to replace the analytical model in the forward direction. It is based on a regular grid designed in the printer *CMY* color space, and the same LUT is used in the inverse direction, simply in switching the domain and co-domain. Note that in displays, the forward model is usually computationally simple and that we need only to use a 3-D LUT for the inverse model. The uniform mapping of the *CMY* space leads to a nonuniform mapping in *CIELAB* space for the inverse direction, and it is common now to resample this space to create a new LUT. To do that, a new grid is usually designed in *CIELAB* and is inverted after gamut mapping of the points located outside the gamut of the printer. Several algorithms can be used to redistribute the data [24, 31, 41] and to fill the grid [9, 77, 88].

Returning to displays, let us call source space the independent color space (typically *CIELAB* or alternatively *CIEXYZ*), the domain from where we want to move, and destination space, the *RGB* color space, the co-domain, where we want to move to. If we want to build a grid, we then have two classical approaches to distribute the patches in the source space, using the forward model. One can use directly a regular distribution in *RGB* and transform it to *CIELAB* using the forward model; this approach is the same as used by [81] for printers, and leads to a non-uniform mapping of the *CIELAB* space, which can lead to a lack of homogeneity of the inverse model depending on the interpolation method used (See Fig. 4.5). An other approach can be to distribute the patches regularly in *CIELAB*, following a given pattern, such as an hexagonal structure [80] or any of the methods used in printers [24, 31, 41]. Then, an optimization process using the forward model can be performed for each point to find the corresponding *RGB* values. The main idea of the method and the notation used in this document are the following:

- One can define a regular 3-D grid in the destination color space (*RGB*).
- This grid defines cubic voxels. Each one can be split into five tetrahedra (See Fig. 4.6).
- This tetrahedral shape is preserved within the transform to the source space (either *CIEXYZ* or *CIELAB*).
- Thus, the model can be generalized to the entire space, using tetrahedral interpolation [53]. It is considered in this case that the color space has a linear behavior within the tetrahedron (e.g., the tetrahedron is small enough).

The most used way to define such a grid is to take directly a linear distribution of points on each digital d_r, d_g, and d_b axis as seeds and to fill up the rest of the destination space. A tetrahedral structure is then built with these points. The built structure is used to retrieve any *RGB* value needed to display a specific color inside the device's gamut. The more points are used to build the grid, the more the tetrahedra will be small and the interpolation accurate. Each vertex is defined by $V_{i,j,k} = (R_i, G_j, B_k)$, where $R_i = d_i$, $G_j = d_j$, $B_k = d_k$, and d_i, d_j, $d_k \in [0,1]$ are the possible normalized digital values, for a linear distribution. $i \in [0, N_r - 1]$, $j \in [0, N_g - 1]$, and $k \in [0, N_b - 1]$ are the indexes (integers) of the seeds of the grid along each primary, and N_r (resp. N_b, N_g) is the number of steps along channel R (resp. G, B).

Once this grid has been built, we define the tetrahedral structure for the interpolation following [53]. Then, we use the forward model to transform the structure into *CIELAB* color space. An inverse model has been built. According to the nonlinearity of the *CIELAB* transform, the size of the tetrahedra is not anymore the same as it was in *RGB*. In the following section, a modification of this framework is proposed that makes this grid more homogeneous in the source color space where we perform the interpolation; this should lead to a better accuracy, following [41].

Let us consider the PLVC model inversion as an example. This model inversion is not as straightforward as the matrix-based models previously defined. For a three primaries display, according to [68], it can be performed defining all subspaces defined by the matrices of each combinations of measured data (note that the

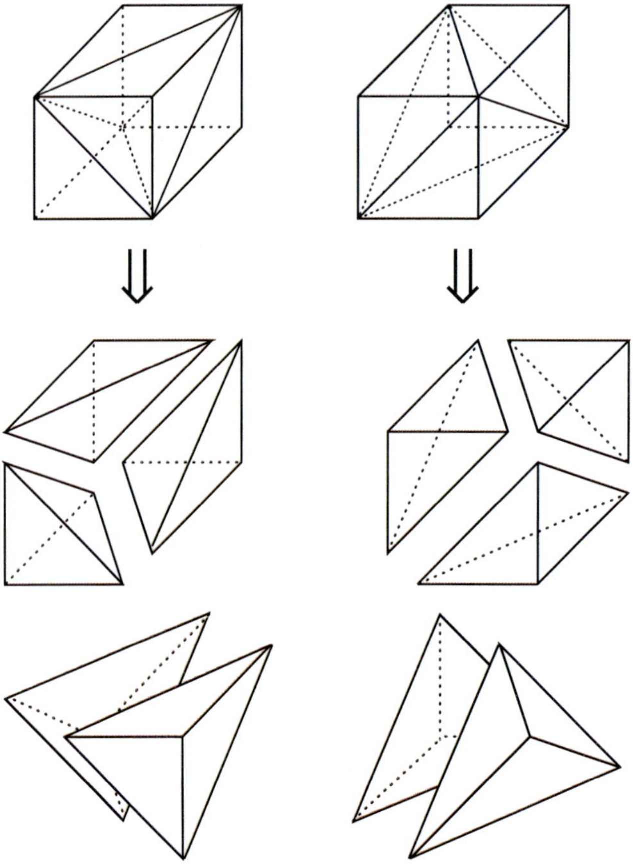

Fig. 4.6 The two ways to split a cubic voxel in 5 tetrahedra. These two methods are combined alternatively when splitting the cubic grid to guarantee that no coplanar segments are crossing

intercepts have to be subtracted, and once all the contributions are known, they have to be added). One can perform an optimization process for each color [50], or define a grid in *RGB*, such as described above, which will allow us to perform the inversion using 3D interpolation. Note that Post and Calhoun have proposed to define a full LUT considering all colors. They said themselves that it is inefficient. Defining a reduced regular grid in *RGB* leads to the building of an irregular grid in *CIELAB* due to the nonlinear transform. This irregular grid could lead to inaccuracy or a lack of homogeneity in interpolation, especially if it is linear. Some studies addressed this problem [84, 85]. They built an optimized LUT, based on a customized *RGB* grid.

4.5 Quality Evaluation

Colorimetric characterization of a color display device is a major issue for the accurate color rendering of a scene. We have seen several models that could possibly be used for this purpose, each of these models have their own advantages and weaknesses.

This section discusses the choice of a model in relation with the technology and the purpose. We first address the problem of defining adequate requirements and constraints, then we discuss the appropriate corresponding model evaluation approach. Before concluding, we propose a qualitative comparison of some display characterization methods.

4.5.1 Purpose

Like any image-processing technique, a display color characterization model has to be chosen considering needs and constraints. For color reproduction, the need is mainly the expected level of accuracy. The constraints depend mainly on two things: the time and the measurement. The time is a major issue, because one may need to minimize the time of establishment of a model, or its application to an image (computational cost). The measurement process is critical because one may need to have access to a special device to establish the model. The constraint of money is distributed on the time, the software, and hardware cost, and particularly on the measurement device. We do not consider here some other features of the device, such as spatial uniformity, gamut size, etc. but only the result of the point-wise colorimetric characterization.

In the case of displays, the combination needs vs constraints seem to be in agreement. Let us expose two situations:

- The person who needs an accurate color characterization (such as a designer or a color scientist) has often a color measurement device available, is working in a more or less controlled environment, and does not mind to spend 15–20 min every day to calibrate his/her monitor/projector. This person may typically want to use an accurate method, an accurate measurement device, to take care of the temporal stability of the device, etc.
- The person who wants to display some pictures in a party or in a seminar, using a projector in an uncontrolled environment does not need a very accurate colorimetric rendering. That is fortunate, because he/she does not have any measurement device, does not have much time to perform a calibration or to properly warm up the projector. However, this person needs the colors not to betray the meaning she/he intends. In this case, a fast end-user characterization should be precise enough. This person might use a visual calibration, or even better, a visual/camera-based calibration. The method should be coupled to a user-friendly software for making it easy and fast.

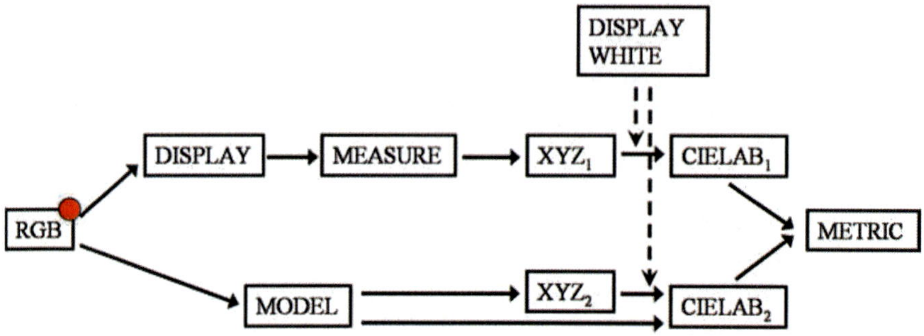

Fig. 4.7 Evaluation of a forward model scheme. A digital value is sent to the model and to the display. A value is computed and a value is measured. The difference between these values represents the error of the model in a perceptually pseudo-uniform color space

We can see a duality between two types of display characterization methods and goals: the consumer, end-user purpose, which intends only to keep the meaning and aesthetic unchanged through the color workflow, and the accurate professional one, which aims to have a very high colorimetric fidelity through the color workflow. We see also through these examples that the constraints and the needs are not necessarily going in the opposite direction.

In the next section, we will relate the quality of a model with colorimetric objective indicators.

4.5.2 Quality

Once a model is set up, there is a need to evaluate its quality to confirm we are within the accuracy we wanted. In this section, we discuss how to use objective indicators for assessing quality.

4.5.2.1 Evaluation

A point-wise quality evaluation process is straightforward. We process a digital value with the model to obtain a result and compare it in a perceptually pseudo-uniform color space, typically *CIELAB*, with the measurement of the same input. Figure 4.7 illustrates the process.

The data set used to evaluate the model should be obviously different than the one used to yield the model. The distribution of this data can be either distributed regularly or following a random distribution. Often, authors are choosing an evaluation data set distributed homogeneously in the *RGB* device space. This is

Table 4.1 This table shows the set of thresholds one can use to assess the quality of a color characterization model, depending on the purpose

ΔE^*_{ab}	Professional		Consumer
	Mean ΔE^*_{ab}	Max ΔE^*_{ab}	Mean ΔE^*_{ab}
$- < 1$	Good	Good	Good
$1 \leq - < 3$	Acceptable		
$3 \leq - < 6$	Not acceptable	Acceptable	Acceptable
$6 \leq -$		Not acceptable	Not acceptable

a good choice, since it will cover the whole device possibility. This can also be a good choice for the comparison of one method over different devices. However, if one wants to relate the result to the visual interpretation of the signal throughout the whole gamut of the device, it might be judicious to select an equiprobably distributed data set in a perceptual color space. This means that most of the data will fall into low digital values.

4.5.2.2 Quantitative Evaluation

Once we have an estimation of the model failure, we would like to be able to say how it is good or not for a given purpose. The ideal colorimetric case is to have an error below the just noticeable difference[3](JND). Kang [52] stated on page 167 of his book that the JND is of 1 ΔE^*_{ab} unit. Mahy et al. [60] study assessed that the JND is of 2.3 ΔE^*_{ab} units. Considering that the *CIELAB* color space is not perfectly uniform, it is impossible to give a perfect threshold with an euclidean metric.[4] Moreover, these thresholds have been defined for simultaneous pair comparison of uniform color patches. This situation almost never fit with a display use, it may then not be the best choice when comparing color display devices.

In the case of ΔE^*_{ab} thresholds for color imaging devices, many thresholds have been used [1, 42, 70, 78]. Stokes et al. [82] found a perceptibility acceptance for pictorial images of an average of 2.15 units. Catrysse et al. [23] used a threshold of 3 units. Gibson and Fairchild [39] found acceptable a characterized display that has a prediction error average of 1.98 and maximum of 5.57, while the non-acceptable has at the best an average of 3.73 and a maximum of 7.63 using ΔE^*_{94}.

Following is a set of thresholds that could be used to quantify the success of the color control depending on the purpose. In Table 4.1, we distinguish between accurate professional color characterization, which purpose is to ensure a really high

[3] A JND is the smallest detectable difference between a starting and secondary level of a particular sensory stimulus, in our case two color samples.

[4] The JND while using ΔE^*_{00} should be closer to one than with other metrics but has still been defined for simultaneous pair comparison of uniform color patches.

quality color reproduction, and a consumer color reproduction, which aims only at the preservation of the intended meaning, and relate the purpose with objective indicators.

Considering the professional reproduction, let us consider the following rule of thumb. If we want to reach a good accuracy, we need to consider two indicators: the average and the maximum error. Let us consider the average: from 0 to 1 is considered good, from 1 to 3 acceptable, and over 3 not acceptable. If now we consider the maximum, from 0 to 3 is good, from 3 to 6 is acceptable, over is not acceptable. If we compare this scale with the rule of thumb used by [42], it makes sense since below three it is hardly perceptible, the same if we look at the work of [1]. If we look at the JND proposed by [52] or [60] it seems to make sense since in both cases, the *good* is under the JND. In this case we would prefer results to be good, and it may be possible to discard a couple model/display if it does not satisfy this condition. In the case of this professional reproduction, it could be better to use the maximum error to discard a couple model/display. Considering the consumer prediction, we propose to consider that from 0 to 3 it is good, from 3 to 6 it is acceptable, and over 6 it is not acceptable. In this case we would rather accept methods that shows average results up to 6, since it should not spoil the meaning of the reproduction. This is basically the same as the rule of thumb proposed by [42], *perceptible but acceptable* being the basic idea of preserving the intended meaning.

4.5.3 Color Correction

The different approaches presented in the previous sections are characterized by different parameters, such as the accuracy on a given technology, the computational cost, the number of measurements required, etc.

The accuracy of the color rendering depends on the choice of both the method and the display technology and features.

Display characteristics, such as temporal stability or spatial uniformity have to be taken into account. Some of these parameters are studied in the literature, for instance, in [87]. However, Table 4.2 presents a qualitative summary of different display colorimetric characterization models based on quality thresholds from Table 4.1, and on the experimentation on several displays of different models in relation with the nature and number of measurements needed. The complete quantitative analysis of these models are presented in the literature [26, 62, 86].

We only focus on five models that are a representative sampling of existing ones: The PLVC model [50, 68, 86], Bala's model [8, 62], an optimized polyharmonic splines-based model [26], the offset-corrected piecewise linear assuming chromaticity constancy model (PLCC*) and the GOGO [13, 14, 29, 50, 54, 55, 68].

Table 4.2 Qualitative interpretation of different models based on Table 4.1. The efficiency of a model is dependent on several factors: the purpose, the number of measurements, the nature of the data to measure, the computational cost, its accuracy, etc. All these parameters depend strongly on each display

Model	PLVC	Bala	PLCC*	Polyharmonic splines	GOGO
Type of measurement	54 (*CIEXYZ*) measures	1–3 visual tasks for 1–3 pictures	54 (Y) measures 3 (*CIEXYZ*)	216 (*CIEXYZ*) measures	3–54 (Y) measures 3 (*CIEXYZ*)
Technology	Dependent	Dependent	Dependent	Independent	CRT
Purpose	Professional or Consumer	Consumer	Professional or Consumer	Professional	Consumer

4.6 Conclusion and Perspectives

Successful color-consistent cross-media color reproduction depends on a multitude of factors. In this chapter we have reviewed briefly the state of the art of this field, focusing specifically on displays.

Device colorimetric characterization is based on a model, which can successfully predict the relationship between the media value and the color itself. A model can be based on knowledge on the device technology, then a few measurement or evaluation is necessary. A numerical model based on measure only can be used too, which requires usually more measurement, and requires to take care of more aspects, such as an interpolation method and the distribution on the training data set.

Point-wise colorimetric characterization of displays is something that is working considering objective indicators. Within this chapter, we reviewed different means to achieve this result. Display technology is evolving really fast. New technology might requires the definition of other types of models. For instance, this happened with the emergence of multi-primaries devices, which means more than three primaries, there are some works that address the transform from a set of N-primaries and a 3-D colorimetric data.

This chapter only treated on point-wise, static models. A research direction could be to define dynamic models, which could take into account the spatial uniformity, the temporal stability, etc.

Within the last section of this chapter, we mainly wanted to show how we can evaluate the quality of a couple display/color characterization model with the tools we have in hands and to give an idea of how to select a model for a given purpose.

In summary, the choice of a couple display/color characterization model depends on the purpose. However, all the considerations we discussed are taking into account colorimetric objective indicators. In the case of complex images, indicators based on pointwise colorimetry show their limit. As far as we know, there is no comprehensive work addressing color fidelity and quality for complex color images on displays based on more human indicators. But there is some research initiated in this direction.

Furthermore, to reach an efficient perceived quality of displayed images, we need to relate the work on image quality metric and the display color rendering quality. That means to define an objective indicator for color image quality viewed on displays related to the accuracy of the color rendering.

This point of view could be of benefit, particularly while considering new "intelligent" displays that adapt backlight to the image content. Such displays makes a static model inefficient.

References

1. Abrardo A, Cappellini V, Cappellini M, Mecocci A (1996) Art-works colour calibration using the vasari scanner. In: Color imaging conference, IS&T, pp 94–97
2. Akima H (1970) A new method of interpolation and smooth curve fitting based on local procedures. J ACM 17:589–602
3. Alleysson D, Susstrunk S (2002) Color characterisation of the digital cinema chain. In: Research report, part II, EPFL, Switzerland
4. Alsam A, Farup I (2009) Colour gamut mapping as a constrained variational problem. In: Salberg A-B, Hardeberg JY, Jenssen R (eds) Image analysis, 16th Scandinavian conference, SCIA, 2009, vol 5575 of lecture notes in computer science, pp 109–117
5. Amidror I (2002) Scattered data interpolation methods for electronic imaging systems: a survey. J Electron Imag 11(2):157–176
6. Arslan O, Pizlo Z, Allebach JP (2003) CRT calibration techniques for better accuracy including low-luminance colors. In: Eschbach R, Marcu GG (eds) Color imaging IX: processing, hardcopy, and applications. SPIE Proceedings, vol 5293, pp 286–297
7. Bakke AM, Farup I, Hardeberg JY (2010) Evaluation of algorithms for the determination of color gamut boundaries. Imag Sci Tech 54(5):050,502–11
8. Bala R, Klassen RV, Braun KM (2007) Efficient and simple methods for display tone-response characterization. J Soc Inform Disp 15(11):947–957
9. Balasubramanian R, Maltz MS (1996) Refinement of printer transformations using weighted regression. In: Bares J (ed) Color imaging: device-independent color, color hard copy, and graphic arts. SPIE Proceedings, vol 2658, pp 334–340
10. Barnard K, Funt B (2002) Camera characterization for color research. Color Res Appl 27(3):152–163
11. Bastani B, Cressman B, Funt B (2005) Calibrated color mapping between LCD and CRT displays: a case study. Color Res Appl 30(6):438–447
12. Berns R (1996) Methods for characterizing CRT displays. Displays 16(4):173–182
13. Berns RS, Gorzynski ME, Motta RJ (1993) CRT colorimetry. part II: metrology. Color Res Appl 18(5):315–325
14. Berns RS, Motta RJ, Gorzynski ME (1993) CRT colorimetry. part I: metrology. Color Res Appl 18(5):299,314
15. Berns RS, Fernandez SR, Taplin L (2003) Estimating black-level emissions of computer-controlled displays. Color Res Appl 28(5):379–383
16. Blondé L, Stauder J, Lee B (2009) Inverse display characterization: a two-step parametric model for digital displays. J Soc Inform Disp 17(1):13–21
17. Bonnier N, Schmitt F, Brettel H, Berche S (2008) Evaluation of spatial gamut mapping algorithms. In: The proceedings of the IS&T/SID fourteenth color imaging conference, pp 56–61
18. Bookstein F (1989) Principal warps: thin-plate splines and the decomposition of deformations. IEEE Trans Pattern Anal Mach Intell 11(6):567–585

19. Brainard DH (1989) Calibration of a computer-controlled color monitor. Color Res Appl 14:23–34
20. Brainard DH, Pelli DG, Robson T (2002) Display characterization. Wiley, New York
21. Braun G, Ebner F, Fairchild M (1998) Color gamut mapping in a hue linearized cielab colorspace. In: The proceedings of the IS&T/SID sixth color imaging conference: color science, systems and applications, Springfield (VA), pp 346–350
22. Braun GJ, Fairchild MD (1999) Image lightness rescaling using sigmoidal contrast enhancement functions. J Electron Imag 8(4):380
23. Catrysse PB, Wandell BA, *a PBC, W BA, B E, Gamal AE (1999) Comparative analysis of color architectures for image sensors. In: Sampat N, Yeh T (eds) proceedings of SPIE, vol 3650, pp 26–35
24. Chan JZ, Allebach JP, Bouman CA (1997) Sequential linear interpolation of multidimensional functions. IEEE Trans Image Process 6:1231–1245
25. Choi SY, Luo MR, Rhodes PA, Heo EG, Choi IS (2007) Colorimetric characterization model for plasma display panel. J Imag Sci Technol 51(4):337–347
26. Colantoni P, Thomas JB (2009) A color management process for real time color reconstruction of multispectral images. In: Lecture notes in computer science, 16th Scandinavian conference, SCIA, vol 5575, pp 128–137
27. Colantoni P, Stauder J, Blonde L (2005) Device and method for characterizing a colour device. European Patent 05300165.7
28. Cowan WB (1983) An inexpensive scheme for calibration of a colour monitor in terms of CIE standard coordinates. SIGGRAPH Comput Graph 17(3):315–321
29. Cowan W, Rowell N (1986) On the gun independency and phosphor constancy of color video monitor. Color Res Appl 11:S34–S38
30. Day EA, Taplin L, Berns RS (2004) Colorimetric characterization of a computer-controlled liquid crystal display. Color Res Appl 29(5):365–373
31. Dianat S, Mestha L, Mathew A (2006) Dynamic optimization algorithm for generating inverse printer map with reduced measurements. Proceedings of international conference on acoustics, speech and signal processing, IEEE 3
32. Dugay F, Farup I, Hardeberg JY (2008) Perceptual evaluation of color gamut mapping algorithms. Color Res Appl 33(6):470–476
33. Fairchild M, Wyble D (1998) Colorimetric characterization of the Apple Studio display (flat panel LCD). Munsell color science laboratory technical report
34. Farley WW, Gutmann JC (1980) Digital image processing systems and an approach to the display of colors of specified chrominance. Technical report HFL-80-2/ONR-80, Virginia Polytechnic Institute and State University, Blacksburg, VA
35. Farup I, Gatta C, Rizzi A (2007) A multiscale framework for spatial gamut mapping. Image Process, IEEE Trans 16(10):2423–2435
36. Finlayson G, Hordley S, Hubel P (1998) Recovering device sensitivities with quadratic programming. In: The proceedings of the IS&T/SID sixth color imaging conference: color science, systems and applications, Springfield (VA): The society for imaging science and technology, pp 90–95
37. Gatt A, Morovic J, Noriega L (2003) Colorimetric characterization of negative film for digital cinema post-production. In: Color imaging conference, IS&T, pp 341–345
38. Gerhardt J (2007) Spectral color reproduction: model based and vector error diffusion approaches. PhD thesis, Ecole Nationale Superieure des Telecommunications and Gjøvik University College, URL http://biblio.telecomparistech.fr/cgi-bin/download.cgi?id=7849
39. Gibson JE, Fairchild MD (2000) Colorimetric characterization of three computer displays (LCD and CRT). Munsell color science laboratory technical report
40. Green P (ed) (2010) Color management: understanding and using ICC profiles. Wiley, Chichester
41. Groff RE, Koditschek DE, Khargonekar PP (2000) Piecewise linear homeomorphisms: The scalar case. In: IJCNN (3), pp 259–264

42. Hardeberg J (1999) Acquisition and reproduction of colour images: colorimetric and multispectral approaches. These de doctorat, Ecole Nationale Superieure des Telecommunications, ENST, Paris, France
43. Hardeberg JY, Seime L, Skogstad T (2003) Colorimetric characterization of projection displays using a digital colorimetric camera. In: Projection displays IX,, SPIE proceedings, vol 5002, pp 51–61
44. Hardeberg J, Bando E, Pedersen M (2008) Evaluating colour image difference metrics for gamut-mapped images. Coloration Technology 124:243–253
45. Hong G, Luo MR, Rhodes PA (2001) A study of digital camera colorimetric characterization based on polynomial modeling. Color Res Appl 26(1):76–84
46. IEC:61966–3 (1999) Color measurement and management in multimedia systems and equipment, part 3: equipment using CRT displays. IEC
47. International Color Consortium (2004) Image technology colour management – architecture, profile format, and data structure. Specification ICC.1.2004–10
48. International Commission on Illumination (1996) The relationship between digital and colorimetric data for computer-controlled CRT displays. CIE, Publ 122
49. Ishii A (2002) Color space conversion for the laser film recorder using 3-d lut. SMPTE J 16(11):525–532
50. Jimenez Del Barco L, Diaz JA, Jimenez JR, Rubino M (1995) Considerations on the calibration of color displays assuming constant channel chromaticity. Color Res Appl 20:377–387
51. Johnson T (1996) Methods for characterizing colour scanners and digital cameras. Displays 16(4)
52. Kang HR (ed) (1997) Color technology for electronic imaging devices. SPIE Press, iSBN: 978-0819421081
53. Kasson LM, Nin SI, Plouffe W, Hafner JL (1995) Performing color space conversions with three-dimensional linear interpolation. J Electron Imag 4(3):226–250
54. Katoh N, Deguchi T, Berns R (2001) An accurate characterization of CRT monitor (i) verification of past studies and clarifications of gamma. Opt Rev 8(5):305–314
55. Katoh N, Deguchi T, Berns R (2001) An accurate characterization of CRT monitor (II) proposal for an extension to CIE method and its verification. Opt Rev 8(5):397–408
56. Kimmel R, Shaked D, Elad M, Sobel I (2005) Space-dependent color gamut mapping: A variational approach. IEEE Trans Image Process 14(6):796–803
57. Klassen R, Bala R, Klassen N (2005) Visually determining gamma for softcopy display. In: Procedeedings of the thirteen's color imaging conference, IS&T/SID, pp 234–238
58. Kwak Y, MacDonald L (2000) Characterisation of a desktop LCD projector. Displays 21(5):179–194
59. Kwak Y, Li C, MacDonald L (2003) Controling color of liquid-crystal displays. J Soc Inform Disp 11(2):341–348
60. Mahy M, Eycken LVV, Oosterlinck A (1994) Evaluation of uniform color spaces developed after the adoption of cielab and cieluv. Color Res Appl 19(2):105–121
61. Marcu GG, Chen W, Chen K, Graffagnino P, Andrade O (2001) Color characterization issues for TFT-LCD displays. In: Color imaging: Device-independent color, color hardcopy, and applications VII, SPIE, SPIE Proceedings, vol 4663, pp 187–198
62. Mikalsen EB, Hardeberg JY, Thomas JB (2008) Verification and extension of a camera-based end-user calibration method for projection displays. In: CGIV, pp 575–579
63. Morovic J (2008) Color gamut mapping. Wiley, Chichester
64. Neugebauer HEJ (1937) Die theoretischen Grundlagen des Mehrfarbendruckes. Zeitschrift für wissenschaftliche Photographie, Photophysik und Photochemie 36(4):73–89
65. Neumann A, Artusi A, Zotti G, Neumann L, Purgathofer W (2003) Interactive perception based model for characterization of display device. In: Color imaging IX: processing, hardcopy, and applications IX, SPIE Proc., vol 5293, pp 232–241
66. Nielson GM, Hagen H, Müller H (eds) (1997) Scientific visualization, overviews, methodologies, and techniques. IEEE Computer Society, see http://www.amazon.com/Scientific-Visualization-Overviews-Methodologies-Techniques/dp/0818677775

67. Noriega L, Morovic J, Lempp W, MacDonald L (2001) Colour characterization of a digital cine film scanner. In: CIC 9, IS&T / SID, pp 239–244
68. Post DL, Calhoun CS (1989) An evaluation of methods for producing desired colors on CRT monitors. Color Res Appl 14:172–186
69. Post DL, Calhoun CS (2000) Further evaluation of methods for producing desired colors on CRT monitors. Color Res Appl 25:90–104
70. Schläpfer K (1993) Farbmetrik in der reproduktionstechnik und im Mehrfarbendruck. In: Schweiz SG (ed) 2. Auflage UGRA
71. Schläpfer K, Steiger W, Grönberg J (1998) Features of color management systems. UGRA Report 113/1, Association for the promotion of research in the graphic arts industry
72. Seime L, Hardeberg JY (2003) Colorimetric characterization of LCD and DLP projection displays. J Soc Inform Disp 11(2):349–358
73. Sharma G (2002) LCDs versus CRTs: color-calibration and gamut considerations. Proc IEEE 90(4):605–622
74. Sharma G (2003) Digital color imaging handbook. CRC Press, iSBN: 978-0849309007
75. Sharma G, Shaw MQ (2006) Thin-plate splines for printer data interpolation. In: Proc. "European Signal Proc. Conf.", Florence, Italy, http://www.eurasip.org/Proceedings/Eusipco/Eusipco2006/papers/1568988556.pdf
76. Sharma G, Trussell HJ (1996) Set theoretic estimation in color scanner characterization. J Electron Imag 5(4):479–489
77. Shepard D (1968) A two-dimensional interpolation function for irregularly-spaced data. In: Proceedings of the 1968 23rd ACM national conference, New York, NY, USA, pp 517–524
78. Stamm S (1981) An investigation of color tolerance. In: TAGA proceedings, TAGA proceedings, pp 156–173
79. Stauder J, Colantoni P, Blonde L (2006) Device and method for characterizing a colour device. European Patent WO/2006/094914, EP1701555
80. Stauder J, Thollot J, Colantoni P, Tremeau A (2007) Device, system and method for characterizing a colour device. European Patent WO/2007/116077, EP1845703
81. Stokes M (1997) Method and system for analytic generation of multi-dimensional color lookup tables. United States Patent 5612902
82. Stokes M, Fairchild MD, Berns RS (1992) Precision requirements for digital color reproduction. ACM Trans Graph 11(4):406–422, DOI http://doi.acm.org/10.1145/146443.146482
83. Tamura N, Tsumura N, Miyake Y (2003) Masking model for accurate colorimetric characterization of LCD. J Soc Inform Disp 11(2):333–339
84. Thomas JB, Colantoni P, Hardeberg JY, Foucherot I, Gouton P (2008) A geometrical approach for inverting display color-characterization models. J Soc Inform Disp 16(10):1021–1031
85. Thomas JB, Colantoni P, Hardeberg JY, Foucherot I, Gouton P (2008) An inverse display color characterization model based on an optimized geometrical structure. In: Color imaging XIII: processing, hardcopy, and applications, SPIE, SPIE proceedings, vol 6807, pp 68, 070A–1–12
86. Thomas JB, Hardeberg JY, Foucherot I, Gouton P (2008) The PLVC color characterization model revisited. Color Res Appl 33(6):449–460
87. Thomas JB, Bakke AM, Gerhardt J (2010) Spatial nonuniformity of color features in projection displays: a quantitative analysis. J Imag Sci Technol 54(3):030,403
88. Viassolo DE, Dianat SA, Mestha LK, Wang YR (2003) Practical algorithm for the inversion of an experimental input-output color map for color correction. Opt Eng 42(3):625–631
89. Wen S, Wu R (2006) Two-primary crosstalk model for characterizing liquid crystal displays. Color Res Appl 31(2):102–108
90. Wyble DR, Berns RS (2000) A critical review of spectral models applied to binary color printing. Color Res Appl 25:4–19
91. Wyble DR, Rosen MR (2004) Color management of DLP projectors. In: Proceedings of the twelfth color imaging conference, IS&T, pp 228–232
92. Wyble DR, Zhang H (2003) Colorimetric characterization model for DLP projectors. In: Procedeedings of the eleventh color imaging conference, IS&T, pp 346–350
93. Yeh P, Gu C (1999) Optics of liquid crystal display. Wiley, New York, ISBN: 978-0471182016

94. Yoshida Y, Yamamoto Y (2002) Color calibration of LCDs. In: Tenth color imaging conference, IS&T - The society for imaging science and technology, pp 305–311, Scottsdale, Arizona, USA
95. Yule JAC, Nielsen WJ (1951) The penetration of light into paper and its effect on halftone reproductions. In: TAGA Proceedings, vol 3, p 65

Chapter 5
Dihedral Color Filtering

Reiner Lenz, Vasileios Zografos, and Martin Solli

The color is a body of flesh where a heart beats

Malcolm de Chazal

Abstract Linear filter systems are used in low-level image processing to analyze the visual properties of small image patches. We show first how to use the theory of group representations to construct filter systems that are both steerable and are minimum mean squared error solutions. The underlying groups are the dihedral groups and the permutation groups and the resulting filter systems define a transform which has many properties in common with the well-known discrete Fourier transform. We also show that the theory of extreme value distributions provides a framework to investigate the statistical properties of the vast majority of the computed filter responses. These distributions are completely characterized by only three parameters and in applications involving huge numbers of such distributions, they provide very compact and efficient descriptors of the visual properties of the images. We compare these descriptors with more conventional methods based on histograms and show how they can be used for re-ranking (finding typical images in a class of images) and classification.

Keywords Linear filtering • Low-level image processing • Theory of group representations • Dihedral groups • Permutation groups • Discrete Fourier transform • Image re-ranking and classification

R. Lenz (✉) • M. Solli
Department of Science and Technology, Linköping University, Linköping, Sweden
e-mail: Reiner.lenz@liu.se; Martin.Solli@flir.se

V. Zografos
Computer Vision Laboratory, Linköping University, Linköping, Sweden
e-mail: zografos@isy.liu.se

C. Fernandez-Maloigne (ed.), *Advanced Color Image Processing and Analysis*,
DOI 10.1007/978-1-4419-6190-7_5,

5.1 Introduction

Linear filter systems are used in low-level image processing to analyze the visual properties of small image patches. The patches are analyzed by computing the similarity between the patch and a number of fixed filter functions. These similarity values are used as descriptors of the analyzed patch. The most important step in this approach is the selection of the filter functions. Two popular methods to select them are the minimum mean squared error (MMSE) criterion and the invariance/steerable filter approach. The MMSE method, like the jpeg transform coding, selects those filters that allow a reconstruction of the original patch with a minimal statistical error. The invariance, or the more general steerable, filter system assumes that the patches of interest come in different variations and that one fixed selection of filters can detect both, if a given patch is of interest and if this is the case which variant of these patches it is. A typical example is an edge detector that is used to detect if a given patch was an edge and in the case of an edge it gives a possible estimate of its orientation (see [4, 15] for a comparison of different types of local descriptors). For the case of digital color images, we will show how the theory of group representations can be used to construct such steerable filter systems and that, under fairly general conditions, these filters systems are of the MMSE type.

The general results from representation theory show that the filter functions implement a transform which has many properties in common with the well-known discrete Fourier transform. One of these properties that is of interest in practical applications is the existence of fast transforms. As is the case for the Fourier transform where the DFT can be computed efficiently by using the FFT, it can be shown here that the basic color filter operations can be optimized by computing intermediate results. Apart of the speedup achievable by reducing the number of necessary arithmetic operations, we will also see that the bulk of the computations are simple additions and subtractions which make these filters suitable for hardware implementations and applications where a huge number of images have to be processed. A typical example of such a task is image retrieval from huge image databases. Such databases can contain millions or billions of images that have to be indexed and often it is also necessary to retrieve images from such a database at very high speed. We will therefore illustrate some properties of these filter systems by investigating properties of image databases harvested from websites.

In our illustrations, we will show that the theory of extreme value distributions provides a framework to investigate the statistical properties of the vast majority of the computed filter responses. These distributions are completely characterized by only three parameters and in applications involving huge numbers of such distributions, they provide very compact and efficient descriptors of the visual properties of the images. We will also compare these descriptors with more conventional methods based on histograms and show how they can be used for re-ranking (finding typical images in a class of images) and classification.

5.2 Basic Group Theory

In the following, we will only consider digital color images defined on a square grid. This is the most important case in practice, but similar results can be derived for images on hexagonal grids. We also assume that the pixel values are RGB vectors. Our first goal is to identify natural transformations that modify spatial configurations of RGB vectors.

We recall that a group is a set of elements with a combination rule that maps a pair of elements to another element in the set. The combination rule is associative, every element has an inverse and there is a neutral element. More information about group theory can be found in every algebra textbook and especially in [3]. In the following, we will only deal with dihedral groups. Such a group is defined as the set of all transformations that map a regular polygon into itself. In the following, we will use the dihedral group D_4 of the symmetry transformations of the square to describe the transformation rules of the grid on which the images are defined. We will also use the group D_3 formed by the symmetry transformations of a triangle to describe the modifications of RGB vectors. A description of the usage of the dihedral groups to model the geometric properties of the sensor grid can be found in [8, 9]. This was then extended to the investigation of color images in [10, 11]. Here we will not describe the application of the same ideas in the study of RGB histograms which can be found in [13].

It can be shown that the dihedral group D_n of the n-sided regular polygon consists of $2n$ elements. The symmetry group D_4 of the square grid has eight elements: the four rotations ρ_k with rotation angles $k = k\pi/2, k = 0, \ldots, 3$ and the reflection σ on one of the diagonals combined with one of the four rotations. The elements are thus given by $\rho_0, \ldots, \rho_3, \rho_0\sigma, \ldots, \rho_3\sigma$. This is a general property of the dihedral groups where we find for the hexagonal grid the corresponding transformation group D_6 consisting of twelve elements, six rotations and six rotations combined with a reflection. For the RGB vectors, we consider the R, G, and B channel as represented by corner points of an equilateral triangle. The symmetry group of the RGB space is thus the group D_3. It has six elements $\rho_0, \rho_1, \rho_2, \rho_0\sigma, \rho_1\sigma, \rho_2\sigma$ where the ρ_k now represent rotations with rotation angle $k \cdot 120°$ and σ is the reflection on one fixed symmetry axis of the triangle. Algebraically, this group is identical to the permutation group $S(3)$ of three elements.

We introduce the following notation to describe the general situation: For a group G with elements g and a set $\mathbf{Z}$ with elements $\mathbf{z}$ we say that G operates on $\mathbf{Z}$ if there is a mapping $(G, \mathbf{Z}) \to \mathbf{Z}$ such that $(\mathrm{g}, \mathbf{z}) \mapsto \mathrm{g}\mathbf{z}$ and $(\mathrm{g}_2\mathrm{g}_1)\mathbf{z} = \mathrm{g}_2(\mathrm{g}_1\mathbf{z})$; we also say that G is a transformation group. Furthermore, $|\mathbf{Z}|$ denotes the number of elements in the set and $|G|$ is the number of the group elements. In the context of the dihedral groups the set $\mathbf{Z}$ consists of the n corner points of the regular polygon, the transformation group is D_n and $|D_n| = 2n$. In the following, we need more general point configurations than the corner points of the polygon and therefore we introduce $\mathbf{X}$ as a set of points in the 2D-plane and we will use $\mathbf{x}$ for its elements. Such a set $\mathbf{X}$ is the collection of points on which the filter functions are defined. As a second set we introduce $\mathbf{Y}$ consisting of the three points of an equilateral triangle. Sometimes we write $\mathbf{y}$ for a general element, $\mathbf{y}_i$ for a given element and R, G, or B if we consider the three points as locations of the R, G, or B channel of a color image. For the elements in the dihedral groups, we use g or g_i to denote elements in D_4 and h or h_i for elements in D_3. For a point $\mathbf{x} \in \mathbf{X}$ on the grid and an element $\mathrm{g} \in \mathrm{D}_4$, we define the point $\mathrm{g}\mathbf{x}$ as the point obtained by applying the transformation g on the point $\mathbf{x}$. In the same way, we define for an element h and an RGB vector $\mathbf{y} = (R, G, B)$ the permuted RGB vector by $\mathrm{h}\mathbf{y}$.

We now combine the spatial transformations of the grid and the permutation of the color channels defining the product of groups. The pairs (g, h) form another group $\mathrm{D}_4 \otimes \mathrm{D}_3$ under componentwise composition $((\mathrm{g}_1, \mathrm{h}_1)(\mathrm{g}_2, \mathrm{h}_2) = (\mathrm{g}_1\mathrm{g}_2, \mathrm{h}_1\mathrm{h}_2))$.

Finally, we define the orbit $\mathrm{D}_4\mathbf{x}$ of a point $\mathbf{x}$ under the grid operations as the set $\{\mathrm{g}\mathbf{x}, \mathrm{g} \in \mathrm{D}_4\}$. It is easy to show that there exist three types of orbits: the one-point orbit consisting of the origin $\mathbf{x} = (0,0)$, four-point orbits, and eight-point orbits. Four-point orbits consist of the corner points of squares $\{(\pm n, \pm n)\}$.

For finite groups, we can use a matrix–vector notation to describe the action of the abstract transformation group. As an example, consider the case of the RGB vectors and the group D_3. Here we identify the R-channel with the first, the G-channel with the second, and the B-channel with the third unit vector in a three-dimensional space. The $120°$ rotation ρ maps the RGB vector to the vector GBR and the corresponding matrix $h(\rho)$ describing the transformation of the 3D unit vectors is given by the permutation matrix:

$$h(\rho) = \begin{pmatrix} 0 & 0 & 1 \\ 1 & 0 & 0 \\ 0 & 1 & 0 \end{pmatrix}.$$

Using this construction, we mapped abstract group elements to matrices in such a way that concatenation of group operations corresponds to the multiplication of matrices. General transformation groups are of limited use in practice since sets have no internal structure. The connection between an abstract group operation and a matrix is an example of a more interesting construction that is obtained when the set is replaced by a vector space. A general method to construct such a vector space as in our application as follows: We define an RGB patch $\mathbf{p}$ as a linear mapping $\mathbf{X} \otimes \mathbf{Y} \to \mathbb{R}$ where $\mathbf{p}(\mathbf{x}, \mathbf{y})$ is the value at pixel $\mathbf{x}$ in channel $\mathbf{y}$. RGB patches form a vector space $\mathcal{P}$. The elements in $\mathrm{D}_4 \otimes \mathrm{D}_3$ operate on the domain of $\mathbf{p}$ and we can therefore define the transformed patch $\mathbf{p}^{(\mathrm{g},\mathrm{h})}$ as $\mathbf{p}^{(\mathrm{g},\mathrm{h})}(\mathbf{x}, \mathbf{y}) = \mathbf{p}(\mathrm{g}^{-1}\mathbf{x}, \mathrm{h}^{-1}\mathbf{y})$. For example, if g is a $90°$ rotation and e is the identity operation in D_3 then the patch $\mathbf{p}^{(\mathrm{g},e)}$ is the rotated original patch and if e is the identity in D_4 and h interchanges the R and the G channel, leaving B fixed the $\mathbf{p}^{(e,\mathrm{h})}$ is the pattern with the same geometrical configuration but with R and G channel interchanged. The pattern space $\mathcal{P}$ has dimension $3|\mathbf{X}|$ where a basis is given by the set of all functions that have value one for exactly one combination $(\mathbf{x}, \mathbf{y})$ and is zero everywhere.

The mapping $\mathbf{p} \mapsto \mathbf{p}^{(\mathrm{g},\mathrm{h})}$ is a linear mapping of pattern space and we can therefore describe it by a matrix $\mathbf{T}(\mathrm{g},\mathrm{h})$ such that $\mathbf{p}^{(\mathrm{g},\mathrm{h})} = \mathbf{T}(\mathrm{g},\mathrm{h})\mathbf{p}$. We calculate:

$$\mathbf{p}^{(\mathrm{g}_1\mathrm{g}_2,\mathrm{h}_1\mathrm{h}_2)}(\mathbf{x},\mathbf{y}) = \mathbf{p}((\mathrm{g}_1\mathrm{g}_2)^{-1}\mathbf{x},(\mathrm{h}_1\mathrm{h}_2)^{-1}\mathbf{y}) = \mathbf{p}(\mathrm{g}_2^{-1}\mathrm{g}_1^{-1}\mathbf{x},\mathrm{h}_2^{-1}\mathrm{h}_1^{-1}\mathbf{y})$$
$$= \mathbf{p}^{(\mathrm{g}_2,\mathrm{h}_2)}(\mathrm{g}_1^{-1}\mathbf{x},\mathrm{h}_1^{-1}\mathbf{y}) = \left(\mathbf{p}^{(\mathrm{g}_2,\mathrm{h}_2)}\right)^{(\mathrm{g}_1,\mathrm{h}_1)}(\mathbf{x},\mathbf{y})$$

and we see that the matrices satisfy $\mathbf{T}(\mathrm{g}_1\mathrm{g}_2,\mathrm{h}_1\mathrm{h}_2) = \mathbf{T}(\mathrm{g}_1,\mathrm{h}_1)\mathbf{T}(\mathrm{g}_2,\mathrm{h}_2)$. This shows that this rule defines a (matrix) representation of the group which is a mapping from the group into a space of matrices such that the group operation maps to matrix multiplication.

Matrices describe linear mappings between vector spaces in a given coordinate system. Changing the basis in the vector space gives a new description of the same linear transformations by different matrices. Changing the basis in the pattern space $\mathcal{P}$ using a matrix $\mathbf{B}$ will replace the representation matrices $\mathbf{T}(\mathrm{g},\mathrm{h})$ by the matrices $\mathbf{BT}(\mathrm{g},\mathrm{h})\mathbf{B}^{-1}$. It is therefore natural to construct matrices $\mathbf{B}$ that simplify the matrices $\mathbf{BT}(\mathrm{g},\mathrm{h})\mathbf{B}^{-1}$ for all group elements (g,h) simultaneously. The following theorem from the representation theory of finite groups collects the basic results that give a complete overview over the relevant properties of these reduced matrices (see [3, 19, 20] for details):

Theorem 1. – *We can always find a matrix* $\mathbf{B}$ *such that all* $\mathbf{BT}(\mathrm{g},\mathrm{h})\mathbf{B}^{-1}$ *are block-diagonal with blocks* $\mathbf{T}_m$ *of minimum size.*
- *These smallest blocks are of the form* $\mathbf{T}(\mathrm{g},\mathrm{h}) = \mathbf{T}_i^{(4)}(g) \otimes \mathbf{T}_j^{(3)}(h)$ *where* $\otimes$ *denotes the Kronecker product of matrices.*
- *The dimensions of* $\mathbf{T}_i^{(4)}(g)$ *and* $\mathbf{T}_j^{(3)}(h)$ *are one or two.*
- *Both, the* $\mathbf{T}_i^{(4)}$ *and the* $\mathbf{T}_j^{(3)}$*, are representations and from their transformation properties follows that it is sufficient to know them for one rotation and the reflection:* $\mathbf{T}^{(.)}(\rho^k\sigma^l) = (\mathbf{T}^{(.)}(\rho))^k(\mathbf{T}^{(.)}(\sigma))^l)$.

For the group D_3 operating on RGB vectors, it is easy to see that the transformation $(R,G,B) \mapsto R+G+B$ is a projection on a one-dimensional subspace invariant under all elements in D_3. The first block of the matrices is therefore one-dimensional and we have $\mathbf{T}^{(3)}(h) = 1$ for all $h \in D_3$. This defines the trivial representation of the group and the one-dimensional subspace of the RGB space defined by this transformation property is the space of all gray value vectors. The other block is two-dimensional and given by the orthogonal complement to this one-dimensional invariant subspace. This two-dimensional complement defines the space of complementary colors. For the group D_4, a complete list of its smallest representations can be found in Table 5.1.

Given an arbitrary set $\mathbf{X}$ closed under D_4, the tools from the representation theory of finite groups provide algorithms to construct the matrix $\mathbf{B}$ such that the transformed matrices $\mathbf{BT}(\mathrm{g},\mathrm{h})\mathbf{B}^{-1}$ are block-diagonal with minimum-sized blocks. For details, we refer again to the literature [3, 9, 19, 20].

Table 5.1 Representations of D_4

Name	$T^{(4)}(\sigma)$	$T^{(4)}(\rho)$	Space	Dimension
Trivial	1	1	V_t	1
Alternating	−1	−1	V_a	1
p	1	−1	V_p	1
m	−1	1	V_m	1
2D	$\begin{pmatrix} 1 & 0 \\ 0 & -1 \end{pmatrix}$	$\begin{pmatrix} 0 & 1 \\ -1 & 0 \end{pmatrix}$	V_2	2

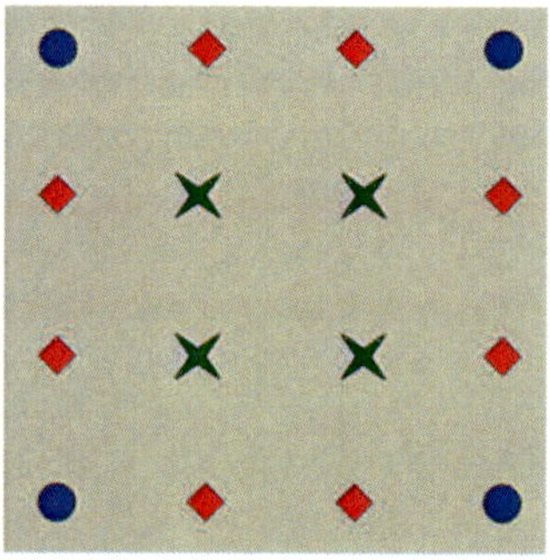

Fig. 5.1 4×4 pattern

Practically we construct the filter functions as follows: We start with a set of points $\mathbf{X}$ which is the union of D_4−orbits. The number of orbits is denoted by K and from the orbit with index k, we select an arbitrary but fixed representative $\mathbf{x}_k$. We thus have $\mathbf{X} = \bigcup_k D_4\mathbf{x}_k$. We denote the number of four-point orbits by K_4, of eight-point orbits by K_8. The number of points in $\mathbf{X}$ is thus $|\mathbf{X}| = 4K_4 + 8K_8 + K_0$ where $K_0 = 1$ if the origin is in $\mathbf{X}$ and zero otherwise. The orbit number k defines a subspace $\mathcal{P}_k$ of pattern space $\mathcal{P}$ of dimension 3, 12, or 24 depending on the number of points in the orbit. These spaces $\mathcal{P}_n$ are then further decomposed into tensor products of the spaces in Table 5.1 and the one-dimensional subspace of the intensities and the two-dimensional space of complementary colors.

As an illustration consider the case of a four-by-four window consisting of 16 grid points and a 48-dimensional pattern space. We construct the spatial filters by first splitting the 16 points in the 4×4 window into two four-point (green cross and blue disks) and one eight-point orbit (red squares) as shown in Fig. 5.1. The four-point orbit splits into the direct sum $V_t \oplus V_a \oplus V_2$. The eight point orbit is the direct sum $V_t \oplus V_a \oplus V_p \oplus V_m \oplus 2V_2$, where $2V_2$ denotes two copies of V_2. Combining these decompositions reveals that the 16-dimensional space V is given by the direct sum

$$V = 3V_t \oplus 3V_a \oplus V_p \oplus V_m \oplus 4V_2. \tag{5.1}$$

Next, we use the tensor representation construction and find for the structure of the full 48-dimensional space the decomposition

$$\begin{aligned} V = &\left(3V_{ti} \oplus 3V_{ai} \oplus V_{pi} \oplus V_{mi} \oplus 4V_{2i}\right) \\ &\oplus \left(3V_{tc} \oplus 3V_{ac} \oplus V_{pc} \oplus V_{mc} \oplus 4V_{2c}\right), \end{aligned} \tag{5.2}$$

Table 5.2 Tensor representations

Space	Dimension
$V_{ti}, V_{ai}, V_{pi}, V_{mi}$	1
$V_{2i}, V_{t2}, V_{a2}, V_{p2}, V_{m2}$	2
V_{22}	4

where $V_{xy} = V_x \otimes V_y$ is the vector space defined by the tensor representation $T_x \otimes T_y$ of the representation $T_x, x = t, a, p, m, 2$ of the group D_4 and the representation T_y, $y = t, 2$ of D_3. The dimensions of the representation spaces are collected in Table 5.2.

5.3 Illustration

We will now illustrate some properties of these filter systems with the help of a single image. We use an image of size 192×128 and filters of size 4×4. The properties of the filter results are of course depending on the relation between the resolution of the original image and the filter size where images with a high resolution will on average contain more homogeneous regions. We selected the combination of a small image size and small filter size since we will later use this combination in an application where we will investigate the statistical properties of databases consisting of very many thumbnails harvested by an image search engine from the internet.

In Fig. 5.2, we see the original image with the two parrots in the upper left corner and the result of the 48 different filters. The first 16 filter results are computed from the intensity channel and the remaining from the 32 color-opponent combinations corresponding to the splitting of the original 48-dimensional vector space described in Table 5.2. The images show the magnitude of the filter responses. In the case of the intensity-based filters, this means that a pattern and its inverted copy will produce the same filter result in this figure. The colormap used in Fig. 5.2 and the following figures is shown below the filter results. From the figure, we can see that the highest filter responses are obtained by the filters that are related to the spatial averaging, i.e., those belonging to vector spaces V_{ti} and V_{ti}. We see also that the intensity filters have in general higher responses than their corresponding color-opponent filters. From the construction (see (5.2)), we saw that these 48 filters come in 24 packages of length one, two, and four where all filters in a package have the same transformation properties and the norm of the filter vectors is invariant under spatial and color transformations. The norm of these 24 filter packages is shown in Fig. 5.3. Again, we see the highest response for the three spatial averaging filters (besides the original) and the three spatial averaging filters combined with the color-opponent colors (last two in the third row and first in the fourth row). In both, Figs. 5.2 and 5.3 the filter results are scaled such that the highest value for all filter responses is one. This makes it possible to see the relative importance of the different filters and filter packages. In Fig. 5.4, we show the magnitude of the

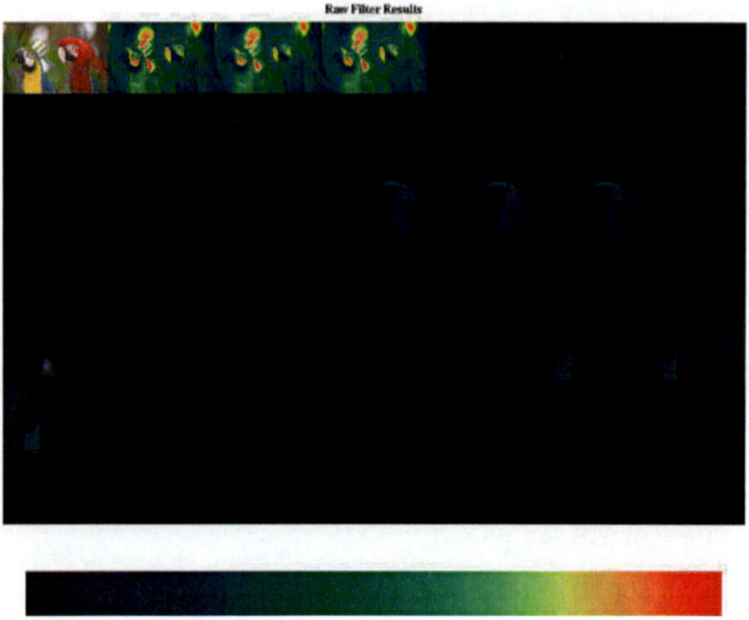

Fig. 5.2 Original image and the 48 filter results

Fig. 5.3 Original image and the 24 magnitude filter images

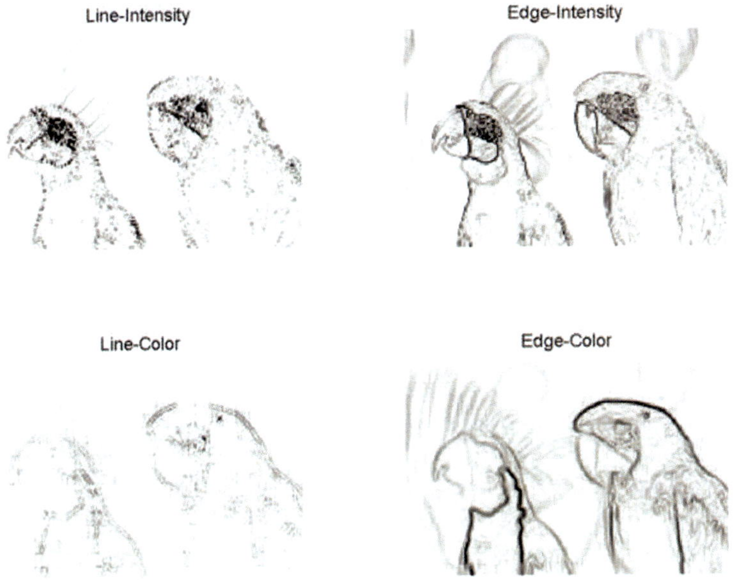

Fig. 5.4 Selected line and edge filters. *Dark* corresponds to large magnitude and *light* corresponds to low magnitude

filter response vectors for four typical filter packages related to the vector spaces of types V_{ai}, V_{2i}, V_{ac}, and V_{2c}. Visually, they correspond to line and edge filters in the intensity and the color-opponent images.

5.4 Linear Filters

Up to now we described and classified the different transformations of the pattern space. Now we will consider functions on these patterns and here especially functions that define projections. We define a linear filter f as a linear map from the pattern space $\mathcal{P}$ into the real (or complex) scalars. The Riesz representation theorem [22] shows that this map is represented by an element in pattern space and that we have: $f : \mathbf{p} \to \mathbb{R}; \mathbf{p} \mapsto f(\mathbf{p}) = \langle f, \mathbf{p} \rangle = f'\mathbf{p}$ where f' is the transposed vector and $\langle f, \mathbf{p} \rangle$ is the scalar product. An L-dimensional linear filter system $\mathbf{F}$ is an L-tupel of linear filters. We write

$$\mathbf{F}(\mathbf{p}) = \langle \mathbf{F}, \mathbf{p} \rangle = (\langle f_1, \mathbf{p} \rangle, \ldots, \langle f_L, \mathbf{p} \rangle)'.$$

We call such a vector a feature vector and the corresponding vector space the feature space.

We now divide the pattern space $\mathcal{P}$ into the smallest subspaces under the transformation group introduced above. It is then easy to see that the filter systems defined as the projection operators onto these invariant subspaces have two properties that are of importance in applications: invariance and steerability. Consider a filter system $\mathbf{F}$ that is a projection on such an invariant subspace. From the construction, we know that a pattern $\mathbf{p}$ in this subspace can be described by a coordinate vector $\mathbf{F}(\mathbf{p})$ and since the subspace is invariant we find for the transformed pattern $\mathbf{p}^{(\mathrm{g},\mathrm{h})}$ a transformation matrix $\mathbf{T}(\mathrm{g},\mathrm{h})$ such that $\mathbf{F}(\mathbf{p}^{(\mathrm{g},\mathrm{h})}) = \mathbf{T}(\mathrm{g},\mathrm{h})\mathbf{F}(\mathbf{p})$. From the general theory, it can also be shown that we can always choose the coordinate system such that the matrices $\mathbf{T}(\mathrm{g},\mathrm{h})$ are orthonormal. The norm $\|\mathbf{F}(\mathbf{p})\|$ of the feature vector is thus invariant under all group transformations. Due to the symmetry of the scalar product we can also apply the transformations of the group to the filter functions and thus modify their behavior. The feature vectors obtained from the modified filter systems are also related via the matrix multiplication with $\mathbf{T}(\mathrm{g},\mathrm{h})$ and we see that we can generate all possible modifications of these feature vectors from any given instance of it. A formal description is the following: a filter system is steerable if it satisfies the following condition:

$$\langle \mathbf{F}, \mathbf{p}^{(\mathrm{g},\mathrm{h})} \rangle = \langle \mathbf{F}, \mathbf{T}(\mathrm{g},\mathrm{h})\mathbf{p} \rangle = \widetilde{\mathbf{T}}(\mathrm{g},\mathrm{h}) \langle \mathbf{F}, \mathbf{p} \rangle \text{ for an } L \times L \text{ matrix } \widetilde{\mathbf{T}}(\mathrm{g},\mathrm{h}).$$

If we collect all filter coefficients in the matrix $\mathbf{F}$, then a steerable filter system satisfies the matrix equations $\mathbf{F}\mathbf{T}(\mathrm{g},\mathrm{h}) = \widetilde{\mathbf{T}}(\mathrm{g},\mathrm{h})\mathbf{F}$.

For a fixed pattern $\mathbf{p}_0$, we can generate its orbit $(\mathrm{D}_4 \otimes \mathrm{D}_3)\mathbf{p}_0$ in pattern space and if $\mathbf{F}$ is a steerable filter system then the feature vectors $\widetilde{\mathbf{T}}(\mathrm{g},\mathrm{h})\mathbf{F}(\mathbf{p}_0)$ define an orbit in feature space. Steerable filters have the advantage that the actual filter vector $\langle \mathbf{F}, \mathbf{p}_0 \rangle$ is computed once all the transformed versions can be computed with the closed form expression $\widetilde{\mathbf{T}}(\mathrm{g},\mathrm{h})\mathbf{F}$.

In summary, we constructed filter systems $\mathbf{F}$ with the following properties:

- $\mathbf{F}(\mathbf{p}^{(\mathrm{g},\mathrm{h})}) = \widetilde{\mathbf{T}}(\mathrm{g},\mathrm{h})\mathbf{F}(\mathbf{p})$ for all transformations $(\mathrm{g},\mathrm{h}) \in \mathrm{D}_4 \otimes \mathrm{D}_3$ and the filter system is covariant.
- $\|\mathbf{F}(\mathbf{p}^{(\mathrm{g},\mathrm{h})})\| = \|\mathbf{F}(\mathbf{p})\|$ for all transformations $(\mathrm{g},\mathrm{h}) \in \mathrm{D}_4 \otimes \mathrm{D}_3$, i.e. the norm of the feature vector is an invariant.
- The filter system is steerable via $\mathbf{T}(\mathrm{g},\mathrm{h})\mathbf{F}(\mathbf{p})$.
- The filter system is irreducible, i.e., the length of the feature vector is minimal.
- Two filter systems that are of the same type, i.e., belonging to the same product $T_x \otimes T_y$ of irreducible representations, have identical transformation properties under the transformations in the group. Especially all feature values computed by the same type of filter systems, applied to different orbits, obey identical transformation rules.

5.5 Statistical Properties

5.5.1 Partial Principal Component Analysis

Up to now we only used the existence of transformation groups to design filter systems. More powerful results can be obtained if we also make assumptions about the statistical properties of the transformations. A reasonable assumption, especially for spatially small patches, is that all transformations are equally likely to be encountered in a sufficiently large set of measurements. In this case, we can compute the second-order moments by first summing up over all transformed versions of a given pattern and then over different patterns. Formally, we consider the patterns $\mathbf{p}$ as output of a stochastic process with variable $\omega : \mathbf{p} = \mathbf{p}_\omega, \omega \in \Omega$. The second-order moment matrix, or correlation matrix, $\mathbf{C}$ is proportional to $\mathbf{C} = \sum_\omega \mathbf{p}_\omega \mathbf{p}_\omega^{'}$. For a general group G with elements g and transformed patterns $\mathbf{p}^{\mathrm{g}} = \mathbf{T}(\mathrm{g})\mathbf{p}$, we find for the correlation matrix $\mathbf{C} = \sum_\Omega \mathbf{p}_\omega \mathbf{p}_\omega^{'} = \sum_G \sum_{\Omega/G} \mathbf{T}(\mathrm{g}) \mathbf{p}_\omega \mathbf{p}_\omega^{'} \mathbf{T}(\mathrm{g})'$. Here, we assumed that all transformations $\mathrm{g} \in G$ have the same probability and we denote by Ω/G the equivalence classes of elements in Ω that are not related via a group transformation. From this follows that the correlation matrix satisfies the equations $\mathbf{C} = \mathbf{T}(\mathrm{g})\mathbf{C}\mathbf{T}(\mathrm{g})'$ for all $\mathrm{g} \in G$. For an orthonormal matrix $\mathbf{T}(\mathrm{g})$, this implies $\mathbf{C}\mathbf{T}(\mathrm{g}) = \mathbf{T}(\mathrm{g})\mathbf{C}$ and a matrix $\mathbf{C}$ with this property is called an intertwining operator. If the matrices $\mathbf{T}(\mathrm{g})$ are those defined by the smallest invariant subspaces then Schur's lemma [3, 7, 19, 20] states that the matrix $\mathbf{C}$ must either be the null-matrix or a scalar multiple of the unit matrix. From this it follows that in the coordinate system based on the smallest invariant subspaces, a general correlation matrix with $\mathbf{C} = \mathbf{T}(\mathrm{g})\mathbf{C}\mathbf{T}(\mathrm{g})'$ for all $\mathrm{g} \in G$ must be block-diagonal. The number of blocks is given by the number of different types of invariant subspaces.

Under the condition that all transformations in the transformation group are equally likely, the matrix of second-order moments is block-diagonal. In reality, this will never be exactly the case and for every collection of images we therefore have to compute how good the block-diagonal approximation describes the current collection. As an example we use the image collections Photo.net and DPChallenge described in [2]. From the Photo.net database, we used twenty three million patches from 18,365 images and from DPChallenge 16,508 images and twenty million patches. We resized the images so that the smallest dimension (height or width) was 128 pixels. The second-order moment matrix computed from the resized images is shown in Fig. 5.5. From the colorbar, it can be seen that all entries in the matrix have approximately the same value but one can also see that the matrix has a certain geometrical structure. For the second-order moment matrix computed from the filtered data, we find that the contributions from the values from the first, averaging, filters are much higher than the contributions from the other filters. We therefore illustrate the properties of these filter magnitudes in logarithmic coordinates.

In Fig. 5.6, we see the logarithms of the diagonal elements of the matrix of second-order moments. The vertical lines mark the boundary between the different blocks given by the dihedral structure. The first blocks (components 1–16) are

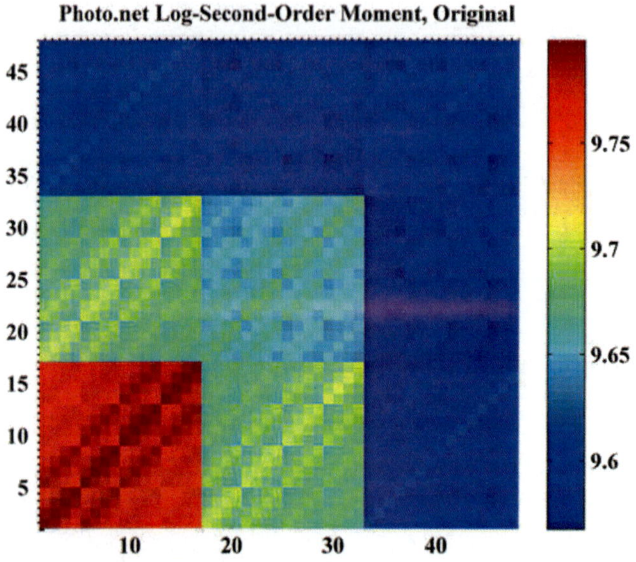

Fig. 5.5 Second-order moment matrix computed from original data

computed from the intensity channel, whereas the remaining (17–48) are related to the two-dimensional complementary color distributions. The first block (1–3 in intensity, 17–22 in complementary color) is computed by spatial averaging, the second (4–6 and 23–28) to line-like structures, the third (7–10, 29–44) to edges, and the last (11–12, 45–48) is related to the one-dimensional representations p and m (see (5.1)) from the inner orbit. The left diagram shows the values computed from the Photo.net images, the right diagram is based on the DPChallenge images. We see that the non-negativity of the first filter results leads to significant correlations with the other filter results. This leads to the structures in the first column and the last rows in the figures in the left column. The structure in the remaining part of the matrices can be clearly seen in the two images in the right column.

In Fig. 5.7, the structure of the full second-order moment matrices of the filtered results is shown. On the left side of the figure the full matrices are shown; in the right column, the rows and columns related to the averanging filters are removed to enhance the visibility of the structure in the remaining part. In the upper row, the results for the Photo.net and in the lower row the DPChallenge database were used.

5.5.2 *Extreme Value Theory*

We may further explore the statistical properties of the filters with the help of the following simple model: consider a black-box unit U with input X the pixel values from a finite-sized window in a digital image (a similar analogy can be applied to the receptive fields of a biological vision system). The purpose of this black box is to

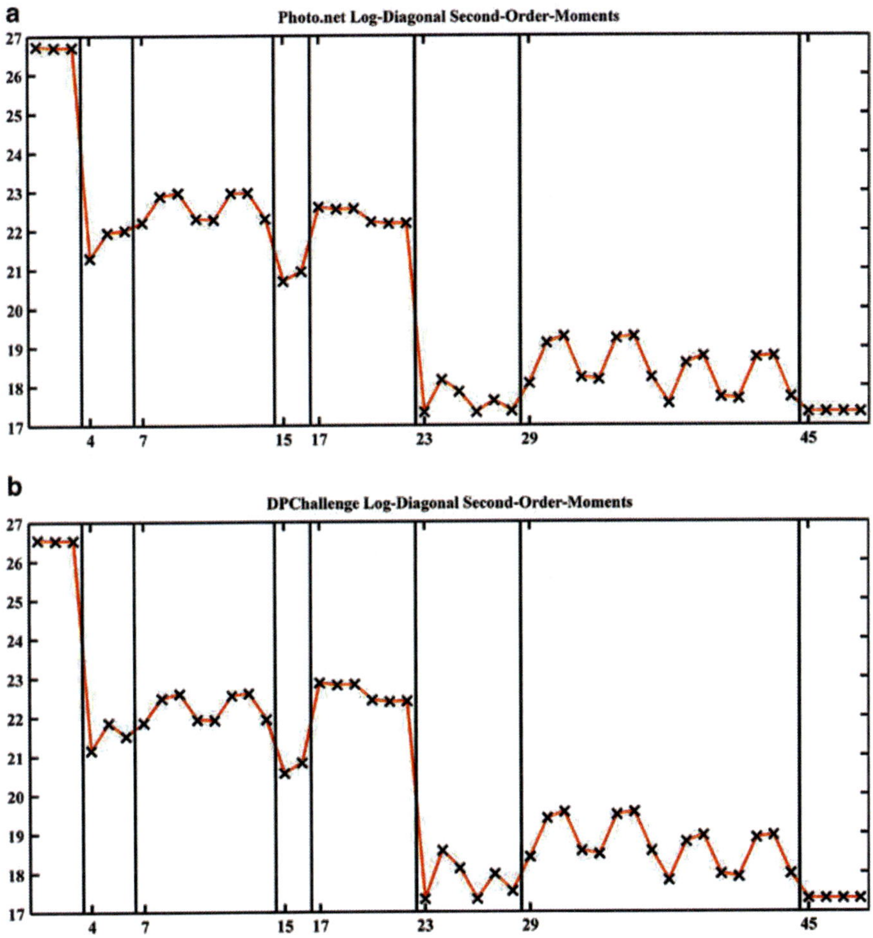

Fig. 5.6 Log Diagonal of the second-order moment matrices after filtering (**a**) Photo.net (**b**) DPChallenge

measure the amount of some non-negative quantity $X(t)$ that changes over time. We write this as $u(t) = U(X(t))$. We also define an accumulator $s(n) = \int_0^n u(t)dt$ that accumulates the measured output from the unit until it reaches a certain threshold $s(n) = \mathrm{Max}^{(n)}(X)$ or a certain period of time, above which the accumulator is reset to zero and the process is restarted.

If we consider $u(t)$, $s(n)$ as stochastic processes and select a finite number N of random samples $u_1, \dots, u_N$, then their joint distribution $J(u_1, \dots, u_N)$ and the distribution $Y(s_N)$ of s_N, depend on the original distribution $F(X_N)$. At this point, we may pose two questions:

1. When $N \to \infty$ is there a limiting form of $Y(s) \to \Phi(s)$?
2. If there exists such a limit distribution, what are the properties of the black-box unit U and of $J(u_1, \dots, u_N)$ that determines the form of $\Phi(s)$?

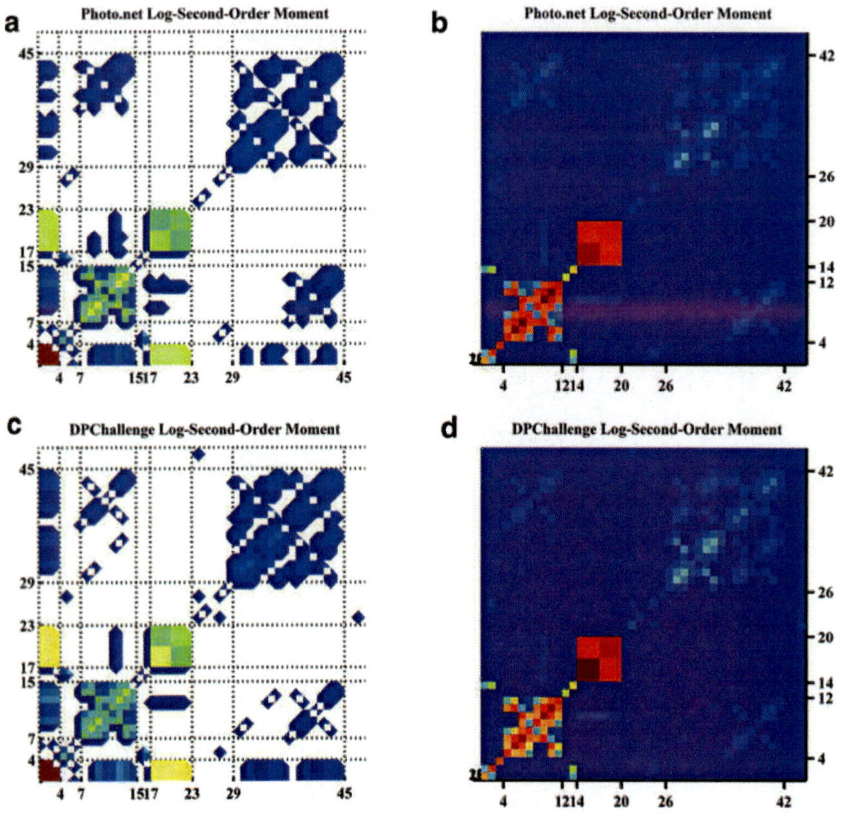

Fig. 5.7 Second-order moment matrices (**a**) Full Matrix Photo.net (**b**) No averaging filters Photo.net (**c**) Full Matrix DPChallenge (**d**) No averaging filters DPChallenge

In [1], the authors have demonstrated that under certain conditions on $Y(s)$ the possible limiting forms of $\Phi(s)$ are given by the three distribution families:

$$\Phi(s) = \exp\left(-\exp\left(\frac{\mu - s}{\sigma}\right)\right), \quad \forall s \quad \text{Gumbel},$$

$$\Phi(s) = 1 - \exp\left(-\left(\frac{s-\mu}{\sigma}\right)^{k}\right), \quad s > \mu \quad \text{Weibull},$$

$$\Phi(s) = \exp\left(-\left(\frac{s-\mu}{\sigma}\right)^{-k}\right), \quad s > \mu \quad \text{Fréchet}, \tag{5.3}$$

where μ, σ, k are the location, scale, and shape parameters of the distributions, respectively. The particular choice between the three families in (5.3) is determined by the tail behavior of $F(X)$ at large X. In this case, we use as units U the black box that computes the absolute value of the filter result vectors from the irreducible

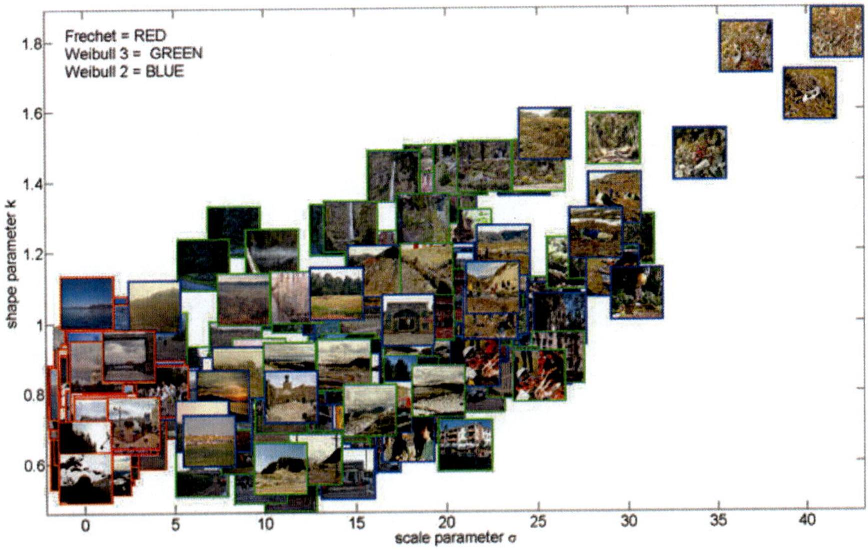

Fig. 5.8 Image type and model distribution in EVT parameter space

representations of the dihedral groups. The filter vectors not associated with the trivial representation are of the form $s = \sum(x_i - x_j)$ where x_i, x_j are pixel values. We can therefore expect that these filter values are usually very small and that high values will appear very seldom. In addition, these sums are calculated over a small, finite neighborhood, and for this reason, the random variables are highly correlated. In short, the output for each filter has a form similar to the sums described in [1], and so it is possible to use the EVT to model their distribution.

We may now analyze which types of images are assigned to each submodel in (5.3). For economy of space, we only illustrate a single filter (an intensity edge filter) on the dataset described in Sect. 5.7.1, but the results generalize to all filters and different datasets. We omit the μ parameter since it usually exhibits very little variation and the most important behavior is observed in the other two parameters. First of all, if we look at Fig. 5.8 we see a correlated dispersion in the two axes, with the Fréchet images spanning only a very small region of the space at low σ, k, and well separated from 2-parameter and 3-parameter Weibull. Also notice how the Fréchet set typically includes images with near-uniform colored regions with smooth transitions between them, or alternatively very coarse-textured, homogeneous regions with sharp boundaries. High frequency textures seem to be relatively absent from the Fréchet, and on average the image intensities seem to be lower in the Fréchet set than in the Weibulls.

On the other hand, the 2-parameter and 3-parameter Weibull clusters are intermixed, with the 2-parameter mostly restricted to the lower portion of the space. For smaller σ, k values, the 2-parameter Weibull images exhibit coarser textures, with the latter becoming more fine-grained as σ, k increase in tandem. Also, there

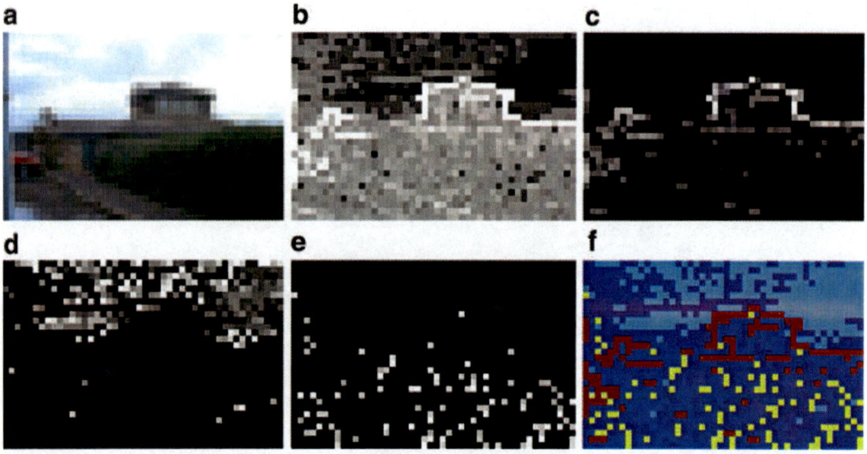

Fig. 5.9 A comparison between the extrema and other regions of a filtered image (**a**) Original image (**b**) Edge filter result (**c**) Tails (maxima) (**d**) Mode (**e**) Median (**f**) Synthesis

seems to be a shift from low-exposure, low-contrast images with shadows (small σ, k) to high-contrast, more illumination, less shadows when σ, k become large. Furthermore, the 2-parameter Weibull set shows a preference for sharp linear edges associated with urban, artificial, or man-made scenes, whereas the 3-parameter Weibull mostly captures the "fractal"-type edges, common in nature images.

Finally, we illustrate the importance of the data at the extrema of a filtered image, as described by the EVT. In Fig. 5.9a, we show an image (rescaled for comparison) and its filtered result using one of the intensity edge filters in Fig. 5.9b. This is essentially a gradient filter in the x- and y-directions. Next is Fig. 5.9c that shows the response at the tails of the fitted distribution. It is immediately obvious that the tails contain all the important edges and boundary outlines that abstract the main objects in the image (house, roof, horizon, diagonal road). These are some (but not necessarily all) of the most salient features that a human observer may focus on, or that a computer vision system might extract for object recognition or navigation. We also show the regions near the mode in Fig. 5.9d. We see that much of it contains small magnitude edges and noise from the almost uniform sky texture. Although this is part of the scene, it has very little significance when one is trying to classify or recognize objects in an image. A similar observation holds for the grass area, which although contains stronger edges than the sky and is distributed near the median (Fig. 5.9e), it is still not as important (magnitude-wise and semantically) as the edges in the tails are. Finally, Fig. 5.9f shows how all the components put together, can describe different regions in the image: the salient object edges in the tails (red); the average response, discounting extreme outliers, (median) in yellow; the most common response in light blue (mode); and the remaining superfluous data in between (dark blue). This is exactly the type of semantic behavior that the EVT models can isolate with their location, scale, and shape parameters, something which is not immediately possible when using histograms.

5.6 Fast Transforms, Orientation, and Scale

From the construction of the filters follows that they can be implemented as a combination of three basic transforms: one operating on the RGB vectors, one for the four-point, and one for the eight-point orbit. These filters are linear and they are therefore completely characterized by three matrices of sizes $3 \times 3, 4 \times 4$, and 8×8. The rows of these matrices define projection operators onto the spaces V_{xy} introduced above and they can be computed using algorithms from the representation theory of finite groups.

We illustrate the basic idea with the help of the transformation matrix related to the RGB transformation. We already noted that the sum R+G+B is invariant under permutations of the RGB channels. It follows that the vector $\begin{pmatrix} 1 & 1 & 1 \end{pmatrix}$ defines a projection onto this invariant subspace. We also know that the orthogonal complement to this one-dimensional subspace defines a two-dimensional subspace that cannot be reduced further. Any two vectors orthogonal to $\begin{pmatrix} 1 & 1 & 1 \end{pmatrix}$ can therefore be used to fill the remaining two rows of the RGB transformation matrix. Among the possible choices we mention here two: the Fourier transform and an integer-valued transform. The Fourier transform is a natural choice considering the interpretation of the RGB vector as three points on a triangle. In this case, the remaining two matrix rows are given by $\cos(2k\pi/3)$ and $\sin(2k\pi/3)$. Since the filters are applied to a large number of vectors, it is important to find implementations that are computationally efficient. One solution is given by the two vectors $\begin{pmatrix} 1 & -1 & 0 \end{pmatrix}$ and $\begin{pmatrix} 1 & 1 & -2 \end{pmatrix}$. The resulting transformation matrix has the advantage that the filters can be implemented using addition and subtraction only. We can furthermore reduce the number of operations by computing intermediate sums. One solution is to compute RG = R+G first and combine that afterward to obtain RG+B and RG-2B. The complete transformation can therefore be computed with the help of five operations instead of the six operations required by a direct implementation of the matrix–vector operation.

For the four- and the eight-point orbit transforms, we can use the general tools of representation theory to find the integer-valued transform matrices:

$$\begin{pmatrix} 1 & 1 & 1 & 1 \\ -1 & 1 & -1 & 1 \\ -1 & -1 & 1 & 1 \\ 1 & -1 & -1 & 1 \end{pmatrix} \text{ and } \begin{pmatrix} 1 & 1 & 1 & 1 & 1 & 1 & 1 & 1 \\ 1 & 1 & -1 & -1 & 1 & 1 & -1 & -1 \\ 1 & -1 & 1 & -1 & 1 & -1 & 1 & -1 \\ -1 & 1 & 1 & -1 & -1 & 1 & 1 & -1 \\ 0 & -1 & -1 & 0 & 0 & 1 & 1 & 0 \\ 1 & 0 & 0 & -1 & -1 & 0 & 0 & 1 \\ -1 & 0 & 0 & -1 & 1 & 0 & 0 & 1 \\ 0 & 1 & -1 & 0 & 0 & -1 & 1 & 0 \end{pmatrix}. \tag{5.4}$$

Also here we see that we can use intermediate sums to reduce the number of operations necessary. This is an example of a general method to construct fast-group theoretical transforms similar to the FFT implementation of the Fourier transform. More information can be found in [18].

One of the characteristic properties of these filter systems is their covariance property described above, i.e., $\mathbf{F}(\mathbf{p}^{(g,h)}) = \widetilde{\mathbf{T}}(g,h)\mathbf{F}(\mathbf{p})$ for all transformations $(g,h) \in D_4 \otimes D_3$. In Fig. 5.3, we used only the fact that the matrix $\widetilde{\mathbf{T}}(g,h)$ is orthonormal and the norm of the filter vectors is therefore preserved when the underlying pixel distribution undergoes one of the transformations described by the group elements. More detailed results can be obtained by analyzing the transformation properties of the computed feature vectors. The easiest example of how this can be done is related to the transformation of the filter result computed from a 2×2 patch with the help of the filter $\begin{pmatrix} -1 & 1 & -1 & 1 \end{pmatrix}$ in the second row of the matrix in (5.4). Denoting the pixels intensity values in the 2×2 patch by a,b,c,d, we get the filter result $\mathbf{F} = (b+d) - (a+c)$. Rotating the patch 90°, gives the pixel distribution d,a,b,c with filter value $(a+c) - (d+b) = -\mathbf{F}$. In the same way, we find that a reflection on the diagonal gives the new pixel vector a,d,c,b with filter result $\mathbf{F} = (d+b) - (a+c)$. Since these two operations generate all possible spatial transforms in D_4, we see that the sign change of the filter result indicates if the original patch was rotated 90° or 270°. We can therefore add the sign of the filter results as another descriptor. Using the same calculations, we see that reflections and 180° rotations cannot be distinguished from the original.

This example can be generalized using group theoretical tools as follows: In the space of all filter vectors $\mathbf{F}$ (in the previous case, the real line $\mathbb{R}$) one introduces an equivalence relation defining two vectors $\mathbf{F}_1, \mathbf{F}_2$ as equivalent if there is an element $(g,h) \in D_4 \otimes D_3$ such that $\mathbf{F}_2 = \widetilde{\mathbf{T}}(g,h)\mathbf{F}_1$. In the previous case, the equivalence classes are represented by the half-axis. Every equivalence class is given by an orbit of a given feature vector and in the general case, there are up to 48 different elements (corresponding to the 48 group elements) in such an orbit. Apart from the norm of the feature vector, we can thus characterize every feature vector by its position on its orbit relative to an arbitrary but fixed element on the orbit. This can be used to construct SIFT-like descriptors [14] where the bins of the histogram are naturally given by the orbit positions. Another illustration is related to the two-dimensional representation V_{2i} representing intensity edge filters. Under the transformation of the original distribution, the computed feature vectors transform in the same way as the dihedral group transforms the points of the square. From this follows directly that the magnitude of the resulting two-dimensional filter vector $(\mathbf{F}_x, \mathbf{F}_y)$ is invariant under rotations and reflections and represents edge-strength. The relation between the vector components $(\mathbf{F}_x, \mathbf{F}_y)$ is related to orientation and we can therefore describe $(\mathbf{F}_x, \mathbf{F}_y)$ in polar coordinates as vector (ρ, θ). In Fig. 5.10, this is illustrated for the edge filters computed from the inner 2×2 corner points. The magnitude image on the left corresponds to the result shown in Fig. 5.4 while the right images encodes the magnitude ρ in the v-component and the angular component θ in the h-component of the hsv-color space.

The results described so far are all derived within the framework of the dihedral groups describing the spatial and color transformation of image patches. Another common transform which is not covered by the framework is scaling. Its systematic analysis lies outside the scope of this description but we mention one strategy that can be used to incorporate scaling. The basic observation is that the different orbits

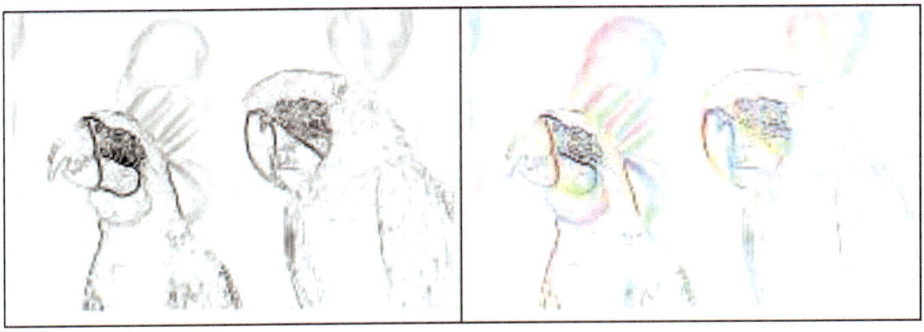

Fig. 5.10 Edge magnitude (*left*) and orientation (*right*)

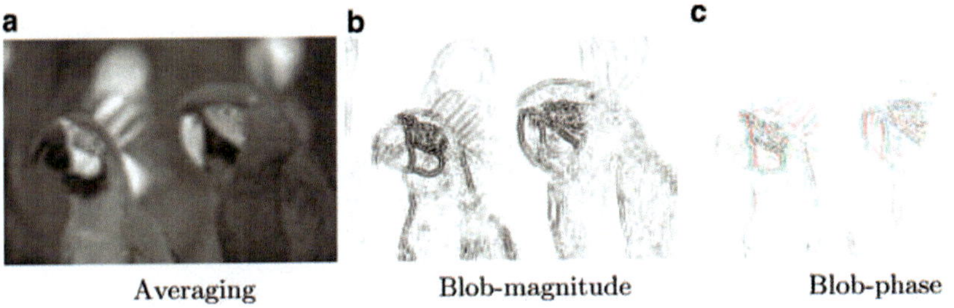

Fig. 5.11 Scale-space

under the group D_4 are related to scaling. In the simplest case of the three averaging intensity filters, one gets as filter results the vector $(\mathbf{F}_1, \mathbf{F}_2, \mathbf{F}_3)$ representing the average intensity over the three different rings in the 4×4 patch. Assuming that the scaling is such that one can, on average, interchange all three orbits then one can treat the three filter results $\mathbf{F}_1, \mathbf{F}_2, \mathbf{F}_3$ as function values defined on the corners of a triangle. In that case, one can use the same strategy as for the RGB components and apply the D_3-based filters to the vectors $(\mathbf{F}_1, \mathbf{F}_2, \mathbf{F}_3)$. The first filter will then compute the average value over three different scales. Its visual effect is a blurring. The second filter computes the difference between the intensities in the inner four-pixel patch and the intensities on the eight points on the next orbit. The visual property is a center-surround filter. Finally, the third filter computes the difference between the two inner and the outer orbit. Also, here we can convert the vectors with the last two filter results to polar coordinates to obtain a magnitude "blob-detector" and an angular phase-like result. We use the same color coding as for the edge-filter illustration in Fig. 5.10 and show the result of these three scale-based filters in Fig. 5.11. We don't describe this construction in detail but we only show its effect on the values of the diagonal elements in the second-order moment matrix. In Fig. 5.12, we show the logarithms of the absolute values of the diagonal elements in the second-order moment matrix computed from the filtered patches as before (marked by crosses) and the result of the scaling operation (given by the circles). For the first three filters, we see that the value of the first component increased

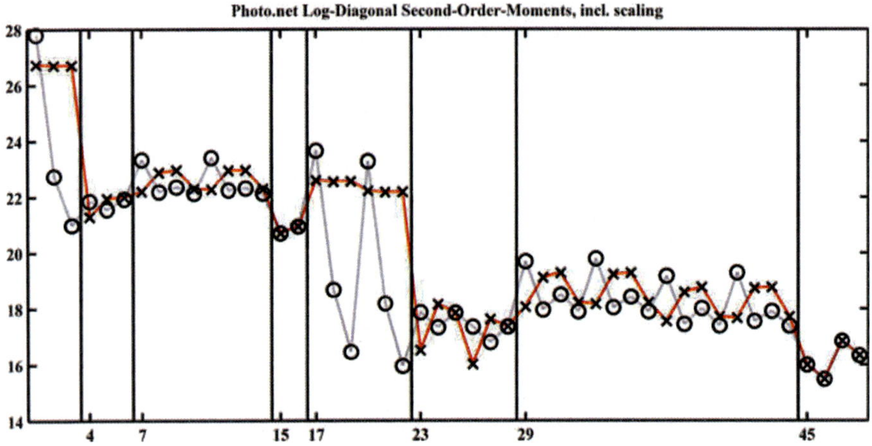

Fig. 5.12 Diagonal elements and scaling operation

significantly while the values of the other two decreased correspondingly. This is a typical effect that can also be observed for the other filter packages. We conclude the descriptions of these generalizations by remarking that this is an illustration showing that one can use the representation theory of the group $D_4 \otimes D_3 \otimes D_3$ to incorporate scaling properties into the framework.

5.7 Image Classification Experiments

Among possible applications that can be based on the presented filter systems we will here illustrate the usefulness in an image classification experiment, where we try to separate classes of images downloaded from the Internet. A popular approach for large-scale image classification is to combine global or local image histograms with a supervised learning algorithm. Here we derive a 16 bins histogram for each filter package, resulting in a 16×24 representation of each image. The learning algorithm is the Support Vector Machine implementation *SVMlight* described in [6]. For simplicity and reproducibility reasons, all experiments are carried out with default settings.

5.7.1 Image Collections

Two-class classification results are illustrated for the keyword pairs *garden–beach* and *andy warhol–claude monet*. We believe these pairs to be representative examples of various tasks that can be encountered in image classification.

Table 5.3 Two-class classification accuracy for various filter packages, and the overall descriptor

	Filter package							
Keyword pair	1:3	4:6	7:10	13:15	16:18	19:22	ALL	ALL+EVT
Garden–beach	0.78	0.77	0.79	0.69	0.68	0.67	0.80	0.79
Andy warhol–claude monet	0.66	0.81	0.81	0.80	0.82	0.83	0.89	0.84

The Picsearch[1] image search service is queried with each keyword, and 500 thumbnail images (maximum size 128 pixels) are saved from each search result. Based on recorded user statistics, we only save the most popular images in each category, which we assume will increase the relevance of each class. The popularity estimate is based on the ratio between how many times an image has been clicked and viewed in the public search interface. For each classification task, we create a training set containing every second image from both keywords in the pair, and remaining images are used for evaluating the classifier.

5.7.2 *Classification Results*

Classification results are given in Table 5.3. The overall result for the entire descriptor (the entire 16×24 representation) is shown in the second to last column, and classification results in earlier columns are based on selected filter packages. The last column summarizes the classification results obtained from the EVT-parameter descriptions of the distributions. The classification accuracy is given by the proportion of correctly labeled images. A value of 0.75 means that 75% of the images were labeled with a correct label. We conclude that the best classification result is obtained when the entire descriptor is used. But as the table indicates, the importance of different filter packages varies with the image categories in use. We see, for instance, that the color content of an image (captured in filter packages 13–24) is more important for the *andy warhol–claude monet* classification result, than for *garden–beach*.

We illustrate the classification result by plotting subsets of classified images. The result based on the entire descriptor, for each keyword pair respectively, can be seen in Fig. 5.13a, b. Each sub-figure shows the 10+10 images than obtained the most positive and most negative score from the Support Vector Machine. Similar plots for selected filter packages are shown in Figs. 5.14–5.19.

In closing, we briefly illustrate the practical application of the Extreme Value theory models in the above classification examples and as an alternative representation to histograms. The input data vector to the SVM in this case contains the three parameters: *location*, *scale*, *shape* estimated by fitting the EVT models to each

[1] http://www.picsearch.com/.

Fig. 5.13 Classification examples based on the entire descriptor (filter results 1–24) (**a**) beach (*top*) vs garden (*bottom*) (**b**) andy warhol (*top*) vs claude monet (*bottom*)

Fig. 5.14 Classification examples based on filter package: 1–3 (intensity mean) (**a**) beach (*top*) vs garden (*bottom*) (**b**) andy warhol (*top*) vs claude monet (*bottom*)

of the 24 filters packages. Compared with the histogram from before, we are now only using a 3×24-dimensional vector for the full filter descriptor, as opposed to a 16×24-dimensional vector. This leads to a much reduced data representation, faster training and classification steps, and no need to optimally set the number of bins. First, in Fig. 5.20 we show the comparative results for a single filter classification on the *andy warhol–claude monet* set. We can see that the EVT, even with its lower dimensionality, is equally or sometimes even more accurate than the histogram representations. In terms of absolute accuracy numbers, the EVT scores for the full-filter descriptor are shown in the last column of Table 5.3. As it is obvious, these scores are very close to the histogram-based results.

Fig. 5.15 Classification examples based on filter package: 4–6 (intensity lines) (**a**) beach (*top*) vs garden (*bottom*) (**b**) andy warhol (*top*) vs claude monet (*bottom*)

Fig. 5.16 Classification examples based on filter package: 7–10 (intensity edges) (**a**) beach (*top*) vs garden (*bottom*) (**b**) andy warhol (*top*) vs claude monet (*bottom*)

Finally, we show an example of image retrieval. More specifically, using the full 3×24-dimensional EVT-based filter descriptor we classified the four classes of 500 thumbnail images each. We trained an SVM with 70% of the images and tested on the remaining 30% using the "One-to-All" classification scheme to build a multi-way classifier. The top SVM-ranked images resulting from this classification (or in other words, the top retrieved images) are shown in Fig. 5.21.

Although the results need not be identical to those using the histograms above, we can see the equally good retrieval and separation in the four different classes. In particular, the vivid, near-constant colors and sharp edges in the *Warhol* set and the less saturated, softer tones, and faint edges of the *Monet* set. In the same way, the *garden* images contain very high frequency natural textures and the *beach* images

Fig. 5.17 Classification examples based on filter package: 13–15 (color mean) (**a**) beach (*top*) vs garden (*bottom*) (**b**) andy warhol (*top*) vs claude monet (*bottom*)

Fig. 5.18 Classification examples based on filter package: 16–18 (color lines) (**a**) beach (*top*) vs garden (*bottom*) (**b**) andy warhol (*top*) vs claude monet (*bottom*)

more homogeneous regions with similarly colored boundaries. These characteristics are the exact information captured by the filters and the EVT models and which can be used very effectively for image classification and retrieval purposes.

5.8 Summary

We started from the obvious observations that the pixels of digital images are located on grids and that, on average, the three color channels are interchangeable. These two properties motivated the application of tools from the representation theory of

Fig. 5.19 Classification examples based on filter package: 19–22 (color edges) (**a**) beach (*top*) vs garden (*bottom*) (**b**) andy warhol (*top*) vs claude monet (*bottom*)

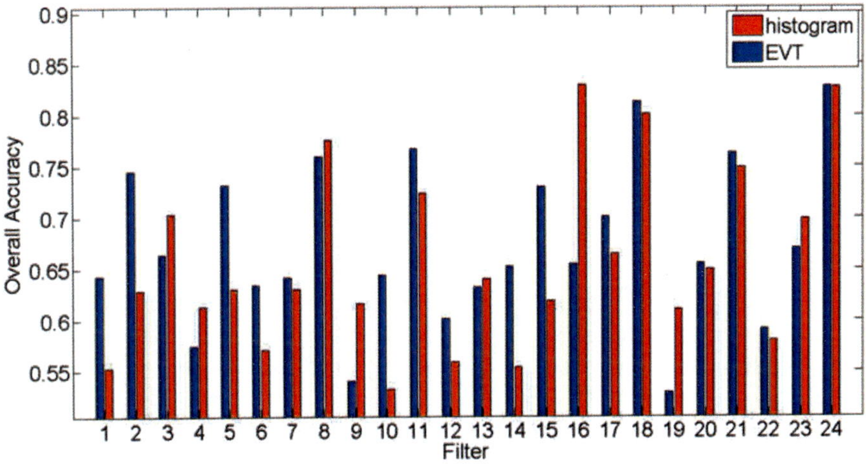

Fig. 5.20 Two-class classification accuracy comparison between the EVT and the histogram representations for the *andy warhol–claude monet* set. We only use a single filter paclage at a time

finite groups and we showed that in this framework, we can explain how steerable filter systems and MMSE-based transform-coding methods are all linked to those group theoretical symmetry properties. Apart from these theoretical properties, the representation theory also provides algorithms that can be used to construct the filter coefficients automatically and it also shows how to create fast filter systems using the same principles as the FFT-implementations of the DFT. We also sketched briefly how the group structure can be used to define natural bins for the histogram descriptors of orientation parameters. A generalization that includes simple scaling properties was also sketched.

Fig. 5.21 Retrieval results from the four classes using EVT (**a**) beach (*top*) vs garden (*bottom*) (**b**) andy warhol (*top*) vs claude monet (*bottom*)

The computational efficiency of these filter systems makes them interesting candidates in applications where huge numbers of images have to be processed at high speed. A typical example of such an application is image database retrieval where billions of images have to be analyzed, indexed, and stored. For image collections, we showed that the statistical distributions of the computed filter values can be approximated by the three-types of extreme value distributions. This results in a descriptor where the statistical distribution of a feature value in an image can be at most three parameters. We used the histograms of the filter results and the three parameters obtained by a statistical parameter estimation procedure to discriminate web images from different image categories. These tests show that most of the feature distributions can indeed by described by the three-parameter extreme-value distribution model and that the classification performance of these parametric model descriptors is comparable with conventional histogram-based descriptors. We also illustrated the visual significance of the distribution types and typical parameter vectors with the help of the scatter plot in Fig. 5.8.

We did not discuss if these filter systems are relevant for the understanding of biological vision systems and we did not compare them in detail with other filter systems like those described in [16]. Their simple structure, their optimality properties (see also [5, 17]), and the fact that they are massively parallel should motivate their further study in situations where very fast decisions are necessary [21]. Finally, we mention that similar techniques can be used to process color histograms [13] and that there are similar algorithms for data defined on three-dimensional grids [12].

Acknowledgements The financial support of the Swedish Science Foundation is gratefully acknowledged. The research leading to these results has received funding from the European Community's Seventh Framework Programme FP7/2007–2013—Challenge 2—Cognitive Systems, Interaction, Robotics—under grant agreement No 247947-GARNICS.

References

1. Bertin E, Clusel M (2006) Generalised extreme value statistics and sum of correlated variables. J Phys A: Math Gen 39(24):7607–7619
2. Datta R, Li J, Wang JZ (2008) Algorithmic inferencing of aesthetics and emotion in natural images: An exposition. In: 2008 IEEE International Conference on Image Processing, ICIP 2008, pp 105–108, San Diego, CA
3. Fässler A, Stiefel EL (1992) Group theoretical methods and their applications. Birkhäuser, Boston
4. Freeman WT, Adelson EH (1991) The design and use of steerable filters. IEEE Trans Pattern Anal Mach Intell 13(9):891–906
5. Hubel DH (1988) Eye, brain, and vision. Scientific American Library, New York
6. Joachims T (1999) Making large-scale support vector machine learning practical. MIT Press, Cambridge, pp 169–184
7. Lenz R (1990) Group theoretical methods in image processing. Lecture notes in computer science (Vol. 413). Springer, Heidelberg
8. Lenz R (1993) Using representations of the dihedral groups in the design of early vision filters. In: Proceedings of international conference on acoustics, speech, and signal processing, pp V165–V168. IEEE
9. Lenz R (1995) Investigation of receptive fields using representations of dihedral groups. J Vis Comm Image Represent 6(3):209–227
10. Lenz R, Bui TH, Takase K (2005) Fast low-level filter systems for multispectral color images. In: Nieves JL, Hernandez-Andres J (eds) Proceedings of 10th congress of the international colour association, vol 1, pp 535–538. International color association
11. Lenz R, Bui TH, Takase K (2005) A group theoretical toolbox for color image operators. In: Proceedings of ICIP 05, pp III–557–III–560. IEEE, September 2005
12. Lenz R, Carmona PL (2009) Octahedral transforms for 3-d image processing. IEEE Trans Image Process 18(12):2618–2628
13. Lenz R, Carmona PL (2010) Hierarchical s(3)-coding of rgb histograms. In: Ranchordas A et al (eds) Selected papers from VISAPP 2009, vol 68 of Communications in computer and information science. Springer, Berlin, pp 188–200
14. Lowe DG (2004) Distinctive image features from scale-invariant keypoints. Int J Comput Vis 60(2):91–110
15. Mikolajczyk K, Schmid C (2005) A performance evaluation of local descriptors. IEEE Trans Pattern Anal Mach Intell 27(10):1615–1630
16. Oliva A, Torralba A (2006) Building the Gist of a Scene: The Role of Global Image Features in Recognition. Progress in Brain Research 155:23-36
17. Olshausen BA, Field DJ (1996) Emergence of simple-cell receptive field properties by learning a sparse code for natural images. Nature 381(6583):607–609
18. Rockmore D (2004) Recent progress and applications in group FFTs. In: Byrnes J, Ostheimer G (eds) Computational noncommutative algebra and applications. Kluwer, Dordrecht
19. Serre J-P (1977) Linear representations of finite groups. Springer, New York
20. Terras A (1999) Fourier analysis on finite groups and applications. Cambridge University Press, Cambridge
21. Thorpe S, Fize D, Marlot C (1996) Speed of processing in the human visual system. Nature 381(6582):520–522
22. Yosida K (1980) Functional analysis. Springer, Berlin

Chapter 6
Color Representation and Processes with Clifford Algebra

Philippe Carré and Michel Berthier

The color is stronger than the language

Marie-Laure Bernadac

Abstract In the literature, colour information of pixels of an image has been represented by different structures. Recently algebraic entities such as quaternions or Clifford algebras have been used to perform image processing for example. We propose to review several contributions for colour image processing by using the Quaternion algebra and the Clifford algebra. First, we illustrate how this formalism can be used to define colour alterations with algebraic operations. We generalise linear filtering algorithms already defined with quaternions and review a Clifford color edge detector. Clifford algebras appear to be an efficient mathematical tool to investigate the geometry of *nD* images. It has been shown for instance how to use quaternions for colour edge detection or to define an hypercomplex Fourier transform. The aim of the second part of this chapter is to present an example of applications, namely the Clifford Fourier transform of Clifford algebras to colour image processing.

Keywords Clifford • Quaternion • Fourier transform • Algebraic operations • Spatial filtering

P. Carré (✉)
Laboratory XLIM-SIC, UMR CNRS 7252, University of Poitiers, France
e-mail: philippe.carre@univ-poitiers.fr

M. Berthier
Laboratory MIA (Mathématiques, Images et Applications), University of La Rochelle, France
e-mail: michel.berthier@univ-lr.fr

C. Fernandez-Maloigne (ed.), *Advanced Color Image Processing and Analysis*,
DOI 10.1007/978-1-4419-6190-7_6,

6.1 Introduction

Nowadays, as multimedia devices and internet are becoming accessible to more and more people, image processing must take colour information into account because colour processings are needed everywhere for new technologies. In this idea, several approaches have been submitted to deal with colour images, one of the oldest is to process any greyscale algorithm on each channel of the colour image to get the equivalent result. Implementing such programs often creates artefacts, so researchers have come to deal differently with colour image information. A quite recent manner to process colour algorithms is to encode the three channel components on the three imaginary parts of a quaternion as proposed by S.T. Sangwine and T. Ell in [1–3]. Recently algebraic entities such as Clifford algebras have been used to perform image processing. We propose in this chapter to review several contributions for colour image processing by using the Quaternion algebra and the Clifford algebra.

The first section describes the spatial approach and illustrates how this formalism can be used to define colour alterations with algebraic operations. By taking this approach, linear filtering algorithms can be generalized with Quaternion and Clifford algebra. To illustrate what can be done, we will conclude this part by present strategies for colour edge detector.

The second part introduces how to use quaternions to define an hypercomplex Fourier transform, and after reviews a Clifford Fourier transform that is suitable for colour image spectral analysis. There have been many attempts to define such a transformation using quaternions or Clifford algebras. We focus here on a geometric approach using group actions. The idea is to generalize the usual definition based on the characters of abelian groups by considering group morphisms from R^2 to spinor groups Spin(3) and Spin(4). The transformation is parameterized by a bivector and a quadratic form, the choice of which is related to the application to be treated.

6.2 Spatial Approach

In this first section, we start with a brief description of basic concepts of Quaternion and Clifford algebra.

6.2.1 *Quaternion Concept*

6.2.1.1 Definition of Quaternion

A quaternion $q \in \mathbb{H}$ is an extension of complex numbers with multiplication rules as follows: $i^2 = j^2 = k^2 = -1$, $k = ij = -ji$, $i = jk = -kj$, and $j = ki = -ik$ and is defined as $q = q_r + q_i i + q_j j + q_k k$ where:

- q_r, q_i, q_j, and q_k are real numbers.
- i, j, and k are two new imaginary numbers, asserting:

$$i^2 = j^2 = k^2 = -1, \quad ij = -ji = k, \quad jk = -kj = i, \quad ki = -ik = j. \quad (6.1)$$

We see that the quaternion product is anti-commutative.

With $q = q_r + q_i i + q_j j + q_k k$ any quaternion, we review a similar vocabulary:

- $\overline{q} = q_r - q_i i - q_j j - q_k k$ is q's *conjugate*.
- If $q \neq 0$, then $q^{-1} = \frac{\overline{q}}{|q|^2}$ is q's *inverse*.
- $\Re(q) = q_r$ is q's *real part*. If $\Re(q) = q$, then q is *real*.
- $\Im(q) = ib + jc + kd$ is q's *imaginary part*. If $\Im(q) = q$, then q is *pure*.
- q's *modulus* or *norm* is $\sqrt{q_r^2 + q_i^2 + q_j^2 + q_k^2} = \sqrt{q\overline{q}}$ noted $|q|$.
- $\mathbb{P} = \{q \in \mathbb{H} \,|\, q = \Im(q)\}$ is the *Pure Quaternion* group.
- $\mathbb{S} = \{q \in \mathbb{H} \,|\, |q| = 1\}$ is the *Unitary Quaternion* group.

As we have said Quaternion can be expressed in a scalar $S(q)$ an vector part $V(q)$,

$$q = S(q) + V(q)$$

with $S(q) = q_r$ and $V(q) = q_i i + q_j j + q_k k$.

6.2.1.2 $\mathbb{R}^3$ Transformations with Quaternions

As pure quaternion are used analogously to $\mathbb{R}^3$ vectors, the classical $\mathbb{R}^3$ transformations: translations, reflections, projections, rejections, and rotations can be defined with only additions and multiplications.

With $(q_1, q_2) \in \mathbb{P}^2$ two pure quaternions, the *translation* vector is supported by the quaternion $q_{\text{trans}} = q_1 + q_2$. If $q \in \mathbb{P}$ and $\mu \in \mathbb{S} \cap \mathbb{P}$ then $q_{\text{refl}} = -\mu q \mu$ is the q's *reflection* vector with μ axis.

If $q \in \mathbb{P}$ and $\mu \in \mathbb{S} \cap \mathbb{P}$ then $q_{\text{proj}} = \frac{1}{2}(q - \mu q \mu)$ is the q's *projection* vector on μ axis.

If $q \in \mathbb{P}$ and $\mu \in \mathbb{S} \cap \mathbb{P}$ then $q_{\text{rej}} = \frac{1}{2}(q + \mu q \mu)$ is the q's orthogonal projection vector on μ axis's orthogonal plan or the q's *rejection* of the μ axis.

And if $q \in \mathbb{P}$, $\phi \in \mathbb{R}$ and $\mu \in \mathbb{S} \cap \mathbb{P}$ then $q_{\text{rot}} = e^{\mu \frac{\phi}{2}} q e^{-\mu \frac{\phi}{2}}$ is the q's *rotation* vector around μ axis with ϕ angle.

All these definitions of transformation allowed us the understand how to interact on colour encoded in quaternion using just operations like additions, substractions and multiplications. Furthermore Sangwine used them to create his spatial quaternionic filters

Sangwine and Ell were the first to use this vector part of quaternion to encode colour images. They took the three imaginary parts to code the colour components r (red), g (green), and b (blue) of an image. A colour image is then considered as a map $f : \mathbb{R}^2 \longrightarrow \mathbb{H}_0$ defined by

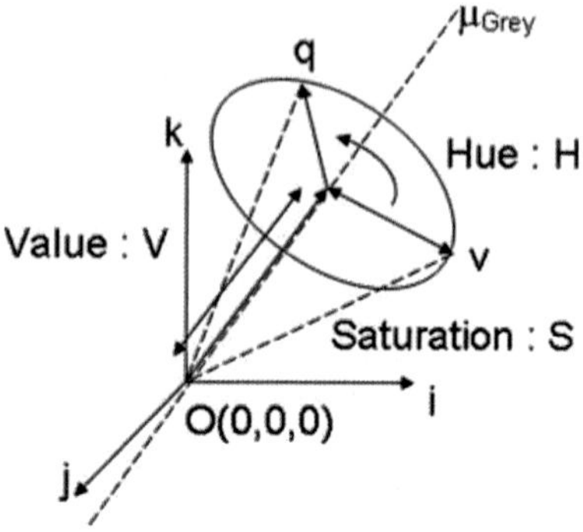

Fig. 6.1 Hue, saturation, and value given with reference μ_{grey} and $H(v) = 0$

$$f[m,n] = r[m,n]i + g[m,n]j + b[m,n]k,$$

where $r(x,y)$, $g(x,y)$, and $b(x,y)$ are the red, green, and blue components of the image.

From a colour described in RGB colour space with a quaternion vector $q \in \mathbb{P}$, HSV colour space coordinates can be found as well with operations on quaternions. We consider that Value is the norm of the colour's orthogonal projection vector $(q.\mu_{\text{grey}})\mu_{\text{grey}}$ on the grey axis μ_{grey} (this axis can be defined such that $\mu_{\text{grey}} = \frac{i+j+k}{\sqrt{3}}$. Saturation and Hue are represented on the orthogonal plan to the grey axis which crosses $(q.\mu_{\text{grey}})\mu_{\text{grey}}$. The Saturation is the distance between the colour vector q and the grey axis μ_{grey}, and Hue is the angle between the colour vector q and a colour vector v taken anywhere on the plan orthogonal to μ_{grey} and which sets the reference zero Hue angle. This reference Hue value is often taken to represent the red colour vector, so we decided arbitrarily to associate the red colour vector or any other one to the v vector and gave it a zero Hue value (Fig. 6.1). Hue is the angle between this reference colour vector and the colour vector q.

If q is a colour vector, then Value V, Saturation S, and Hue H can be given with the grey-axis $\mu_{\text{grey}} \in \mathbb{S} \cap \mathbb{P}$ and the reference colour vector $v \in \mathbb{S} \cap \mathbb{P}$ with elementary quaternionic operations as below

$$\begin{cases} H = \tan^{-1} \frac{|q - \mu v q v \mu|}{|q - v q v|} \\ S = |\frac{1}{2}(q + \mu q \mu)| \\ V = |\frac{1}{2}(q - \mu q \mu)| \end{cases} . \tag{6.2}$$

6.2.1.3 Quaternionic Filtering

After a first view of several possible manipulations on colour image encoded with quaternions, we focus on applications which can be done with. We propose to study low level operations on colour images like filtering operations for instance.

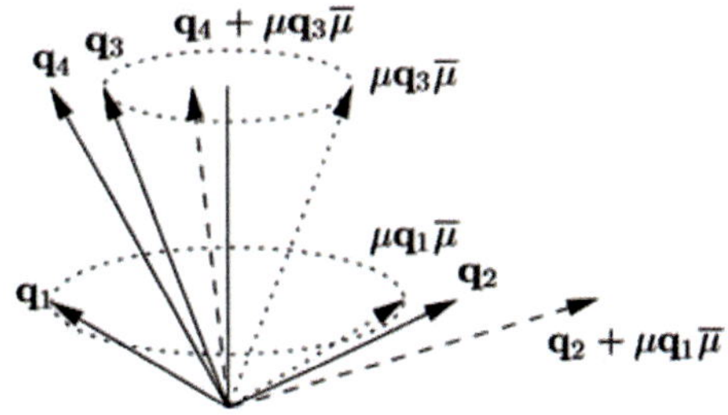

Fig. 6.2 Sangwine edge detector scheme

We propose to review the quaternionic colour gradient defined by Sangwine based on the possibility to use quaternion to represent $\mathbb{R}^3$ transformations and applied to colour geometry.

As a lot of gradient detector, which are of first order, the following method uses convolution filters. Sangwine proposed the convolution on a colour image can be defined by:

$$q_{\text{filtered}}(s,t) = \sum_{\tau_1=-n_1}^{n_1} \sum_{\tau_2=-m_1}^{m_1} h_l(\tau_1,\tau_2)q((s-\tau_1)(t-\tau_2))h_r(\tau_1,\tau_2), \tag{6.3}$$

where h_l and h_r are the two conjugate filters of dimension $N_1 \times M_1$ where $N_1 = 2n_1 + 1 \in \mathbb{N}$ and $M_1 = 2m_1 + 1 \in \mathbb{N}$.

From this definition of the convolution product, Sangwine proposed a colour edge detector in [2]. In this method, the two filters h_1 and h_2 are conjugated in order to fulfill a rotation operation of every pixel around the greyscale axis by an angle of π and compare it to its neighbours (Fig. 6.2).

The filter composed by a pair of quaternion conjugated filters is defined as follow

$$h_l = \frac{1}{6}\begin{bmatrix} 1 & 1 & 1 \\ 0 & 0 & 0 \\ Q & Q & Q \end{bmatrix} \quad \text{and} \quad h_r = \frac{1}{6}\begin{bmatrix} 1 & 1 & 1 \\ 0 & 0 & 0 \\ \overline{Q} & \overline{Q} & \overline{Q} \end{bmatrix}, \tag{6.4}$$

where $Q = e^{\mu \frac{\pi}{2}}$ and $\mu = \mu_{\text{grey}} = \frac{i+j+k}{\sqrt{3}}$ the greyscale axis.

The filtered image is a greyscale image almost everywhere, because the vector sum of one pixel to its neighbours rotated by π around the grey axis has a low saturation (q_3 *and* q_4). However, pixels in colour opposition (q_1 *and* q_2, *for example, representing a colour edge*) present a vector sum far from the grey axis, so edges are coloured due to this high distance (Fig. 6.3).

After this description of some basic issues of colour processing by using Quaternion, we introduce the second mathematical tool: Clifford algebra.

Fig. 6.3 Sangwine edge detector result

6.2.2 *Clifford Algebra*

6.2.2.1 Definition

The Clifford algebra framework allows to encode geometric transformations via algebraic formulas. Let us fix an orthonormal basis (e_1, e_2, e_3) of the vector space $\mathbb{R}^3$. We embed this space in a larger 8-dimensional vector space, denoted $\mathbb{R}_{3,0}$, with basis given by a unit 1, the three vectors e_1, e_2, e_3 and the formal products e_1e_2, e_2e_3, e_1e_3, and $e_1e_2e_3$. The key point is that elements of $\mathbb{R}_{3,0}$ can be multiplied: the product of e_i and e_j is, for example, e_ie_j. The rules of multiplication are given by:

$$e_i^2 = 1, \quad e_ie_j = -e_je_i.$$

The *geometric product* of two vectors u and v of $\mathbb{R}^3$ is given by the formula

$$uv = u \cdot v + u \wedge v, \tag{6.5}$$

where $\cdot$ denotes the scalar product and $u \wedge v$ is the *bivector* generated by u and v. Since the e_i's are orthogonal, then

$$e_ie_j = e_i \wedge e_j.$$

The linear combinations of the elements $e_i \wedge e_j$ are called grade-2 entities termed bivectors. They encode pieces of two-dimensional vector subspaces of $\mathbb{R}^3$ with a magnitude and an orientation. In these algebras, multivectors which are the extension of vectors to higher dimensions, are the basic elements. One example is that we often represent vectors as one-dimensional directed quantities (also represented by arrows), they are thus represented in geometric algebras by 1-vectors. As their dimension is one, they are said 1-graded. By extension, in geometric algebra, there are grade-2 entities termed bivectors which are plane segments endowed with orientation. In general, a k-dimensional oriented entity is known as a k-vector. For an overview on geometric algebras see [4–7] for instance. In geometric algebra, oriented subspaces are basic elements, as vectors in a m-dimensional linear vector space $\mathcal{V}^m$. These oriented subspaces are called blades, and the term k-blade is used to describe a k-dimensional homogeneous subspace. A multivector is then a linear combination of blades.

Any multivector $\mathbf{M}$[1] $\in \mathbb{R}_{n,0}$ is so described by the following equation:

$$\mathbf{M} = \sum_{k=0}^{n} \langle M \rangle_k \tag{6.6}$$

with $\langle M \rangle_k$ the k-vector part of any multivector $\mathbf{M}$ i.e., the grade k operator.

The Geometric product is an associative law and distributive over the addition of multivectors. In general, the result of the geometric product is a multivector. As we have said, if used on 1-vectors a and b, this is the sum of the inner and outer product: $ab = a.b + a \wedge b$. Note also that the geometric product is not commutative.

This product is used to construct k-dimensional subspace elements from independent combinations of blades. For example, multiplying with this product the two independent 1-vectors e_1 and e_2 gets the bi-vector e_{12}. And if you multiply again this bivector by the third 1-vector in $\mathcal{V}^3$, e_3, you get the trivector e_{123}. The basis of $\mathbb{R}_{3,0}$ algebra is so given by $(e_0, e_1, e_2, e_3, e_{23}, e_{31}, e_{12}, e_{123})$. Here e_0 stands for the element of grade 0, it is so the scalar part.

- External or Wedge product: The wedge product is denoted by $\wedge$.

  Wedge product can then be described using geometric product as follow with $\mathbf{A}$ a s-graded multivector and $\mathbf{B}$ a r-graded multivector, both in $\mathbb{R}_{n,0}$:

$$A_r \wedge B_s = \langle A_r B_s \rangle_{r+s}\,. \tag{6.7}$$

- Inner Product: This product also called interior product denoted by ., is used to give the notion of orthogonality between two multivectors [4]. Let $\mathbf{A}$ be a a-vector and $\mathbf{B}$ be a b-vector, then $\mathbf{A}.\mathbf{B}$ is the subspace of $\mathbf{B}$, with a $(b-a)$ dimension, orthogonal to subspace $\mathbf{A}$. If $b < a$ then $\mathbf{A}.\mathbf{B} = 0$ and $\mathbf{A}$ is orthogonal to $\mathbf{B}$. For 1-vectors, the inner product equals the scalar product used in linear algebra $\mathcal{V}^m$.

$$A_r.B_s = \langle A_r B_s \rangle_{|r-s|}\,. \tag{6.8}$$

- Scalar Product: This product denoted by $*$, is used to define distances and modulus.

$$\mathbf{A} * \mathbf{B} \equiv \langle AB \rangle_0\,.$$

[1]In the following, little letters will be used to represent 1-vectors whereas bolded capital letters will stand for any multivectors.

As said before, geometric algebras allows to handle geometric entities with the formalism of algebras. In our case, we study the representation of colours by 1-vectors of $\mathbb{R}_{3,0}$. Here is shown the translation of classical geometric transformations into the algebraic formalism $\mathbb{R}_{3,0}$.

Let v_1, v_2, v_t, $v_\|$, $v_\perp$ and v_r be 1-vectors of $\mathbb{R}_{3,0}$, the transcription into algebraic formalism is given here for:

- Translation: $v_t = v_1 + v_2$ is the result of the translation of v_1 by v_2.
- Projection: $v_\| = v_\|^{v_2} = (v_1 \wedge v_2)v_2^{-1}$ is the v_1's projection on v_2.
- Rejection: $v_\perp = v_\perp^{v_2} = (v_1.v_2)v_2^{-1}$ is the v_1's rejection with respect to v_2.
- Reflection: $v_r = v_r^{v_2} = v_2 v_1 v_2^{-1}$ is the v_1's reflection with respect to v_2.

6.2.2.2 Clifford Colour Transform

We will now introduce how geometric algebras can be associated with colour image processing, but first of all, we give a survey on what have been done already linking image processing and geometric algebra.

We propose to use the geometric transformations to express hue, saturation and value colour information from any colour pixel m incoded as a 1-vector of $\mathbb{R}_{3,0}$ in *RGB* colour space. This resolution is performed using only algebraic expressions in $\mathbb{R}_{3,0}$ and is a generalization of what was already done in the quaternionic formalism. A colour image is seen as a function f from $\mathbb{R}^2$ with values in the vector part of the Clifford algebra $\mathbb{R}_{3,0}$:

$$f : [x,y] \longmapsto f_r[x,y]e_1 + f_2[x,y]e_2 + f_3[x,y]e_3. \tag{6.9}$$

Let ϑ be the 1-vector carrying the grey level axis, r carries the pur red vector and m represents any colour vector.

- Value is the modulus of the projection of m with respect to ϑ, it is then expressed by:

$$V = |(m.\vartheta)\vartheta^{-1}|. \tag{6.10}$$

- Saturation is the distance from the vector m to the grey level axis ϑ, it is then the modulus of the rejection of m with respect to ϑ:

$$S = |(m \wedge \vartheta)\vartheta^{-1}|. \tag{6.11}$$

- To reach the hue, we need to define a colour which hue is zero, let ϑ_2 be the 1-vector that represents $H = 0$. A general agreement is to say that pur red has a null hue, ϑ_2 is then r's rejection with respect to ϑ. Therefore, H is the angle between ϑ_2 and $m_\perp^\vartheta = (m \wedge \vartheta)\vartheta^{-1}$ which is the m's rejection with respect to ϑ .

  The hue can then be given by:

$$H = \cos^{-1}\left(\frac{m_\perp}{|m_\perp|} \cdot \vartheta\right).$$

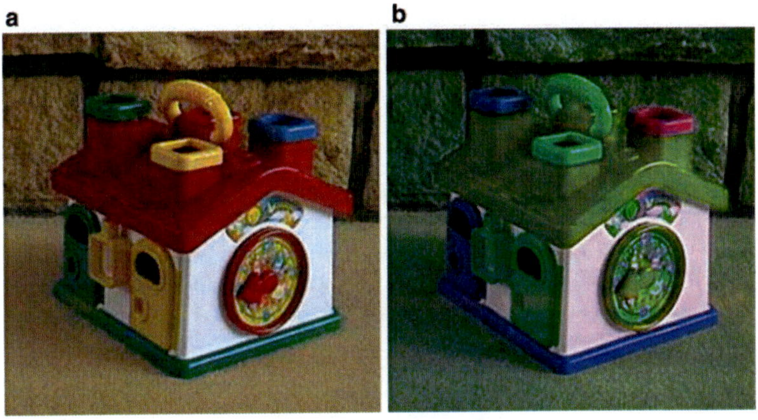

Fig. 6.4 Hue's modification: (**a**) original image (**b**) modified image

We thus have formulated the hue, saturation and value of a RGB colour vector using only algebraic expressions. From these concepts, we can define colour transform by using only algebraic expressions.

Performing the translation operator along the grey axis ϑ with coefficient $\alpha \in \mathbb{R}$ on every pixel of an image will result on an alteration of the general value or brightness of the original image. The result of such an alteration of the brightness seems more contrasted and warmer than the original image.

$$m' = m + \alpha\vartheta = I\vartheta + Se^{\vartheta^* T}\vartheta_2 + \alpha\vartheta \rightarrow I' = I + \alpha.$$

Here the rotation operator is applied on each pixels $f[m,n]$ of the image around the grey axis ϑ, this is shown in Fig. 6.4. The result is an alteration of the hue of the original image.

$$m' = e^{-\vartheta\frac{\theta}{2}} m e^{\vartheta\frac{\theta}{2}} = Ie^{-\vartheta\frac{\theta}{2}}\vartheta e^{\vartheta\frac{\theta}{2}} + Se^{-\vartheta\frac{\theta}{2}} e^{\vartheta T}\vartheta_2 e^{\vartheta\frac{\theta}{2}}$$

$$m' = I\vartheta + Se^{\vartheta(T+\theta)}\vartheta_2 \longrightarrow T' = T + \theta.$$

Figure 6.4 shows this kind of hue modification. The original image (Fig. 6.4a) has been modified by the rotation around the greyscale axis with an angle of $\pi/3$. So the red roof is now green, the green threshold is now blue, etc. Note that one can imagine to choose any other colour vector for the rotation axis than the grey one to perform an other operation than hue alteration.

The translation operator, associated with the weight $\beta \in \mathbb{R}$ on its saturation axis $f_\perp[x,y]$, is applied to perform an alteration of an image's saturation. The saturation axis $f_\perp[x,y]$ is the rejection of $f[x,y]$ with respect to ϑ.

$$m' = I\vartheta + S'e^{\vartheta^* T}\vartheta_2 = m + \beta e^{\vartheta^* T}\vartheta_2$$

$$m' = I\vartheta + (S+\beta)e^{\vartheta^* T}\vartheta_2 \longrightarrow S' = S + \beta.$$

Fig. 6.5 Saturation's modification: (**a**) original image (**b**) modified image

Figure 6.5 illustrates this saturation's alteration where the original image (Fig. 6.4a) has been altered as described in the precedent equation to get the result (Fig. 6.4b). The operation is equivalent to:

- Get pale or washed colours in opposition to the original image when the saturation level is lowered as in the Fig. 6.5.
- Whereas when the saturation level is uppered, colours seem to be more vivid than in the original image.

Note that all of these colour transformations can be done because colours are encoded on the vector part of a $\mathbb{R}_{3,0}$ multivector. In fact, reflections, translations, rotations and rejections are defined only for simple multivectors that is to say information included in such multivectors is described on one and only one grade.

In this section, we introduced geometric algebras and studied several of their properties which will help us through the next section where we will describe how to use them to process digital images. We also showed that embedding colours into $\mathbb{R}_{3,0}$ algebra allows to perform image alteration using algebraic formalisation only. From these concepts, we now propose to study colour edge detection with Clifford algebra.

6.2.2.3 Spatial Filtering with Clifford Algebra

The formalisation of colour information into $\mathbb{R}_{3,0}$ allows to define more complex colour processings. Here, we describe how to use the geometric concepts studied before to define algebraically spatial approaches such as colour edge detection.

As we have seen, Sangwine [8] defined a method to detect colour edges by using colour pixels incoded in pur quaternions. The idea is to define a convolution operation for quaternions and apply specific filters to perform the colour edge detection. This method can be written with the geometric algebra formalism where colours are represented by 1-vectors of $\mathbb{R}_{3,0}$:

$$Sang[x,y] = (h_1 * I * h_2)[x,y], \tag{6.12}$$

and where h_1 and h_2 are a couple of filters which are used to perform a reflection of every colour with respect to the 1-vector associated with the greyscale axis $\vartheta = (e_1 + e_2 + e_3)/\sqrt{3}$ defined as:

$$h_1 = \frac{1}{6}\begin{bmatrix} 1 & 1 & 1 \\ 0 & 0 & 0 \\ \vartheta & \vartheta & \vartheta \end{bmatrix} \quad \text{and} \quad h_2 = \frac{1}{6}\begin{bmatrix} 1 & 1 & 1 \\ 0 & 0 & 0 \\ \vartheta^{-1} & \vartheta^{-1} & \vartheta^{-1} \end{bmatrix}. \tag{6.13}$$

Note that these are vertical filters to detect horizontal edges.

As for the Quaternion formalism, this fulfills an average vector of every pixel in the filters neighbourhood reflected with respect to the greyscale axis. The filtered image is a greyscale image almost everywhere, because in homogeneous regions the vector sum of one pixel to its neighbours reflected by the grey axis has a low saturation in other words low distance from the greyscale axis and edges are thus coloured due to this high distance. The limit of this method is that the filters can be applied horizontally from left to right and from right to left for example depending on the filters definitions but without giving the same results.

To counter this side effect involving the direction applied to the filters from Sangwine's method, we propose to illustrate a simple modification [9] of the filtering scheme, with the computation of the distance between the Sangwine's comparison vector to the greyscale axis. This time, it gives the same results if the filters are applied clockwise or anticlockwise. Because we are looking for a distance between Sangwine's comparison vector which is a colour vector to the greyscale axis, the result is a gradient of saturation. Thus the expression of the saturation gradient is given by:

$$SatGrad[x,y] = |(Sang[x,y] \wedge \vartheta)\vartheta^{-1}|. \tag{6.14}$$

Figure 6.6 illustrates the saturation filtering scheme. We observe the detection of all chromatic edges. But, the main drawback of this method is still that it is based on a saturation measurement only. In fact, when edges contain achromaticity information only, this approach is not able to detect them properly.

The geometric product allow us to fulfill the drawback of the previous method by describing geometrically every colour pixel $f[x,y]$ with respect to the greyscale axis. This description is given by this geometric product $f[x,y]\vartheta$ where $f[x,y]\vartheta$ is broken into two terms for every pixels:

$$\begin{aligned} f[x,y]\vartheta &= \underbrace{f[x,y].\vartheta}_{\text{scalar}} + \underbrace{f[x,y] \wedge \vartheta}_{\text{bivector}} \\ &= Scal[x,y] + Biv[x,y]. \end{aligned} \tag{6.15}$$

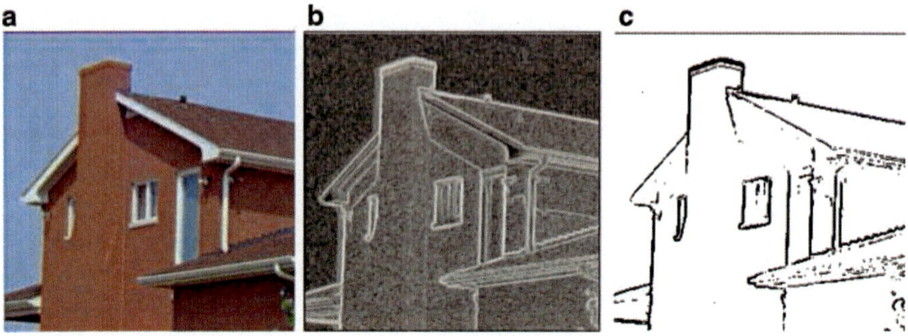

Fig. 6.6 Saturation gradient approach: (**a**) original image, (**b**) saturation gradient (**c**) edges selected by maxima extraction

In order to compute the geometric product and the Sangwine filtering, we use this couple of filters

$$v = \begin{pmatrix} 1 & 1 & 1 \\ 0 & 1 & 0 \\ \vartheta & \vartheta & \vartheta \end{pmatrix} \quad \text{and} \quad u = \begin{pmatrix} 1 & 1 & 1 \\ 0 & \vartheta & 0 \\ \vartheta^{-1} & \vartheta^{-1} & \vartheta^{-1} \end{pmatrix}.$$

The result of the filtering is:

$$g[m,n] = \{[\vartheta f[m+1,n+1]\vartheta^{-1} + f[m+1,n-1]] + [\vartheta f[m,n+1]\vartheta^{-1} + f[m,n-1]] + [\vartheta f[m-1,n+1]\vartheta^{-1} + f[m-1,n-1]]\} + f[m,n]\vartheta. \quad (6.16)$$

Figure 6.7 shows the AG filtering scheme.

We disjoin the scalar and bivectorial parts. The modulus of the bivectorial part is calculated to discover where information is achromatic only (Fig. 6.7b). In fact, the less the modulus of this bivector is, the less the saturation of the pixel is so information is achromatic.

Next step is to apply a threshold on this modulus map to get a mask. This mask will emphasise the areas where the modulus is low so where information is mainly intensity in contrast to elsewhere where the colour pixels are full of chromaticity information.

As the scalar part is the projection of every pixel on the greyscale axis μ, it thus represents intensity information. A Prewitt filtering is applied on this scalar part for every pixels in horizontal, vertical and diagonal directions to get an intensity or value gradient.

$$P_{1,2,3,4}[x,y] = (G_{1,2,3,4} * Scal[x,y]) \quad (6.17)$$

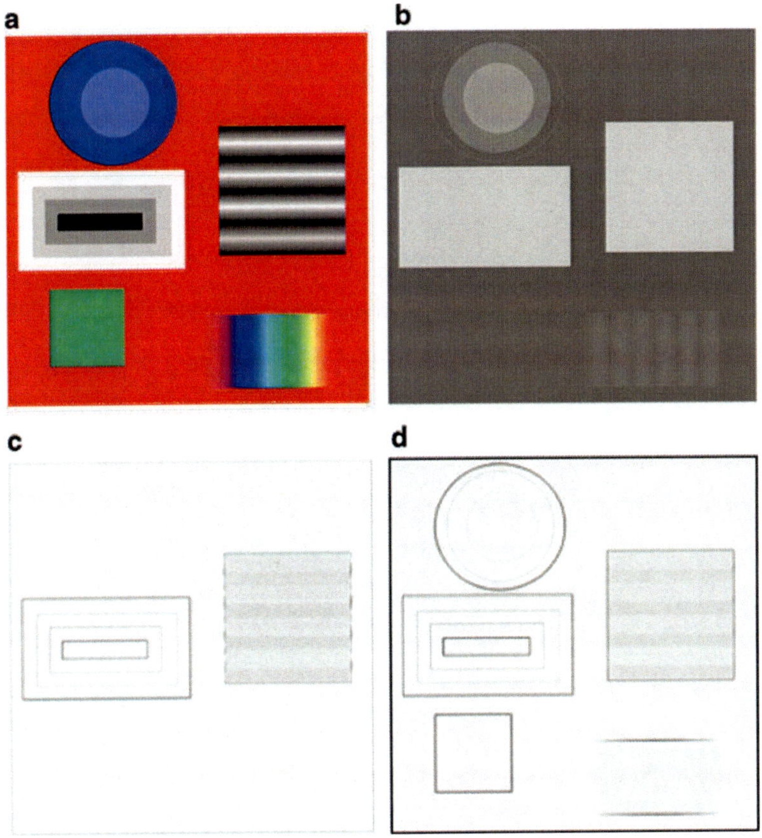

Fig. 6.7 (**a**) Original image with chromatic and achromatic information; (**b**) bivector part $|f[x,y] \wedge \vartheta|$; (**c**) Prewitt filtering applied on the scalar part and combined with the achromatic mask; (**d**) final result

with

$$G_1 = \begin{bmatrix} -1 & -1 & -1 \\ 0 & 0 & 0 \\ 1 & 1 & 1 \end{bmatrix}$$

and $G_{2,3,4}$ rotated version of G_1.

Then, to store just achromatic information, we use the mask defined with the modulus of the bivectial part. The result gives the gradient of pixels which do not contain chromaticity information (Fig. 6.7c). The last step is to combine this value gradient to the saturation one defined before. We can use different techniques to merge the two gradients as the maximum operator to preserve only the maximum value between those two gradients (Fig. 6.7d).

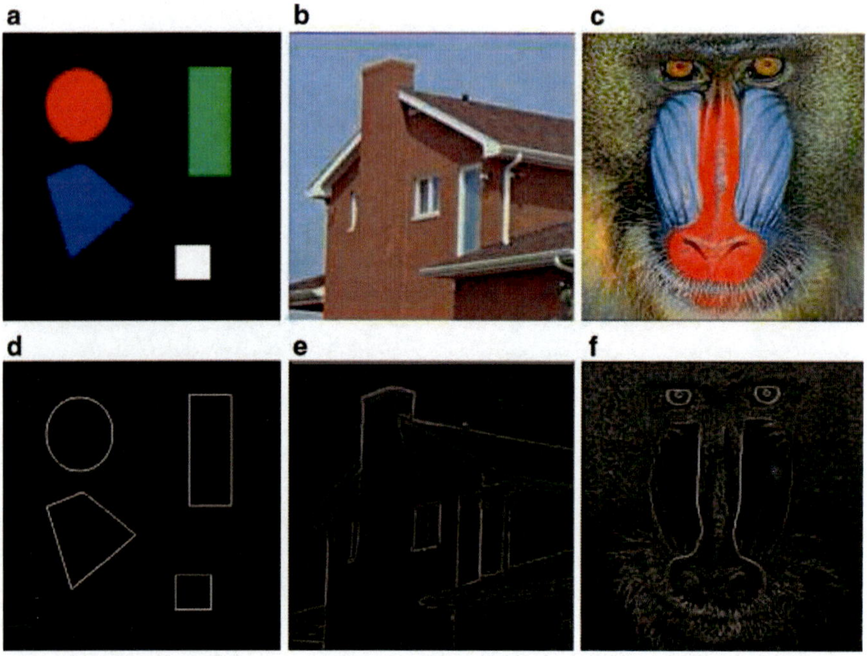

Fig. 6.8 Gradient examples on colour images: (**a**) (**b**) (**c**) original images; (**d**) (**e**) (**f**) final gradient

Figure 6.8 shows results on classical digital colour processing images. One can note that the computer graphics image (Fig. 6.8a) points up that achromatic regions are here well detected (Fig. 6.8d). The following house image (Fig. 6.8b) also includes achromatic areas such as the gutters and the windows frame which appear in the calculated gradient (Fig. 6.8e).

We now describe the use of these new concepts for the definition of a colour Fourier transform.

6.3 The Colour Fourier Transform

The Fourier transform is well known to be an efficient tool for analyzing signals and especially grey level images. When dealing with *nD* images, as colour images, it is not so clear how to define a Fourier transform which is more than *n* Fourier transforms computed marginally. The first attempt to define such a transform is due to S. Sangwine and T. Ells [10] who proposed to encode the colour space *RGB* by the space of *imaginary quaternions* $\mathbb{H}_0$. For a function f from $\mathbb{R}^2$ to $\mathbb{H}_0$ representing a colour image, the Fourier Transform is given by

$$\mathcal{F}_\mu f(U) = \int_{\mathbb{R}^2} f(X)\exp(-\mu\langle X,U\rangle)\mathrm{d}X, \tag{6.18}$$

where $X = (x_1, x_2)$, $U = (u_1, u_2)$ and μ is a unit imaginary quaternion. In this expression μ (satisfying $\mu^2 = -1$) is a parameter which corresponds to a privileged direction of analysis (typically μ is the unit quaternion giving the grey axis in [10]). It is a fact that will be discussed later that such a choice must be taken into account.

It appears that quaternions is a special case of the more general mathematical notion of *Clifford Algebras*. As we have said, these algebras, under the name of *Geometric Algebras* are widely used in computer vision and robotics [11, 12]. One of the main topics of this part of the chapter is to describe a rigorous construction of a Fourier Transform in this context.

There are in fact many constructions of Clifford Fourier transforms with several different motivations. In this chapter we focus on applications to colour image processing. The Clifford definition relies strongly on the notion of *group actions* which is not the case for the already existing transforms. This is precisely this viewpoint that justifies the necessity of choosing an analyzing direction. We illustrate how to perform frequencies filtering in colour images.

6.3.1 First Works on Fourier Transform in the Context of Quaternion/Clifford Algebra

To our knowledge, the only generalizations of the usual Fourier transform using quaternions and concerning image processing are those proposed by Sangwine et al. and by Bülow. The first one is clearly motivated by colour analysis and the second one aims at detecting two-dimensional symmetries in grey-level images.

Several constructions have been proposed in the context of Clifford algebras. In [13], a definition is given using the algebras $\mathbb{R}_{2,0}$ and $\mathbb{R}_{3,0}$ in order to introduce the concept of $2D$ analytic signal. A definition appears also in [14] which is mainly applied to analyse frequencies of vector fields. With the same Fourier kernel Mawardi and Hitzer in [15] establish in [15] an uncertainty principle for multivector functions. The reader may find in [16] a construction using the Dirac operator and applications to Gabor filters. Let us also mention, from a different viewpoint, reference [17] where generalized Fourier descriptors are defined by considering the action of the motion group of $\mathbb{R}^2$ that is the semidirect product of the groups $\mathbb{R}^2$ and $SO(2)$.

The final part of this section describes the generalization of the Fourier Transform by using Bivectors of the Clifford Algebra $\mathbb{R}_{3,0}$. We start by illustrating how Quaternion are used to define colour Fourier Transform.

6.3.1.1 Quaternionic Fourier Transforms: Definition

As already mentioned the idea of [10] is to encode colour information through imaginary quaternions. A colour image is then considered as a map $f : \mathbb{R}^2 \longrightarrow \mathbb{H}_0$ defined by

$$f(x,y) = r(x,y)i + g(x,y)j + b(x,y)k,$$

where $r(x,y)$, $g(x,y)$, and $b(x,y)$ are the red, green and blue components of the image. R To define a quaternionic Fourier transform the authors of [10] propose to replace the imaginary complex i by some imaginary unit quaternion μ. It is easily checked that $\mu^2 = -1$ and a typical choice for μ is $\mu = (i + j + k)/\sqrt{3}$ which corresponds in *RBG* to the grey axis. The transform is given by formula:

$$\mathcal{F}_\mu f(U) = \int_{\mathbb{R}^2} f(X)\exp(-\mu\langle X,U\rangle)\mathrm{d}X. \tag{6.19}$$

The quaternionic Fourier coefficients are decomposed with respect to a *symplectic decomposition* associated to μ, each one of the factors being expressed in the polar form:

$$\mathcal{F}_\mu f = A_{\parallel}\exp[\mu\theta_{\parallel}] + A_{\perp}\exp[\mu\theta_{\perp}]v$$

with v an imaginary unit quaternion orthogonal to μ. The authors propose a spectral interpretation from this decomposition (see [10] for details).

Bülow's approach is quite different since it concerns mainly the analysis of symmetries of a signal f from $\mathbb{R}^2$ to $\mathbb{R}$ given for example by a grey-level image. For such a signal, the quaternionic Fourier transform proposed by Bülow reads:

$$\mathcal{F}_{ij}f(U) = \int_{\mathbb{R}^2} \exp(-2\pi i x_1 u_1) f(X)\exp(-2\pi j u_2 x_2)\mathrm{d}X. \tag{6.20}$$

Note that i and j can be replaced by arbitrary pur imaginary quaternions. The choice of this formula is justified by the following equality:

$$\mathcal{F}_{ij}f(U) = \mathcal{F}_{cc}f(U) - i\mathcal{F}_{sc}f(U) - j\mathcal{F}_{cs}f(U) + k\mathcal{F}_{ss}f(U),$$

where

$$\mathcal{F}_{cc}f(U) = \int_{\mathbb{R}^2} f(X)\cos(2\pi u_1 x_1)\cos(2\pi u_2 x_2)\mathrm{d}X$$

and similar expressions involving sinus and cosinus for $\mathcal{F}_{sc}f$, $\mathcal{F}_{cs}f$ and $\mathcal{F}_{ss}f$. We refer the reader to [18] for details and applications to analytic signals.

6.3.1.2 Quaternionic Fourier Transforms: Numerical Analysis

In order to understand what the Fourier coefficients stand for, we studied the digital caracterization of the Discrete Quaternionic Fourier Transform (DQFT). The colour

Fourier spectrum presented some symmetries due to zero scalar spatial part of any colour image exactly as it was well known that the spectrum of a real signal by a complex Fourier transform (CFT) had hermitian properties of symmetry. Even if the spatial information of a colour image is using pure quaternions only, applying a DQFT on an image results in full quaternions (i.e., with scalar part non zero). We wanted to find, after Inverse DQFT, a space where scalar part is zero in order to avoid any loss of information as the spatial colour image is coded on a pure quaternion matrix which real part automatically set to zero.
Let

$$F[o,p] = F_r[o,p] + F_i[o,p]i + F_j[o,p]j + F_k[o,p]k \tag{6.21}$$

be the spectral quaternion at coordinates $(o,p) \in ([\frac{-N}{2}+1..\frac{N}{2}],[\frac{-M}{2}..\frac{M}{2}])$ and

$$f[m,n] = \frac{1}{\sqrt{MN}} \sum_{o} \sum_{p} e^{2\mu\pi(\frac{om}{M}+\frac{pn}{N})} F[o,p] \tag{6.22}$$

the Inverse DQFT quaternion of (m,n) spatial coordinates.
Developing this, with $\mu = \mu_i i + \mu_j j + \mu_k k$, the cartesian real part form of the spatial domain leads to

$$\begin{aligned} f_r[m,n] = \frac{1}{\sqrt{MN}} \sum_{o} \sum_{p} \Big[&\cos\left(2\pi\left(\frac{om}{M}+\frac{pn}{N}\right)\right) F_r[o,p] \\ &- \sin\left(2\pi\left(\frac{om}{M}+\frac{pn}{N}\right)\right) (\mu_i F_i[o,p] + \mu_j F_j[o,p] + \mu_k F_k[o,p]) \Big] \end{aligned} \tag{6.23}$$

$f_r(s,t)$ is null when $F_r[0,0] = F_r[\frac{M}{2},\frac{N}{2}] = 0$ and for all $o \in [1;\frac{M}{2}-1]$ and $p \in [1;\frac{N}{2}-1]$:

$$F_r[-o,-p] = -F_r[o,p] \tag{6.24}$$

Moreover for all $o \in [0;\frac{M}{2}]$ and $p \in [0;\frac{N}{2}]$:

$$\begin{aligned} F_i[-o,-p] &= F_i[o,p] \\ F_j[-o,-p] &= F_j[o,p] \\ F_k[-o,-p] &= F_k[o,p]. \end{aligned} \tag{6.25}$$

We can see that the real part must be odd and all the imaginary parts must be even. This is a direct extension of the antihermitian property of the complex Fourier transform of imaginary signal.

When studying the complex spectrum domain, several notions are helpful such as the modulus and the angle.

A Fourier coefficient, $F[o,p] = q_0 + iq_1 + jq_2 + kq_3$ can be written as:

$$F[o,p] = q = |q|e^{\nu\varphi}$$

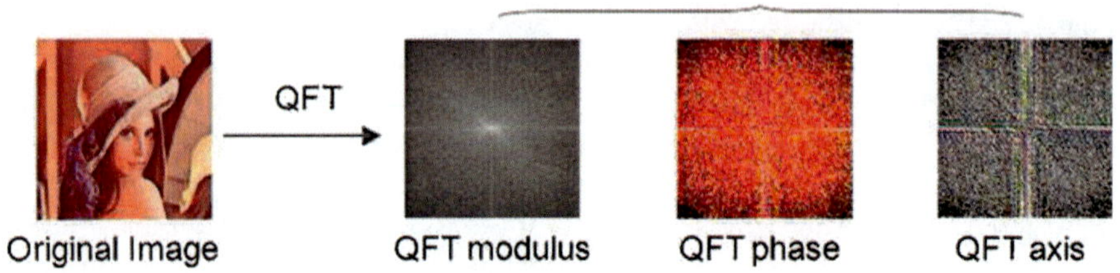

Fig. 6.9 Polar representation of the quaternionic Fourier transform

with $|q|$ the QFT modulus, $\varphi \in \mathbf{R}$ the QFT phase and $\nu \in \mathbb{H}_0 \cap \mathbb{H}^1$ the QFT axis. Figure 6.9 illustrates this polar representation of the Fourier coefficient for the image Lena. We can see that the Fourier coefficient has a modulus similar to that Greyscale image. It is more difficult to give interpretation of the information contained in the angle or the axis.

In order to try to give an interpretation of the information contained in the quaternionic spectrum of colour images, we can study spatial atoms associated with a pulse (Dirac) in the frequency domain.

Initialization could be done in two different ways:

- $F[o,p] = K_r.\delta_{o_0,p_0}[o,p] - K_r\delta_{-o_0,-p_0}[o,p]$. Initialization done on the real part of the spectrum, leading to odd oscillations on the spatial domain linked to the direction μ parameter of the Fourier transform. Complex colours are obtained in the RGB colour space after modifying this μ parameter and normalising it because it always needs to stay a pure unit quaternion.

  $F_r[o_0,p_0] = K_r$ and $F_r[-o_0,-p_0] = -K_r$ is associated with:

$$f[m,n] = 2\mu(K_r)\sin\left(2\pi\left(\frac{o_0 m}{M} + \frac{p_0 n}{N}\right)\right). \tag{6.26}$$

  Initializing a pair of constants on the real component leads to a spatial oscillation following the same imaginary component(s) as those included in the direction μ (Fig. 6.10b).
- $F[o,p] = e.(K_e.\delta_{o_0,p_0}[o,p] + K_e\delta_{-o_0,-p_0}[o,p])$ with $e = i,j$ or k. Initialization done on the imaginary part of the spectrum, leading to even oscillations on the spatial domain independently from the μ parameter of the Fourier transform. Complex colours in the RGB colour space are reached by initialization on several imaginary components weighted as in the additive colour synthesis theory.

  With $e = i,j,k$, $F_e[o_0,p_0] = F_e[-o_0,-p_0] = K_e$ is associated with:

$$f[m,n] = e\left[2(K_e)\cos\left(2\pi\left(\frac{o_0 m}{M} + \frac{p_0 n}{N}\right)\right)\right]. \tag{6.27}$$

  Initializing a pair of constants on any imaginary component with any direction μ leads to a spatial oscillation on the same component (Fig. 6.10a).

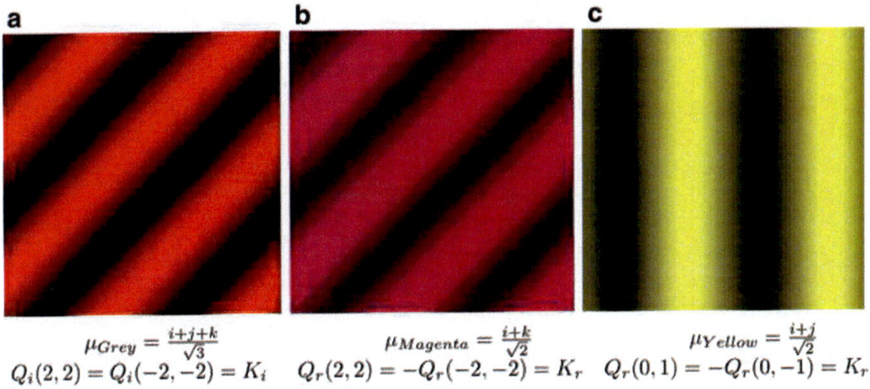

Fig. 6.10 Spectrum initialization examples

The coordinates (o_0, p_0) and $(-o_0, -p_0)$ of the two initialization points in the Fourier domain affect the orientation and the frequency of the oscillations in the spatial domain as it does so with greyscale image in complex Fourier domain. Orientation of the oscillations can be changed as shown in Fig. 6.10c.

Below we outline the different ways of defining a Clifford Fourier Transform.

6.3.1.3 Clifford Fourier Transforms

In [13], the Clifford Fourier transform is defined by

$$\mathcal{F}_{e_1e_2}f(U) = \int_{\mathbb{R}^2} \exp(-2\pi e_1e_2\langle U,X\rangle)f(X)\mathrm{d}X \tag{6.28}$$

for a function $f(X) = f(xe_1) = f(x)e_2$ from $\mathbb{R}$ to $\mathbb{R}$ and

$$\mathcal{F}_{e_1e_2e_3}f(U) = \int_{\mathbb{R}^3} \exp(-2\pi e_1e_2e_3\langle U,X\rangle)f(X)\mathrm{d}X \tag{6.29}$$

for a function $f(X) = f(x_1e_1 + x_2e_2) = f(x_1,x_2)e_3$ from $\mathbb{R}^2$ to $\mathbb{R}$. The coefficient e_1e_2, resp. $e_1e_2e_3$, is the so-called pseudoscalar of the Clifford algebra $\mathbb{R}_{2,0}$, resp. $\mathbb{R}_{3,0}$. These transforms appear naturally when dealing with the analytic and monogenic signals.

Scheuermann et al. also use the last kernel for a function $f : \mathbb{R}^3 \longrightarrow \mathbb{R}_{3,0}$:

$$\mathcal{F}_{e_1e_2e_3}f(U) = \int_{\mathbb{R}^3} f(X)\exp(-2\pi e_1e_2e_3\langle U,X\rangle)\mathrm{d}X. \tag{6.30}$$

Note that if we set

$$\begin{aligned} f = {} & f_0 + f_1e_1 + f_2e_2 + f_3e_3 + \\ & + f_{23}i_3e_1 + f_{31}i_3e_2 + f_{12}i_3e_3 + f_{123}i_3 \end{aligned}$$

with $i_3 = e_1e_2e_3$, this transform can be written as a sum of four complex Fourier transforms by identifying i_3 with the imaginary complex i. In particular, for a function f with values in the vector part of the Clifford algebra, this reduces to marginal processing. This Clifford Fourier transform is used to analyse frequencies of vector fields and the behavior of vector valued filters.

The definition proposed in [16] relies on Clifford analysis and involves the so-called *angular Dirac operator* Γ. The general formula is

$$\mathcal{F}_{\pm}f(U) = \left(\frac{1}{\sqrt{2\pi}}\right)^n \int_{\mathbb{R}^n} \exp\left(\mp i\frac{\pi}{2}\Gamma_U\right) \times \exp(-i < U,X >) f(X)\mathrm{d}X. \quad (6.31)$$

For the special case of a function f from $\mathbb{R}^2$ to $\mathbb{C}_2 = \mathbb{R}_{0,2} \otimes \mathbb{C}$, the kernel can be made explicit and

$$\mathcal{F}_{\pm}f(U) = \frac{1}{2\pi}\int_{\mathbb{R}^2} \exp(\pm U \wedge X) f(X)\mathrm{d}X. \quad (6.32)$$

Let us remark that $\exp(\pm U \wedge X)$ is the exponential of a bivector, i.e., a spinor. This construction allows to introduce two-dimensional Clifford Gabor filters (see [16] for details).

As the reader may notice, there are many ways to generalize the usual definition of the complex Fourier transform. In all the situations mentioned above the multiplication is non commutative and as a consequence the position of the kernel in the integral is arbitrary. We may in fact distinguish two kinds of approaches: the first ones deal with so called *bivectors* (see below) and the second ones involve the *pseudoscalar* $e_1e_2e_3$ of the Clifford algebra $\mathbb{R}_{3,0}$. The rest of this paper focus on the first approaches. The purpose of the last part of this chapter is to propose a well founded mathematical definition that explains why it is necessary to introduce those bivectors and their role in the definition. Before going into details, we recall some mathematical notions.

6.3.2 Mathematical Background

6.3.2.1 Mathematical Viewpoint on Fourier Transforms

We start by some considerations about the theory of abstract Fourier transform and then introduce basic notions on Clifford algebras and spinor groups. The main result of this section is the description of the $Spin(3)$ and $Spin(4)$ characters. From the mathematical viewpoint, defining a Fourier Transform requires to deal with group actions. For example, in the classical one-dimensional formula

$$\mathcal{F}f(u) = \int_{-\infty}^{+\infty} f(x)\exp(-iux)\mathrm{d}x \quad (6.33)$$

the involved group is the additive group $(\mathbb{R},+)$. This is closely related to the well-known *Shift Theorem*

$$\mathcal{F}f_{\alpha}(u)=\mathcal{F}f(u)\exp(iu\alpha) \tag{6.34}$$

where $f_{\alpha}(x)$ denotes the function $x\longmapsto f(x+\alpha)$, which reflects the fact that a translation of a vector α produces a multiplication by $\exp(iu\alpha)$. The correspondance $\alpha\longmapsto\exp(iu\alpha)$ is a so-called *character* of the additive group $(\mathbb{R},+)$.

More precisely, a character of an abelian group G is a map $\varphi: G\longrightarrow\mathbb{S}^1$ that preserves the composition laws of both groups. Here $\mathbb{S}^1$ is the multiplicative group of unit complex numbers. It is a special case, for abelian groups, of the notion of irreducible unitary representations, [19]. The abstract definition of a Fourier Transform for an (abelian) additive group G and a function f from G to $\mathbb{C}$ is given by

$$\mathcal{F}f(\varphi)=\int_G f(x)\varphi(-x)\mathrm{d}\nu(x), \tag{6.35}$$

where φ is a character and ν is a measure on G.

The characters of the group $(\mathbb{R}^n,+)$ are the maps

$$X=(x_1,\ldots,x_n)\longrightarrow\exp(u_1x_1+\cdots+u_nx_n)$$

parametrized by $U=(u_1,\ldots,u_n)$. They form the group $(\mathbb{R}^n,+)$. Applying the above formula to this situation leads to the usual Fourier Transform

$$\mathcal{F}f(U)=\int_{\mathbb{R}^n} f(X)\exp(-i\langle U,X\rangle)\mathrm{d}X. \tag{6.36}$$

It is classical, see [19], that considering the group of rotations $SO(2,\mathbb{R})$, resp. the group $\mathbb{Z}_n$, and the corresponding characters yields to the Fourier series theory, resp. the discrete Fourier Transform.

One of the ingredients of the construction of the Colour Fourier transform is the notion of *Spin characters* which extends the notion of characters to maps from $\mathbb{R}^2$ to spinor groups representing rotations.

6.3.2.2 Rotation

In the same way that characters of the Fourier transform for grey-level images are maps form $\mathbb{R}^2$ to the rotation group $\mathbb{S}^1$ of the complex plane $\mathbb{C}$, we want to define characters for colour images as maps from $\mathbb{R}^2$ to the rotation group acting on the space of colours, chosen in the sequel to be *RGB*. The Clifford algebra framework is particulary well adapted to treat this problem since it allows to encode geometric transformations via algebraic formulas.

Rotations of $\mathbb{R}^3$ correspond to specific elements of $\mathbb{R}_{3,0}$, namely those given by

$$\tau=a1+be_1e_2+ce_2e_3+de_1e_3$$

with $a^2+b^2+c^2+d^2=1$. These are called *spinors* and form a group $Spin(3)$ isomorphic to the group of unit quaternions. The image of a vector v under a rotation given by some spinor τ is the vector

$$\tau \perp v := \tau^{-1} v \tau. \tag{6.37}$$

As it is more convenient to define the usual Fourier transform in the complex setting, it is more convenient for the following to consider the colour space as embedded in $\mathbb{R}^4$. This simplifies, in particular, the implementation through a double complex FFT. The Clifford algebra $\mathbb{R}_{4,0}$ is the vector space of dimension 16 with basis given by the set

$$\{e_{i_1} \cdots e_{i_k},\ i_1 < \cdots < i_k \in [1,\ldots,4]\}$$

and the unit 1 (the vectors e_1, e_2, e_3 and e_4 are elements of an orthonormal basis of $\mathbb{R}^4$). As before, the multiplication rules are given by $e_i^2 = 1$ and $e_i e_j = -e_j e_i$. The corresponding spinor group $Spin(4)$ is the cross product of two copies of $Spin(3)$ and acts as rotations on vectors of $\mathbb{R}^4$ by formula (6.37).

One fundamental remark is that every spinor τ of $Spin(3)$, resp. $Spin(4)$, can be written as the exponential of a bivector of $\mathbb{R}_{3,0}$, resp. $\mathbb{R}_{4,0}$, i.e.,

$$\tau = \sum_{i\geq 0} \frac{1}{i!} B^i \tag{6.38}$$

for some bivector B. This means precisely that the *Lie exponential map* is onto (see [20] for a general theorem on compact connected Lie groups). As an example, the spinor

$$\tau = \frac{(1+n_2 n_1)}{\sqrt{2(1+n_1 \cdot n_2)}} = \exp\left[(\theta/2)\frac{n_2 \wedge n_1}{|n_2 \wedge n_1|}\right]$$

is the rotation of $\mathbb{R}^3$ that sends by formula (6.37) the unit vector n_1 to the unit vector n_2 leaving the plane (n_1, n_2) globally invariant. In the above expression, θ is the angle between n_1 and n_2 and $|n_2 \wedge n_1|$ is the magnitude of the bivector $n_2 \wedge n_1$.

6.3.2.3 Spin Characters

The aim here is to compute the group morphisms (i.e., maps preserving the composition laws) from the additive group $\mathbb{R}^2$ to the spinor group of the involved Clifford algebra. We don't detail the proofs since they need specific tools on Lie algebras (see [21] for explanations). In the sequel, we denote $\mathbb{S}^2_{3,0}$, resp. $\mathbb{S}^2_{4,0}$, the set of unit bivectors of the algebra $\mathbb{R}_{3,0}$, resp. $\mathbb{R}_{4,0}$. Let us first treat the case of $Spin(3)$ characters.

Theorem 1 (*Spin (3)* Characters). *The group morphisms of the additive group* $\mathbb{R}^2$ *to* $Spin(3)$ *are given by the maps that send* (x_1,x_2) *to:*

$$\exp\left[\frac{1}{2}(x_1u_1+x_2u_2)B\right], \tag{6.39}$$

where B *belongs to* $\mathbb{S}^2_{3,0}$ *and* u_1 *and* u_2 *are reals.*

It is important to notice that if the $Spin(3)$ characters are parametrized as usual by two frequencies u_1 and u_2, they are also parametrized by a unit bivector B of the Clifford algebra $\mathbb{R}_{3,0}$. We have already mentionned that, for implementation reasons, it is preferable to deal with the Clifford algebra $\mathbb{R}_{4,0}$ and the corresponding $Spin(4)$ group. Using the fact that this one is the cross product of two copies of $Spin(3)$, one can prove the following result.

Theorem 2 (*Spin (4)* Characters). *The group morphisms of the additive group* $\mathbb{R}^2$ *to* $Spin(4)$ *are given by the maps that send* (x_1,x_2) *to:*

$$\exp\left[\frac{1}{2}[x_1(u_1+u_3)+x_2(u_2+u_4)]D\right]\exp\left[\frac{1}{2}[x_1(u_1-u_3)+x_2(u_2-u_4)]I_4D\right], \tag{6.40}$$

where D *belongs to* $\mathbb{S}^2_{4,0}$ *and* u_1, u_2, u_3, *and* u_4 *are reals. In this expression* I_4 *denotes the pseudo scalar* $e_1e_2e_3e_4$ *of the algebra* $\mathbb{R}_{4,0}$.

Let us make few comments. The first one is that $Spin(4)$ characters are parametrized by four frequencies and a bivector of $\mathbb{S}^2_{4,0}$. This is not really suprising in view of the classification of rotation in $\mathbb{R}^4$ (see below). The second one concerns the product I_4D of the pseudo scalar I_4 by the bivector D. A *simple bivector* D (i.e., the exterior product of two vectors) represents a piece of a two-dimensional vector space of $\mathbb{R}^4$ with a magnitude and an orientation. Multiplying this one by I_4 consists in fact to consider the element of $\mathbb{S}^2_{4,0}$ which represents the piece of vector space orthogonal to D in $\mathbb{R}^4$ (see [22]). The spinor is written as a product of two *commuting* spinors each one acting as a rotation (the first one in the D plane, the second one in the I_4D plane). Finally, note that these formulas are quite natural and generalize the usual formula since the imaginary complex i can be viewed as the unit bivector coding the complex plane.

We denote $\varphi_{(u_1,u_2,u_3,u_4,D)}$ the morphisms given by equation (6.40).

6.3.2.4 About Rotations of $\mathbb{R}^4$

The reader may find in [21] the complete description of the rotations in the space $\mathbb{R}^4$. The classification is given as follows.

- *Simple rotations* are exponential of *simple bivectors* that is exterior products of two vectors. These rotations turn only one plane.

- *Isoclinic rotations* are exponential of simple bivectors multiplied by one of the elements $(1 \pm I_4)/2$. An isoclinic rotation has an infinity of rotation planes.
- *General rotations* have two invariant planes which are completely orthogonal with different angle of rotation.

In the Clifford algebra $\mathbb{R}_{3,0}$ every bivector is simple, i.e., it represents a piece of plane. Formula (6.39) describes simple rotations. Formula (6.40) describes general rotations in $\mathbb{R}^4$. In the next section, we make use of the following special $Spin(4)$ characters:

$$(x_1, x_2) \longmapsto \varphi_{(u_1,u_2,0,0,D)}(x_1, x_2). \tag{6.41}$$

They correspond to isoclinic rotations.

6.3.3 *Clifford Colour Fourier Transform with Spin Characters*

Before examining Clifford Colour Fourier, it is useful to rewrite the usual definition of the complex Fourier transform in the language of Clifford algebras.

6.3.3.1 The Usual Transform in the Clifford Framework

Let us consider the usual Fourier formula when n equals 2:

$$\mathcal{F}f(u_1,u_2) = \int_{\mathbb{R}^2} f(x_1,x_2) \exp(-i(x_1u_1 + x_2u_2)) \mathrm{d}x_1 \mathrm{d}x_2. \tag{6.42}$$

The involved characters are the maps

$$(x_1, x_2) \longmapsto \exp(i(x_1u_1 + x_2u_2))$$

with values in the group of unit complex numbers which is in fact the group $Spin(2)$ of the Clifford algebra $\mathbb{R}_{2,0}$. Considering the complex valued function $f = f_1 + if_2$ as a map in the vector part of this algebra, i.e.,

$$f(x_1,x_2) = f_1(x_1,x_2)e_1 + f_2(x_1,x_2)e_2$$

the Fourier transform may be written

$$\begin{aligned}\mathcal{F}f(u_1,u_2) = \int_{\mathbb{R}^2} \{&[\cos((x_1u_1 + x_2u_2)/2) + \sin((x_1u_1 + x_2u_2)/2)e_1e_2] \\ &[f_1(x_1,x_2)e_1 + f_2(x_1,x_2)e_2]\,[\cos(-(x_1u_1 + x_2u_2)/2) \\ &+ \sin(-(x_1u_1 + x_2u_2)/2)e_1e_2]\}\, \mathrm{d}x_1 \mathrm{d}x_2. \end{aligned} \tag{6.43}$$

If we consider the action $\perp$ introduced in formula (6.37), we obtain:

$$\mathcal{F}f(u_1,u_2) = \int_{\mathbb{R}^2} [f_1(x_1,x_2)e_1 + f_2(x_1,x_2)e_2] \perp \varphi_{(u_1,u_2,e_1e_2)}(-x_1,-x_2)\mathrm{d}x_1\mathrm{d}x_2, \tag{6.44}$$

where

$$\varphi_{(u_1,u_2,e_1e_2)}(x_1,x_2) = \exp\left[\frac{1}{2}(x_1u_1 + x_2u_2)e_1e_2\right]$$

since as said before, the imaginary complex i corresponds to the bivector e_1e_2. We now describe the generalization of this definition.

6.3.3.2 Definition of the Clifford Fourier Transform

We give first a general definition for a function f from $\mathbb{R}^2$ with values in the vector part of the Clifford algebra $\mathbb{R}_{4,0}$:

$$f : (x_1,x_2) \longmapsto f_1(x_1,x_2)e_1 + f_2(x_1,x_2)e_2 + f_3(x_1,x_2)e_3 + f_4(x_1,x_2)e_4. \tag{6.45}$$

Definition 1 (General Definition). The Clifford Fourier transform of the function f defined by (6.45) is given by

$$\mathcal{CF}f(u_1,u_2,u_3,u_4,D) = \int_{\mathbb{R}^2} f(x_1,x_2) \perp \varphi_{(u_1,u_2,u_3,u_4,D)}(-x_1,-x_2)\mathrm{d}x_1\mathrm{d}x_2. \tag{6.46}$$

It is defined on $\mathbb{R}^4 \times \mathbb{S}^2_{4,0}$.

Let us give an example. The vector space $\mathbb{H}$ of quaternions can be identified with the vector space $\mathbb{R}^4$ under the correspondance: $e_1 \leftrightarrow i$, $e_2 \leftrightarrow j$, $e_3 \leftrightarrow k$, and $e_4 \leftrightarrow 1$. It then can be shown that

$$\mathcal{F}_{ij}f(u_1,u_2) = \mathcal{CF}f(2\pi u_1, 0, 0, 2\pi u_2, D_{ij}),$$

where $\mathcal{F}_{ij}$ is the quaternioninc transform of Bülow and D_{ij} is the bivector

$$D_{ij} = -\frac{1}{4}(e_1 + e_2)(e_3 - e_4).$$

For most of the applications to colour image processing that will be investigated below, it is sufficient to consider a transform that can be applied to functions with values in the vector part of the algebra $\mathbb{R}_{3,0}$. Such a function is given by

$$f : (x_1,x_2) \longmapsto f_1(x_1,x_2)e_1 + f_2(x_1,x_2)e_2 + f_3(x_1,x_2)e_3 + 0e_4 \tag{6.47}$$

just as a real function is a complex function with 0 imaginary part.

Definition 2 (Definition for Colour Images). The Clifford Fourier transform of the function f defined by (6.47) in the direction D is given by

$$\mathcal{CF}_D f(u_1,u_2) = \int_{\mathbb{R}^2} f(x_1,x_2) \perp \varphi_{(u_1,u_2,0,0,D)}(-x_1,-x_2)\mathrm{d}x_1\mathrm{d}x_2. \tag{6.48}$$

It is defined on $\mathbb{R}^2$.

As an example, let us mention that (under the above identification of $\mathbb{H}$ with $\mathbb{R}^4$)

$$\mathcal{F}_\mu f(u_1,u_2) = \mathcal{CF}_{D_\mu} f(u_1,u_2),$$

where $\mathcal{F}_\mu$ is the quaternionic transform of Sangwine et al. and D_μ is the bivector

$$D_\mu = (\mu_1 e_1 + \mu_2 e_2 + \mu_3 e_3) \wedge e_4$$

with $\mu = \mu_1 i + \mu_2 j + \mu_3 k$ a unit imaginary quaternion.

Both definitions involve bivectors of $\mathbb{S}^2_{4,0}$ (as variable and as parameter). We give now some of the properties satisfied by the Clifford Fourier transform.

6.3.3.3 Properties of the Clifford Fourier Transform

A parallel and orthogonal decomposition, very closed to the symplectic decomposition used by Sangwine, is used to study the properties of the colour Fourier Transform.

Parallel and Orthogonal Decomposition

The function f given by equation (6.47) can be decomposed as

$$f = f_{\parallel D} + f_{\perp D}, \tag{6.49}$$

where $f_{\parallel D}$, resp. $f_{\perp D}$, is the parallel part, resp. the orthogonal part, of f with respect to the bivector D. Simple computations show that

$$\begin{aligned}
&\mathcal{CF}f(u_1,u_2,u_3,u_4,D) \\
&\quad = \int_{\mathbb{R}^2} f_{\parallel D}(x_1,x_2)\exp\left[-(x_1(u_1+u_3)+x_2(u_2+u_4))D\right]\mathrm{d}x_1\mathrm{d}x_2 \\
&\qquad + \int_{\mathbb{R}^2} f_{\perp D}(x_1,x_2)\exp\left[-(x_1(u_1+u_3)+x_2(u_2+u_4))I_4 D\right]\mathrm{d}x_1\mathrm{d}x_2.
\end{aligned} \tag{6.50}$$

Applying this decomposition to colour images, leads to the following result.

Proposition 1 (Clifford Fourier Transform Decomposition for Colour Images). *Let f be as in (6.47), then*

$$\mathcal{CF}_D f = \mathcal{CF}_D(f_{\|_D}) + \mathcal{CF}_D(f_{\perp_D}) = (\mathcal{CF}_D f)_{\|_D} + (\mathcal{CF}_D f)_{\perp_D}. \tag{6.51}$$

In practice, the decomposition is obtained as follows. Let us fix a *simple bivector* $D = v_1 \wedge v_2$ of $\mathbb{S}^2_{4,0}$. There exists a vector w_2 of $\mathbb{R}^4$, namely $w_2 = v_1^{-1}(v_1 \wedge v_2) = v_1^{-1}D$, such that

$$D = v_1 \wedge v_2 = v_1 \wedge w_2 = v_1 w_2.$$

In the same way, if v_3 is a unit vector such that $v_3 \wedge I_4 D = 0$, then the vector $w_4 = v_3^{-1} I_4 D$ satisfies

$$I_4 D = v_3 \wedge w_4 = v_3 w_4.$$

This precisely means that if v_1 and v_3 are chosen to be unit vectors (in this case, $v_1^{-1} = v_1$ and $v_3^{-1} = v_3$), the set (v_1, w_2, v_3, w_4) is an orthonormal basis of $\mathbb{R}^4$ adapted to D and $I_4 D$. We can then write a function f

$$\begin{aligned} f(x_1,x_2) &= [(f(x_1,x_2)\cdot v_1)v_1 + (f(x_1,x_2)\cdot(v_1 D))v_1 D] \\ &\quad + [(f(x_1,x_2)\cdot v_3)v_3 + (f(x_1,x_2)\cdot(v_3 I_4 D))v_3 I_4 D] \end{aligned} \tag{6.52}$$

or equivalently

$$\begin{aligned} f(x_1,x_2) &= v_1\,[(f(x_1,x_2)\cdot v_1) + (f(x_1,x_2)\cdot(v_1 D))D] \\ &\quad + v_3\,[(f(x_1,x_2)\cdot v_3) + (f(x_1,x_2)\cdot(v_3 I_4 D))I_4 D] \\ &= v_1\,[\alpha(x_1,x_2) + \beta(x_1,x_2)D] + v_3\,[\gamma(x_1,x_2) + \delta(x_1,x_2)I_4 D] \end{aligned} \tag{6.53}$$

Since $D^2 = (I_4 D)^2 = -1$, the terms in the brackets can be identified with complex numbers $\alpha(x_1,x_2) + i\beta(x_1,x_2)$ and $\gamma(x_1,x_2) + i\delta(x_1,x_2)$ on which a usual complex FFT can be applied. Let us denote $\widehat{\alpha}(u_1,u_2) + i\widehat{\beta}(u_1,u_2)$ and $\widehat{\gamma}(u_1,u_2) + i\widehat{\delta}(u_1,u_2)$ the results. The Clifford Fourier transform of f in the direction D is given by

$$\mathcal{CF}_D f(u_1,u_2) = v_1\left[\widehat{\alpha}(u_1,u_2) + \widehat{\beta}(u_1,u_2)D\right] + v_3\left[\widehat{\gamma}(u_1,u_2) + \widehat{\delta}(u_1,u_2)I_4 D\right]. \tag{6.54}$$

For the applications treated below, it will be clear how to choose the unit vectors v_1 and v_3.

Inverse Clifford Fourier Transform

The Clifford Fourier transform defined by equation (6.46) is left invertible. Its inverse is given by

$$\begin{aligned} &\mathcal{CF}^{-1} g(x_1,x_2) \\ &= \int_{\mathbb{R}^4 \times \mathbb{S}^2_{4,0}} g(u_1,u_2,u_3,u_4,D) \perp \varphi_{(u_1,u_2,u_3,u_4,D)}(x_1,x_2)\mathrm{d}u_1\mathrm{d}u_2\mathrm{d}u_3\mathrm{d}u_4\mathrm{d}v(D), \end{aligned} \tag{6.55}$$

where ν is a measure on the set $\mathbb{S}^2_{4,0}$. The inversion formula for the Clifford Fourier transform (6.48) (colour image definition) is much more simpler.

Proposition 2 (Inverse Clifford Fourier Transform for Colour Images). *The Clifford Fourier transform defined by equation (6.48) is invertible. Its inverse is given by*

$$\mathcal{CF}_D^{-1} g(x_1,x_2) = \int_{\mathbb{R}^2} g(u_1,u_2) \perp \varphi_{(u_1,u_2,0,0,D)}(x_1,x_2) \mathrm{d}u_1 \mathrm{d}u_2. \tag{6.56}$$

Remark the analogy (change of signs in the spin characters) with the usual inversion formula.

Since this chapter is mainly devoted to colour image processing and for sake of simplicity, we describe now properties concerning the only transformation (6.48).

Shift Theorem

It is important here to notice that the transform $\mathcal{CF}_D$ satisfies a natural Shift Theorem which results in fact from the way it has been constructed. Let us denote $f_{(\alpha_1,\alpha_2)}$ the function defined by

$$f_{(\alpha_1,\alpha_2)}(x_1,x_2) = f(x_1+\alpha_1, x_2+\alpha_2), \tag{6.57}$$

where f is as in (6.47).

Proposition 3 (Shift Theorem for Colour Images). *The Clifford Fourier Transform of the function $f_{(\alpha_1,\alpha_2)}$ in the direction D is given by*

$$\mathcal{CF}_D f_{(\alpha_1,\alpha_2)}(u_1,u_2) = \mathcal{CF}_D f(u_1,u_2) \perp \varphi_{(u_1,u_2,0,0,D)}(\alpha_1,\alpha_2). \tag{6.58}$$

Generalized Hermitian Symmetry

It is well known that a function f defined on $\mathbb{R}^2$ is real if and only if the usual Fourier coefficients satisfy

$$\mathcal{F}f(-u_1,-u_2) = \overline{\mathcal{F}f(u_1,v_2)}, \tag{6.59}$$

where $\mathcal{F}$ is the usual Fourier transform and the overline denotes the complex conjugacy. This property, called *hermitian symmetry* is important when dealing with frequencies filtering. Note that the precedent equation implies that

$$\Im\left[\mathcal{F}f(u_1,u_2)\exp(i(x_1u_1+x_2u_2)) + \mathcal{F}f(-u_1,-u_2)\exp(-i(x_1u_1+x_2u_2))\right] = 0, \tag{6.60}$$

where $\Im$ denotes the imaginary part and thus that the function f is real.

With the quaternionic Fourier transform, we noted that the colour Fourier coefficients satisfied an anti-hermitian symmetry.

The next proposition generalizes this hermitian property to the Clifford Fourier transform for color images.

Proposition 4 (Generalized Hermitian Symmetry for Colour Images). *Let f be given as in (6.47), then the e_4 term in*

$$\mathcal{CF}_D f(u_1,u_2) \perp \varphi_{(u_1,u_2,0,0,D)}(x_1,x_2) + \mathcal{CF}_D f(-u_1,-u_2) \perp \varphi_{(-u_1,-u_2,0,0,D)}(x_1,x_2) \tag{6.61}$$

is zero. Moreover, the expression does not depend on D.

This proposition justifies the fact that the masks used for filtering in the frequency domain are chosen to be invariant with respect to the transformation $(u_1,u_2) \longmapsto (-u_1,-u_2)$.

Energy Conservation

The following statement is an analog of the usual Parceval equality satisfied by the usual Fourier transform.

Proposition 5 (Clifford Parceval Equality). *Let f be given as in (6.47), then*

$$\int_{\mathbb{R}^2} (\mathcal{CF}_D f(u_1,u_2))^2 \mathrm{d}u_1 \mathrm{d}u_2 = \int_{\mathbb{R}^2} (f(x_1,x_2))^2 \mathrm{d}x_1 \mathrm{d}x_2 \tag{6.62}$$

whenever one term is defined (and thus both terms are defined).

Let us recall that for a vector u of the algebra $\mathbb{R}_{4,0}$, $u^2 = Q(u)$ where Q is the Euclidean quadratic form on $\mathbb{R}^4$.

6.3.3.4 Examples of the Use of the Colour Spectrum by Frequency Windowing

The definition of the Colour Fourier transform involves explicitly a bivector D of $\mathbb{R}_{4,0}$ which, as already said, corresponds to an analyzing direction. We precise now what kinds of bivectors may be considered. We only deal here with simple bivectors, i.e., bivectors that are wedge products of two vectors of $\mathbb{R}^4$. These one correspond to pieces of two-dimensional subspaces of $\mathbb{R}^4$ with a magnitude and an orientation. In the Fourier definition, the bivector D is of magnitude 1.

Colour Bivector

Let $\mu = \mu_1 e_1 + \mu_2 e_2 + \mu_3 e_3$ be a unit colour of the cube RGB. The bivector corresponding to this colour is given by

$$D_\mu = \mu \wedge e_4. \tag{6.63}$$

The parallel part of the Clifford Fourier transform $\mathcal{CF}_{D_\mu}$ can be used to analyse the frequencies of the colour image in the direction μ, whereas the orthogonal part can be used to analyse frequencies of colours that are orthogonal to μ (see examples below).

Hue Bivector

Proposition 6. *Let H be the set of bivectors*

$$H = \{(e_1 + e_2 + e_3) \wedge \mu,\ \mu \in RGB\} \tag{6.64}$$

with the quivalence relation

$$D_1 \simeq D_2 \Longleftrightarrow D_1 = \lambda D_2 \ for\ some\ \lambda > 0. \tag{6.65}$$

Then, $H/\simeq$ is in bijection with the set of hues.

It appears that choosing a unit bivector $D_{\tilde{\mu}}$ of the form $(e_1 + e_2 + e_3) \wedge \mu$ makes it possible to analyse the frequencies of a colour image (through the parallel part of the Clifford Fourier transform) with respect to the hue of the colour μ.

Note that it is also possible to choose a unit bivector which is the wedge product of two colours μ_1 and μ_2.

Figure 6.11 shows the result of a directional filtering applied on a colour version of the classical Fourier house. The original image is on the left. The bivector used on this example is the one coding the red colour, i.e., $D = e_1 \wedge e_4$. The mask is defined in the Fourier domain by 0 on the set $\{|\arg(z) - \pi/2| < \varepsilon\} \cup \{|\arg(z) + \pi/2| < \varepsilon\}$ and by 1 elsewhere. It can be seen on the right image that the horizontal red lines have disappeared, whereas the green horizontal lines remain unchanged.

Figure 6.12 gives an illustration of the influence of the choice of the bivector D which, as said before, corresponds to an analizing direction. The filter used in this case is a low-pass filter in the parallel part. In the left image, D is once again the bivector $e_1 \wedge e_4$ coding the red color. For the right image, D is the bivector $(1/\sqrt{2})(e_2 + e_3) \wedge e_1$ coding the red hue.

On the left, both green and cyan stripes are not modified. This comes from the fact that these colours belong to the orthogonal part given by $I_4 D$. The result is different on the right image. Cyan stripes are blurred since the bivectors representing the red and cyan hues are opposite and thus generate the same plane. Green stripes are no more invariant since the vector e_2 of the green colour is no longer orthogonal to $D = (1/\sqrt{2})(e_2 + e_3) \wedge e_1$.

Fig. 6.11 Original image—directional filtering in the red color

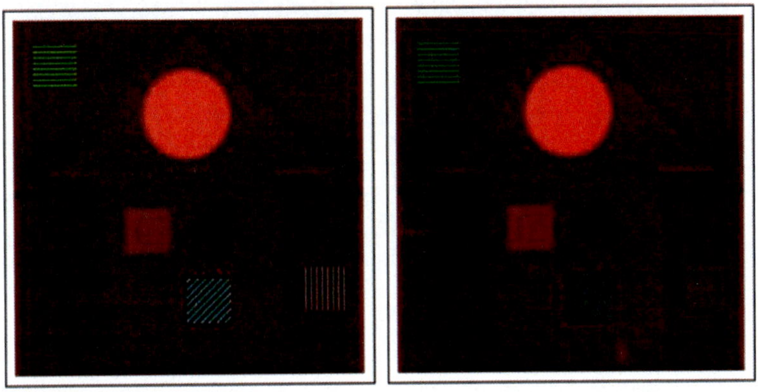

Fig. 6.12 Colour filtering–Hue filtering

Let us emphasize one fundamental property of the Clifford Fourier transform defined with Spin characters. The preceeding definition can be extended by considering any positive definite quadratic form on $\mathbb{R}^4$. We do not enter into details here and just give an illustration in Fig. 6.13.

The right image is the original image on which is applied a low-pass filter in the part orthogonal to the bivector $D = e_1 \wedge e_4$. The middle image corresponds to the usual Euclidean quadratic form while the image on the right involves the quadratic form given by the identity matrix in the basis $(e_1, (1/\sqrt{2})(e_1 + e_2), i_\alpha/\|i_\alpha\|, e_4)$ of $\mathbb{R}^4$, i_α being the vector coding the colour α of the background leaves. This precisely means that, in this case, the red, yellow colours, and α are considered as orthogonal. Note that these ones are the dominant colours of the original image. The bivector $I_4 D$ is given by $(1/\sqrt{2})(e_1 + e_2) \wedge (i_\alpha/\|i_\alpha\|)$. It contains the yellow colour and α.

Fig. 6.13 Original image—Eclidean metric–adapted metric

In the middle image, the green and blue high frequencies are removed while the red ones are preserved ($I_4D = e_2 \wedge e_3$). The low-pass filter removes all high frequencies of the right image excepted those of the red petals.

6.4 Conclusion

Hypercomplex or quaternions numbers have been used recently for both greyscale and colour image processing. Geometric algebra allows to handle geometric entities such as scalars, vectors, or bivectors independently. These entities are handled with the help algebraic expressions such as products (inner, outer, geometric, ...) for instance and rules over these products allow to affect or modify entities.

This chapter presents how quaternion and geometric algebra is used as a new formalism to perform colour image processing.

The first section reminds us how to use quaternions to process colour information, and how the three components of a colour pixel split to the vectorial part of $\mathbb{R}_{3,0}$ multivector. This condition is required to apply and define geometric operations algebraically on colour vectors such as translations and rotations for instance. After that, we illustrate that the $\mathbb{R}_{3,0}$ algebra is convenient to analyse and/or alter geometrically colour in images with operations tools defined algebraically. For that, we gave examples with alteration of the global hue, saturation, or value of colour images. After this description of some basic issues of colour manipulations, we shows different existent filtering approaches for using quaternions with colour images, and we proposed to generalize approaches already defined with quaternions and enhanced them with this new formalism. Illustrations proved it gave more accurate colour edge detection.

The second section introduces the discrete quaternionic Fourier transform proposed by Sangwine and by Bülow, and the conditions on the quaternionic spectrum to enable manipulations into this frequency domain without loosing information when going back to the spatial domain. This parts gives some interpretation of the quaternionic Fourier space. We conclude on a geometric approach using group actions for the Clifford colour fourier transform. The idea is to generalize the usual definition based on the characters of abelian groups by considering group morphisms from R^2 to spinor groups Spin(3) and Spin(4). The transformation is parameterized by a bivector and a quadratic form, the choice of which is related to the application to be treated.

References

1. Sangwine SJ (1996) Fourier transforms of colour images using quaternion, or hypercomplex, numbers. Electron Lett 32(21):1979–1980
2. Sangwine SJ (1998) Colour image edge detector based on quaternion convolution. Electron Lett 34(10):969–971
3. Moxey CE, Sangwine SJ, Ell TA (2002) Vector correlation of colour images. In: First European conference on colour in graphics, imaging and vision (CGIV 2002), pp 343–347
4. Dorst L, Mann S (2002) Geometric algebra: a computational framework for geometrical applications (part i: algebra). IEEE Comput Graph Appl 22(3):24–31
5. Hestenes D, Sobczyk G (1984) Clifford algebra to geometric calculus: a unified language for mathematics and physics. Reidel, Dordrecht
6. Hestenes D (1986) New foundations for classical mechanics, 2nd edn. Kluwer Academic Publishers, Dordrecht
7. Lasenby J, Lasenby AN, Doran CJL (2000) A unified mathematical language for physics and engineering in the 21st century. Phil Trans Math Phys Eng Sci 358:21–39
8. Sangwine SJ (2000) Colour in image processing. Electron Comm Eng J 12(5):211–219
9. Denis P, Carré P (2007) Colour gradient using geometric algebra. In EUSIPCO2007, 15th European signal processing conference, Poznań, Poland
10. Ell TA, Sangwine SJ (2007) Hypercomplex fourier transform of color images. IEEE Trans Signal Process 16(1):22–35
11. Sochen N, Kimmel R, Malladi R (1998) A general framework for low level vision. IEEE Trans Image Process 7:310–318
12. Batard T, Saint-Jean C, Berthier M (2009) A metric approach to nd images edge detection with clifford algebras. J Math Imag Vis 33:296–312
13. Felsberg M (2002) Low-level image processing with the structure multivector. Ph.D. thesis, Christian Albrechts University of Kiel
14. Ebling J, Scheuermann G (2005) Clifford fourier transform on vector fields. IEEE Trans Visual Comput Graph 11(4):469–479
15. Mawardi M, Hitzer E (2006) Clifford fourier transformation and uncertainty principle for the clifford geometric algebra cl3,0. Advances in applied Clifford algebras 16:41–61
16. Brackx F, De Schepper N, Sommen F (2006) The two-dimensional clifford-fourier transform. J Math Imaging Vis 26(1–2):5–18
17. Smach F, Lemaitre C, Gauthier JP, Miteran J, Atri M (2008) Generalized fourier descriptors with applications to objects recognition in svm context. J Math Imaging Vis 30:43–71
18. Bülow T (1999) Hypercomplex spectral sinal representations for the processing and analysis of images. Ph.D. thesis, Christian Albrechts University of Kiel
19. Vilenkin NJ (1968) Special functions and the theory of group representations, vol 22. American Mathematical Society, Providence, RI
20. Helgason S (1978) Differential geometry, Lie groups and symmetric spaces. Academic Press, London
21. Lounesto P (1997) Clifford algebras and spinors. Cambridge University Press, Cambridge
22. Hestenes D, Sobczyk G (1987) Clifford algebra to geometric calculus: a unified language for mathematics and physics. Springer, Berlin

Chapter 7
Image Super-Resolution, a State-of-the-Art Review and Evaluation

Aldo Maalouf and Mohamed-Chaker Larabi

The perfumes, the colors and the sounds are answered

Charles Baudelaire

Abstract Image super-resolution is a popular technique for increasing the resolution of a given image. Its most common application is to provide better visual effect after resizing a digital image for display or printing. In recent years, due to consumer multimedia products being in vogue, imaging and display device become ubiquitous, and image super-resolution is becoming more and more important. There are mainly three categories of approaches for this problem: interpolation-based methods, reconstruction-based methods, and learning-based methods.

This chapter is aimed, first, to explain the objective of image super-resolution, and then to describe the existing methods with special emphasis on color super-resolution. Finally, the performance of these methods is studied by carrying on objective and subjective image quality assessment on the super-resolution images.

Keywords Image super-resolution • Color super-resolution • Interpolation-based methods • Reconstruction-based methods • Learning-based methods • Streaming video websites • HDTV displays • Digital cinema

7.1 Introduction

Image super-resolution is the process of increasing the resolution of a given image [1–3]. This process has also been referred to in the literature as resolution enhancement. One such application to image super-resolution can be found in streaming

A. Maalouf • M.-C. Larabi (✉)
Laboratory XLIM-SIC, UMR CNRS 7252, University of Poitiers, France
e-mail: chaker.larabi@univ-poitiers.fr

C. Fernandez-Maloigne (ed.), *Advanced Color Image Processing and Analysis*,
DOI 10.1007/978-1-4419-6190-7_7,

video websites, which often store video at low resolutions (e.g., 352×288 pixels CIF format) for various reasons. The problem is that users often wish to expand the size of the video to watch at full screen with resolutions of $1{,}024 \times 768$ or higher, and this process requires that the images be interpolated to the higher resolution. Another application comes from the emergence of HDTV displays. To better utilize the display technical prowess of the existing viewing devices, input signals coming from a low-resolution source must first be converted to higher resolutions through interpolation. Moreover, filmmakers today are increasingly turning toward an all-digital solution, from image capture to postproduction and projection. Due to its fairly recent appearance, the digital cinema chain still suffers from limitations which can hamper the productivity and creativity of cinematographers and production companies. One of these limitations is that the cameras used for high resolutions are expensive and the data files they produce are large. Because of this, studios may chose to capture some sequences at lower resolution (2K, for example). These sequences can later be interpolated to 4K sequences by using a super-resolution technique and projected in higher resolution display devices.

Increasing the resolution of the imaging sensor is clearly one way to increase the resolution of the acquired images. This solution, however, may not be feasible due to the increased associated cost and the fact that the shot noise increases during acquisition as the pixel size becomes smaller. Furthermore, increasing the chip size to accommodate the larger number of pixels increases the capacitance, which in tern reduces the data transfer rate. Therefore, image-processing techniques, like the ones described in this chapter, provide a clear alternative for increasing the resolution of the acquired images.

There are various possible models for performing resolution enhancement. These models can be grouped in three categories: interpolation-based methods, reconstruction-based methods, and learning-based methods.

The goals of this chapter can be summarized as follows:

- To review the super-resolution techniques used in the literature from the image quality point of view.
- To evaluate the performance of each technique.

The rest of the chapter is organized as follows: first, interpolation-based super-resolution techniques are described. Then, we present the reconstruction-based and the learning-based methods. Thereafter, these methods are evaluated by using objective and subjective image quality assessment metrics. Finally, we summarize the chapter and comment on super-resolution problems that still remain open.

7.2 Interpolation-Based Super-Resolution Methods

Interpolation-based methods generate a high-resolution image from its low-resolution version by estimating the pixel intensities on an upsampled grid. Suppose the low-resolution image is $I_{i,j}$ of size $W \times H$ and its corresponding

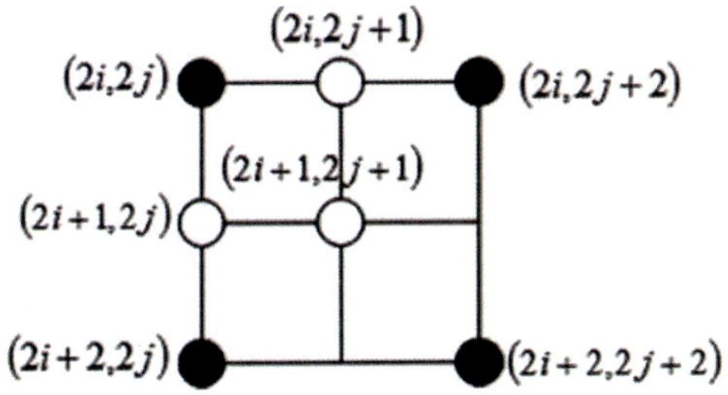

Fig. 7.1 Pixels in the high-resolution image

high-resolution image is $I'_{i,j}$ of size $aW \times bH$, here a and b are magnification factors of the width and height, respectively. Without the loss of generality, we assume $a = b = 2$. We can easily get pixel value of $I'_{2i,2j}$ from the low resolution just because $I'_{2i,2j} = I_{i,j}$ $(i = 0, 1, \ldots\ldots, H; j = 0, 1, \ldots\ldots, W)$, and the interpolation is needed when we settle the problem that how we get the pixel values of $I'_{2i+1,2j}$, $I'_{2i,2j+1}$, and $I'_{2i+1,2j+1}$. As is shown in Fig. 7.1, the black nodes denote the pixels which can be directly obtained from the low-resolution image, we call them original pixels; and the white nodes denote the pixel which are unknown and can be gained by interpolation method; this is image interpolation.

The most common interpolation methods used in practice are the bilinear and bicubic interpolation methods [4, 5], requiring only a small amount of computation. However, because they are based on an oversimplified slow varying image model, these simple methods often produce images with various problems along object boundaries, including aliasing, blurring, and zigzagging edges.

To cope with these problems, various algorithms have been proposed to improve the interpolation-based approaches and reduce edge artifacts, aiming at obtaining images with regularity (i.e., smoothness) along edges. In one of the earliest papers on the subject, Jensen and Anastassiou [6] propose to estimate the orientation of each edge in the image by using projections onto an orthonormal basis and the interpolation process is modified to avoid interpolating across the edge. To this end, Jensen and Anastassiou modeled the edge by a function of four parameters taking the following form:

$$S(i, j, A, B, \rho, \theta) = \begin{cases} A \text{ if } i\cos\theta + j\sin\theta \geq \rho \\ B \text{ if } i\cos\theta + j\sin\theta < \rho \end{cases}, \tag{7.1}$$

where $i\cos\theta + j\sin\theta = \rho$ is a straight line separating two regions.

Looking from a distance $R > \rho$, they showed that the edge model (7.1) is none other than a periodic function of the angular coordinate ϕ. Then, they defined an orthogonal basis set $B_n(\phi)$ by:

$$B_0(\phi) = \frac{1}{\sqrt{2\pi}},$$

$$
\begin{aligned}
B_1(\phi) &= \frac{1}{\sqrt{\pi}} \cos(\phi), \\
B_2(\phi) &= \frac{1}{\sqrt{\pi}} \sin(\phi), \\
B_3(\phi) &= \frac{1}{\sqrt{\pi}} \cos(2\phi), \\
B_4(\phi) &= \frac{1}{\sqrt{\pi}} \sin(2\phi).
\end{aligned} \tag{7.2}
$$

The projection of the edge model S onto $B_n(\phi)$ is given by:

$$
a_n = \langle S(\phi, A, B, \rho, \theta), B_n(\phi) \rangle = \int_0^{2\pi} S(\phi, A, B, \rho, \theta) B_n(\phi)\, \mathrm{d}\phi. \tag{7.3}
$$

Equation (7.3) yields to a set of spectral coefficients a_n from which we can compute the edge orientation θ by:

$$
\theta = \tan^{-1}\left(\frac{a_2}{a_1}\right). \tag{7.4}
$$

In order to make the interpolation selective, i.e., avoiding interpolation on edges, Jensen et al. proposed to find not only edges orientation but also the step height of the edges so as to perform edge continuation to preserve the image's geometry and interpolation in homogeneous regions. The step height of edges is estimated as follows:

Let $W[i,j]$ be a 3×3 window of image data. For each W in I, we compute the edge step height at the center pixel of W by using the following equation:

$$
\lambda_k = \sum_{m=1}^{3} \sum_{n=1}^{3} W[m,n] M_k[m,n], \tag{7.5}
$$

where M_k's are the operators shown in Fig. 7.2 with the weightings $\alpha = \frac{\sqrt{\pi}}{4\sqrt{2}}$ and $\beta = \frac{\sqrt{\pi}}{4}$.

Finally, the value of the missing pixels A and B on either side of the edge are computed by:

$$
B = \frac{1}{2\pi}\left(\sqrt{2\pi}\lambda_0 - 2\sigma\delta\right), \tag{7.6}
$$

$$
A = \delta + B, \tag{7.7}
$$

Fig. 7.2 Operators

$$M_0 = \begin{bmatrix} \alpha & \alpha & \alpha \\ \alpha & 0 & \alpha \\ \alpha & \alpha & \alpha \end{bmatrix} \quad M_1 = \begin{bmatrix} -\alpha & 0 & \alpha \\ -\beta & 0 & \beta \\ -\alpha & 0 & \alpha \end{bmatrix} \quad M_3 = \begin{bmatrix} \alpha & \beta & \alpha \\ 0 & 0 & 0 \\ -\alpha & -\beta & -\alpha \end{bmatrix}$$

$$M_4 = \begin{bmatrix} 0 & -\beta & 0 \\ \beta & 0 & \beta \\ 0 & -\beta & 0 \end{bmatrix} \quad M_5 = \begin{bmatrix} -\beta & 0 & \beta \\ 0 & 0 & 0 \\ \beta & 0 & -\beta \end{bmatrix}$$

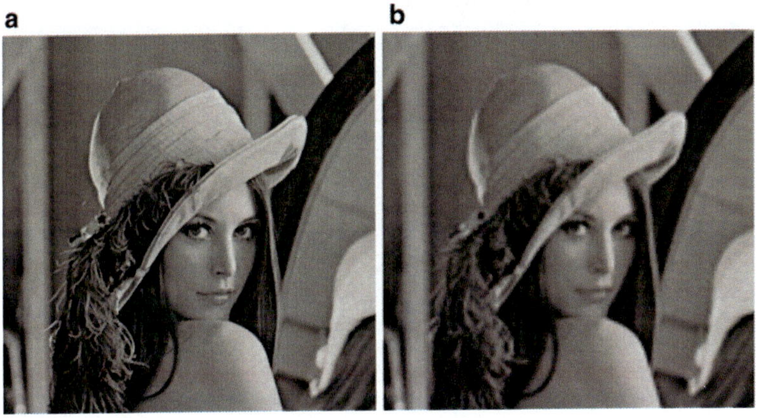

Fig. 7.3 (a) Original Lena image and (b) result obtained by using the method of Jensen et al. [6]

where,

$$\delta = \frac{\sqrt{\pi}\,[\lambda_1 \cos\theta + \lambda_2 \sin\theta]}{2\sin\sigma} \tag{7.8}$$

and

$$\sigma = \cos^{-1}\left(\frac{\lambda_3 \cos 2\theta + \lambda_4 \sin 2\theta}{\lambda_1 \cos\theta + \lambda_2 sin\theta}\right). \tag{7.9}$$

Figures 7.3 and 7.4 show an example of applying the method of Jensen et al. [6] on "Lena" and "Lighthouse" images, respectively. For all the experiments in this chapter, the original images were first downsampled by a factor of four and then interpolated back to their original size. This provides a better comparison than a factor of two interpolation, and is well justified if one compares the ratio of NTSC scan lines (240 per frame) to state-of-the-art HDTV (1080 per frame), which is a

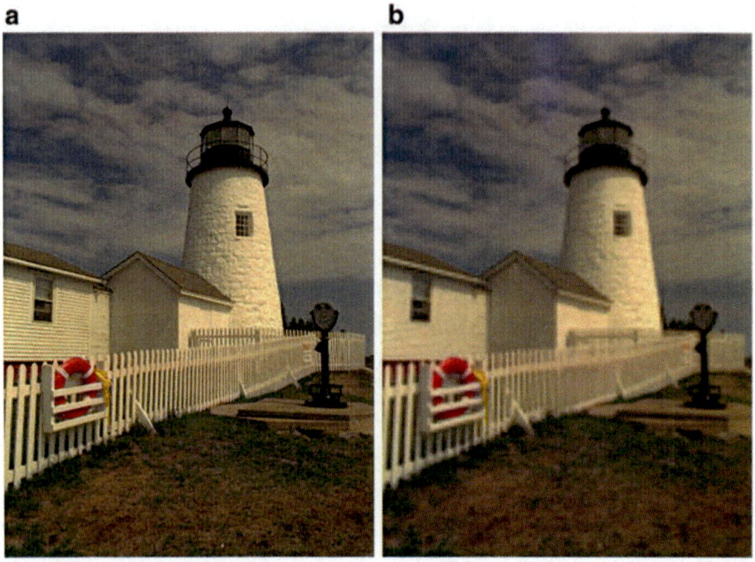

Fig. 7.4 (**a**) Original lighthouse image and (**b**) result obtained by using the method of Jensen et al. [6]

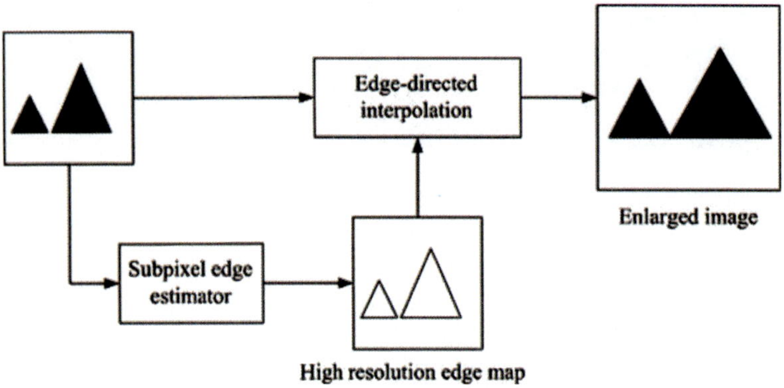

Fig. 7.5 Framework of the edge-directed interpolation method

factor of 4.5. As we can see from the reconstructed images, the method of Jensen et al. performs well in homogeneous regions; however, the reconstructed images are blurred especially near fine edges.

Rather than modeling the edges of the low-resolution image and avoiding interpolation on these edges, Allebach et al. proposed in [7] to generate a high-resolution edge map from the low-resolution image, and then use the high-resolution edge map to guide the interpolation. Figure 7.5 shows the framework within with the edge-directed interpolation method works. First, a subpixel edge estimation technique is

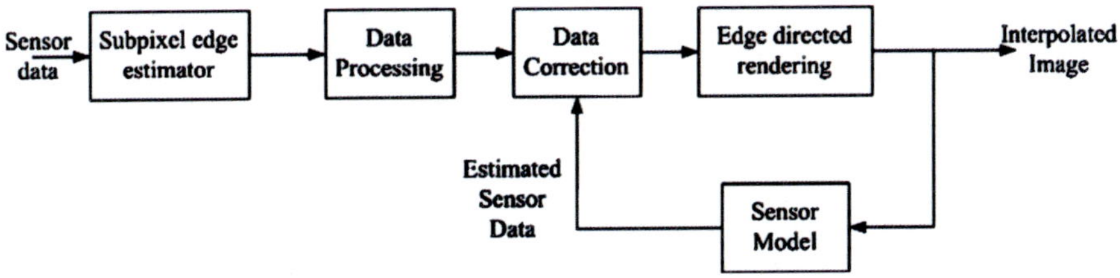

Fig. 7.6 Architecture of the edge directed interpolation method

used to generate a high-resolution edge map from the low-resolution image. Then, the obtained high-resolution edge map is used to guide the interpolation of the low-resolution image to the high-resolution version. Figure 7.6 shows the architecture of the edge-directed interpolation technique itself. It consists of two steps: rendering and data correction. Rendering is none other than a modified form of bilinear interpolation of the low-resolution image data. An implicit assumption underlying bilinear interpolation is that the low-resolution data consists of point samples from the high-resolution image. However, most sensors generate low-resolution data by averaging the light incident at the focal plane over the unit cell corresponding to the low-resolution sampling lattice. After that, Allebach et al. proposed to iteratively compensate for this effect by feeding the interpolated image back through the sensor model and using the disparity between the resulting estimated sensor data and the true sensor data to correct the mesh values on which the bilinear interpolation is based.

To estimate the subpixel edge map, the low-resolution image is filtered with a rectangular center-on-surround-off (COSO) filter with a constant positive center region embedded within a constant negative surround region. The relative heights are chosen to yield zero DC response. The COSO filter coefficients are given by:

$$h^{\mathrm{COSO}}(i,j)=\begin{cases} h_{\mathrm{c}}, & |i|,|j|\le N_{\mathrm{c}} \\ h_{\mathrm{s}}, & \begin{matrix} N_{\mathrm{c}}<|i|\le N_{\mathrm{s}} \text{ and } |i|\le N_{\mathrm{s}} \\ N_{\mathrm{c}}<|i|\le N_{\mathrm{s}} \text{ and } |j|\le N_{\mathrm{s}} \end{matrix}, \\ 0, & \text{otherwise} \end{cases} \tag{7.10}$$

where h_{c} and h_{s} are computed at the center and the sides, respectively, by using the point-spread function for the Laplacian-of-Gaussian (LOG) given by:

$$h^{\mathrm{LoG}}(i,j)=\frac{2}{\sigma^2}\left[1-\left(i^2+j^2\right)/2\sigma^2\right]\mathrm{e}^{-\left(i^2+j^2\right)/2\sigma^2} \tag{7.11}$$

and N_{c} and N_{s} are the width of the center and side respectively.

The COSO filter results in a good approximation to the edge map. To determine the high-resolution edge map, the COSO filter output is linearly interpolated between points on the low-resolution lattice to estimate zero-crossing positions on the high-resolution lattice.

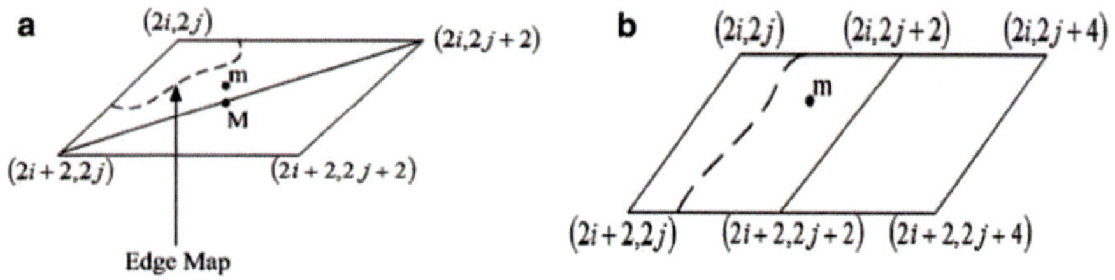

Fig. 7.7 Computation of replacement values for the low-resolution corner pixels to be used when bilinearly interpolating the image value at high-resolution pixel m. The cases shown are (**a**) replacement of one pixel, and (**b**) replacement of two adjacent pixels

Now let us turn our attention to Fig. 7.6. The essential feature of the rendering step is that we modify bilinear interpolation on a pixel-by-pixel basis to prevent interpolation across edges. To illustrate the approach, let us consider interpolation at the high-resolution pixel m in Fig. 7.7. We first determine whether or not any of the low-resolution corner pixels $I(2i+2,2j), I(2i,2j), I(2i,2j+2)$, and $I(2i+2,2j+2)$ are separated from m by edges. For all those pixels that are, replacement values are computed according to a heuristic procedure that depends on the number and geometry of the pixels to be replaced. Figure 7.7a shows the situation in which a single corner pixel $I(2i,2j)$ is to replaced. In this case, linearly interpolate is used to compute the value of the midpoint M of the line $I(2i+2,2j)-I(2i,2j+2)$, and then an extrapolation along the line $I(2i+2,2j+2)-M$ is performed to yield the replacement value of $I(2i,2j)$. If two corner pixels are to be replaced, they can be either adjacent or not adjacent. Figure 7.7b shows the case in which two adjacent pixels $I(2i+2,2j)$ and $I(2i,2j+2)$ must be replaced. In this case, we check to see if any edges cross the lines $I(2i,2j+4)-I(2i,2j+6)$ and $I(2i+2,2j+4)-I(2i+2,2j+6)$. If none does, we linearly extrapolate along the lines $I(2i,2j+4)-I(2i,2j+6)$ and $I(2i+2,2j+4)-I(2i+2,2j+6)$ to generate the replacement values of $I(2i+2,2j)$ and $I(2i,2j)$, respectively. If an edge crosses $I(2i,2j+4)-I(2i,2j+6)$, we simply let $I(2i,2j+2)=I(2i,2j+4)$. The cases in which two nonadjacent pixels are to be replaced or in which three pixels are to be replaced are treated similarly. The final case to be considered is that which occurs when the pixel m to be interpolated is separated from all four corner pixels $I(2i+2,2j), I(2i,2j), I(2i,2j+2)$, and $I(2i+2,2j+2)$. This case would only occur in regions of high spatial activity. In such areas, it is assumed that it is not possible to obtain a meaningful estimate of the high-resolution edge map from just the four low-resolution corner pixels; so the high-resolution image will be rendered with unmodified bilinear interpolation.

Figures 7.8 and 7.9 shows the results of four times interpolation using the edge-directed interpolation algorithm. As we can see, edge-directed interpolation yields a much sharper result than the method of Jensen et al. [6]. While some of the aliasing artifacts that occur with pixel replication can be seen in the edge-directed interpolation result, they are not nearly as prominent.

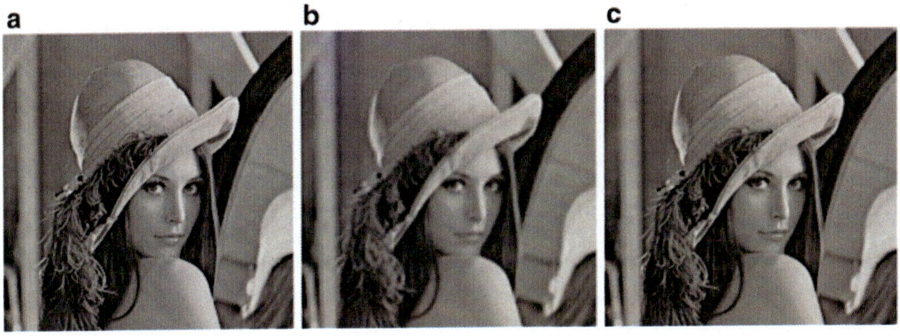

Fig. 7.8 (**a**) Original Lena image and results obtained by using the method of: (**b**) Jensen et al. [6] and (**c**) Allebach et al. [7]

Fig. 7.9 (**a**) Original Lighthouse image and results obtained by using the method of: (**b**) Jensen et al. [6] and (**c**) Allebach et al. [7]

Other proposed methods perform interpolation in a transform (e.g., wavelet) domain [8, 9]. These algorithms assume the low-resolution image to be the low-pass output of the wavelet transform and utilize dependence across wavelet scales to predict the "missing" coefficients in the more detailed scales.

In [9], Carey et al. made use of the wavelet transform because it provides a mean by which the local smoothness of a signal may be quantified; the mathematical smoothness (or regularity) is bounded by the rate of decay of its wavelet transform coefficients across scales. The algorithm proposed in [9] creates new wavelet subbands by extrapolating the local coefficient decay. These new, fine scale subbands

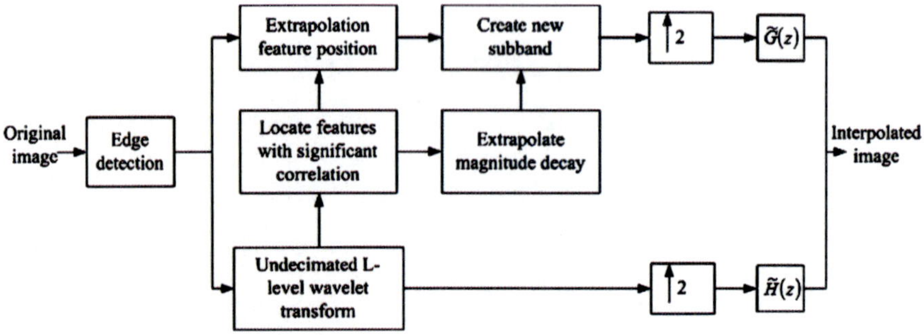

Fig. 7.10 Block diagram of the method proposed in [9]

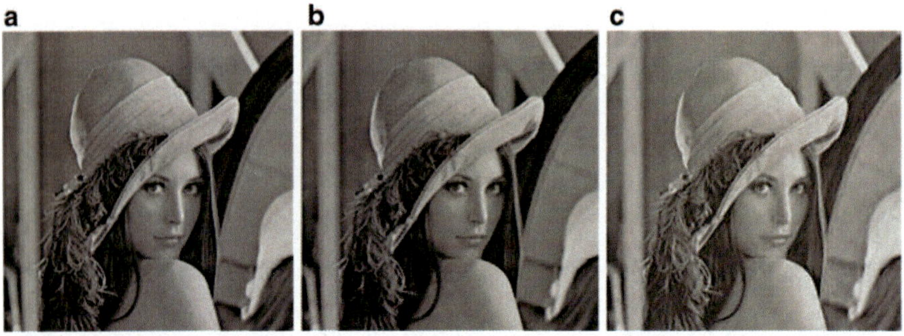

Fig. 7.11 (**a**) Original Lena image and results obtained by using the method of: (**b**) Allebach et al. [7] and (**c**) Carey et al. [9]

are used together with the original wavelet subbands to synthesize an image of twice the original size. Extrapolation of the coefficient decay preserves the local regularity of the original image, thus avoiding the over smoothing problems.

The block diagram of the method proposed in [9] is shown in Fig. 7.10.

The original image is considered to be the low-pass output of a wavelet analysis stage. Thus, it can be input to a single wavelet synthesis stage along with the corresponding high-frequency subbands to produce an image interpolated by a factor of two in both directions. Creation of these high-frequency subbands is therefore required for this interpolation strategy. After a non-separable edge detection, the unknown high-frequency subbands are created separately by a two-step process. First, edges with significant correlation across scales in each row are identified. The rate of decay of the wavelet coefficients near these edges is then extrapolated to approximate the high-frequency subband required to resynthesize a row of twice the original size. The same procedure is then applied to each column of the row-interpolated image.

Figures 7.11 and 7.12 show two examples of image interpolation using the method of Carey et al.. As we can see, the method of Carey et al. showed more edge artifacts when compared to the method of Allebach et al.. This is due to the reconstruction by the wavelet filters.

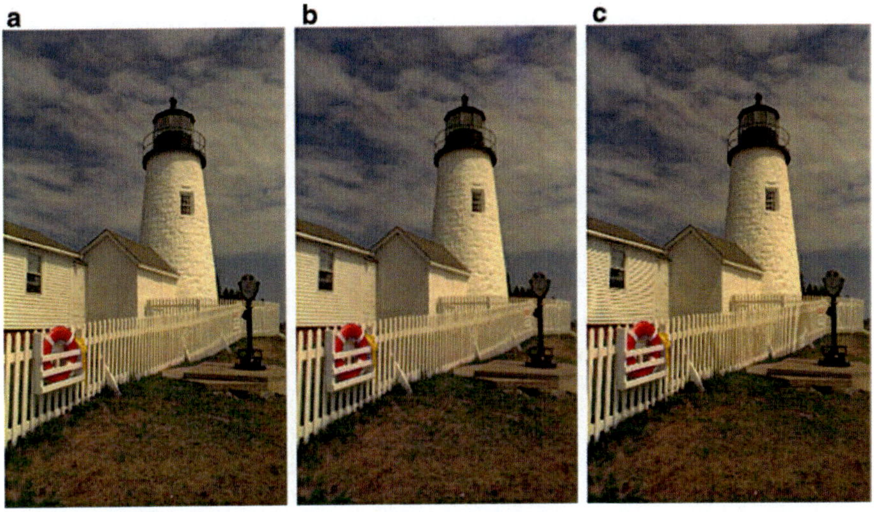

Fig. 7.12 (**a**) Original Lighthouse image and results obtained by using the method of: (**b**) Allebach et al. [7] and (**c**) Carey et al. [9]

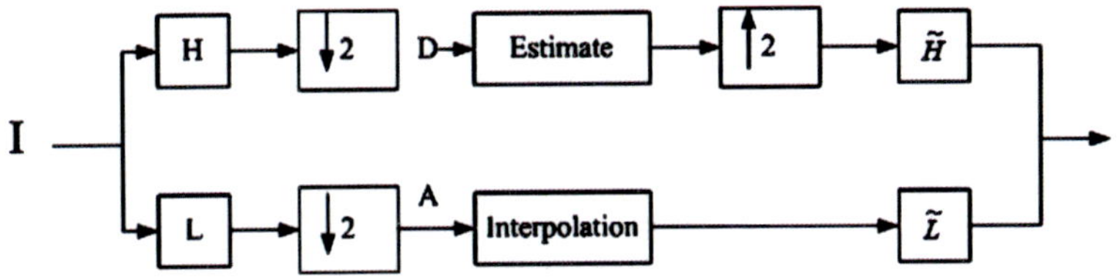

Fig. 7.13 Block diagram of the method proposed in [8]

Carey et al. [9] exploited the Lipschitz property of sharp edges in wavelet scales. In other words, they used the modulus maxima information at coarse scales to predict the unknown wavelet coefficients at the finest scale. Then, the HR image is constructed by inverse wavelet transform. Muresan and Parks [8] extended this strategy by using the entire cone influence of a sharp edge in wavelet scale space, instead of only the modulus maxima, to estimate the finest scale coefficients through optimal recovery theory. The approach to image interpolation of Muresan and Parks which is called *Prediction of Image Detail* can be explained with the help of Fig. 7.13. In Fig. 7.13, the high-resolution image is represented as the signal I at the input to the filter bank. Muresan et al. assumed that the low resolution, more coarsely sampled image is the result of a low-pass filtering operation followed by decimation to give the signal A. The low-pass filter, L, represents the effects of the image acquisition system. It will be possible to reconstruct the original image if we were able to filter the original high-resolution signal with the high-pass filter H to

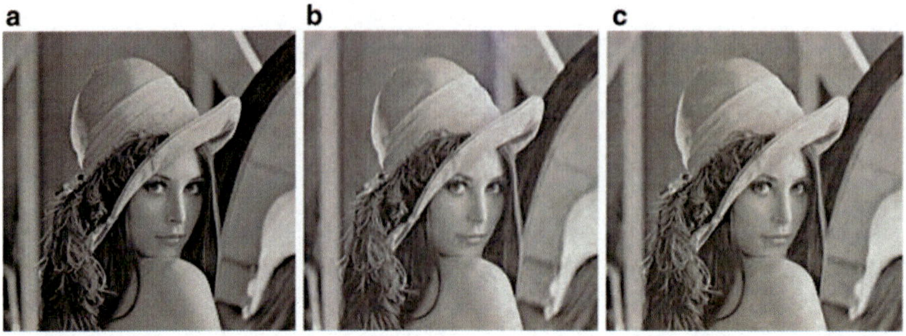

Fig. 7.14 (**a**) Original Lena image and results obtained by using the method of: (**b**) Carey et al. [9] and (**c**) Muresan et al. [8]

obtain the detail signal D, and if we had a perfect reconstruction filter bank, it would then be possible to reconstruct the original image. It is not possible to access to the detail signal D. Therefore, it must be estimated or predicted.

The approach followed by Muresan et al. to add image detail is based on the behavior of edges across scales in the scale-space domain. The approach of Carey et al. was to use only the modulus maxima information to estimate the detail coefficients at finest level. However, in practice, there may be a lot more details to add, than just using the modulus maxima information. Murasen et al. suggested using the entire cone of influence, from the coarser scales, for adding details to the finest scale.

Particularly, they used the energy of the wavelet coefficients around edges concentrated inside the cone of influence. They used this observation together with the theory of optimal recovery for estimating the coefficients of the fine scale from the known coefficients, inside the cone of influence, at the coarser scales.

Figures 7.14 and 7.15 show some results of applying the method of Murasen et al. for generating super-resolution images. It is clear that the method of Muresan et al. outperforms the method proposed by Carey et al. by reducing the edge artifacts.

Another adaptive image interpolation has been proposed by Li et al. in [4]. Their method is edge directed and is called NEDI. They used the duality between the low-resolution and high-resolution covariance for super-resolution. The covariance between neighboring pixels in a local window around the low-resolution source is used to estimate the covariance between neighboring pixels in the high-resolution target. An example covariance problem is represented in Fig. 7.16.

The covariance of the b_0 relation in Fig. 7.16 is estimated by the covariance of neighboring a_0 relations in the local window. The open circles in the figure represent the low-resolution pixels and the closed circle represents a high-resolution pixel to be estimated. The covariance used is that between pixels and their four diagonal neighbors. The covariance between low-resolution pixels and their four diagonals in a $m \times m$ local window is calculated. This covariance determines the optimal way

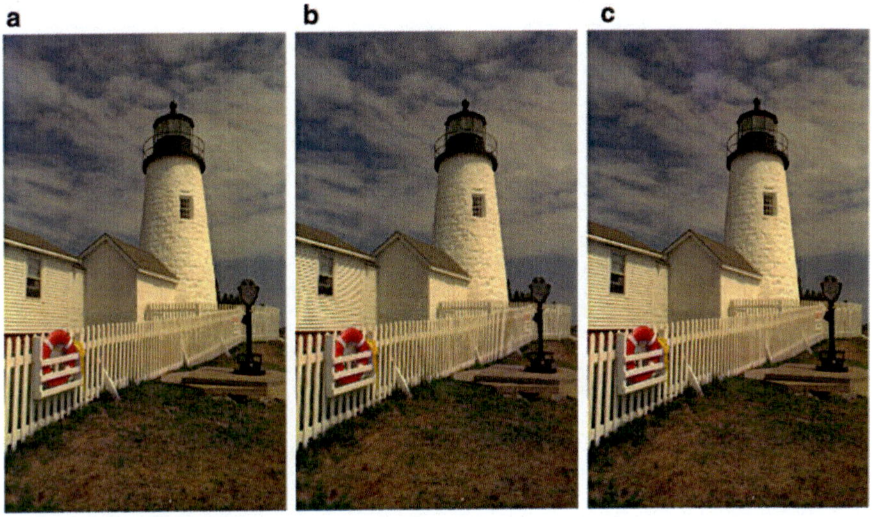

Fig. 7.15 (**a**) Original Lighthouse image and results obtained by using the method of: (**b**) Carey et al. [9] and (**c**) Muresan et al. [8]

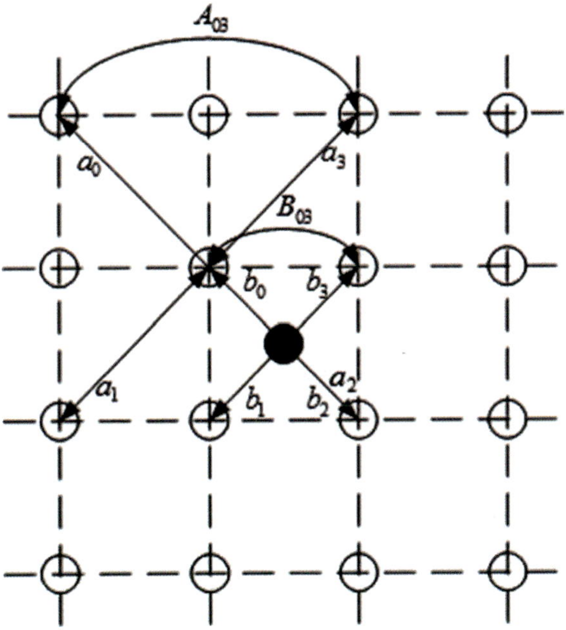

Fig. 7.16 Local covariance

of blending the four diagonals into the center pixel. This optimal value in the low-resolution window is used to blend a new pixel in the super-resolution image. By using the local covariance, the interpolation can adhere to arbitrarily oriented edges to reduce edge blurring and blocking.

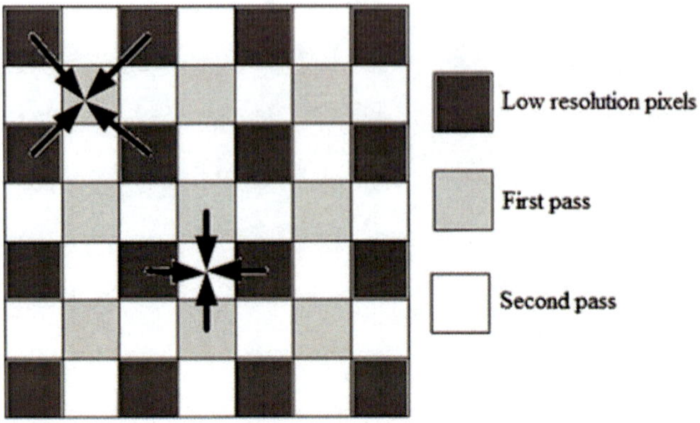

Fig. 7.17 Two steps in NEDI

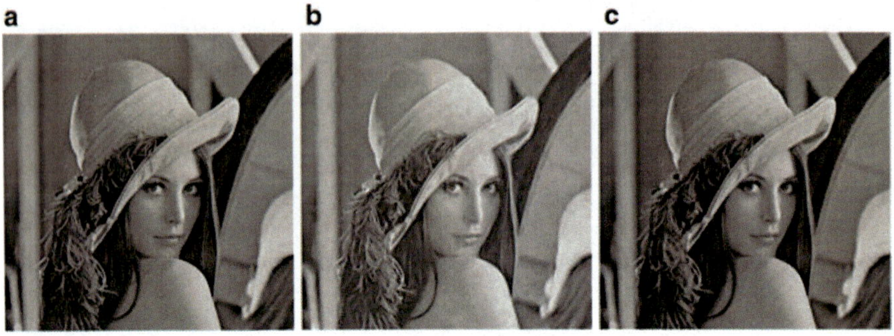

Fig. 7.18 (**a**) Original Lena image and results obtained by using the method of: (**b**) Muresan et al. [8] and (**c**) Li et al. [4]

Let A be a vector of size four containing the diagonal neighbors of the target pixel o, X a vector of size m^2 containing the pixels in the $m \times m$ window and C a $4 \times m^2$ matrix containing the diagonal neighbors of the pixels in X. The equation for enlargement using this method is shown in (7.12):

$$o = \left(C^{\mathrm{T}} \cdot C\right)^{-1} \left(C^{\mathrm{T}} \cdot X\right) \cdot A^{\mathrm{T}}. \tag{7.12}$$

The NEDI algorithm uses two passes to determine all high-resolution pixels. The first pass uses the diagonal neighbors to interpolate the high-resolution pixels with both coordinates odd and the second pass uses the horizontal and vertical neighbors to interpolate the rest of the high-resolution pixels as shown in Fig. 7.17.

Figures 7.18 and 7.19 show two examples of image interpolation using the NEDI algorithm for the Lena and lighthouse images, respectively. Visually comparing the results, the quality of the reconstructed images using the method of Lee et al. is better compared to the results obtained by the other interpolation techniques. This confirms the hypothesis of Li et al. that a better interpolation can be obtained by

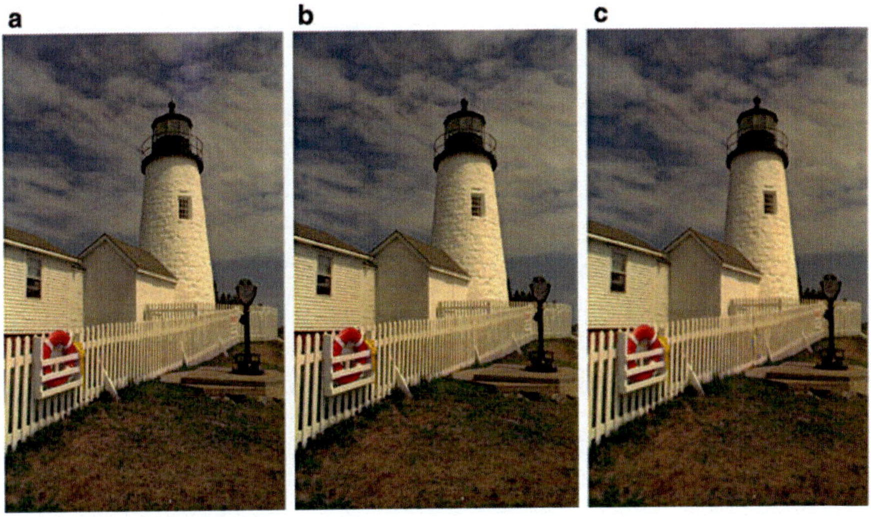

Fig. 7.19 (**a**) Original Lighthouse image and results obtained by using the method of: (**b**) Muresan et al. [8] and (**c**) Li et al. [4]

integrating information from edges pixels and their neighboring into the super-resolution process. More subjective and objective comparisons will be made in the evaluation section.

Another super-resolution approach has been proposed by Irani et al. in [10]. The approach by Irani et al. is based on generating a set of simulated low-resolution images. The image differences between this set of images and the actual observed low-resolution images are back projected, using a back-projecting kernel, onto an initial estimate of the high-resolution image. Figure 7.20, adapted from Irani and Peleg [10], illustrates the super-resolution process.

The generation of each observed image is the result of simulating an imaging process, which is the process where the observed low-resolution images are obtained from the high-resolution image. The imaging process can be modeled by the following equation:

$$g_k(m,n) = \alpha_k\left(h\left(T_k\left(I'(i,j)\right)\right)\right) + \eta_k(i,j), \tag{7.13}$$

where

- g_k is the kth observed image.
- I' is the high-resolution image that the algorithm is trying to find.
- T_k is the 2D transformation that maps I' to g_k.
- h is a blurring function that is dependent on the point spread function (PSF) of the sensor.
- η_k is an additive noise term.
- α_k is a down-sampling operator.

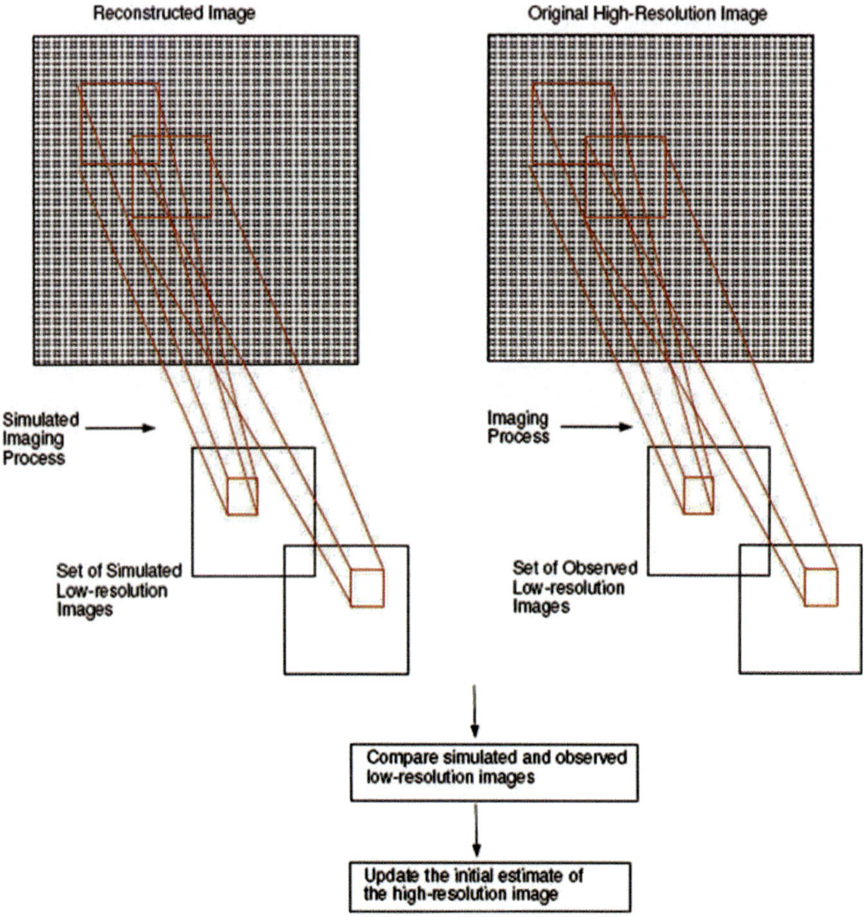

Fig. 7.20 The super-resolution process proposed by Irani and Peleg. The initial estimate of the high-resolution image is iteratively updated so that the simulated low-resolution images are as close as possible to the observed low-resolution images

The initial stages of the super-resolution algorithm involve creating an initial estimate $f^{(0)}$ of the high-resolution image, and then simulating a set of low-resolution images. This set of low-resolution images $\{g_k^{(0)}\}_{k=1}^{K}$ correspond to the set of observed images $\{g_k\}_{k=1}^{K}$. The process that yields these simulated low-resolution images can be expressed by the following equation:

$$g_k^{(n)} = \left(T_k\left(I^{(n)}\right) * h\right) \downarrow s, \tag{7.14}$$

where

- $\downarrow$ is a down-sampling operation according to a scale factor s.
- n is the nth iteration.
- $*$ is the convolution operator.

The differences between each simulated image and its corresponding observed image are now used to update the initial estimate image. If the initial estimate image $I^{(0)}$ is the correct high-resolution image, then the set of simulated low-resolution images $\{g_k^{(0)}\}_{k=1}^K$ should be identical to the set of observed low-resolution images $\{g_k\}_{k=1}^K$. Therefore, these image differences $\{g_k - g_k^{(0)}\}_{k=1}^K$ can be used to improve the initial guess image $I^{(0)}$ in order to obtain a high-resolution image $I^{(1)}$. Each value in the difference images is back projected onto its receptive field in the initial guess image $I^{(0)}$.

The above process is repeated iteratively in order to minimize the following error function:

$$e^{(n)} = \sqrt{\frac{1}{K}\sum_{k=1}^{K}\left\|g_k - g_k^{(n)}\right\|_2^2}. \tag{7.15}$$

The iterative update scheme for the super-resolution process can now be expressed as follows:

$$I^{(n+1)} = I^{(n)} + \frac{1}{K}\sum_{k=1}^{K} T_k^{-1}\left(\left(g_k - g_k^{(n)}\right)\uparrow s\right) * p, \tag{7.16}$$

where

- K is the number of low-resolution images.
- $uparrow$ is an up-sampling operation according to a scale factor s.
- p is the back-projection kernel used to deblur the image.
- $*$ is the convolution operator.

Figures 7.21 and 7.22 show some examples of image interpolation using the method of Irani et al. If we inspect both images Fig. 7.22b, c carefully, the details on the lighthouse are sharper and slightly clearer in the image obtained by the method of Irani et al.

The above-listed super-resolution methods have been designed to increase the resolution of a single channel (monochromatic) image. To date, there is very little work addressing the problem of color super-resolution. The typical solution involves applying monochromatic super-resolution algorithms to each of the color channels independently [11, 12], while using the color information to improve the accuracy. Another approach is transforming the problem to a different color space, where chrominance layers are separated from luminance, and super-resolution is applied only to the luminance channel [10]. Both of these methods are suboptimal as they do not fully exploit the correlation across the color bands.

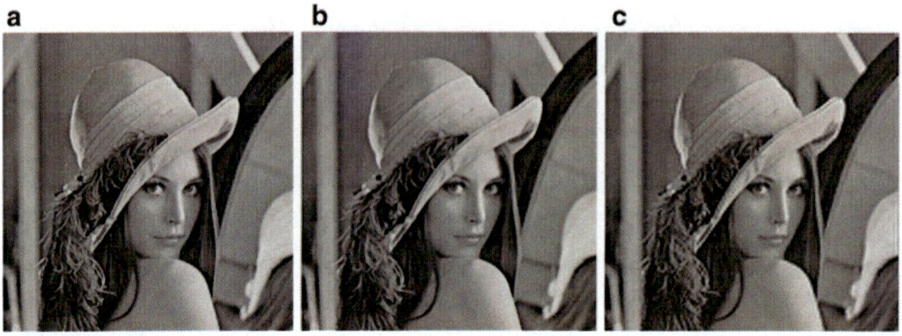

Fig. 7.21 (**a**) Original Lena image and results obtained by using the method of: (**b**) Li et al. [4] and (**c**) Irani et al. [10]

Fig. 7.22 (**a**) Original Lighthouse image and results obtained by using the method of: (**b**) Li et al. [4] and (**c**) Irani et al. [10]

To cop this problem, Maalouf et al. proposed in [13] a super-resolution method that is defined on the geometry of multispectral images. In this method, image geometry is obtained via the grouplet transform [14], then, the captured geometry is used to orient the interpolation process. Figure 7.23 shows the geometric flow for the lenna image computed by using the association field of the grouplet transform.

In order to define the contours of color images, Maalouf et al. used the model proposed by Di Zenzo in [15]. In fact, the extension of the differential-based operations to color or multi-valued images is hindered by the multi-channel nature of color images. The derivatives in different channels can point in opposite directions; hence, cancellation might occur by simple addition. The solution to this problem is given by the structure tensor for which opposing vectors reinforce each other.

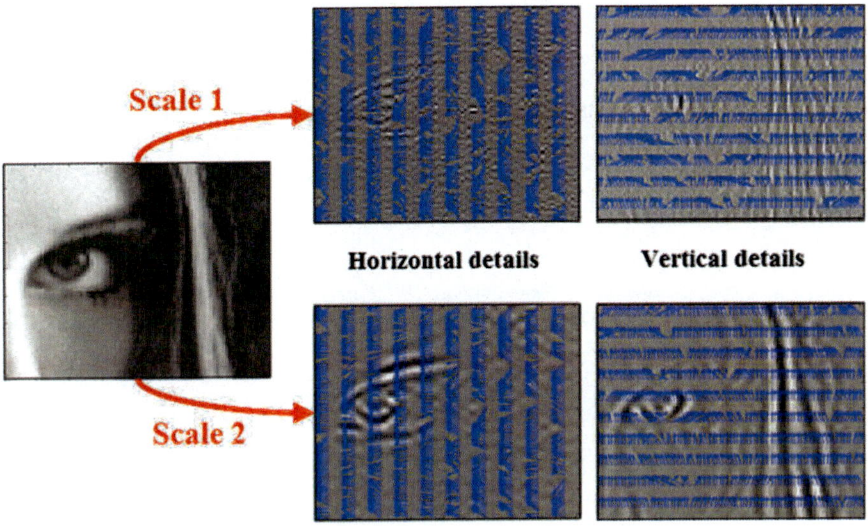

Fig. 7.23 Geometric flow of Lena image on different grouplet scales

In [15], Di Zenzo pointed out that the correct method to combine the first-order derivative structure is by using a local tensor. Analysis of the shape of the tensor leads to an orientation and a gradient norm estimate. For a multichannel image $I = \left(I^1, I^2, \ldots\ldots, I^n\right)^{\mathrm{T}}$, the structure tensor is given by

$$M = \begin{pmatrix} I_x^{\mathrm{T}} I_x & I_x^{\mathrm{T}} I_y \\ I_y^{\mathrm{T}} I_x & I_y^{\mathrm{T}} I_y \end{pmatrix}. \tag{7.17}$$

The multichannel structure tensor describes the 2D first-order differential structure at a certain point in the image.

The motivation of the method proposed in [13] is to make the interpolation oriented by the optimal geometry direction captured by the grouplet transform in order to synthesize fine structures for the super-resolution image. For that purpose, a multiscale multistructure grouplet-oriented tensor for an m-valued ($m = 3$ for color images and $m = 1$ for gray images) image is defined by:

$$G^q = \begin{bmatrix} \sum\limits_{r=1}^{m} \left(\frac{\partial}{\partial x}\tilde{h}_r^q \cos\theta_r\right)^2 & \sum\limits_{r=1}^{m} \frac{\partial}{\partial x}\tilde{h}_r^q \cos\theta_r \frac{\partial}{\partial y}\tilde{h}_r^q \sin\theta_r \\ \sum\limits_{r=1}^{m} \frac{\partial}{\partial x}\tilde{h}_r^q \cos\theta_r \frac{\partial}{\partial y}\tilde{h}_r^q \sin\theta_r & \sum\limits_{r=1}^{m} \left(\frac{\partial}{\partial y}\tilde{h}_r^q \sin\theta_r\right)^2 \end{bmatrix} \tag{7.18}$$

for $r = 1, 2, \ldots\ldots m$,

where q is the grouplet scale.

The norm of G^q is defined in terms of its eigenvalues λ_+ and λ_-, $||G^q|| = \sqrt{\lambda_+ + \lambda_-}$. The angle θ_r represents the angle of the direction of the grouplet association field. q is the scale of the grouplet transform. g_r^q is the corresponding grouplet coefficient. r designates the image channel ($r = 1, 2, \ldots, m$).

After characterizing edges and the geometrical flow (the association field) of the image, Maalouf et al. presented a variational super-resolution approach that is oriented by these two geometric features.

Their variational interpolation approach is formulated as follows,

$$\begin{aligned} I_r' = \min_{I_r} \big(\textstyle\int_\Omega \big(\|\tilde{\nabla} I_r(x,y)\| + \|\nabla G^q(x,y)\| \\ + \lambda \|G^q(x,y)\| \big) \mathrm{d}\Omega \big), \end{aligned} \tag{7.19}$$

subjected to the following constraints,

$$I'(xs\Delta, ys\Delta) = I(x,y) \quad \begin{matrix} 0 \le x \le \left\lfloor \frac{w}{s\Delta} \right\rfloor \\ 0 \le y \le \left\lfloor \frac{h}{s\Delta} \right\rfloor \end{matrix}, \tag{7.20}$$

where $I(x,y)$ is the original image before interpolation, Δ is the grid size of the upsampled image, w and h are the width and the height of the image, respectively, and s is the scaling factor.

$\tilde{\nabla}$ is the directional gradient with respect to the grouplet geometric direction θ and λ is a constant.

The first term in (7.19) is a marginal regularization term oriented by the directions of the geometrical flow defined by the association fields of the grouplet transform. The second is a multispectral regularization term while the third is edge driven, which aims at the orientation of the interpolation process to the orientation of the color edges. In fact, the norm $\|G^q(x,y)\|$ is a weighting factor such that more influence is given to points where the color gradient is high in the interpolation process.

The Euler equation of (7.19) is

$$\tilde{\nabla} \cdot \left(\frac{\nabla I_r(x,y)}{\|\nabla I_r(x,y)\|} \right) + \tilde{\nabla} \cdot \left(\frac{\nabla \|G^q(x,y)\|}{\|\nabla I_r(x,y)\|} \right) + \lambda \frac{\tilde{\nabla} \|G^q(x,y)\|}{\|\nabla I_r(x,y)\|} = 0. \tag{7.21}$$

By expanding (7.21), we obtain after simplification,

$$\begin{aligned} & I_{r_{xx}} \cos\theta + I_{r_{yy}} \sin\theta - \left(I_{r_{xy}} + I_{r_{yx}} \right) \cos\theta \sin\theta \\ & + \|G^q(x,y)\|_{xx} \cos\theta + \|G^q(x,y)\|_{yy} \sin\theta \\ & + \lambda \|G^q(x,y)\|_x \cos\theta + \lambda \|G^q(x,y)\|_y \sin\theta = 0, \end{aligned} \tag{7.22}$$

where $\|G^q(x,y)\|_x$ and $\|G^q(x,y)\|_y$ are, respectively, the horizontal and vertical derivatives of the norm matrix $\|G^q(x,y)\|$ computed at a scale q to extract the horizontal and vertical details of the color image.

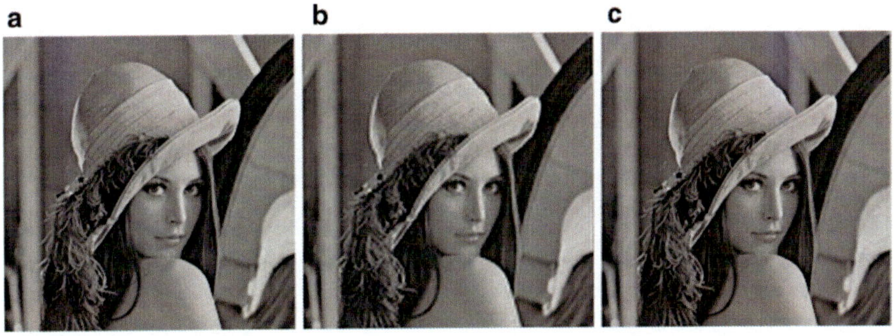

Fig. 7.24 (**a**) Original Lena image and results obtained by using the method of: (**b**) Irani et al. [10] and (**c**) Maalouf et al.

Fig. 7.25 (**a**) Original Lighthouse image and results obtained by using the method of: (**b**) Irani et al. [10] and (**c**) Maalouf et al.

Equation (7.22), which yields to a factor-of-two interpolation scheme, is applied to each color band r.

Figures 7.24 and 7.25 show some results of applying the method of Maalouf et al. for generating super-resolution images. From these two figures, we can see that the method of Maalouf et al. better conserves textures and edges in the reconstructed image especially for the color image. This confirms the hypothesis of Maalouf et al. that considering the multispectral geometry of color images in the interpolation process can improve the visual quality of the interpolated images.

7.3 Reconstructed-Based Methods

Reconstruction-based algorithms compute high-resolution images by simulating the image-formation process and usually first form a linear system,

$$I = PI' + E, \tag{7.23}$$

where I is the column vector of the irradiance of all low-resolution pixels considered, I' is the vector of the irradiance of the high-resolution image, P gives the weights of the high-resolution pixels in order to obtain the irradiance of the corresponding low-resolution pixels, and E is the noise. To solve (7.23), various methods, such as maximum a posteriori (MAP) [16, 17], regularized maximum likelihood (ML) [17], projection onto convex sets (POCS) [18], and iterative back projection [10], have been proposed to solve for the high-resolution image.

In the MAP approach, super-resolution is posed as finding the maximum a posteriori super-resolution image I': i.e., estimating $\underset{I'}{\arg\max} \Pr\left[I'/I_k\right]$. Bayes law for this estimation problem is

$$\Pr\left[I'/I_k\right] = \frac{\Pr\left[I_k/I'\right] \cdot \Pr\left[I'\right]}{\Pr\left[I_k\right]}, \tag{7.24}$$

where I is the low-resolution image.

Since $\Pr[I_k]$ is a constant because the images I_k are inputs (and so are known), and since the logarithmic function is a monotonically increasing function, we have:

$$\underset{I'}{\arg\max} \Pr\left[I'/I_k\right] = \underset{I'}{\arg\min}\left(-\ln \Pr\left[I_k/I'\right] - \ln \Pr\left[I'\right]\right). \tag{7.25}$$

The first term in this equation $-\ln \Pr\left[I_k/I'\right]$ is the negative log probability of reconstructing the low-resolution images I_k, given that the super-resolution image is I'. It is therefore set to be a quadratic (i.e., energy) function of the error in the reconstruction constraints:

$$-\ln \Pr\left[I_k/I'\right] = \frac{1}{2\sigma^2} \sum_{m,k} \left[I_k(m) - \sum_p I'(p) \,.\, \int_p PSF_k\left(r_k(z) - m\right) . \left|\frac{\partial r_k}{\partial z}\right| \mathrm{d}z \right]^2, \tag{7.26}$$

where $\left|\frac{\partial r_k}{\partial z}\right|$ is the determinant of the Jacobian of the registration transformation $r_k(.)$ that is used to align the low-resolution images I_k. In using (7.26), it is implicitly assumed that the noise is independently and identically distributed (across both the images I_k and the pixels m) and is Gaussian with covariance σ^2. PSF is the PSF. Minimizing the expression in (7.26) is then equivalent to finding the (unweighed) least-squares solution of the reconstruction constraints.

In the ML approach, the total probability of the observed image I_k, given an estimate of super-resolution image I' is:

$$\Pr\left(I_k/I'\right) = \prod_{\forall x,y} \frac{1}{\sigma\sqrt{2\pi}} \mathrm{e}^{\frac{-(g_k(x,y)-I_k(x,y))^2}{2\sigma^2}} \tag{7.27}$$

and the associated log-likehood function is

$$L(I_k) = -\sum_{\forall x,y} \left(g_k(x,y) - I_k(x,y)\right)^2, \tag{7.28}$$

where g_k is an image model defined by (7.14).

To find the maximum likehood estimate, s_{ML}, we need to maximize $L(I_k)$ over all images:

$$s_{\mathrm{ML}} = \arg\max_{s} \sum_{k} L(I_k). \tag{7.29}$$

By adding some noise E to (7.14), it can be written as

$$g_k^{(n)} = \left(T_k\left(I^{(n)}\right) * h\right) \downarrow s + E. \tag{7.30}$$

If we combine h, T, and s in a matrix M_k, (7.30) becomes

$$g_k^{(n)} = M_k s + E. \tag{7.31}$$

Equation (7.31) can be rewritten by using matrix notation as

$$\begin{bmatrix} g_0 \\ g_1 \\ \cdot \\ \cdot \\ \cdot \\ g_{K-1} \end{bmatrix} = \begin{bmatrix} M_0 \\ M_1 \\ \cdot \\ \cdot \\ \cdot \\ M_{K-1} \end{bmatrix} + \begin{bmatrix} E_0 \\ E_1 \\ \cdot \\ \cdot \\ \cdot \\ E_{K-1} \end{bmatrix}, \tag{7.32}$$

$$g = Ms + E. \tag{7.33}$$

By using (7.33), we have

$$\sum_{k} L(g_k) = -\sum_{k} \|M_k s - g_k\|^2 = -\|Ms - g\|^2. \tag{7.34}$$

The maximization above, becomes equivalent to:

$$s_{\mathrm{ML}} = \arg\min_{s} \|Ms - g\|^2, \tag{7.35}$$

which is none other than a standard minimization problem:

$$s_{\mathrm{ML}} = \left(M^{\mathrm{T}}M\right)^{-1}M^{\mathrm{T}}m. \tag{7.36}$$

In the POCS super-resolution reconstruction approach, the unknown signal I' is assumed to be an element of an appropriate Hilbert space. Each a priori information or constraint restricts the solution to a closed convex set in H. Thus, for K pieces of information, there are K corresponding closed convex sets $C_k \in H,\ k = 1,2,\ldots\ldots,K$ and $I' \in C_0 = \bigcap_{k=1}^{K} C_k$, provided that the intersection C_0 is nonempty. Given the constraint sets C, and their respective projection operators P_k, the sequence generated is therefore given by:

$$I'_{k+1} = P_K P_{K-1} \ldots\ldots. P_1 I_k. \tag{7.37}$$

Let $T_k = (1-\lambda_k)I + \lambda_k P_k,\ 0 < \lambda_k < 2$ be the relaxed projection operator that converges weakly to a feasible solution in the intersection C_0 of the constraint sets. Indeed, any solution in the intersection set is consistent with the a priori constraints, and therefore, it is a feasible solution. Note that the T_k's reduce to P_k's for unity relaxation parameters, i.e., $\lambda_k = 1$. The initialization I'_0 can be arbitrarily chosen from H. Translating the above description to an analytical model, we get:

$$I'_{k+1} = P\left[I'_k + \sum_{k=1}^{K} \lambda_k w_k P\left(g_k - H_k I_k\right)\right]. \tag{7.38}$$

Where P represents band-limited operator of image, g_k represents the ith measured low-resolution image, λ_k represents the relaxed operator, H represents a blurring operator determined by the PSF, downsampling and transformation of the ith measured low-resolution image, w_k represents the weights.

Figures 7.26 and 7.27 show the interpolation results obtained by the reconstruction-based super-resolution methods proposed by Hardie et al. [16], Elad et al. [17], and Patti et al. [18].

7.4 Learning-Based Methods

In learning-based super-resolution methods, a training set, which is composed of a number of high-resolution images, is used to later predict the details of lower resolution images.

In [19], Super-resolution is defined as a process whose goal is to increase the resolution and at the same time adding appropriate high-frequency information. Therefore, [19] employs a large database of image pairs, which stores a rectangular patch of high-frequency component, a high-resolution image, and corresponding smoothed and downsampled low-level counterpart. The relationship between middle

Fig. 7.26 (**a**) Original Lena image and the results obtained by using the method of: (**b**) Hardie et al. [16], (**c**) Elad et al. [17], and (**d**) Patti et al. [18]

and high frequency of natural images are captured and used to super resolve low-resolution static images and movies. Although a zoom factor of 4 is established for static images, direct application of the approach is not successful in video.

In [20], spatio-temporal consistencies and imageformation/degradation processes are employed to estimate or hallucinate high-resolution video. Since learning-based approaches are more powerful when their application is limited to a specific domain, a database of facial expressions is obtained using a sequence (video) of high-resolution images. Their low-resolution counterpart is acquired using a local smoothing and downsampling process. Images are divided into patches, and spatial (via single image) and temporal (via time) consistencies are established among these image patches. After the training database is constructed, they try to find Maximum A Priori high-resolution image and illumination offset by finding out a template image. Template image is constructed from high-resolution patches in the database, maximizing some constraints.

Fig. 7.27 (**a**) Original Lighthouse image and the results obtained by using the method of: (**b**) Hardie et al. [16], (**c**) Elad et al. [17], and (**d**) Patti et al. [18]

In general, the learning-based methods can be summarized as follows:

The estimation process is based on finding a unique template from which high-resolution image is extracted. In the work presented in [19, 20], high-resolution image and an intensity offset were two unknowns of the problem. The super-resolution problem maps to finding MAP high-resolution image I':

$$I'_{\mathrm{MAP}} = \arg\max_{I'} \log P\left(I'/I\right). \tag{7.39}$$

Above equation is marginalized over unknown template image $Temp$, which is composed of image patches in the training database. Therefore,

$$P\left(I'/I\right) = \sum_{Temp} P\left(I', Temp/I\right). \tag{7.40}$$

If chain rule is applied to the above probabilistic formula,

$$P\left(I'/I\right) = \sum_{Temp} P\left(I'/Temp, I\right) P\left(Temp/I\right) \tag{7.41}$$

is obtained. Bayes rules is used to obtain,

$$P\left(I'/I\right) = \sum_{Temp} \left\{ \frac{P\left(I/I', Temp\right) P\left(I'/Temp\right)}{P\left(I/I'\right)} P\left(Temp/I\right) \right\}. \tag{7.42}$$

Since Markov Random Field model is used to relate different nodes, and there is no relation between two nodes which are not linked via MRF, no direct conditioning exists among such nodes. As a result $P\left(I/Temp, I\right) = P\left(I/I'\right)$,

$$P\left(I'/I\right) = \sum_{Temp} \left\{ P\left(I'/I, Temp\right) P\left(I'/I\right) P\left(Temp/I\right) \right\}. \tag{7.43}$$

Since $P\left(Temp/I\right)$ has its maximum value around true high-resolution solution, a unique (peak) template *Temp* is computed. to maximize the posterior using low-resolution images and database entries. If posterior of $P\left(Temp/I\right)$ is approximated to highly concentrate around $Temp^* = Temp^*\left(I\right)$, original posterior which is tried to be maximized, $P\left(I'/I\right)$ becomes,

$$P\left(I'/I\right) = P\left(I/I'\right) P\left(I'/Temp^*\right). \tag{7.44}$$

Therefore, MAP high-resolution image will be computed by,

$$I'_{\text{MAP}} = \arg\max \left\{ \log P\left(I/I'\right) \log P\left(I'/Temp^*\right) \right\}. \tag{7.45}$$

In the above equation, it is described that in order to maximize $P\left(Temp/I\right)$, a unique template should be constructed from patches in database. Since nodes in I are conditionally dependent, Bayes rule can be applied to in maximization of $P\left(Temp/I\right)$,

$$P\left(Temp/I\right) \propto P\left(I/Temp\right) P\left(Temp\right) = \prod_{p=1}^{N} P\left(I_p/Temp_p\right) P\left(Temp\right). \tag{7.46}$$

The peak template will be computed according to above formulation. By maximizing the first term of right hand side of the equation, the difference between low-resolution observation and downsampled unknown template is minimized. Second term of right hand side provides a consistent template, where spatial consistency in the data is established by means of MRF modeling.

7.5 Evaluation Using Objective and Subjective Quality Metrics

In this section, we propose to evaluate subjectively (panel of observers) and objectively (metrics) the studied algorithms in order to classify their results in terms of visual quality

7.5.1 *Subjective Evaluation*

Subjective experiments consist in asking a panel of subjects to watch a set of images or video sequences and to score their quality. The main output of these tests is the mean opinion scores (MOS) computed using the values assigned by the observer. In order to obtain meaningful and useful values of MOS, the experiment needs to be constructed carefully by selecting rigorously the test material and defining scrupulously the subjective evaluation procedure. The most important recommendations have been published by ITU [21, 22] or described in VQEG test plans [23].

7.5.1.1 Test Material

For this subjective evaluation, we have selected five images partly coming from a state-of-the-art database such as Lena (512×512), Lighthouse (512×768), Iris (512×512), Caster (512×512), and Haifa (512×512). These images have been chosen because of their content and their availability for comparisons. Additionally, Lena has been also used in its grayscale version in order to perform the subjective evaluation on its structural content rather than color. Figure 7.28 gives an overview of the used test material.

All images described above have been downsampled by a ratio of 2 in width and height. Then it has been provided as an input of the super-resolution algorithms. This process allows to compare algorithms with regards to the original image.

7.5.1.2 Environment Setup

The subjective experiments took place in a normalized test room built with respect to ITU standards [21] (cf. Fig. 7.29). It is very important to control accurately the environment setup in order to ensure the repeatability of the experiments and to be able to compare results between different test locations.

Only one observer per display has been admitted during the test session. This observer is seated at a distance between 2H and 4H; H being the height of the

Fig. 7.28 Overview of the test material. (**a**) Lena, (**b**) Lena (grayscale), (**c**) Lighthouse, (**d**) Iris, (**e**) Caster and (**f**) Haifa

Fig. 7.29 A Synthetized view of the used test room

displayed image. His vision is checked for acuity and color blindness. Table 7.1 provides the most important features of the used display. The ambient lighting of the test room has been chosen with a color temperature of 6,500 K.

Table 7.1 Display characteristics for the subjective evaluation

Type	Dell 3008WFP
Diagonal size	30 in.
Resolution	2,560 × 1,600 (native)
Calibration tool	EyeOne Display 2
Gamut	sRGB
White point	D65
Brightness	370 cd/m^2
Black level	lowest

Fig. 7.30 Discrete quality scale used to score images during the test

7.5.1.3 Subjective Evaluation Procedure

In order to compare super-resolution algorithms from the point of view of subjective evaluation, we used a single stimulus approach. This means that processed images are scored without any comparison with the original image (reference image). This latter is used as a result image and is scored in order to study the reliability and accuracy of the observer results.

The test session starts by a training allowing to show to the observer the types of degradation and the way to score impaired images. The results for these items are not registered by the evaluation software but the subject is not told about this. Then, each image is displayed for 10 s, three times, to the observer to stabilize his judgment. At the end of each presentation, a neutral gray is displayed with a GUI containing a discrete scale as shown in Fig. 7.30. This scale corresponding to a quality range from bad to excellent ([0 - 5]) allows to affect a score to each image. Of course, numbers shown on Fig. 7.30 are given here for illustration and do not exist on the GUI.

Each subjective experiment is composed of 198 stimulus: 6 images × 11 (10 algorithms + reference image) × 3 repetitions in addition to 5 stabilizing images (training). A panel of 15 observers has participated to the test. Most of them were naive subjects. The presentation order for each observer is randomized. To better explain the aim of the experiment and the scoring scale, we give the following description to the observers: Imagine you receive an image as an attachment of an email. The resolution of this latter does not fit with the display and you want to see it in full screen. A given algorithm performs the interpolation and you have to score the result as: Excellent: the image content does not present any noticeable artifact; Good: The global quality of the image is good even if a given artifact is noticeable; Fair: several noticeable artifacts are noticeable all over the image; Poor: many noticeable artifacts and strong artifacts corrupt the visual quality of the image; Bad: strong artifacts are detected and the image is unusable.

7.5.1.4 Scores Processing

The raw subjective scores have been processed in order to obtain the final MOS presented in the results section.

The MOS $\bar{u}_{jkr}$ is computed for each presentation:

$$\bar{u}_{jkr} = \frac{1}{N} \sum_{i=1}^{N} u_{ijkr}, \tag{7.47}$$

where u_{ijkr} is the score of the observer i for the impairment j of the image k and the rth iteration. N represents the number of observers. In a similar way, we can calculate the global average scores, $\bar{u}_j$ and $\bar{u}_k$, respectively, for each test condition (algorithm) and each test image.

7.5.2 *Objective Evaluation*

Objective quality measurement is an alternative of a tedious and time-consuming subjective assessment. In literature, there is plenty of metrics (Full reference, Reduced reference and no reference) that models or not the human visual system (HVS). Most of them are not very popular due to their complexity, difficult calibration, or lack of freely available implementation. This is why metrics like PSNR and Structural SIMilarity (SSIM) [24] are widely used to compare algorithms.

PSNR is the most commonly used metrics and its calculation is based on the mean squared error (MSE).

$$PSNR(x,y) = 20log_{10}\frac{255}{MSE(x,y)}. \tag{7.48}$$

SSIM works under the assumption that human visual perception is highly adapted for extracting structural information from a scene. So, it directly evaluates the structural changes between two complex-structured signals.

$$SSIM(x,y) = l(\mu_x,\mu_y)^{\alpha} c(\sigma_x,\sigma_y)^{\beta} s(\sigma_x,\sigma_y)^{\gamma}. \tag{7.49}$$

7.5.3 Evaluation Results

Ten super-resolution algorithms coming for the state of the art have been evaluated objectively and subjectively: *A* for Allebach [7] , *C* for Chang [9], *E* for Elad [17], *H* for Hardie [16], *I* for Irani [10], *J* for Jensen [6], *L* for Li [4], *Ma* for Maalouf [13], *Mu* for Muresan [8], *P* for Patti [18].

Graphics a, b, c, d, and e of Fig. 7.31 show, for each image from the test material, the MOS values obtained after the processing applied to the subjective scores. It shows also the confidence interval associated with each algorithm. From the subjective scores, one can notice that the evaluated algorithms can be grouped in three classes: Low-quality algorithms (*E*, *H*, *I*, and *J*), Medium-quality algorithms (*A* and *C*), and High-quality algorithms (*Ma*, *Mu*, *P*). Only one algorithm, *L*, seems to be content dependent and provide results that can be put in medium and high-quality groups.

For the subjective experiments, we inserted the original images with the test material without giving this information to the observers. The scores obtained for these images are high, approximatively ranging within the 20% highest scores. However, The difference between images is relatively high. This is due to the acquisition condition of the image itself. Figure 7.32 gives the MOS values and the associated confidence interval for the original images. Obviously, Haifa and Lighthouse are around 5 and have a very small confidence interval because these images are relatively sharp and colourful. The worst was Lena because its background contains some acquisition artifacts that can be considered by the observers as generated by the super-resolution algorithm. One important thing that we can exploit from these results is to use them as an offset to calibrate the MOS of the test images.

The test material contains two versions of Lena, i.e., color and grayscale. This has been used to study the effect of the super-resolution algorithms on colors and on human judgment. Figure 7.33 shows MOS values and their confidence intervals for color and grayscale versions. First of all, the scores are relatively close and it is impossible to draw a conclusion about which is the best. Then, the confidence intervals are of the same size approximatively. Finally, these results leads to the conclusion that the used algorithms either allow to conserve the color information or do not deal with the color information in their conception. Hence, for the evaluation of super-resolution algorithms (those used here at least) one can use the Luminance information rather than the three components.

The PSNR and the SSIM have been used to evaluate the quality of the test material used for the subjective evaluation. Figures 7.34 and 7.35 show, respectively, the results for PSNR and SSIM. It is difficult to draw the same conclusion than the subjective assessment because the categories are not clearly present especially for PSNR. This confirms its lack of correlation with the human perception. However, from Fig. 7.34 the low-quality category (PSNR lower than 32dB) is confirmed for algorithms *E*, *H*, *I*, and *J*.

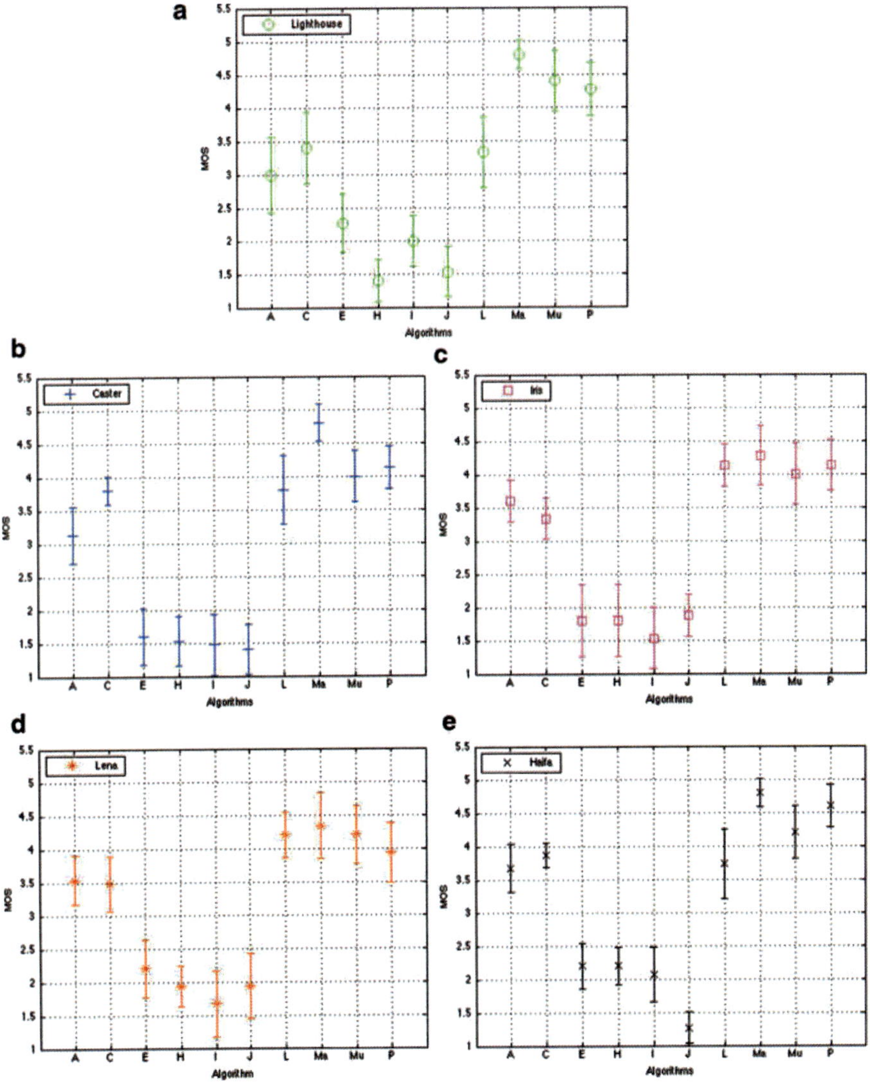

Fig. 7.31 Mean opinion scores (MOS) values and 95% confidence interval obtained for the different algorithms for: (**a**) Lighthouse, (**b**) Caster, (**c**) Iris, (**d**) Lena, and (**e**) Haifa

One can notice that the other algorithms perform better especially *Ma*. Results of Fig. 7.35 are more correlated to human perception than the PSNR because, on the one hand, we can retrieve the same group of high-quality algorithms with values very close to one. On the other hand, the medium quality group can be considered at values between 0.96 and 0.98. For low-quality algorithms, it is really difficult to have a clear range of scores.

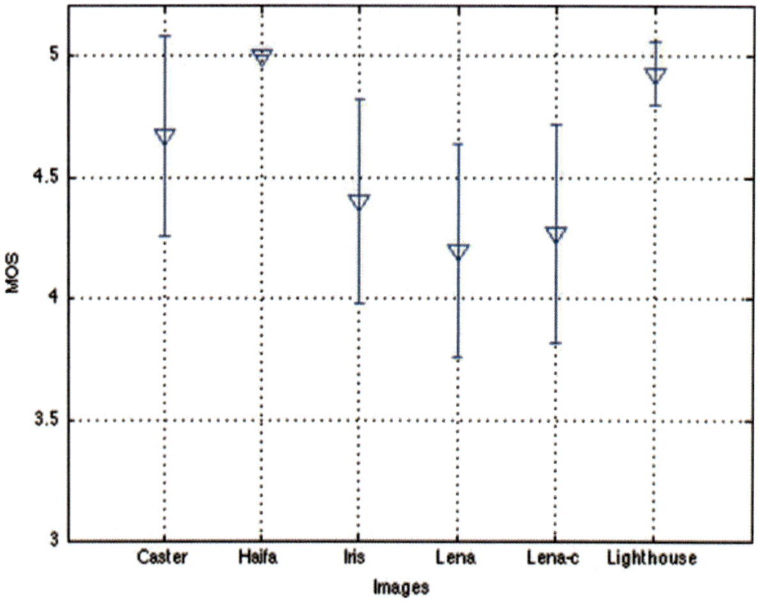

Fig. 7.32 Mean opinion scores (MOS) values and 95% confidence interval obtained for the original images

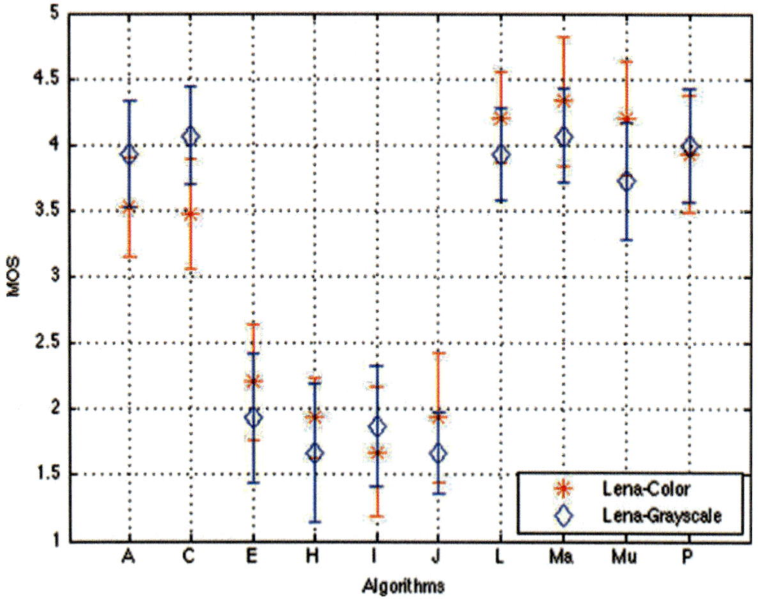

Fig. 7.33 Mean opinion scores (MOS) values and 95% confidence interval obtained for Lena in color and grayscale

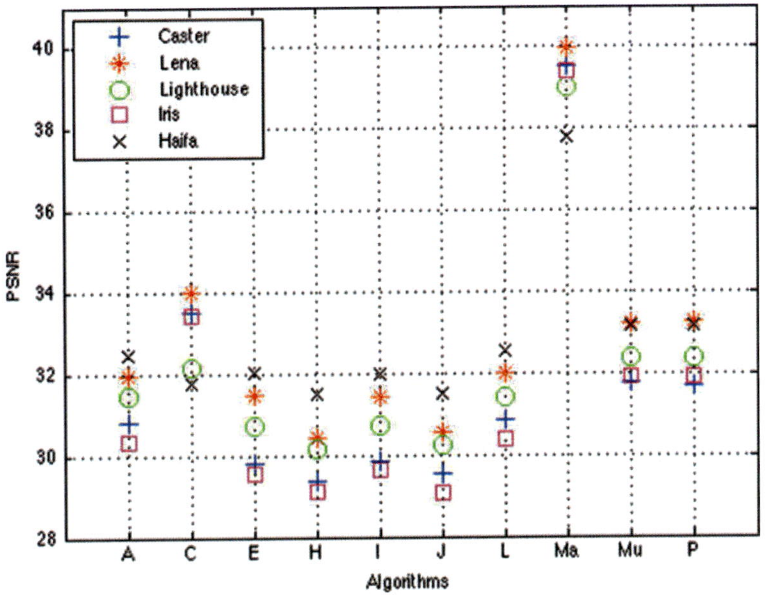

Fig. 7.34 PSNR results for the five images and the ten algorithms

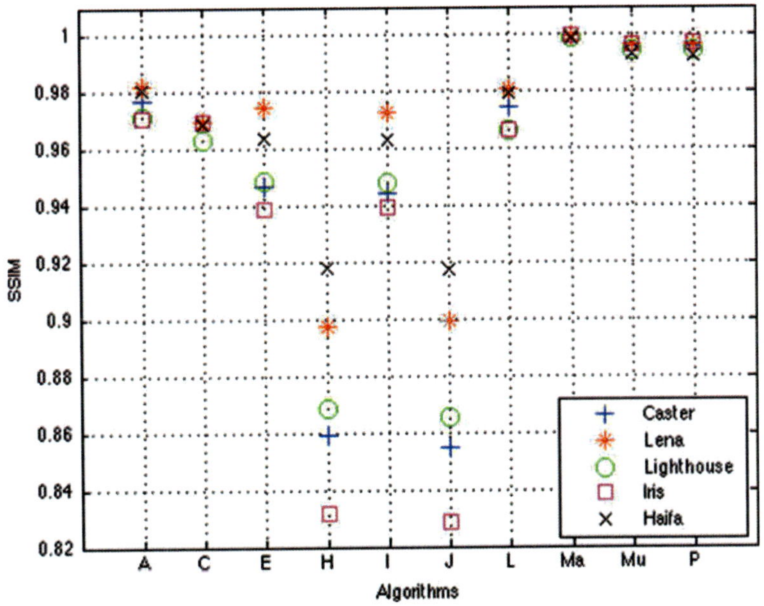

Fig. 7.35 SSIM results for the five images and the ten algorithms

Table 7.2 Pearson correlation coefficient between objective metrics (PSNR, SSIM) and subjective scores

Image	SSIM	PSNR
Caster	0,8224	0,7486
Haifa	0,8758	0,6542
Iris	0,7749	0,6232
Lena	0,7223	0,6465
Lighthouse	0,8963	0,7510
Global	0,7866	0,6745

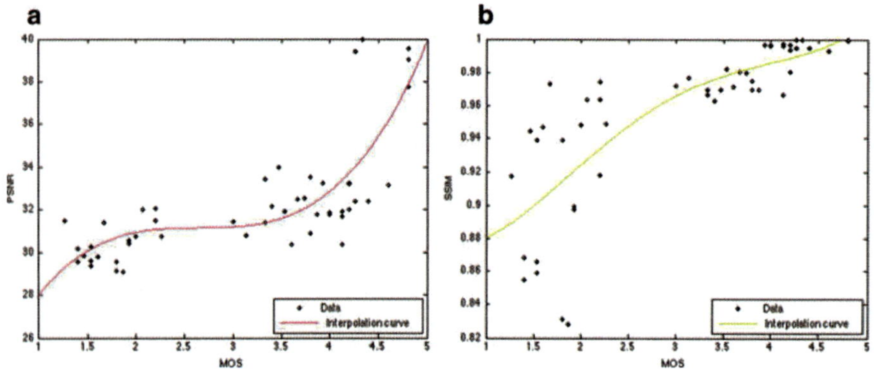

Fig. 7.36 Scatter plots between the MOS values collected for all images and the PSNR (**a**) and the SSIM (**b**)

In order to confirm the declaration about the more or less correlation of the PSNR and the SSIM with the human perception, we computed the Pearson correlation coefficient (PCC). Table 7.2 gives the PCC values first for each image and then for the global data. The PCC values show clearly that SSIM is more correlated than PSNR but the correlation is not very high.

Figure 7.36a, b give scatter plots for the correlation of the PSNR and the SSIM. It easy to notice that the correlation of the first is lower than the second and both are low with regards to human perception. This means that the use of these metrics to replace the human judgment for super-resolution algorithms evaluation is to a certain extent incorrect.

7.6 Summary

In this chapter, a state of the art of super-resolution techniques has been presented through three different conceptions: interpolation-based methods, reconstruction-based methods, and learning-based methods. This research field has seen much progress during the last two decades. The described algorithms have been evaluated subjectively by psychophysical experiments allowing to quantify the human judgment and objectively by using two common metrics: PSNR and SSIM. Finally, image super-resolution appears quite promising for new applications such as digital cinema where it can be used at different stages (acquisition, postproduction. . . projection). Another promising field but somehow related to the previous one is the exploitation of motion information to improve the quality of the results for image sequences.

References

1. Borman S, Stevenson RL (1998) Super-resolution from image sequences – A Review. Midwest Symp Circ Syst, 374–378
2. Park SC, Park MK, Kang MG (2003) Super-resolution image reconstruction: a technical overview. IEEE Signal Process Mag 20(3):21–36
3. Farsiu S, Robinson D, Elad M, Milanfar P (2004) Advances and challenges in super-resolution. Int J Imag Syst Tech 14(2):47–57
4. Li X, Orchard MT (2001) New edge-directed interpolation. IEEE Trans Image Process 10(10):1521–1527
5. Blu T Thevenaz P, Unser M (2000) Image interpolation and resampling. Handbook of medical imaging, processing and analysis. Academic, San Diego
6. Jensen K, Anastassiou D (1995) Subpixel edge localization and the interpolation of still images. IEEE Trans Image Process 4:285–295
7. Allebach J, Wong PW (1996) Edge-directed interpolation. Proc IEEE Int Conf Image Proc 3:707–710
8. Muresan DD, Parks TW (2000) Prediction of image detail. Proc IEEE Int Conf Image Proc, 323–326
9. Chang DB Carey WK, Hermami SS (1999) Regularity-preserving image interpolation. Proc IEEE Int Conf Image Proc, 1293–1297
10. Irani M, Peleg S (1991) Improving resolution by image registration. CVGIP: Graph Models Image Process 53:231–239
11. Shah NR, Zakhor A (1999) Resolution enhancement of color video sequence. IEEE Trans Image Process 6(8):879–885
12. Tom BC, Katsaggelos A (2001) Resolution enhancement of monochrome and color video using motion compensation. IEEE Trans Image Process 2(10):278–287
13. Maalouf A, Larabi MC (2009) Grouplet-based color image super-resolution . EUSIPCO2009, 17th European signal processing conference, Glasgow, Scotland
14. Mallat S (2009) Geometrical grouplets. Appl Comput Harmon Anal 26(2):161–180
15. DiZenzo S (1986) A note on the gradient of multi images. Comput Vis Graph Image Process 33(1):116–125
16. Hardie R, Barnard K, Amstrong E (1997) Joint map registration and high-resolution image estimation using a sequence of undersampled images. IEEE Trans Image Process 6 (12):1621–1633
17. Elad M, Feuer A (1997) Restoration of single super-resolution image from several blurred, noisy and down-sampled measured images. IEEE Trans Image Process 6(12):1646–1658
18. Patti AJ, Sezan MI, Tekalp AM (1997) Superresolution video reconstruction with arbitrary sampling lattices and nonzero aperture time. IEEE Trans Image Process 6(8):1064–1076
19. Bishop CM, Blake A, Marthi B (2003) Super-resolution enhancement of video. In: Bishop CM, Frey B (eds) Proceedings artificial intelligence and statistics. Society for Artificial Intelligence and Statistics, 2003

20. Dedeoglu G, Kanade T, August J (2004) High-zoom video hallucination by exploiting spatio-temporal regularities. In: Proceedings of the 2004 IEEE computer society conference on computer vision and pattern recognition (CVPR 04), June, 2004
21. ITU-T (2000) Recommendation ITU-R BT500-10. Methodology for the subjective assessment of the quality of the television pictures, March 2000
22. ITU-T (1999) Recommendation ITU-R P910. Subjective video quality assessment methods for multimedia applications, September 1999
23. VQEG, Video Quality recommendations, VQEG testplans, ftp://vqeg.its.bldrdoc.gov
24. Wang Z, Bovik AC, Sheikh HR, Simoncelli EP (2004) Image quality assessment: from error visibility to structural similarity. IEEE Trans Image Process 13(4):600–612

Chapter 8
Color Image Segmentation

Mihai Ivanovici, Noël Richard, and Dietrich Paulus

Colors, like features, follow the changes of the emotions

Pablo Picasso

Abstract Splitting an input image into connected sets of pixels is the purpose of image segmentation. The resulting sets, called regions, are defined based on visual properties extracted by local features. To reduce the gap between the computed segmentation and the one expected by the user, these properties tend to embed the perceived complexity of the regions and sometimes their spatial relationship as well. Therefore, we developed different segmentation approaches, sweeping from classical color texture to recent color fractal features, in order to express this visual complexity and show how it can be used to express homogeneity, distances, and similarity measures. We present several segmentation algorithms, like JSEG and color structure code (CSC), and provide examples for different parameter settings of features and algorithms. The now classical segmentation approaches, like pyramidal segmentation and watershed, are also presented and discussed, as well as the graph-based approaches. For the active contour approach, a diffusion model for color images is proposed. Before drawing the conclusions, we talk about segmentation performance evaluation, including the concepts of closed-

M. Ivanovici (✉)
MIV Imaging Venture Laboratory, Department of Electronics and Computers, Transilvania University Braşov, Brasov, România
e-mail: mihai.ivanovici@unitbv.ro

N. Richard
Laboratory XLIM-SIC, UMR CNRS 7252, University of Poitiers, France
e-mail: noel.richard@univ-poitiers.fr

D. Paulus
Computervisualistik, Universität Koblenz-Landau, D-56070 Koblenz, Germany
e-mail: paulus@uni-koblenz.de

C. Fernandez-Maloigne (ed.), *Advanced Color Image Processing and Analysis*,
DOI 10.1007/978-1-4419-6190-7_8,

loop segmentation, supervised segmentation and quality metrics, i.e., the criteria for assessing the quality of an image segmentation approach. An extensive list of references that covers most of the relevant related literature is provided.

Keywords Segmentation • Region • Neighborhood • Homogeneity • Distance • Similarity measure • Feature • Texture • Fractal • Pyramidal segmentation • CSC • Watershed • JSEG • Active contour • Graph-based approaches • Closed-loop segmentation • Supervised segmentation • Quality metric

8.1 Introduction

Image segmentation is, roughly speaking, the process of dividing an input image into *regions* according to the chosen *criteria*. Those regions are called *segments*, thus the name of the operation. Segmentation is often considered to be at the border between image processing and image analysis, having the ingrate role of preparing the content of an image for the subsequent "higher-level" specialized operations, e.g., object detection or recognition. Being an early stage of the analysis phase, "errors in the segmentation process almost certainly lead to inaccuracies in any subsequent analysis" [143]. It is thus worthwhile to produce an image segmentation that is as *accurate* as possible with respect to application requirements. In addition, ideally, it is desired that each resulting region or segment represents an object in the original image, in other words each segment is semantically meaningful, which greatly facilitates the image content analysis and interpretation. A learning phase or classification can follow in order to associate the segments to terms describing the content of the image, like annotations, or, in other words to map the *pixelic* content to the *semantic* image description. Another way would be to match the set of segmented regions to an a priori model represented by a semantic graph [93] in order to interpret the image.

From the point of view of the existing image segmentation frameworks, we can say segmentation is one of the most complex operations performed on images. An argument for this last statement, the finding of [102], p. 579, is quite pessimistic: "There is no theory of image segmentation. As a consequence, no single standard method of image segmentation has emerged. Rather, there are a collection of ad hoc methods that have received some degree of popularity". However, several general paths can be identified, paths that are usually followed by the authors of various segmentation approaches.

Historically, image segmentation was the center point of computer vision—the processing and the analysis procedures aimed at helping the robots to detect simple geometrical objects based on line and circle detection. In addition, we should mention the fact that the image spatial resolution and number of quantification levels were small, quite inferior to the human capabilities. The computers had also low computational power. Since the world of image became colored, the applications, and consequently the segmentation approaches, became more sophisticated. In most of the current books, approaches for gray-level image segmentation are presented by authors who afterward claim that most of the techniques can be extended to multi-spectral, 3D or color images. However, the direct extension of operations on scalar values to vectors is not straightforward; therefore, the same approaches for gray-scale images should not be applied *as they are* to the color domain. Then an avalanche of questions follow: what is the approapriate color space to use, which are the properties of the chosen color space that should be respected, who judges the segmentation result and how...

For gray-scale images, the segmentation techniques were always divided in two major classes: contour- and region-oriented, followed recently by more elaborated techniques based on features. Sometimes segmentation means just detecting certain points of interest, like corners, for instance [53, 88]. Surprisingly enough, even in books edited in the recent years, like [143], the classification of the segmentation approaches still is in *region-* and *boundary-based* segmentation. According to Fu [45], the segmentation techniques can be categorized into three classes: (1) characteristic feature thresholding or clustering, (2) edge detection, and (3) region extraction. In [94], there are six categories identified, which are finally reduced to the same three already mentioned. The reader is advised to read the chapters on segmentation from a couple of classical books in image processing, i.e., [49, 62] for a complete understanding of the basics on image segmentation. The theoretical concepts that form the ground for all segmentation approaches, e.g., similarity, discontinuity, and pixel connectivity [45], [49, 152], constitute the prerequisites for the easy reading of this chapter. We will focus on region segmentation, i.e., we will not treat edge detection and clustering in the context of color images.

Segmentation evolved in the last two decades, from the initial exploratory approaches mostly in the pixel value space to feature space-based techniques, and, in addition, it became multiresolution and multistage. The image is analyzed at various resolutions, from a rough or coarse view, to a fine and detailed one (see Fig. 8.1). In addition, the image segmentation process is performed in several stages, starting with a preprocessing phase whose purpose is to reduce noise (e.g., smoothing), thus reducing the complexity of the color information in the image, followed by a computation of a local descriptor, a *feature*—a characterization of the color information and texture.

The widely used techniques—region growing and edge detection [16]—are nowadays performed on the feature images, in order to detect ruptures in the color texture, for instance. Last but not least, the refinement of the segmentation may take place, in a closed loop, based on a chosen segmentation quality metric (SQM): the low-level parameters (like thresholds) are tuned according to rules imposed by high-level parameters. According to some authors [93] the *classical* dichotomy in *low* and *high* levels should be abandoned since the segmentation process should not be considered as being completely disjunct to the interpretation process [9].

The purpose of the current chapter is to present the current stage in image segmentation and in the same time to identify the trends, as well as the evolution of the segmentation techniques, in the present context of color image processing. We

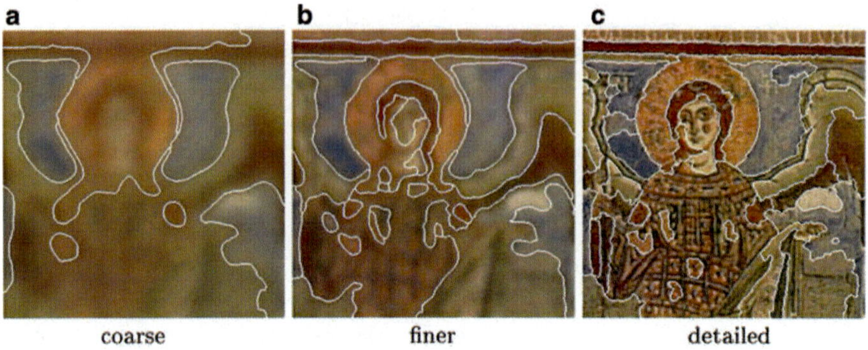

Fig. 8.1 Images at various resolutions and a possible segmentation using JSEG

also try to give some hints regarding the possible answers to the open questions of the topic, for which, unfortunately, a clear answer still does not exist. In [152], the author identifies four paths for the future trends in image segmentation: mathematical models and theories, high-level studies, incorporating human factors and application-oriented segmentation. The first two are definitely intrinsic to any development; therefore, we consider that the latter two are really indicating the ways to go.

It is not our purpose to focus on formalisms and classifications, however, given that the context of this book—color image processing—the taxonomy that comes to our minds would be the following: *marginal*—per component processing—and *purely vectorial* approaches. Vectorial approaches are usually desired, in order to preserve the intrinsic correlation between components, i.e., the multidimensional nature of the color information. This classification would be in fact the answer to a more fundamental question about color: are the properties of the chosen color space considered in the development of the segmentation approach? One may argue that the question stands even for any color image-processing technique, not just for segmentation. From the application point of view—another question is related to the human visual system and human perception of the images: do the approaches take into account the human perception of color information? Is it required for a certain application to integrate the human perception or the energetical point of view should be enough?

Last but not least, the question on the performance of a segmentation approach still remains, albeit the fact that various attempts were made to assess the quality of the segmentation response. Pratt [102] also had a similar observation "Because the methods are ad hoc, it would be useful to have some means of assessing their performance." Haralick and Shapiro [52] have established qualitative guidelines for a good image segmentation which will be discussed in detail in Sect. 8.2.1. Developing quantitative image segmentation performance metrics is, however, a delicate and complex task, and it should take into account several aspects, like human perception and the application point of view.

The chapter is organized as follows: a formal description and fundamental notions of segmentation in Sect. 8.2, including pixel neighborhoods and various distances to be used as similarity measures. The color and color texture features are discussed in Sect. 8.3, including color fractal features. Then in Sect. 8.4, the major segmentation frameworks are presented. Finally, we describe the SQMs in Sect. 8.5 and then draw our conclusions.

8.2 Formalism and Fundamental Concepts

8.2.1 Formalisms

A digital image I is modeled from a mathematical point of view as a function $I(x,y)$ which maps the locations (x,y) in space to the pixel value $I(x,y)=v$. Traditionally, images were black and white or gray and values were discrete from 0 to 255; in this case, v will be a scalar. Since the world of images became colored, color images are used everywhere and RGB images are very common. Each color channel can be represented by an integer or by a floating point number; in both cases, v will be a vector (r,g,b).

A discrete *image* I is a function $I:\mathbb{N}^2\rightarrow\mathbb{V}$. *Locations* P belong to the image support, i.e., a finite rectangular grid, i.e., $D=[0,\ldots M]\times[0,\ldots N]\subseteq\mathbb{N}^2$. For *gray-scale* images $\mathbb{V}=[0,\ldots,255]\subseteq\mathbb{N}$; for *color images*, we (usually) have $\mathbb{V}=[0,\ldots,255]^3\subseteq\mathbb{N}^3$. An image element X is called a *pixel* which has a *pixel location* $\Lambda(X)=P$ and a *pixel value* $\Upsilon(X)=I(\Lambda(X))=v\in\mathbb{V}$. If we enumerate the pixels in an image as $\{X_1,\ldots,X_{N_\mathrm{P}}\}$, we use $N_\mathrm{P}=M\cdot N$ as the number of pixels in an image.

From a mathematical point of view, for an image I, the segmentation operation formalism states that the image is decomposed into a number N_R of regions R_i, with $i=1..N_\mathrm{R}$, which are disjoint nonempty sections of I, like in Fig. 8.2.

Regions are connected sets of pixel locations that exhibit some similarity in the pixel values which can be defined in various ways. For the notion of connectedness we need the definition of neighborhoods, which we will define in Sect. 8.2.2.

The segmentation of an image I into regions R_i is called *complete*, if the regions exhibit the properties listed in [45] which we formalize in the following:

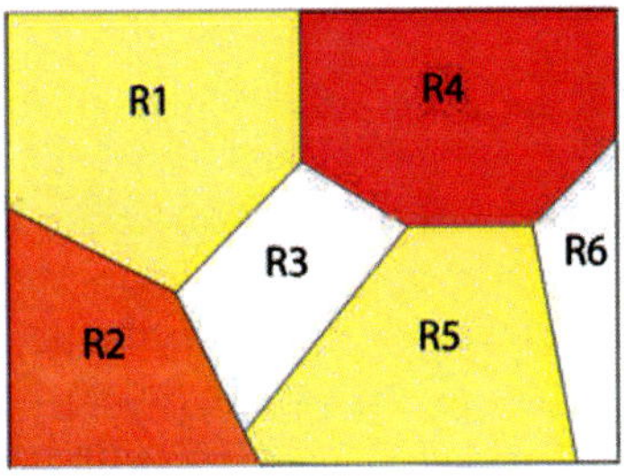

Fig. 8.2 Theoretical example of segmentation

- $\bigcup_{i=1}^{N_R} R_i = D$, i.e., the union of all regions should give the entire image, or in other words, all the pixels should belong to a region at the end of segmentation.
- $R_i \bigcap R_j = \emptyset \ \forall \, i \neq j$, i.e., the regions should not overlap.
- Each segment R_i is a connected component or compact, i.e., the pixel locations $P \in R_i$ in a region R_i are connected; we will define different notions of connectivity in the following paragraphs.
- $\forall i$, a certain criterion of uniformity $\gamma(R_i)$ is satisfied, $(\gamma(R_i) = TRUE)$, i.e., pixels belonging to the same region have similar properties.
- $\forall i \neq j$, the uniformity criterion for $R_i \bigcup R_j$ is not satisfied $(\gamma(R_i \bigcup R_j) = FALSE)$, i.e., pixels belonging to different regions should exhibit different properties.

The result of segmentation is a set of regions $\{R_i\}, i \in \{1, \ldots, N_R\}$ which can be represented in several ways. The simple solution used frequently is to create a so-called *region label image* (I_R) which is a feature image where each location contains the index of the region that this location is assigned to, i.e., $I_R : \mathbb{N}^2 \rightarrow \{1, \ldots, N_R\}$. This label image is also called a *map*.

Haralick and Shapiro state in [52] the guidelines for achieving a *good* segmentation: (1) regions of an image segmentation should be uniform and homogeneous with respect to some characteristic such as gray tone or texture; (2) region interiors should be simple and without many small holes; (3) adjacent regions of a segmentation should have significantly different values with respect to the characteristic on which they are uniform; and (4) boundaries of each segment should be simple, not ragged, and must be spatially accurate. Partially, these guidelines are met by the formal properties mentioned above. The others will be used for the development of image SQMs in Sect. 8.5.4.

If a segmentation is complete, the result of segmentation is a partitioning of the input image, corresponding to the choice of homogeneity criterion γ for the segmentation. There are usually two distinguished cases: *oversegmentation* and *undersegmentation*. The oversegmentation means that the number of regions is larger than the number of objects in the images, or it is simply larger than desired. This case is usually preferred because it can be fixed by a post-processing stage called *region merging*. The undersegmentation is the opposite case and usually less satisfying.

8.2.2 Neighborhoods

The authors of [143] emphasize the need to introduce the concept of *pixel connectivity* as a fundamental concept within the context of image segmentation. The pixels can be adjacent or touching [110], and there are mainly two types of connectivity: *4-connectivity* and *8-connectivity*, the latter one being mostly used, but there are also variants like the *6-connectivity* which is used in segmentation approach proposed by [104]. A region which is 4-connected is also 8-connected. The various pixel neighborhoods are illustrated in Fig. 8.3.

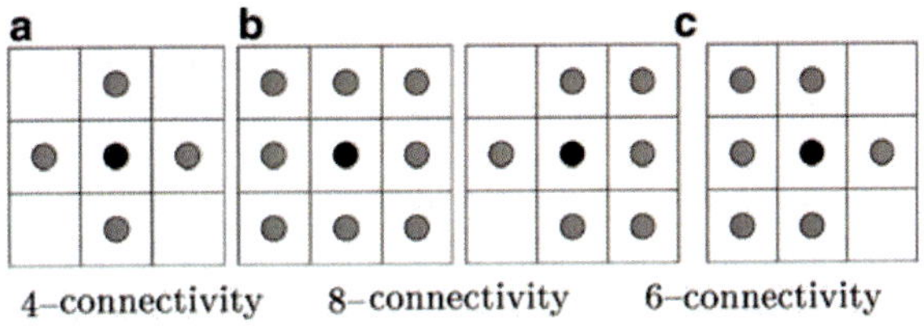

Fig. 8.3 Pixel neighborhoods

Rectangular or squared tessellation of digital images induces the problem of how to define a neighborhood of a pixel for a discrete set, as an approximation for the case of continuous metric space, when the neighborhood usually represents an open ball with a certain centre and radius [40]. The choice of the neighborhood is of extreme importance for region-growing-based segmentation methods. Each version has advantages and disadvantages, when sets of similar pixels are searched in segmentation that should result in a connected region.

8.2.3 *Homogeneity*

Since Haralick [52], the notion of homogeneity is inseparable from the segmentation purpose. Apparently, content homogeneity seems to describe a simple concept: that visual content is visually and physically into an inseparable whole. And right behind this definition, authors simplify the purpose as a problem concerning only the distribution of a variable. If we assume that we could define one information feature which explain this phenomena, we need to define the measure that could indicate if the content is homogeneus or heterogeneus upon this feature. In a context of landscape complexity analysis, Feagin lists several possible criteria [43], upon the fact that the variables are distributed in a qualitatively patchy form [134] or quantitatively defined [76] by an index such as lacunarity [101] or wavelet analysis [117]. We could define a binary criterion $\gamma(R_i)$ for the homogeneity that for a region R_i defines if the region is homogeneous or not:

$$\gamma(R_i) = \begin{cases} \text{TRUE} & \text{if } \forall P \in R_i : ||I(P) - \mu(R_i)|| \leq \theta \\ \text{FALSE} & \text{otherwise} \end{cases}, \tag{8.1}$$

where $\mu(R)$ is the mean $\mu(R) = \frac{1}{||R||} \sum_{P' \in R} I(P')$ of the pixel values in the region and where θ is some threshold.

This describes regions where no pixel differs from the mean of the pixel values inside the region by more than a threshold. This measure requires that the range $\mathbb{V}$ is an algebra that supports addition and a norm (see also the discussion in Sect. 8.2.5). Such definitions do not necessarily lead to a unique segmentation; different processing strategies and algorithms will thus yield different regions. Classically in image processing, variables are chosen from gray level or color distribution, more rarely from color texture.

Nevertheless, such definitions are too simple for the actual challenge of segmentation, in particular, with the increasing of size, resolution, and definition of color images. We need to enhance such definitions. For example, Feagin defines the homogeneity criteria from the shape of the distribution by several parameters like the relative richness, entropy, difference, and scale-dependent variance [43]. Is interesting to note that texture definition used by Feagin is defined as a multiscale one, for which the homogeneity is linked to the stationarity of the distribution along the scale. But the more interesting conclusion of this work dedicated to the notion of homogeneity and heterogeneity is around the fact that the homogeneity perception depends on the perceived scale and could be homogeneous for large scales and heterogeneous for fine scales for the same feature. In [77], the authors explored the heterogeneity definition and they defined it as *the complexity and/or variability of a system property in space* with two subquestions: the structural heterogeneity, i.e., *the complexity or variability of a system property measured without reference to any functional effects* and the heterogeneity as a function of scale [70]. As it is for the definition of homogeneity or heterogeneity, the purpose is clearly expressed as a multiscale complexity and as a result, the question of the uniqueness of the analysis parameter along the scale is asked. And by extension, as features are specialized in color distribution, texture parameters, or wavelets analysis, the true question becomes for the next years: how to merge dynamically several features in function of the analysis scale.

8.2.4 Distances and Similarity Measures

The segmentation process usually searches to maximize the heterogeneity between regions and the homogeneity inside each regions, with a penalizing factor proportional to the region number (see the section on segmentation quality for such writings). So several issues are related to the question of similarity measure of content or distances between content seen through a particular feature. In a first consideration at the lowest possible level, the considered feature could be the pixel color and the question about color distances, then in superior levels the question reach features linked to the color distribution, then to the content description under a texture aspect.

In each case, authors question the choice of the best color space to compute the features and then the distance or similarity measure. For distances between color coordinates, the question is application dependent. If humans are present or active in the processing chain (for decision, for example), perceptual distances are required and thresholds should be expressed in JNDs, (multiple of Just Noticeable Difference) and upon the application a value of three JNDs is usually chosen as threshold for the eye to be able to distinguish between two different colors.

When human perception is disregarded, the difference between colors can be judged from a purely energetic point of view; therefore, any distance between vectors can be used to express the similarity between colors. Moreover, when local features are used for segmentation, the perceptual distances make no sense for the assessment of the similarity between two features. The perceptual distances can be, however, used in the definition of the local features.

8.2.4.1 Color-Specific Distances: The ΔE Family

We should start our section on similarity between colors with the remark made in the third edition of the CIE standard on colorimetry [24]: "*The three-dimensional color space produced by plotting CIE tristimulus values* (X,Y,Z) *in rectangular coordinates is not visually uniform, nor is the* (x,y,Y) *space nor the two-dimensional CIE* (x,y) *chromaticity diagram. Equal distances in these spaces do not represent equally perceptible differences between color stimuli.*". For this reason, the CIE introduced and recommended the L*a*b* and Luv spaces to compute color differences [98, 120, 121]. But the choice of the appropriate color space would be the subject of another chapter or even an entire book.

These *more uniform* color spaces have been constructed so that the Euclidean distance ΔE computed between colors will be in accordance to the perceived differences by humans. Several improvements have emerged from the CIE: ΔE_{94} [25] and ΔE_{2000} [26]. Other distances that are worth mentioning are CMC [27] and DIN99 [36].

The CIELAB and CIELUV color spaces are created to attempt to linearize the perceptibility of color differences. The color distance associated, called ΔE, is an Euclidian distance. For two color values (L_1^*, a_1^*, b_1^*) and (L_2^*, a_2^*, b_2^*), the difference ΔE is

$$\Delta E = \sqrt{(L_1^* - L_2^*)^2 + (a_1^* - a_2^*)^2 + (b_1^* - b_2^*)^2}. \tag{8.2}$$

But equidistant coordinates in these spaces do not define colors perceptually similar [79]. Physiologically, the eye is more sensitive to hue differences than chroma and lightness and ΔE does not take this aspect into account. Then the CIE recommendations for color distance evolved from the initial ΔE, to the ΔE_{94} and finally ΔE_{2000}. The equation of color distance ΔE_{94} published in 1995 [25] is the following:

$$\Delta E_{94} = \sqrt{\left(\frac{\Delta L}{K_L S_L}\right)^2 + \left(\frac{\Delta C}{K_C S_C}\right)^2 + \left(\frac{\Delta H}{K_H S_H}\right)^2}, \tag{8.3}$$

where ΔL, ΔC, and ΔH are the lightness, chroma, and hue differences, respectively. The parameters K_L, K_C, and K_H are weights depending on the conditions of observation, usually $K_L = K_C = K_H = 1$. The S_L, S_C, and S_H parameters adjust the color difference with the chroma value of colors to assess: $S_C = 1 + K_1 C_1$, $S_H = 1 + K_2 C_1$, with $C_1 = \sqrt{a_1^2 + b_1^2}$ and $C_2 = \sqrt{a_2^2 + b_2^2}$. There exists two variants of the writing of the ΔE_{94}. The first one gives a non-symmetric metric, where the weight S_C and S_H are in function of C_1 which depends of the first color, called the *reference* color. The second one uses the geometric mean chroma, but is less robust than the first one.

The equation of color distance ΔE_{2000} published in 2001 [26] is:

$$\Delta E_{2000} = \sqrt{\left(\frac{\Delta L'}{K_L S_L}\right)^2 + \left(\frac{\Delta C'}{K_C S_C}\right)^2 + \left(\frac{\Delta H'}{K_H S_H}\right)^2 + \frac{R_T \Delta C'}{K_C S_C}\frac{R_T \Delta H'}{K_H S_H}}. \quad (8.4)$$

The parameters K_L, K_C, and K_H weight the formula depending on the conditions of observation. The following terms were added in ΔE_{2000} in order to bring several corrections, and ultimately the ΔE_{2000} has a better behavior that suits the human vision than Δ_{E} and ΔE_{94} for small color differences: S_L, S_C, S_H, and R_T. The term S_L realizes a compensation for lightness, corrects for the fact that Δ gives the predictions larger than the visual sensation for light or dark colors; S_C is the compensation for chroma and mitigates for the significant elongation of the ellipses with the chromaticity; S_H is the compensation for hue, that corrects for the magnification of ellipses with the chromaticity and hue; and finally, the term R_T is to take into account the rotation of the ellipses in the blue.

8.2.4.2 General Purpose Distances

As noted above, regions resulting from segmentation should exhibit homogeneity properties. To compute such properties we often need quite the opposite—to define dissimilarities or distances. We can the minimized such differences inside a region. We first introduce distance measures that are commonly used for that purpose.

One of the most used distances, the Minkowski distance of order p between two points $X = (x_1, x_2, \ldots, x_n)$ and $Y = (y_1, y_2, \ldots, y_n)$ is defined as:

$$d(X,Y) = \left(\sum_{i=1}^{n} |x_i - y_i|^p\right)^{\frac{1}{p}}. \quad (8.5)$$

Minkowski distance is typically used with p being 1 or 2. The latter is the Euclidean distance, while the former is sometimes known as the Manhattan distance. In the limiting case of p reaching infinity, we obtain the Chebyshev distance:

$$\lim_{p\to\infty} \left(\sum_{i=1}^{n} |x_i - y_i|^p\right)^{\frac{1}{p}} = \max_{i=1}^{n} |x_i - y_i|. \quad (8.6)$$

In information theory, the Hamming distance between two vectors of equal length is the number of positions at which the corresponding symbols are different. In other words, it measures the minimum number of substitutions required to change one vector into the other, or the number of errors that transformed one vector into the other. A fuzzy extension of this distance is proposed in [59] to quantify the similarity of color images.

The Mahalanobis distance is a useful way of determining the similarity of an unknown sample set to a known one. The fundamental difference from the Euclidean distance is that it takes into account the correlation between variables, being also scale-invariant. Formally, the Mahalanobis distance of a multivariate vector $X = (x_1, x_2, \ldots, x_n)^{\mathrm{T}}$ from a group of values with mean $\mu = (\mu_1, \mu_2, \ldots, \mu_n)^{\mathrm{T}}$ and covariance matrix Σ is defined as:

$$D_{\mathrm{M}}(X,\mu) = \sqrt{(X-\mu)^{\mathrm{T}}\Sigma^{-1}(X-\mu)}. \tag{8.7}$$

The Mahalanobis distance can also be defined as a dissimilarity measure between two random vectors X and Y of the same distribution with the covariance matrix Σ:

$$d_{\mathrm{M}}(X,Y) = \sqrt{(X-Y)^{\mathrm{T}}\Sigma^{-1}(X-Y)}. \tag{8.8}$$

If the covariance matrix is the identity matrix, the Mahalanobis distance reduces to the Euclidean distance. If the covariance matrix is diagonal, then the resulting distance measure is called the normalized Euclidean distance:

$$d_{\mathrm{M}}(X,Y) = \sqrt{\sum_{i=1}^{N} \frac{(x_i - y_i)^2}{\sigma_i^2}}, \tag{8.9}$$

where σ_i is the standard deviation of the x_i over the sample set.

The difference measures introduced in (8.5)–(8.9) can be used for differences in pixel locations or differences in pixel values. Often we want to compare sets of pixels to each other. We will introduce histograms for such sets later. As histograms can be seen as empirical distributions, statistical measures of differences can be applied to histograms, which we introduce next. The following distances are used to quantify the similarity between two distribution functions and they can be used in the case when the local features that are extracted are either multidimensional or represent a distribution function. They all come from the information theory.

The Bhattacharyya distance measures the similarity of two discrete or continuous probability distributions, $p(x)$ and $q(x)$, and it is usually used in classification to measure the separability of classes. It is closely related to the Bhattacharyya coefficient which is a measure of the amount of overlap between two statistical samples or populations, and can thus be used to determine the relative closeness of the two samples being considered:

$$D_{\mathrm{B}}(p(x),q(x)) = -\ln\left[BC(p(x),q(x))\right], \tag{8.10}$$

where

$$BC(p(x),q(x)) = \int_{-\infty}^{+\infty} \sqrt{p(x)q(x)}\mathrm{d}x. \tag{8.11}$$

The Hellinger distance is also used to quantify the similarity between two probability distributions. It is defined in terms of the "Hellinger integral," being a type of *f-divergence*. The Kullback–Leibler divergence (also known as information divergence, relative entropy or KLIC) is a non-symmetric measure of the difference between two probability distributions $p(x)$ and $q(x)$. The KL distance from $p(x)$ to $q(x)$ is not necessarily the same as the one from $q(x)$ to $p(x)$. For the distributions $p(x)$ and $q(x)$ of a continuous random variable, KL-divergence is defined as the integral:

$$D_{\mathrm{KL}}(p(x),q(x)) = \int_{-\infty}^{\infty} p(x)\log\frac{p(x)}{q(x)}\mathrm{d}x. \tag{8.12}$$

The Kolmogorov–Smirnov distance measure is defined as the maximum discrepancy between two cumulative size distributions. It is given by the formula:

$$D(X,Y) = \max_i |F(i,X) - F(i,Y)|, \tag{8.13}$$

where $F(i,X)$ is entry number i of the cumulative distribution of X.

The chi-square statistic (χ^2) can be used to compare distributions. A symmetric version of this measure is often used in computer vision as given by the formula

$$D(X,Y) = \sum_{i=1}^{n} \frac{(f(i,X)_i - f(i,Y)_i)^2}{f(i,Y)_i + f(i,X)_i}. \tag{8.14}$$

The histogram intersection [128] is very useful for partial matches. It is defined as:

$$D(X,Y) = 1 - \frac{\sum_{i=1}^{n} \min(x_i,y_i)}{\sum_{i=1}^{n} x_i}. \tag{8.15}$$

In probability theory, the earth mover's distance (EMD) is a measure of the distance between two probability distributions over a region [114, 115]. From the segmentation point of view, the EMD is used for the definition of OCCD—the optimal color composition distance defined in [87] and improved and used in the context of perceptual color-texture image segmentation [22].

8.2.5 Discussion

So far we remembered several well-known and widely used distances. In this subsection, we draw the attention on the differences in behavior between some of the existing distances and also on the aspect of adapting them to the purpose of image segmentation.

Table 8.1 Color gradient and distances of adjacjent pixels

R	G	B	ΔE	ΔE_{94}	ΔE_{2000}	Euclidian	Mahalanobis
146	15	59	–	–	–	–	–
139	39	52	11.28	5.14	5.26	12.35	2.85
132	63	45	17.12	9.52	10.43	339.79	77.97
125	87	38	20.44	13.58	15.47	21.38	4.91
118	112	31	21.57	14.14	15.41	22.07	5.06
111	136	24	19.52	11.10	11.03	17.53	4.03
104	160	17	17.64	9.27	8.55	10.10	2.34
97	184	10	15.88	8.34	7.13	6.50	1.52
90	209	3	15.06	8.20	6.47	4.66	1.11

In Table 8.1, we show a color gradient and results of different distances between consecutive pixels: ΔE, ΔE_{94}, ΔE_{2000}, Euclidian, and Mahalanobis distances computed in HLS. Neighboring colors have a constant Euclidian distance of $\approx$26 in RGB, each.

We emphasize that the importance of using the proper distance in the right color space. From the point of view of the definition of color spaces, it makes no sense to use the Euclidian distance in other color spaces than CIE L*a*b*, which was especially designed to enable the use of such a distance. In addition, the L*a*b* color space was created so that the Euclidian distance is consistent from the human perception point of view. Therefore, when extending the existing segmentation approaches to the color domain, one has to make sure the correct choices are made regarding color distances and color spaces.

8.3 Features

Image or signal processing is greatly dependent on the definition of features and associated similarity measures. Through these definitions, we search to restore the complex information embedded in image or regions into some values, vectors or small multidimensional values. In addition, for the segmentation purpose, the problem is exposed under a generic form as the partitioning of the image in homogeneous regions or upon the most heterogeneous boundaries. So the right question is how to estimate these criteria, in particular, for color images. Since the 1970s, a variety of features constructions have been made. To develop this section, we chose to organize this question around the analysis scale. As the estimation of this homogeneity or heterogeneity criterion did not take the same sense and consequently the same mathematical formulation in a closed neighborhood, a extended neighborhood or in a complete perspective through multiresolution approaches was chosen.

We start with local information for closed neighborhoods defined by color distributions. When this spatial organization could be reached, the features are integrated into the texture level of regions or zones. As we will see, these approaches are always extended in a multiresolution purpose to be more independent of the analysis scale. In this sense, we present a formulation based on the fractal theory, directly expressed as a multiscale feature and not a set of mono-scale features. Nevertheless, we should draw reader's attention on the conclusions of the authors of [4, 50] who claim that many color spaces—like HLS and HSV—are not suited for image processing and analysis, despite the fact that they were developed for human discussion about color and they are, however, widely used in image processing and analysis approaches. So for each case, the question of the right color space for processing is developed without a unanimous adopted solution.

8.3.1 Color Distribution

Color distributions are a key element in color image segmentation, offering ways to characterize locally the color feeling and some particular objects. The color itself can be a robust feature, exhibiting the invariance to image size, resolution or orientation. However, sometimes the color property may be regarded as useless, as everybody knows that the sky is blue and the grass is green, but let us bear in mind the Latin maxim that is "de gustisbus et coloribus non est disputandum," thus *the blue* is not the same one for all of us and definitely this may not be disregarded for an image-processing application.

In addition, understanding the physics behind color for a given application may be extremely useful for choosing the appropriate way to characterize a color impression. Nevertheless, a physical description of color requires several models for illumination, surface reflection, material properties, and sensor characteristics. When these informations are managed, the image-processing chain deals with the initial discretized continuous variable (energy, frequency, exact localization in three-dimensional spaces) and moves into the vision processing, through typically the solution of an inverse problem. However, most of these properties will be unknown uncontrolled environments as they are often the case for image analysis. These additional knowledge on the image-formation process may lack to improve the results of analysis, and the processing chain will try to overcome these limits. In the sequel, we only use the output of the physical input chain, i.e., the pixel data that result from the image-formation process.

Several features are constructed based on the mathematical assumption that the analyzed image is a particular realization of a random field [95], the image being statistically modeled by a spatially organized random multi-variate variable. For the color domain in particular, thus excluding the multi-spectral or hyper-spectral domain, there will be a tri-variate random variable. The histogram of a color region is an estimate of the probability density function in this case. For arbitrary color spaces or gray-level images, a histogram is defined as follows:

A *histogram* H of a (discrete) image I is defined as:

$H : \mathbb{V} \to \mathbb{N}$

Fig. 8.4 Original images used in the following examples in this chapter

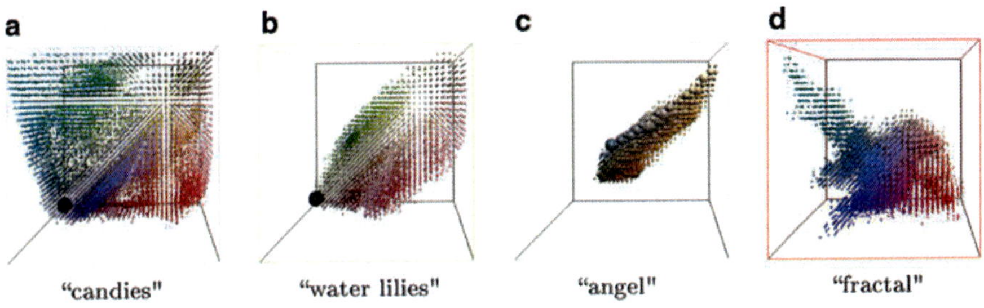

Fig. 8.5 RGB color histograms

$H(v) : \frac{1}{M \times N} \sum_{n=1}^{N} \sum_{m=1}^{M} \delta(||I(x,y) - v||), \quad v \in \mathbb{V}$ where $M \times N$ is the size of the image I and δ is the Dirac function. If the values $v \in \mathbb{V}$ are not discrete, we would have to use an empirical distribution function instead. Similarly, we can define histograms of sets of pixels or of regions.

The number of colors in a color space can be extremely large; thus, the information provided by the histogram may be too large to deal with properly (see in Fig. 8.5 the 3D RGB color histograms of the three images from Fig. 8.4 used in this chapter to illustrate the results of different segmentation approaches). All the approaches based on the histogram thresholding techniques are almost impossible to extend to the color domain. In addition, the spatial information is missing and the histogram is sensitive to changes of illumination. Very often a reduction of the number of colors is used, i.e., quantization. Funt and Finlayson [46] improved the robustness of histogram with respect to illumination changes. In Fig. 8.5, the histograms of the images from Fig. 8.4, which are used along this chapter to illustrate different segmentation approaches, are depicted. Depending on the content of the image and the color information, the histograms may be very complex, but somtimes very coarse because of the reduced number of colors compared to the available number of bins.

Statistical moments are very often preferred, reducing in this way the information offered by a histogram to some scalar values. An example is depicted in Fig. 8.6, where the mean value of all the CIE Lab ΔE distances between the pixels in a region is used as the color feature. As will be seen later, the extension of this feature in a

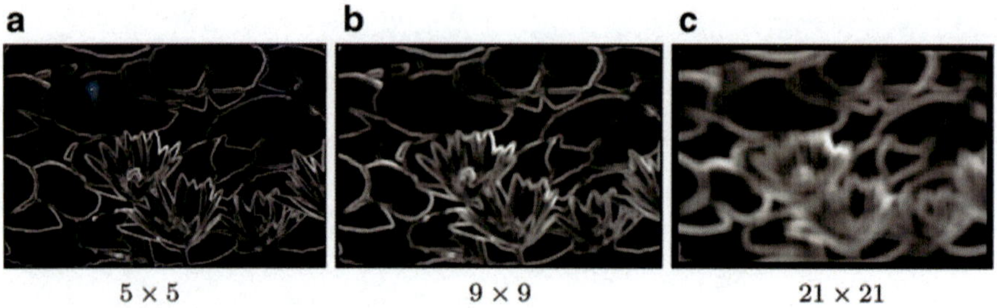

Fig. 8.6 Pseudo images of mean ΔE distance for three sizes of the local window

multiresolution scheme allow to compute the color correlation fractal dimension, and further used to illustrate the watershed segmentation approach.

Based on the color cumulative histogram—an estimate of the cumulative density function—Stricker and Orengo [126] proposed a color feature representation, i.e., a similarity measure of two color distributions, based on the L_1, L_2, and L_∞ distances of histograms. The authors use only the first three statistical moments for each color channel; in their famous article, they use the HSV color space, and obtain good results for image indexation, showing a successful application of statistical moments as local criteria, even if the color space choice could be discussed after 20 years of research in the color domain.

Given the limitations of the histogram, especially the fact that the spatial information is discarded, Pass et al. [97] proposed something close to the histogram, but taking into account the spatial information: the color coherence vectors. Smith and Chang proposed the use of so-called *color sets* [125].

So far we have analyzed color, but the color itself is not enough for a *good* segmentation. See, for instance, the fractal image in Fig. 8.4: the color information may look extremely complex; however, the texture is not complex because it exhibits small variations in the color information for the neighbor pixels. Therefore, the spatial or topological information has to be taken into consideration. This is also important for the merging phase for an oversegmented image. In other words, the local features should be spatial-colorimetric and that is the case of the following features.

8.3.2 *Color Texture Features*

The J-factor is defined as a measure of *similarity* between colors texture inside a window of analysis. Deng wrote the J-criterion to identify region homogeneity upon a color texture point of view [34], with the assumption that the color information in each region of the image could be represented by a few representative colors. In this sense, J expresses the normalized variance of spatial distances of class centers. The

Fig. 8.7 *J*-values upon the spatial organization of colors

homogeneity induced by this metric is of order 1, given that the center of the classes are finally the centers of a region in the case of an homogeneous region.

For a set Q of N pixel locations $Q = \{P_1, P_2, \ldots, P_N\}$ let m be the mean position of all pixels: $m = \frac{1}{N}\sum_{i=1}^{N} P_i$. If Q is classified into C classes Q_i according to the color values at those locations, then let m_i be the mean position of the N_i points of class Q_i: $m_i = \frac{1}{N_i}\sum_{P \in Q_i} P_i$. Then let $S_T = \sum_{q \in Q} ||q - m||^2$, the total spatial variance and

$$S_W = \sum_{i=1}^{C} S_i = \sum_{i=1}^{C} \sum_{q \in Q_i} ||q - m_i||^2, \tag{8.16}$$

the spatial variance relative to the Q_i classes. The measure J is defined as:

$$J = \frac{S_B}{S_W} = \frac{S_T - S_W}{S_W}, \tag{8.17}$$

where J basically measures the distances between different classes S_B over the distances between the members within each class S_W: a high value of J indicates that the classes are more separated from each other and the members within each class are closer to each other, and vice versa. As the spatial variance relative to the color classes S_W depends on the color class of each window of analysis, this formulation is not ideal, then Dombre modifies the expression to be stable on the complete image:

$$\widetilde{J} = \frac{S_B}{S_W} = \frac{S_T - S_W}{S_T}. \tag{8.18}$$

The *J-image* is a gray-scale pseudo-image whose pixel values are the J values calculated over local windows centered on each pixel position (Fig. 8.7). The higher the local J value is, the more likely that the pixel is near region boundaries. In Fig. 8.8, we show some examples of *J-images* at three different resolutions:

In fact, the color information is not used by the J-criterion. Behind Deng's assumption, there is an unspoken idea that all colors are really different, far in term of distance. The color classes Q_i used in (8.16) are obtained by color quantization.

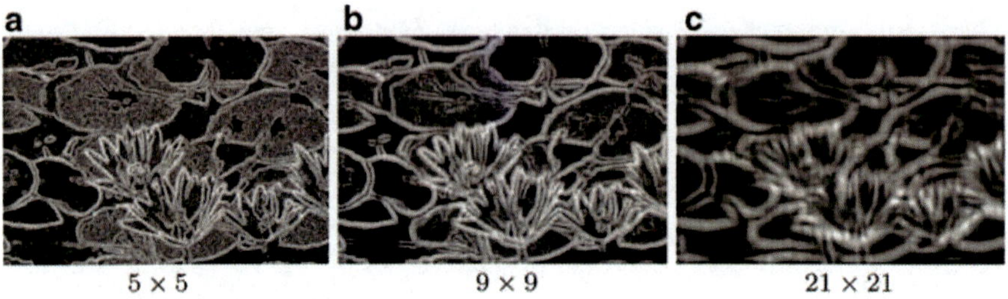

Fig. 8.8 *J*-images for three sizes of the local window

So the choice of the quantization scheme is really important in this kind of approach to produce the right reduced color set. Another important aspect is the number of color classes, to obtain good results; this number should be reduced to obtain some color classes in the analysis window. The J-factor is only useful if the number of quantized colors is smaller than the number of pixels in the window.

Another idea to describe textures was introduced by Huang et al. who defines color correlograms [57] which practically compute the probability $\gamma^{(k)}_{v_i,v_j}$ that a pixel at location P_1 of color v_i has a neighbor P_2 of color v_j, the position of the neighbor being defined by the distance k:

$$\gamma^{(k)}_{v_i,v_j} = p\left(v_i, v_j \,|\exists X_1 \exists X_2 : (\Upsilon(X_1) = v_i) \,\text{and}\, (\Upsilon(X_2) = v_2)\right.$$
$$\left.\text{and}\, (||\Lambda(X_1) - \Lambda(X_2)|| = k)\right). \quad (8.19)$$

As a generalization, the cooccurrence matrices are used to characterize textures by computing the probability $\chi^{(k,\theta)}_{v_i,v_j}$ that a pixel P_1 of color v_i has a neighbor P_2 of color v_j, the position of the neighbor being defined by the distance k and the direction θ:

$$\chi^{(k,\theta)}_{v_i,v_j} \stackrel{\Delta}{=} p(v_i, v_j|\exists X_1 \exists X_2 : (\Upsilon(X_1) = v_i) \,\text{and}\, (\Upsilon(X_2) = v_2)$$
$$\text{and}\, (||\Lambda(X_1) - \Lambda(X_2)|| = k) \,\text{and}\, (\sin^{-1}(\Lambda(X_1) \cdot \Lambda(X_2) = \theta)) \quad (8.20)$$

For color images, we propose the use overlaid cooccurrence images computed independently on each color channel. This is definitely a marginal analysis, but the vectorial approach would require the representation of a 6-dimensional matrix. In Fig. 8.9, we show the overlaid RGB cooccurrence matrices for various images, for the right neighbor of the pixel ($k = 1$ and $\theta = 90°$): the larger spread for the image "angel" is the consequence of a less correlated information in the texture, as well as a lower resolution of the image.

The Haralick texture features (Table 8.2) are defined based on the co-occurrence matrix. The element (i, j) of the square matrix of size $G \times G$ (where G is the number of levels in gray-scale images), represents the number of times a pixel with value i is

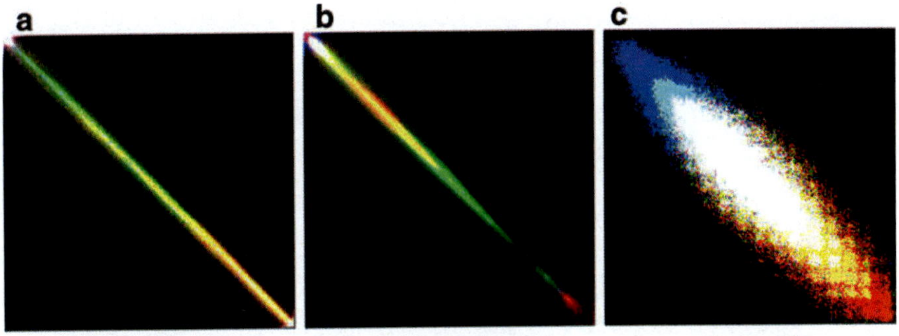

Fig. 8.9 Overlaid RGB cooccurrence matrices

Table 8.2 The Haralick texture features

Angular second moment	$f_1 = \sum_{i=1}^{G}\sum_{j=1}^{G}\{p(i,j)\}^2$
Contrast	$f_2 = \sum_{g=0}^{G-1} g^2 \{\underbrace{\sum_{i=1}^{G}\sum_{j=1}^{G}}_{\lvert i-j\rvert=g} p(i,j)\}$
Correlation	$f_3 = \frac{\sum_{i=1}^{G}\sum_{j=1}^{G}(ij)p(i,j)-\mu_x\mu_y}{\sigma_x\sigma_y}$
Sum of squares: variance	$f_4 = \sum_{i=1}^{G}\sum_{j=1}^{G}(i-\mu)^2 p(i,j)$
Inverse difference moment	$f_5 = \sum_{i=1}^{G}\sum_{j=1}^{G}\frac{1}{1+(i-j)^2}p(i,j)$
Sum average	$f_6 = \sum_{i=2}^{2G} i p_{x+y}(i)$
Sum variance	$f_7 = \sum_{i=2}^{2G}(i-f_8)^2 p_{x+y}(i)$
Sum entropy	$f_8 = -\sum_{i=2}^{2G} p_{x+y}(i)\log(p_{x+y}(i))$
Entropy	$f_9 = -\sum_{i=1}^{G}\sum_{j=1}^{G} p(i,j)\log(p(i,j))$
Difference variance	$f_{10} =$ variance of p_{x-y}
Difference entropy	$f_{11} = -\sum_{i=0}^{G-1} p_{x-y}(i)\log(p_{x-y}(i))$
Information measures of correlation	$f_{12} = \frac{\mathrm{HXY}-\mathrm{HXY}_1}{\max(\mathrm{HX},\mathrm{HY})}$
	$f_{13} = (1-\exp[-2.0(\mathrm{HXY}_2-\mathrm{HXY})])^{\frac{1}{2}}$
Maximal correlation coefficient	$f_{14} =$ (second largest eigenvalue of $Q)^{\frac{1}{2}}$

adjacent to a pixel with value j. The matrix is normalized, thus obtaining an estimate of the probability that a pixel with value i will be found adjacent to a pixel of value j. Since adjacency can be defined, for instance, in a neighborhood of $4-connectivity$, four such co-occurrence matrices can be calculated.

If $p(i,j)$ is the normalized cooccurrence matrix, then $p_x(i) = \sum_{j=1}^{G} p(i,j)$ and $p_y(j) = \sum_{i=1}^{G} p(i,j)$ are the marginal probability matrices. The following two expressions are used in the definition of the Haralick texture features:

$$\begin{cases} p_{x+y}(g) = \underbrace{\sum_{i=1}^{G}\sum_{j=1}^{G}}_{i+j=g} p(i,j), g = 2,3,\ldots,2G \\ p_{x-y}(g) = \underbrace{\sum_{i=1}^{G}\sum_{j=1}^{G}}_{|i-j|=g} p(i,j), g = 0,1,\ldots,G-1 \end{cases} . \tag{8.21}$$

In Table 8.2, μ_x, μ_y, σ_x, and σ_y are the means and standard deviations of p_x and p_y, respectively. Regarding f_8, since some of the probabilities may be zero and $\log(0)$ is not defined, it is recommended that the term $\log(p+\varepsilon)$ (ε an arbitrarily small positive constant) be used in place of $\log(p)$ in entropy computations. HX and HY are entropies of p_x and p_y. The matrix $Q \in \mathbb{R}^{G,G}$ is defined by $Q_{i,j} = \sum_{g=1}^{G} \frac{p(i,g)p(j,g)}{p_x(i)p_y(j)}$.

$$\begin{cases} \text{HXY} &= -\sum_{i=1}^{G}\sum_{j=1}^{G} p(i,j)\log[p(i,j)] \\ \text{HXY}_1 &= -\sum_{i=1}^{G}\sum_{j=1}^{G} p(i,j)\log[p_x(i)p_y(j)] \\ \text{HXY}_2 &= -\sum_{i=1}^{G}\sum_{j=1}^{G} p_x(i)p_y(j)\log[p_x(i)p_y(j)] \end{cases} . \tag{8.22}$$

The so-called *run-length matrix* is another widely used texture description [47, 129].

8.3.3 Color Fractal Features

In the context of this chapter, we illustrate in this section an example of features that naturally integrate, by definition, the multiresolution view. The fractal dimension and lacunarity are the two widely used complexity measures from the fractal geometry [80]. Fractal dimension is a measure that characterizes the complexity of a fractal, indicating how much of the space is filled by the fractal object. Lacunarity is a mass distribution function indicating how the space is occupied (see [60, 61] for an extension to the color domain of the two fractal measures). The fractal dimension can be used as an indicator whether texture-like surfaces belong to a class or another, therefore and due to its invariance to scale, rotation or translation it is successfully used for classification and segmentation [21, 144]. Komati [71] combined the fractal features with the J-factor to improve segmentation.

There are many specific definitions of fractal dimension; however, the most important theoretical dimensions are the Hausdorff dimension [42] and the Renyi dimension [106]. These dimensions are not used in practice, due to their definition for continuous objects. However, there are several expressions which are directly linked to the theoretical ones and whose simple algorithmic formulations make them very popular. The probabilistic algorithm defined by Voss [68, 138] upon the proposal from Mandelbrot [80], Chap. 34, considers the image as a set S of points in an Euclidian space of dimension E. The spatial arrangement of the set is characterized by the probabilities $\mathbf{P}(m,L)$ probability of having m points included into a hypercube of size L (also called a box), centered in an arbitrary point of S.

The counts are normalized to probabilities, so that $\sum_{m=1}^{N_\mathrm{p}} \mathbf{P}(m,L) = 1$, $\forall L$, where N_p is the number of pixels included in a box of size L. Given the total number of points in the image is M, the number of boxes that contain m points is $(M/m)\mathbf{P}(m,L)$. The total number of boxes needed to cover the image is:

$$N(L) = \sum_{m=1}^{N_\mathrm{p}} \frac{M}{m}\mathbf{P}(m,L) = M\sum_{m=1}^{N_\mathrm{p}} \frac{1}{m}\mathbf{P}(m,L). \tag{8.23}$$

Therefore, we conclude that $N(L) = \sum_{m=1}^{N_\mathrm{p}} \frac{1}{m}\mathbf{P}(m,L) \propto L^{-D}$, where D is the fractal dimension, corresponding to the commonly used mass dimension. A gray-level image is a three-dimensional object, in particular, a discrete surface $z = I(x,y)$ where z is the luminance in every (x,y) point of the space. There is also an extension of the approach to complex three-dimensional objects, but its validation is limited to the Cantor dust [38]. However, there are very few references to a development dedicated to color images, despite the fact that the theoretical background for fractal analysis is based on the Borel set measure in an n-dimensional Euclidian space [42].

A color image is a hypersurface in a color space, like RGB, for instance: $I(x,y) = (r,g,b)$. Therefore in the case of color images we deal with a five-dimensional Euclidian hyperspace and each pixel can be seen as a five-dimensional vector (x,y,r,g,b). The RGB color space was chosen, due to the fact that the RGB space exhibits a cubic organization coherent with the two-dimensional spatial organization of the image. In this way, the constraint of expression in a five-dimensional space was fulfilled.

For gray-level images, the classical algorithm of Voss [138] defines cubes of size L centered in the current pixel $(x,y,z = I(x,y))$ and counts the number of pixels that fall inside a cube characterized by the following corners: $(x-\frac{L}{2}, y-\frac{L}{2}, z-\frac{L}{2})$ and $(x+\frac{L}{2}, y+\frac{L}{2}, z+\frac{L}{2})$. A direct extension of the Voss approach to color images would count the pixels $F = I(x,y,r,g,b)$ for which the Euclidian distance to the center of the hypercube, $F_\mathrm{c} = I(x_\mathrm{c}, y_\mathrm{c}, r_\mathrm{c}, g_\mathrm{c}, b_\mathrm{c})$ would be smaller than L:

$$|F - F_\mathrm{c}| = \sqrt{\sum_{i=1}^{5} |f_i - f_{\mathrm{c}i}|^2} \leq L. \tag{8.24}$$

Given that the Euclidian distance in RGB space does not correspond to the perceptual distance between colors, the use of the Minkowski infinity norm distance was preferred instead:

$$|F - F_\mathrm{c}| = \max_{i=\in\{1\ldots5\}} (|I_i - I_{\mathrm{c}i}|) \leq L\,. \tag{8.25}$$

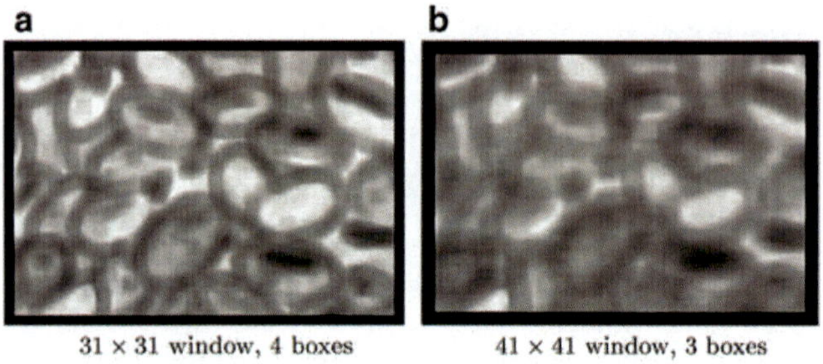

Fig. 8.10 Local color fractal features pseudo-images

Practically, for a certain square of size L in the (x,y) plane, the number of pixels that fall inside a three-dimensional RGB cube of size L, centered in the current pixel, is counted. Once the $\mathbf{P}(m,L)$ is computed, then the measure $N(L)$ is calculated. The slope of the regression line through the points $(log(L), -log(N(L)))$ represents an estimate of the fractal dimension. For details, refer to [60].

In Fig. 8.10, we show the pseudo-images of local color fractal dimensions obtained for the image "candies" when using two sizes for the analysis window and a certain number of boxes, starting with the smallest size of 3. One can see that the multiresolution color feature is able to distinguish between different complexity regions in the image, for further segmentation.

Another approximation of the theoretical fractal dimension is the correlation dimension, introduced in [7]: the correlation integral $C(r)$, whose computation gives us the correlation dimension for a set of points $\{P_1, P_2, \ldots, P_N\}$ is defined as:

$$C(r) = \lim_{N \to \infty} \frac{2q_r}{N(N-1)}, \tag{8.26}$$

where q_r is the number of pairs (i, j) whose color distance is less than r. In [7], the following relationships between the local correlation dimension and the Hausdorff dimension of continuous random fields are proven: For a continuous random vector process P, in certain conditions,[1] the local correlation dimension ν, if it exists, is smaller than the Hausdorff dimension . For the extension to the color domain, the ΔE_{2000} color distance in the CIELab color space can be used, integrating the human perception of color distances into the feature expression.

8.4 Segmentation Approaches

So far we presented several color, color texture, and color fractal features to be chosen and used as local criteria for the implementation of the segmentation

[1]The process must be absolutely continuous with respect to the $\mathbb{R}^d$—Lebesgue measure.

Fig. 8.11 Pyramid structure (original image size: 496×336)

operation. The criteria is usually refined in several major segmentation frameworks. The refinement of the criteria—which can be seen as an optimization of a measure—is usually performed at various scales, in a multiresolution or multiscale approach. The segmentation techniques thus evolved from the open-loop one-step approach to more sophisticated multiple-level optimization-based closed-loop approaches.

There exists a wide spectrum of segmentation approaches. We shall describe several of the mostly-used segmentation frameworks in this section: the pyramid-based approaches, the watershed, the JSEG and JSEG-related approaches, the active contours, the color-structure-code, and graph-based methods. Apart from the ones presented, we should mention also several approaches that are worth mentioning, like the color quantization, the approach proposed by Huang in [58] based on the expectation maximization (EM) technique in the HSV color space; the approach based on Voronoi partitions [6].

8.4.1 Pyramidal Approaches

For image segmentation, pyramid is a hierarchal structure used to represent the image at different resolution levels, each pyramid level being recursively obtained from its underlying level. The bottom level contains the image to be segmented. In a *4-to-1* pyramid, each level is obtained by reducing the resolution of the previous one by a factor of 4 (a factor of 2 on each image axis), usually by means of downsampling and Gaussian smoothing. From the top of the pyramid, the process can be reversed by oversampling the images and gradually increase the resolution in order to refine the rough segmentation obtained on the image on top by analyzing the details in the lower levels of the pyramid (see Fig. 8.11 for an example of a pyramid). Such pyramidal structures allow reducing the computational complexity of the segmentation process: the relationship defined for neighbor pixels between adjacent levels help reduce the time required to analyze the image. There exist various pyramidal approaches for gray-level image segmentation [5, 81, 107, 113].

In Fig. 8.12, the results of pyramidal segmentation are depicted, when the region merging threshold T_2 was varied. The implementation available in the OpenCV library has several input parameters: the number of levels and two thresholds T_1 and T and T_2. The relationship between any pixel X_1 on level i and its candidate-

Fig. 8.12 Pyramid structure segmentation results

father pixel X_2 on the adjacent level is established if $d(\Upsilon(X_1),\Upsilon(X_2)) < T_1$. Once the connected components are determined, they are classified: any two regions R_1 and R_2 belong to the same class, if $d(c(R_1),c(R_2)) < T_2$, where $c(R_)$ is the mean color of the region R_i. For color images, the distances used are $p(c_1,c_2) = 0.3(c_{1r} - c_{2r}) + 0.59(c_{1g} - c_{1g}) + 0.11(c_{1b} - c_{2b})$ computed in RGB.

The pyramids can be either *regular* or *irregular*. The regular pyramids can be *linked* or *weighted*, but they all have various issues because of their inflexibility [13]: most of the approaches are shift-, rotation-, and scale-variant and consequently the resulting segmentation maps are strongly depending on the image data and they are not reproducible. The main reason for all these issues is the subsampling. In addition, the authors of [13] conclude that "multiresolution algorithms in general have a fundamental and inherent difficulty in analyzing elongated objects and ensuring connectivity."

The irregular pyramids were introduced to overcome the issues of regular pyramids. According to the authors of [82], the most referenced are the stochastic and adaptive pyramids [64]. Unfortunately, by renouncing to the rigorous and well-defined neighborhood structure—which was one of the advantages of the regular pyramids—the pyramid size can no longer be bounded; thus, the time to compute the pyramid cannot be estimated [72, 139].

For the color domain, there exist several pyramid-based approaches [82, 135]. The color space used in [82] is HSV due to its correspondence to the human color perception. The homogeneity criterion is defined as follows: $\gamma(x,y,l)$ as a function of pixel position (x,y) and the level l in the pyramid. $\gamma(x,y,l) = 1$ if the four pixels on the lower level have color differences below a threshold T. Unfortunately, neither the distance nor the threshold that are used are specified.

8.4.1.1 Color Structure Code

We now proceed to introduce a variant of the pyramidal approaches, the so-called color structure code (CSC) algorithm which is mainly used to segment color images using color homogeneity but can be used on any other homogeneity predicate. The algorithm was introduced in [104].

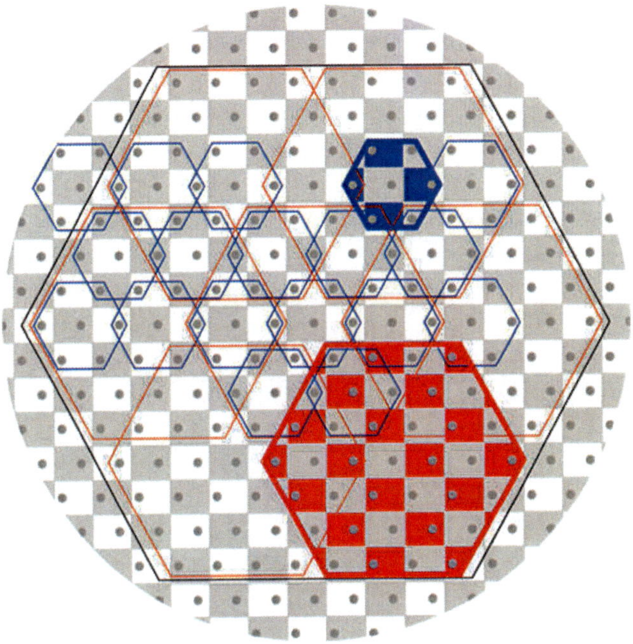

Fig. 8.13 Hexagonal hierarchical island structure for CSC and some code elements

The algorithm combines several strategies: it computes several resolutions and it can use different criteria for homogeneity. It first creates segments by region growing, then it splits them, if other segmentations are better suited. Its input is a color image and its output is a graph that links homogenous regions in various resolutions. The algorithm uses the hexagonal neighborhood shown in Fig. 8.3c. Using this neighborhood, we can map an orthogonal image to an hexagonal image and neighborhood structure as shown in Fig. 8.13.

The algorithm works in four phases: initialization, detection, merging, and splitting.

In the region detection stage, we start by inspecting the islands of level 0. A group of at least two adjacent pixels in an island that fulfill a homogeneity predicate γ are joined to a so-called code element. As islands of level 0 consist of seven pixels, at most three code elements can be defined within an island. For each code element we store its features, such as mean color, and its corresponding elements, i.e., the pixels in the case of level 0 code elements.

In Fig. 8.14 on the left, we see black pixels that are assumed to be similar with respect to a homogeneity predicate γ. Numbers 1–9 denote code elements of level 0 which are always written inside the island of level 0 that they belong to. As pixels may belong to two islands, they may also belong to two code elements. The island on the left contains two code elements.

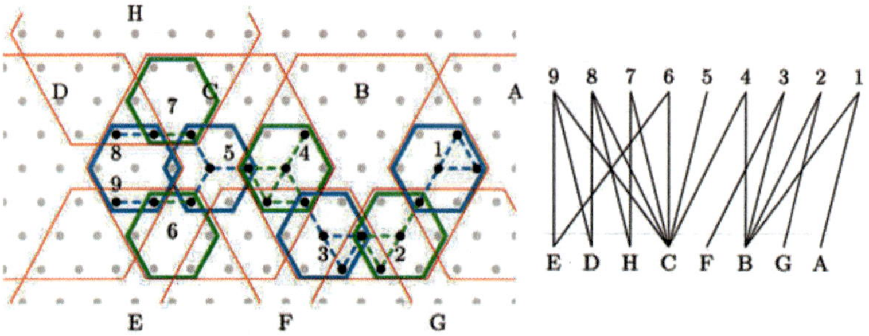

Fig. 8.14 Example of linked code elements and corresponding graph structure

Fig. 8.15 Examples of CSC segmentation on test images

On level $n+1$, we now inspect each island of level $n+1$ separately. This makes the algorithm inherently parallel. For an island of level $n+1$, we check the code elements in its seven subislands. Overlapping code elements are merged; the merging stops at the borders of the island. Two code elements of level $n+1$ are connected, if they share a code element of level n.

If two overlapping code elements would violate the homogeneity criterion, if they were merged, a recursive splitting of the overlapping range is started (see [104] for details).

The association of code elements is stored as a graph which links islands to code elements. The graph can be efficiently stored using bitfields of length 7. The result of segmentation is a region segmentation of the input image represented directly by a region map or by a set of such graphs which each describe a region in several resolutions. Fig. 8.14 right shows such a graph for code elements of level 0 (numbered 1–9) and of islands of level 1 (called A, B, C, D, E, F, G, and H).

In Fig. 8.15, we present the result of the CSC algorithm on several test images where each region is filled with its mean color (in RGB). The homogeneity criterion used here is the Euclidian difference to the mean color of the region computed in intensity-normalized *rgb* as in eq. (8.1). Depending on the threshold θ for the maximum allowed difference, we can obtain oversegmented (image on the left) or undersegmented images (on the right).

8.4.2 Watershed

The watershed is a region-based segmentation approach in which the image is considered to be a relief or a landscape [10, 109]. Historically and by definition, watershed is the approach of the mathematical morphology. Since its first appearance several improvements followed, for instance [32]. The segmentation process is similar—from a metaphorical point of view—to the rain falling on that landscape which will gradually flood the *basins* [11]. The watersheds or dams are determined as the lines between two different flooded basins that will merge. When the topographical relief is flooded step by step, three situations can be observed: (1) a new object is registered if the water reaches a new local minimum. The corresponding pixel location is tagged with a new region label; (2) if a basin extends without merging to another, the new borders have to be assigned to this basin; and (3) if two basins are about to unite, a dam has to be built in between.

Unsupervised approaches of watershed use local minima of the gradient or heterogeneity image as markers and flood the relief from these sources. To separate the different basins, dams are built when flooded zones are meeting. In the end, when the entire relief becomes completely flooded, the resulting set of dams constitutes the *watershed*: (a) the considered basins begin to flood; (b) basin V_3 floods a local minimum; (c) basin V_1 floods another local minimum; (d) a dam is built between valleys V_2 and V_3; and (e) the final stage of the algorithm.

The watershed approach is traditionally applied in the original image domain, but it has a major disadvantage since it fails to capture the global information about the color content of the image. Therefore, in Chapter X of [152], an approach that uses the watershed to find clusters in the feature space is proposed. As an alternative, the gradient or heterogeneity information does not produce closed contours and, hence, does not necessarily provide a partition of the image into regions. As this information could be used as scalar information—classically the norm of the gradient vector—they are well adapted to watershed processing. Having this point of view, the gradient images can be seen as a topographical relief: the gray level of a pixel becomes the elevation of a point, the basins and valleys of the relief correspond to the dark areas, whereas the mountains and crest lines correspond to the light areas. The watershed line may be intuitively introduced as the set of points where a drop of water, falling there, may flow down toward several catchment basins of the relief [31].

An implementation of the classical watershed approach is available in the OpenCV library [85]. In our experiments, we used the implementation available in Matlab. In Fig. 8.16 the evolution of the watershed along the scale is illustrated. Note that usually a merging phase follows.

The critical point of the algorithm is how the dam will be set up. If we imagine a heavy rain falling on the relief, we expect a drop of water to flow to the next lower location (the natural assumption is that raindrops never go uphill); the issue is thus finding the length of the path to the nearest location on the digital grid which is lower than the point where the rain drops on the terrain. A first approach would be to use

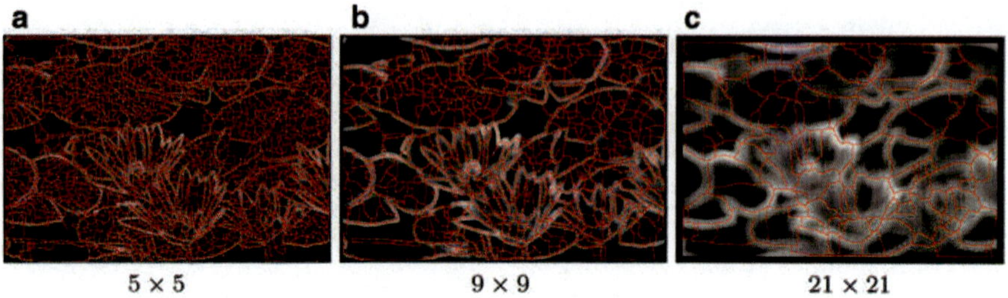

Fig. 8.16 The evolution of the watershed segmentation along the scale

the Euclidean distance; this is, however, computationally too expensive, especially if the line is long and curved. Vincent [136] uses the *city–block* distance in his formal description. However, many points cannot be assigned to a region because of this distance measure.

There are several watershed approaches for color images [20,23,116]. Chanussot et al. extend the watershed approach to the color domain by using the bit mixing technique for multivalued morphology [19]. In [23], the authors use a perceptual color contrast defined in the HSV color space, after a Gaussian low pass filter and a uniform color quantization to reduce the number of colors in the image. According to Itten, there are seven types of color contrast: hue, light-dark, cold-warm, complementary, simultaneous, saturation, and extension [74].

8.4.3 JSEG

In Sect. 8.3.2, the J-criterion was defined as a pseudo spatio-chromatic feature, established upon window size parameters. These parameters lead to a behavior similar to a low-pass filter, i.e., a large window size is well adapted to the extraction of the most important color texture rupture, and on the opposite, a small window size (ws in Fig. 8.17) is better adapted to small details changes in a region. These characteristics allow to establish a specific segmentation algorithm using a multiscale approach in a coarse-to-fine scheme (Fig. 8.17) [33]. In this scheme, all the inner parameters depend on a scale parameter, defined by the i variable. Deng proposes to fix this parameter to an initial value of 128, but in fact these values could be chosen in function of the desired scale number as $i = 2^{S_N}$. Then the window size for the J-criterion computation is established upon this scale parameter as the different used threshold.

The algorithm starts with a computation of the J-image at different scales. Then it performs a segmentation of the J-image starting at the coarsest level. Uniform regions are identified by low J-values; Deng proposes to identify the so-called *valleys* which are the local minima in an empirical way as pixels with values lower than a level predefined threshold T_k established as

$$T_k = \mu_k + a\sigma_k, \tag{8.27}$$

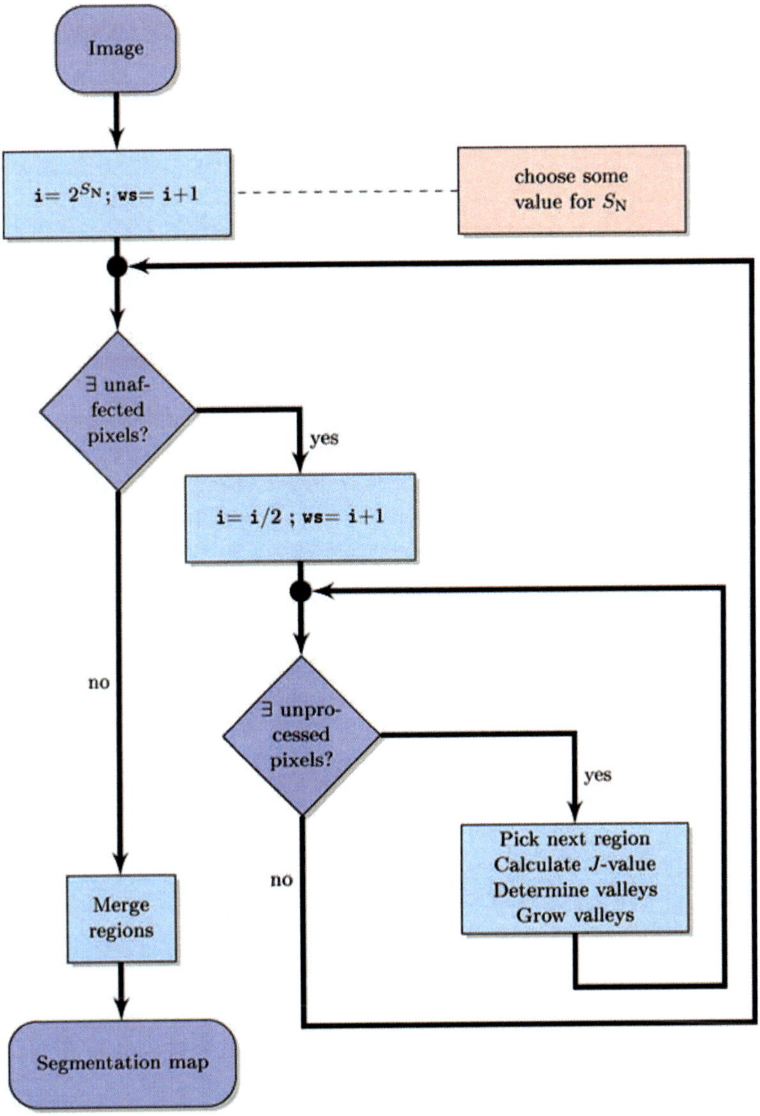

Fig. 8.17 The JSEG based segmentation

where $a \in [-1,1]$ is a parameter that is adjusted to obtain the maximum number of regions[2] and μ_k and σ_k are the mean and standard deviation of the J-values of the k-the region R_k. Then the neighboring pixels with J-values lower than the threshold

[2]To reduce the computational cost, Deng proposes a reduced list of values to try: $[-0.6,-0.4,-0.2,0,0.2,0.4]$.

are aggregated. Only sets of size M_k larger than a fixed value depending on the scales are kept ($|M_k \geq i^2/2|$).[3]

A region growing step is then performed to remove holes in valley, and to construct growing areas around the valley. The J-values are averaged in the remaining unsegmented part of the region, then the pixels with values below this average are affected to the growing areas. If a growing area is adjacent to one and only one valley, it is assigned to that valley. In this way, the $\overline{J}$ average is defined by :

$$\overline{J} = \frac{1}{N}\sum_{k} M_k J_k, \tag{8.28}$$

where J_k is the J-criterion computed for the k-th region R_k with M_k pixels and N the total number of pixels.

To continue the segmentation scheme, the process is repeated for the next lower scales. As the J-value will grow up around the region boundaries, this iterative scheme ensures to enhance the final localization of contours. The result of this iterative scheme is an oversegmented image; a post-processing step is used to reduce this problem. As the first steps of the algorithm are based on the J-criterion, which did not really use the color information, just the notion of color class resulting from quantization, the last step merge the regions upon the similarity between the color histograms. The two most closest regions, in the sense of their color histograms, are merged, then the next closest region, etc. The process stops when the maximum color distance is lower than a threshold λ.

The quantization parameter and region merging threshold are the most important parameters of the JSEG approach, the former determines the minimum distance between two quantized colors, and the latter determines the maximum similarity between two adjacent regions in spatial segmentation. When the quantization parameter is small, oversegmentation will occur and then a large region merging threshold is often needed. When the quantization parameter is large, region merging threshold should be small. An *appropriate* quantization parameter can result in an accurate class-map or region label image, which will lead to a *good* segmentation. However, it is hard to identify a combination of a quantization parameter and a region merge threshold which will make the algorithm more efficient.

In addition, the principal part of the segmentation based on J-criterion calculates the J-values over the class-map, which describes the texture information but does not consider color information of the pixels. In other words, the measure J can only characterize the homogeneity of texture, but fails to represent the discontinuity of color, thus degrading the robustness and discrimination of JSEG. These observations made by the authors of [150] lead to a refined version of the JSEG segmentation approach. Color information is taken into account in the preprocessing step with the color quantization and in the post-processing step with the final region merging.

[3] i arbitrary, but large.

Fig. 8.18 JSEG results

In Fig. 8.18, one can see the influence of the color quantization and the merging threshold on the final result. The result of the approach strongly depends on the color quantization, which in our view is the weak point of the approach. An aggressive quantization, i.e., a small number of colors leads to a small number of regions, while a small merging threshold leads to a larger number of regions (for the same number of colors, $N = 17$ for "candies"). We used the JSEG algorithm implementation from [33, 34] to illustrate how the approach works.

The segmentation scheme proposed by Deng is not so far from a classical watershed algorithm. The third line in Fig. 8.18 show results obtained when the J-image is used as a gradient image in a watershed process, 17 color classes 17 are kept as the second line of results. The segmentation result is quite different and coherent in color content. As the interesting point of the JSEG segmentation algorithm was to be expressed in a multiscale scheme, the same purpose could be developed with a valleys management made by a watershed approach. With this objective, Dombre proposes to combine several scales of segmentation, in a fine-to-coarse expression [37]. As the contours are better localized for small scales (i.e., small

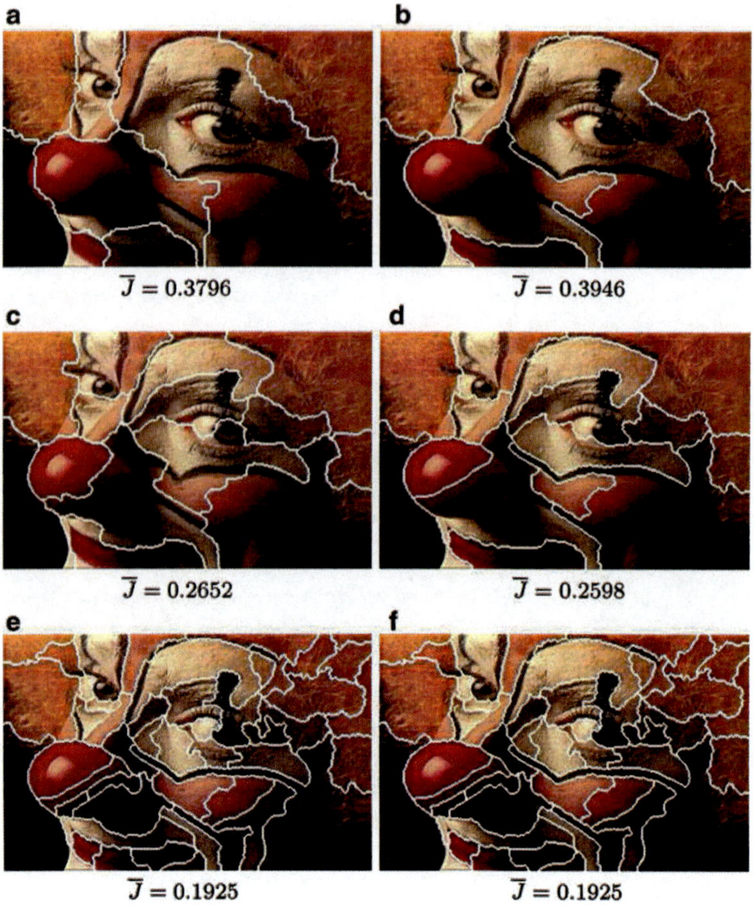

Fig. 8.19 Left column—without and right colum—with repositioning of the edges

analysis window size), but the result often oversegmented. Dombre proposes to project the region obtained for a particular scale in the next one in the scale list. Each small region from low scale is affected to a region in the upper scale; for that, the decision criterion is based on the maximum area of intersection between them. Only outside contours are kept for the low-scale region, but as observed in Fig. 8.19, these contours are now coherent among the scales and well positioned.

The original JSEG uses the J-factor as a criterion for the detection of heterogeneity zones. A continuation of the JSEG proposed by Deng, by integrating the human perception of the color differences for the texture analysis is presented in [108]. For the multiresolution aspect, the contrast sensitivity function (CSF) is used, more precisely the functions described in [91]. The CSF describes the pattern sensitivity of the human visual system as a function of contrast and spatial frequencies.

Fig. 8.20 Image viewed at different distances and the resulting perceptual gradient

Fig. 8.21 Segmentation results

In Fig. 8.20, the perceptual gradient J-images are presented, obtained by CSF filtering, simulating in this way different viewing distances. For the computation of the gradient, the ΔE distance was used.

The final segmentation result is presented in Fig. 8.21. As the emulated viewing distance increases, the segmentation result moves from an oversegmentation to one which is closer to the desired result.

8.4.4 Active Contours

Active contours, also known as *snakes*, were introduced by Kass, Witkin, and Terzopoulos in 1988 [67] and they are defined as "an energy-minimizing spline guided by external constraint forces and influenced by image forces that pull it toward features such as lines or edges." The snakes are successfully used for image segmentation in medical applications [123], especially for computed tomography and brain magnetic resonance imaging. The initial contour is incrementally deformed according to several specified energies. According to the original definition, an active contour is a spline $c(s) = [x(s), y(s)]$, with $s \in [0,1]$, that minimizes the following energy functional [131]:

$$\varepsilon(c) = \varepsilon_{\mathrm{int}}(c) + \varepsilon_{\mathrm{ext}}(c) = \int_0^1 [E_{\mathrm{int}}(c(s)) + E_{\mathrm{ext}}(c(s))]\mathrm{d}s, \tag{8.29}$$

where $\varepsilon_{\mathrm{int}}(c)$ represents the *internal* energy and $\varepsilon_{\mathrm{ext}}(c)$ represents the *external* energy. The internal energy is intrinsic to the spline and the external energies come either from the image or specified by the user, usually as external constraints. The internal energy $\varepsilon_{\mathrm{int}}$ is usually written as:

$$\varepsilon_{\mathrm{int}}(c) = \int_0^1 \tfrac{1}{2}[\alpha(s)\left|c'(s)\right|^2 + \beta(s)\left|c''(s)\right|^2]\mathrm{d}s, \tag{8.30}$$

where $c'(s)$ and $c''(s)$ are the first and the second derivatives, weighted by $\alpha(s)$ and $\beta(s)$, which are usually considered to be constants in most of the implementations.

Xu [148] identifies several issues of the original model [67]: (1) the initialization of the snake has to be close to the edge and (2) the convergence to concave boundaries is poor. These were partially addressed in the original article [67] by using the propagation in the scale space described in [141, 142]. The "balloon model" [28] introduced supplementary external forces for stopping the snakes in the case when the contour was not "visible" enough. The drawbacks of this approach were corrected by the approach of Tina Kapur [66]. Later on, the gradient vector flow (GVF) was introduced by Xu, as well as the generalized gradient vector flow (GGVF) [146–148], the two methods being widely used, despite the fact that they are complex and time consuming. Also, there exist several alternative approaches to GVF: the virtual electrical field (VEF) [96], Curvature Vector Flow in [48], boundary vector field (BVF) [127], Gradient Diffusion Field [69], and Fluid Vector Flow [130]. Active contours have been extended to the so-called *level-set* segmentation which has also been extended to color images in [63].

Here, we present the results of a multiresolution approach extended to the color domain, restricted to a medical application in dermatology (see Fig. 8.22). The external energy forces that drive the active contours are given by the average CIE Lab ΔE distance computed locally at different resolutions based on the original image. Basically, for a certain resolution, the value of one point (x, y) in the energetic surfaces is given by the average CIE Lab ΔE distance computed in a neighborhood

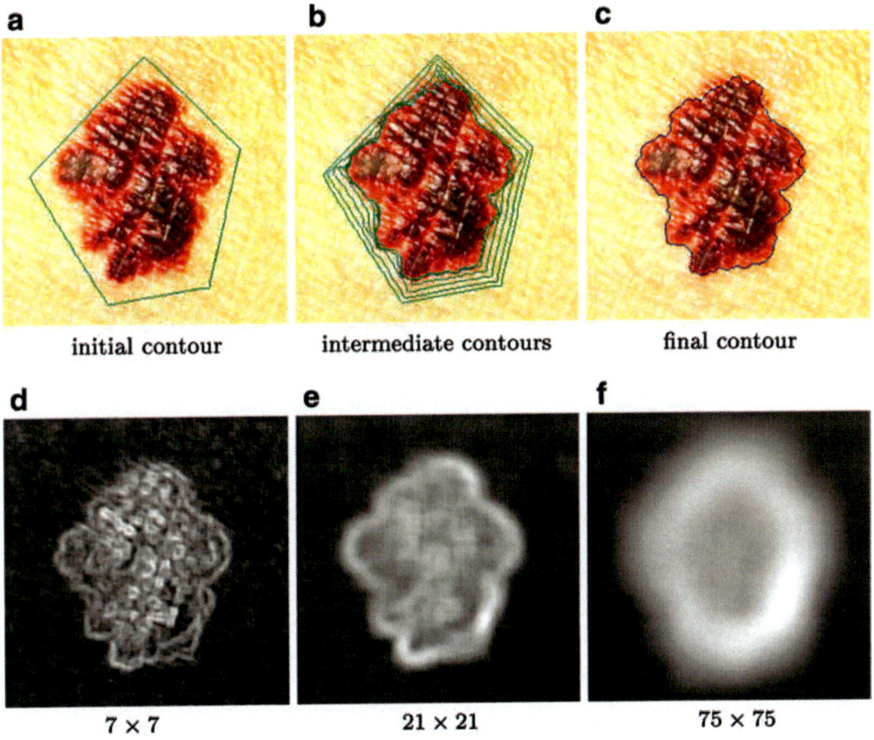

Fig. 8.22 Active contours—initial, intermediate and final snakes, as well as the diffusion at different resolutions

of size $n \times n$ centered in that specific point. If we consider the pixel in the $n \times n$ vicinity as places in a vector of size n^2, we have to compute the average value of $\frac{n^2(n^2-1)}{2}$ distances:

$$E_{\mathrm{ext}}(x,y)|_{n\times n} = \frac{2}{n^2(n^2-1)} \sum_{i=1}^{n^2} \sum_{j=i+1}^{n^2} \Delta E(v_i, v_j). \tag{8.31}$$

In addition, the points of the active contour independently move on the average ΔE surfaces in order to ensure the rapid convergence of the algorithm toward the final solution [133]. The hypothesis that is made is that in such medical images there are two types of textures, exhibiting different complexities: one corresponding to the healthy skin and the other to the lesion (the complexity of the latter one being usually larger).

The proposed external energy is linked to the correlation dimension (practically being the mean value of the $C(r)$ distribution) and also related to the J-factor given that it represents a measure of the heterogeneity in a certain neighborhood at a given resolution.

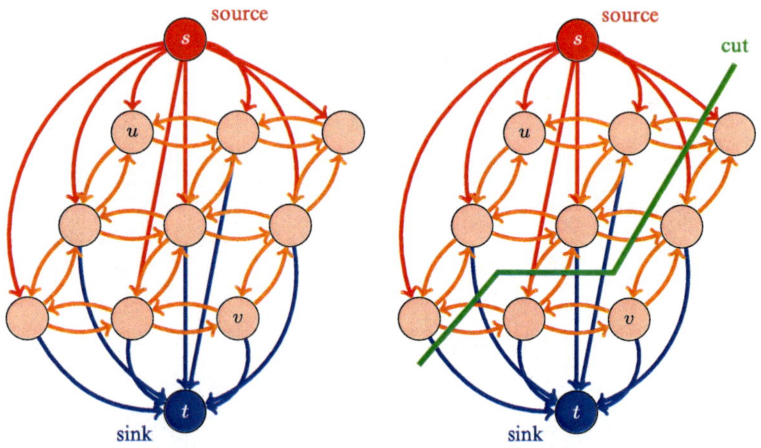

Fig. 8.23 Classical graph construction and cut between nodes according to [15]

8.4.5 Graph-Based Approaches

We will now introduce the segmentation approach the graph-based approaches for image segmentation. We describe graph cuts [73, 124] which use the local features as the information loaded on the edges of a graph, which is initially mapped to the pixel structure of the image.

From an informatics point of view, the segmentation process could be written as the searching of a minimal path between one region to segment and the image rest and be based on a graph representation of the image content. Clearly, the problem is then solved in a spatially discrete space and defined as the partition of an original graph—the image—into several subgraphs—the regions. Among the interest of this consideration, one of them lies in the proximity of this expression to the Gestalt's theory and the principles that govern the human vision of complex scene by perceptual clusters organization.

A graph $\mathcal{G} = \{\mathcal{V}, \mathcal{E}\}$ is composed of a set of nodes $\mathcal{V}$ and a set of directed edges $\mathcal{E} \subset \mathcal{V} \times \mathcal{V}$ that connect the nodes. Each directed edge $(u,v) \in \mathcal{E}$ is assigned to a nonnegative weight $w(u,v) \in \mathbb{R}^+ \cup \{0\}$. Two particular nodes are defined, namely, the *source* denoted by s and the *sink* denoted by t. These nodes are called *terminal* nodes in contrast to the others that are the non-terminal nodes.

The segmentation process separates the nodes into two disjoint subsets $\mathcal{S}$ and $\mathcal{T}$, including, respectively, the source and sink nodes (Fig. 8.23b). The path between the nodes which defines the partition is called the *cut* with $\mathcal{C} \subset \mathcal{E}$ and a cost $C(\mathcal{C})$ is assigned to this cut, which is equal to the sum of all weights assigned to the disconnected edges:

$$C(\mathcal{C}) = \sum_{(u,v) \in \mathcal{C}} w(u,v). \tag{8.32}$$

For the graphcut approaches, the segmentation problem is then posed as a minimization process of this cost to obtain the so-called *Mincut*. Thanks to the theorem of Ford and Fulkerson [44], the minimum cut searching is equivalent to the searching of maximal flow from the source to the sink trough the graph.

The classical scheme that present the purpose is:

The initial formalism is well adapted to binary segmentation of an image I into an object "obj" and a background "bkg." In this case, we have a region map $K = I_{\mathrm{R}}$ with $N_{\mathrm{R}} = 2$. That could be expressed on the minimization of a soft-constraint $E_{\mathrm{R}}(K)$ which includes one term linked to the region property and one term linked to the boundary property of the segmentation K [15]. Under this term, the segmentation is expressed through a binary vector

$$I_{\mathrm{R}} = (I_{\mathrm{R}}(1,1),\ldots,I_{\mathrm{R}}(M,N)) = (\ldots,K(u),\ldots K(v),\ldots) = K$$

with $K(u) \in \{\text{“obj”},\text{“bkg”}\}$.

We now link the graph problem to the segmentation problem. We define a graph with $|\mathcal{V}| = N_{\mathrm{P}}$, i.e., each pixel is assigned to a node. The problem to segment the image into object and background is mapped to the problem to define a graph cut that separates object nodes $\mathcal{O}$ from background nodes $\mathcal{B}$. The choice of source and sink nodes initializes an optimization problem using an energy function expressed the quality of the obtained partition:

$$E_{\mathrm{R}}(K) = \lambda E_{\mathrm{A}}(K) + E_{\mathrm{B}}(K), \tag{8.33}$$

where

$$E_{\mathrm{A}}(K) = \sum_{p\in\mathcal{V}} A(u,K(u)), \tag{8.34}$$

$$E_{\mathrm{B}}(K) = \sum_{(u,v)\in\mathcal{E}} B(u,v)\delta(u,v) \tag{8.35}$$

and $\delta(u,v) = 1$ if $(u,v) \in \mathcal{E}$ and 0 otherwise.

The coefficient $\lambda \leq 0$ is chosen to weight the importance of the region content $E_{\mathrm{A}}(K)$ in front of the boundary information $E_{\mathrm{B}}(K)$. In such writing, it is supposed that the individual penalties $A(u,\text{“obj”})$ (resp. $A(u,\text{“bkg”})$) that assign a location u to the object (resp. background) label could be defined, e.g., in a similarity measure. In the same manner, the term attached to the boundary information $B(u,v)$ could evolve as a similarity measure between the content of the u and v nodes, which is near from zero when the two contents are very different. Generally, $B(u,v)$ is based on gradient or assimilated functions, but Mortensen propose to combine a weighted sum of six features, such as Laplacian zero-crossing, gradient magnitude, and direction [89].

Boykov gives and demonstrates in a theorem that the segmentation defined by a minimal cut minimizes the expression (8.33) distributes the edges weighting functions upon their type [15] as in Table 8.3.

Table 8.3 Weighting functions for different types of edges according to [15]

Edge	Weight w	For
(u,v)	$B(u,v)$	$(u,v) \in \mathcal{C}$
	$\lambda A(u,\text{“}\mathtt{bkg}\text{''})$	$u \in \mathcal{V}, u \notin \mathcal{O} \cup \mathcal{B}$
(s,u)	K	$u \in \mathcal{O}$
	0	$u \in \mathcal{B}$
	$\lambda A(v,\text{“}\mathtt{obj}\text{''})$	$v \in \mathcal{V}, v \notin \mathcal{O} \cup \mathcal{B}$
(v,t)	0	$v \in \mathcal{O}$
	K	$v \in \mathcal{B}$

$K = 1 + \max_{p \in \mathcal{V}} \sum_{q:(u,v) \in \mathcal{E}} B(u,v)$

Among the different form of boundary penalty functions [89, 122], the most popular one is on a Gaussian form:

$$B(u,v) \propto \exp(-\frac{-|I(u)-I(v)|^2}{2\sigma^2}) \cdot \frac{1}{\text{dist}(u,v)}, \tag{8.36}$$

where $|I(u)-I(v)|$ is a color distance at pixel scales, or a color distribution similarity metric for regions scales. Without the nonlinearity of the exponential function, the algorithm could not work properly, because the flow between edges is not sufficiently different. So whatever the boundary penalties function uses, it needs a high dynamical response to allow a good flow saturation.

For the individual penalties, the expression are explained in a negative log-likehoods based on a color histogram similarity metrics too:

$$\begin{aligned} A(u,\text{“}\mathtt{obj}\text{''}) &= -\ln(p(I(u)|\mathcal{O})), \\ A(u,\text{“}\mathtt{bkg}\text{''}) &= -\ln(p(I(u)|\mathcal{B})), \end{aligned} \tag{8.37}$$

where the probabilities are obtained from a learning stage [86].

To understand the relation between the used parameters in the graphcut formulation, we develop an example around the waterlily image (Fig. 8.24). In this example, we chose to start with a pre-segmented image, through a CSF gradient and a watershed segementation. The major information used during the processing is the similarity measure between nodes. As we work with regions of different size, we use a particular similarity metrics based on a cumulative sum of ΔE color distance, randomly extracted from the region $A(u,K(u))$ and $A(v,K(v))$ to be independent from the spatial organization. The histogram in Fig. 8.24c shows that the distribution between small differences to larger is continuous. Ideally, we expected that such histograms present two modes, one for small differences resulting from nodes in the same object or region, and another for larger distances between nodes belonging to two different region or objects.

In a first result, we work only with the boundaries information only, as we set $\lambda = 0$ parameter by modifying only the sigma value. In these results, we consider only one node for the source (a light purple petals of the water lily on the left) and one for the sink (a green leaf on top of the central water lily). Due to the definition

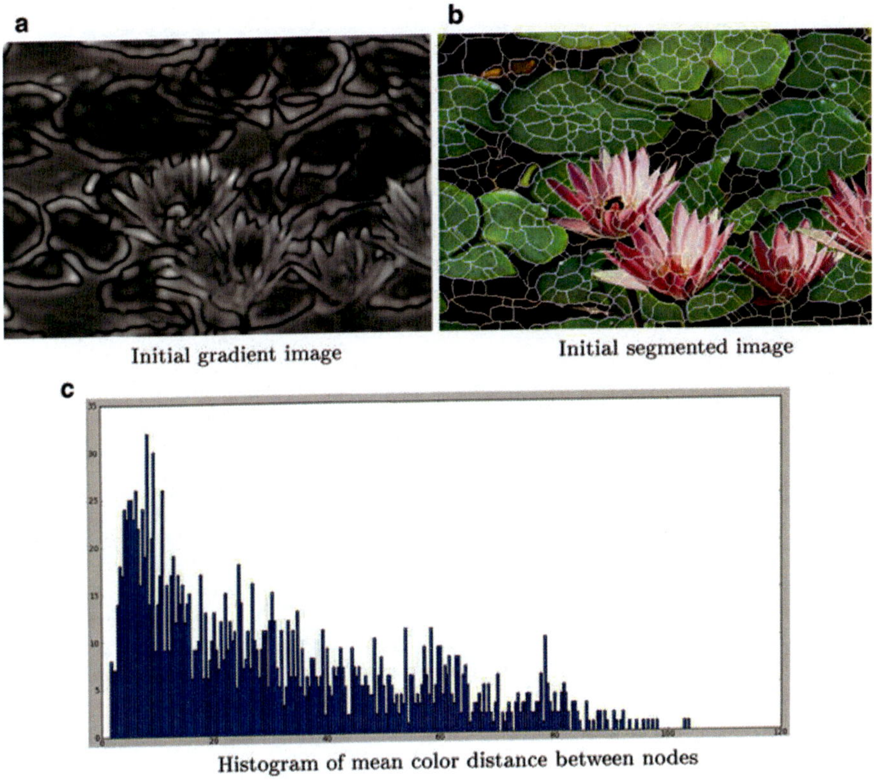

Fig. 8.24 Initial data for the graphcuts segmentation of "waterlilies" image

of the distances distribution, and thus of the weights attached to the boundaries, σ values closed to one allow to separate two connected flowers. When the σ value grows up, only the source node is extracted and labelled as *object*; in contrast, a reduction of the σ value induced that more important distance values are considered in the process, and then the integration of more dissimilar regions (Fig. 8.25).

In a second result, we study the impact of the region information; for that, we chose a value for σ, such that the response of the boundaries information is quite similar for all the distances values between nodes: $\sigma = 1000.0$. For a small value of *lambda*, the cut isolates just the sink node because the λ multiplicative factor does not create sufficiently differences between the edges weights. For more important values of λ, the gap between the nodes affected with high probability and the nodes affected with a low probability is amplified, so the nodes classification is better resolved (Fig. 8.26).

In Fig. 8.27, we show the obtained result for the best parameter set ($\sigma = 1.0$, $\lambda = 10.0$) and for the closest ones, around these values the graphcut behavior tends to be closed from the previous ones. In a second step, we modify the initialization nodes. In Fig. 8.27c, we chose a lighter petal in the central water lily for the source

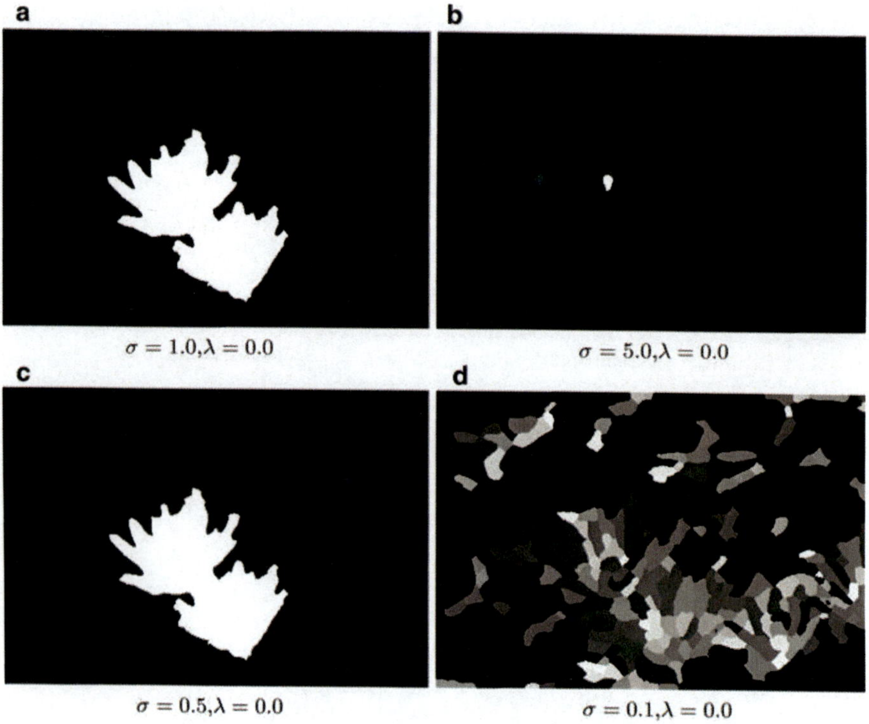

Fig. 8.25 Graphcut results based only on boundaries informations

and a black region at the top of the central water lily. With this initialization and the same parameter set, the region information takes over the boundaries information to extract nodes with the colors with the highest intensity. The results in Fig. 8.27d show that the sink node is the same as the first results (green leaf), but the source is a dark purple petal; the behavior is thus reversed compared to the previous result with the extraction of dark region in image.

There exist lot of variation from the initial graphcut formulation, in particular in the dissimilarity metrics used in the boundaries information or in the computation of the probability to label a node. Nevertheless, from the initial expression we could produce some remarks. First ones appear clearly in this section; the λ parameter exists to control the weight of the region information in front of the boundaries information. But this control is not clear and must be associated to the σ parameter management. A classical formulation like in (8.38) is more adapted to this management.

$$E_{\mathrm{R}}(K) = \lambda E_{\mathrm{A}}(K) + (1-\lambda)E_{\mathrm{B}}(K). \tag{8.38}$$

In addition, a second comment should be made on the dynamic from each part of (8.33) or (8.38), upon the proposed expression in (8.36) and (8.37): the $B(u,v)$ expression is included in the $[0,1]$ interval, but the $A(u,K(u))$ is included

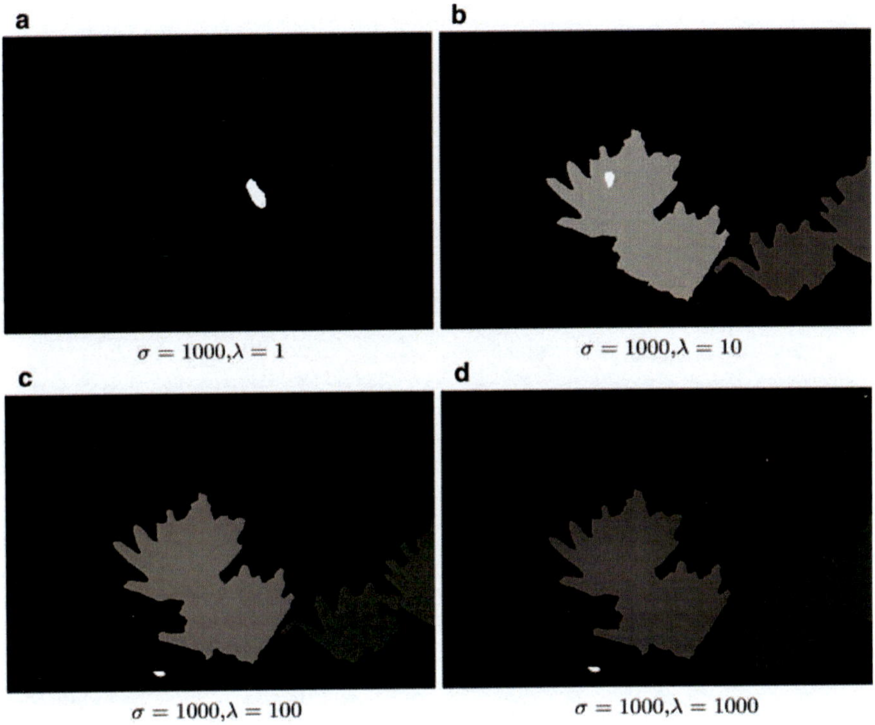

Fig. 8.26 Graphcut results based only on region informations

in the $[0, +\infty]$, so depending on the λ parameter the behavior of the graphcut algorithm should closer than a based-features classification algorithm than a mixed boundaries-content segmentation algorithm. For examples images, this behavior is crucial to obtain good results, and is linked to the important value of the λ parameter, greater than a unit value and consequently inducing the Boykov formulation in (8.33).

There exist several methods for image segmentation that extend the ideas for graphs introduced so far; the reader should refer to normalized cuts [122] or to the classical *mean shift* pattern recognition technique [29].

8.5 Performance Evaluation

Classically, an expert used to compile an appropriate sequence of image-processing operations consisting of preprocessing, feature extraction, image segmentation, and post-processing to solve given problem. He would tune parameters so that the results are as expected. Test cases will verify the behavior.

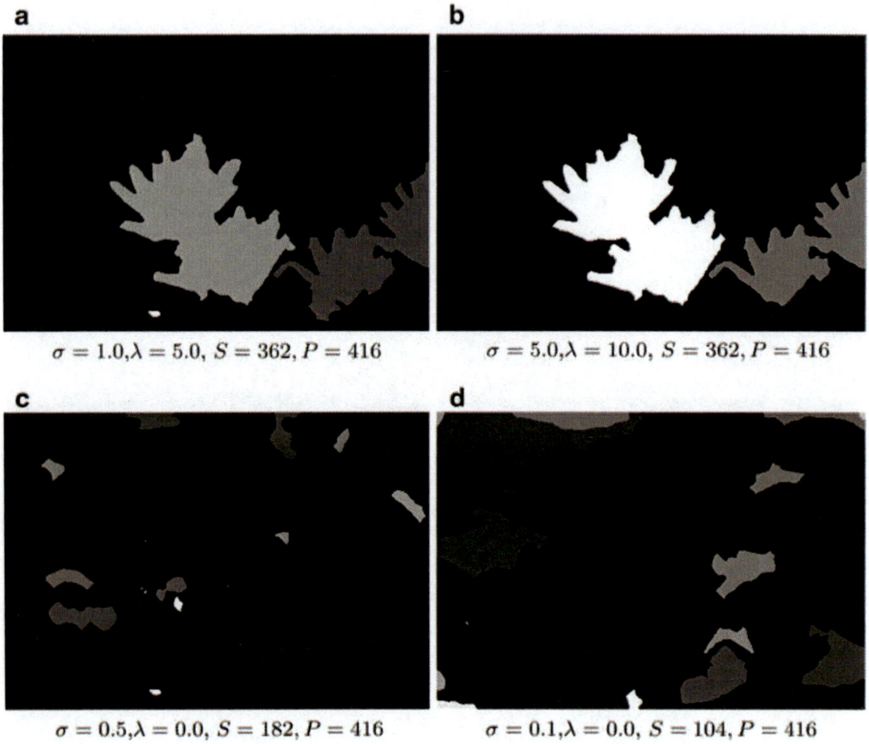

Fig. 8.27 Graphcut results upon different sources and sink nodes initialization

In this sense, segmentation was an open-loop process and the idea of refining the output of this operation appeared some time ago [12]. However, the criteria to use are still to be defined. In our view, such criteria could be defined only from the application perspective, especially from three directions: (1) from the human visual system point of view, by integrating models of human perception; (2) from a metrologic point of view, when the purpose of segmentation is to identify particular objects or regions of interest for the assessment of their features (area, number, and so on); (3) task driven in cognitive vision systems (see, e.g., [17]). In [152], the fact that the segmentation has impact on the measurement of object features (like length or area) is emphasized.

8.5.1 Validation and Strategies

The quality of a segmentation algorithm is usually *evaluated* by comparing it against a ground-truth image, i.e., a manually segmented image. The first observation would be the fact that the evaluation is therefore subjective, and based on human

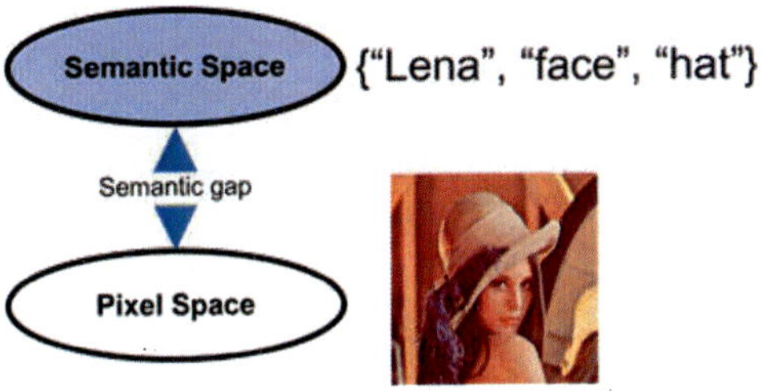

Fig. 8.28 The pixel and semantic spaces for a color image

perception. From a human point of view, segmentation is *correct* if we identify in the image the object that we are looking for: if we see an airplane, the desired segmentation result should indicate only one object for the airplane, regardless the fact that the object airplane may have different color and texture regions. Such a high-level assessment may not take into account the precision at pixel level or the exact matching between the manually segmented region and the automatically-segmented one.

A possible path for the evaluation of segmentation performance would be to find the relationships between the image and its content, i.e., the mapping between a *pixel space* and a *semantic space* (see Fig. 8.28) in order to define appropriate SQMs for the chosen segmentation approach, with respect to the content of the image. In [93], the authors identify two major approaches for accomplishing the task of pattern recognition: "close to the pixels" and "close to the goal" and they ask the question of how to fill the gap between the two approaches. This gap is often called the "semantic gap" (see, e.g., [153]).

Segmentation can be either *bottom-up* or *top-down*. For a bottom-up approach, the starting point is the obtained segmentation map, at the pixel level of the image, and the output is a set of concepts, describing the content of the image. The input data are the pixels of the image and the result is a semantic interpretation, the mapping can be performed by means of object recognition, based on the hypothesis that the regions in the segmentation map are semantically meaningful. For a top-down approach is the vice versa. In general, the gap between the two spaces is filled by a process of learning, either supervised [84] or unsupervised [8, 9, 39]. Sometimes an intermediate level is introduced via graph structures, like region adjacency graphs (RAGs) used to group the small regions obtained in an oversegmentation process, based on rules of the Gestalt theory [151]. See [132] for an example of using RAGs for color image segmentation. In [99], we introduced another graph structure that represents arbitrary segmentation results and that can also represent RAG or the RSE-graph that was used in [51]. The CSC-graph introduced in Sect. 8.4.1.1 is another example of such a graph structure. So-called *T-Graphs* were used to enrich semantical models for image analysis in [140].

We propose a top-down approach based on object detection for linking the two spaces. The object recognition allows us to perform several tasks: (1) automatic annotation, (2) initialization of the segmentation algorithm and (3) supervise the segmentation. In the diagram in Fig. 8.29, the relationship between the degree of

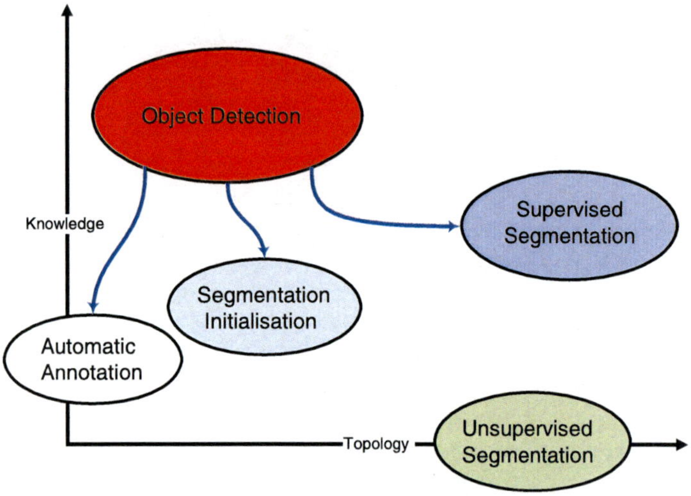

Fig. 8.29 Topology versus knowledge for various levels of segmentation

learned information, or knowledge, and the topology is depicted. For the completely unsupervised segmentation, independent from the high-level content of the image, the results may be nonsatisfactory, or the choice of parameters for tuning the segmentation would be highly complicated and time consuming. As the knowledge about the content of the image increases, segmentation may be improved.

Once the object or the pattern of interest is identified, the result of the segmentation algorithm can be analyzed, for instance, from the human perspective, the two results should be consistent according to some criteria [55, 56]. If there is one face object detected in the image, the ideal segmentation map should indicate one region corresponding to this object, thus determining at least one semantically meaningful region. To illustrate the concept, we take the example of human face detection, based on the approach proposed by Viola and Jones [65, 137]. The implementation of the approach is available in the OpenCV library [1,2]. Any segmentation algorithm may be chosen.

8.5.2 Closed-Loop Segmentation

In general, the result of the segmentation depends on the values of several input parameters, usually used in the definition of the homogeneity criteria (e.g., a threshold representing the maximum value for the distance between colors belonging to the same region). Sometimes the values for such input parameters are chosen based on the experience of the application user or they are determined automatically based on a priori information about the content of the image (e.g., the threshold for the histogram-bases segmentation approaches roughly estimated on

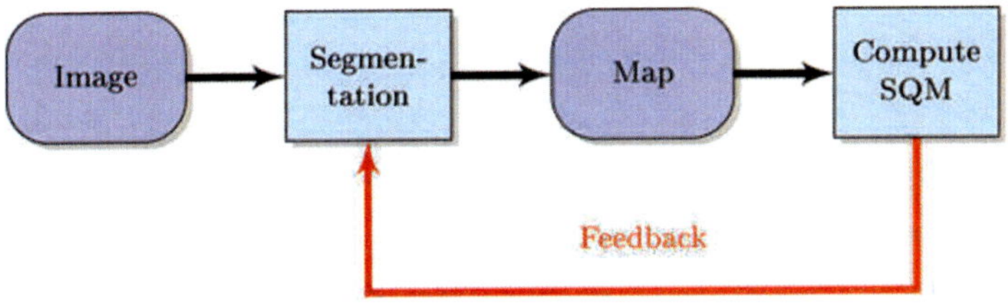

Fig. 8.30 Closed-loop segmentation

the cumulative density function knowing the size of the object and the interval of gray-level values that are characteristic for that particular object). Moreover, the process of segmentation is usually performed in *open loop*, i.e., there is no feedback to adjust the values of the input parameters. The system to be used for closed-loop segmentation may be like the one shown in Fig. 8.30, inspired from [12]. The SQM allows to modify the input parameters in order to refine the result of segmentation.

Refining or enhancing the segmentation is not as simple as it may sound. Normally there exists no gold standard or notion of correctness of a segmentation, as such judgements can only be done for testing as in Sect. 8.5.1. Still we want to tune segmentation parameters, as they were mentioned in Sect. 8.3 for almost all of the methods. The tuning should result in "better" segmentation. As shown in Fig. 8.30, this results in a control loop where a judgment has to be made on the results of segmentation. Based on this judgment, the parameters are modified. This control problem is in general not well posed, as it is unclear, which influence the parameters have on the quality measure of the result. One option is to have a human in the loop who—by its experience— will tune the parameters appropriately. For working systems this is normally not feasible and other solutions are required, as we will describe next.

8.5.3 *Supervised Segmentation*

However, even if subjective, the opinion of the *expert* could be translated into a set of rules, which can be implemented as an algorithm for steering the segmentation approach. In order to search for the *correct* segmentation, we propose to use a *supervised* approach, based on a model. The principle is depicted in Fig. 8.31.

The methodology should comprise three stages: (1) first, the image is segmented using the algorithm under evaluation; (2) secondly, objects of interest are identified in the image using another approach which is not based on segmentation; and (3) third, the quality of segmentation is evaluated based on a metric, by analyzing the segmentation map and the identified objects. Note that the first two stages can actually take place in parallel. Rather than building a graph from an oversegmented image and find rules to merge the regions and simplify the graph [149], the search space should be reduced by adjusting the values of the segmentation parameters and

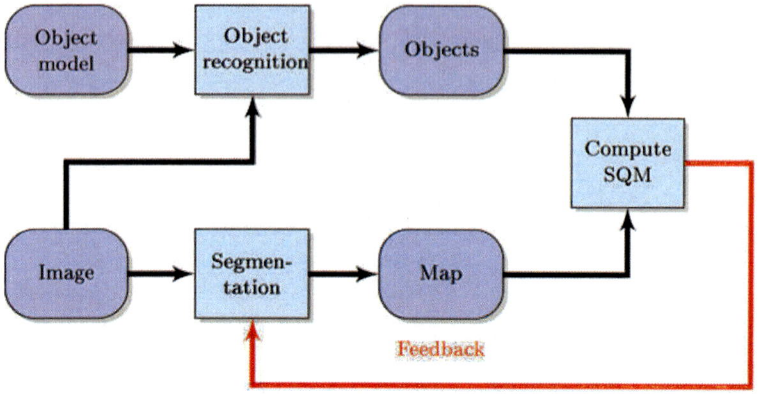

Fig. 8.31 The supervised segmentation

then try to identify the regions, associate high-level information to them, i.e., match them to a model, like in [92].

If the segmented map in the region of the detected object contains more than the expected number of regions, according to the model used, then the image is most likely oversegmented; on contrary, the image is under-segmented. One subregion may be also an indication that the whole image was undersegmented; therefore, the appropriate choice of metric or criteria should be defined. In this case, the number of regions may be a first indication or quality metric for segmentation.

Let us consider the following example of supervised segmentation, when the segmentation approach was pyramidal (see the result of face detection and one possible segmentation in Fig. 8.32). For the pyramidal segmentation algorithms consult [81], but the implementation we used in our experiments was the one available in the OpenCV libraries. As already said, the object recognition in our example was the face detection algorithm of [65].

We experimentally determined the dependency between the number of regions and the threshold representing the input parameter of the chosen segmentation approach. The results are presented in Fig. 8.33. The number of regions is a measure able to indicate if the image is over- or undersegmented. In Fig. 8.33 if the threshold exceeds 70, the number of regions remains more or less constant; therefore, a value larger than 70 is useless. If the threshold is between 1 and 15 the rapid decrease of the number of regions indicate an *unstable*, still to be refined segmentation.

8.5.4 Semantical Quality Metrics

However, the number of regions is not enough to indicate a correct segmentation from the semantic point of view; therefore, a more complex metric should be used to be able to capture some information about the content of the image.

Fig. 8.32 Original image with face detection and one possible result of pyramidal segmentation

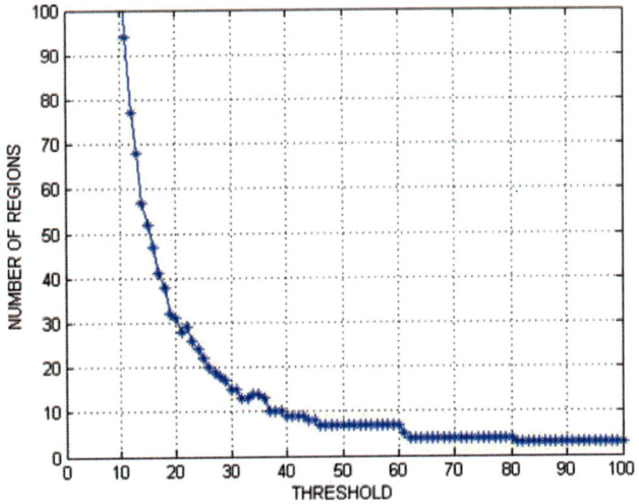

Fig. 8.33 The number of regions versus the input parameter (threshold)

The methodology we envisage is model-based recognition combined with graph matching (see Fig. 8.34 for an example of the desired segmentation map and the associate graph for the case of human face detection in a supervised segmentation: the vertices represent regions and the edges the adjacency relation between regions [111, 112]). As a consequence, the segmentation quality metric will translate into a metric of similarity between two graphs. For the matching of two graphs, there exist several metrics: in [18], the metric is a function that combines the measure of adequacy between the vertex attributes and the edges of the two graphs and in [54]

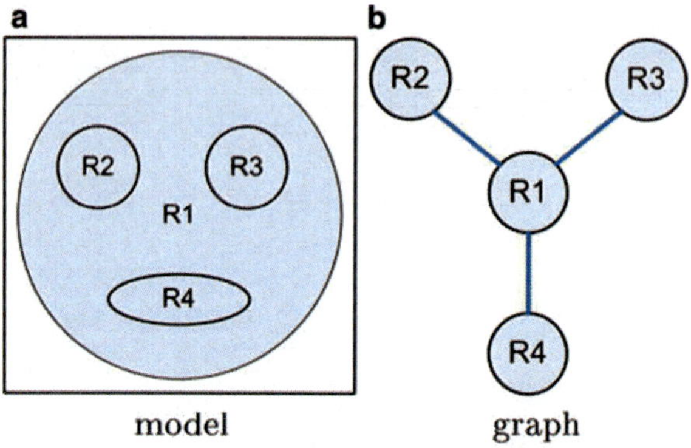

Fig. 8.34 The model of a face and the associated graph

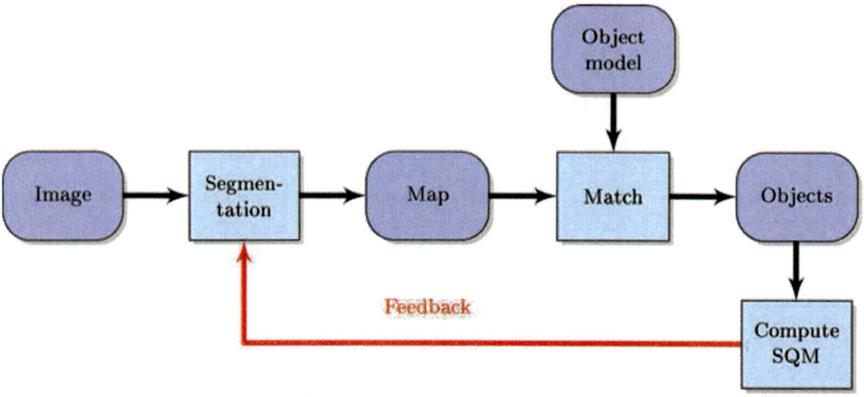

Fig. 8.35 Task-driven segmentation

the metric used is the normalized distance to the points defining the skeleton of the model. Fundamentally, all the approaches reduce to graph isomorphism [30].

Note that the proposed methodology or the chosen quality criteria may not be appropriate for the entire image or for all the objects in the semantic space. In addition, there are still several open questions: if the two graphs are isomorphic, are the objects identical? in the case when the number of regions is the same for two different values of the input parameters, are the regions also the same? Do the regions represent the elements composing the object from the semantic space? Are they pertinent for the semantic characterization of the image?

Figure 8.35 shows the classical approach to model-based image analysis extended by a measuring step based on the result of object recognition: Images are segmented and the segments are matched to object models to obtain object hypotheses, e.g., by graph matching. If expected objects are not found, the segmentation process is tuned. This provides a semantic feedback to the segmentation step.

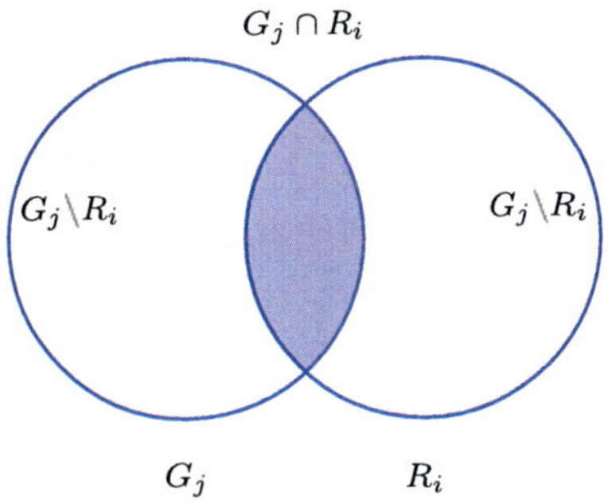

Fig. 8.36 Region segmentation compared to ground truth

8.5.5 *Image-Based Quality Metrics*

There are already several widely used segmentation quality metrics. Most simply, we compare ground truth regions G_j, as obtained by experts (see Sect. 8.5.1) to regions R_j obtained by segmentation. The number of pixels that are correct $(G_j \cup R_i)$, missing $(G_j \backslash R_i)$, or superfluent $(G_j \backslash R_i)$ can be scaled by the size of the region to derive various measures for the fit of that region (Fig. 8.36). Such measures are used, in particular, for segmentation of medical images, as can be seen, e.g., in [3].

The empirical function F (8.39) proposed by Liu and Yang in [78] basically incorporates to some extent three of the four heuristic criteria suggested by Haralick and Shapiro [52] for the objective assessment of the segmentation result, namely (1), (2) and (3) listed on p. 224:

$$F(I) = \frac{1}{1000(N \cdot M)} \sqrt{N_{\mathrm{R}}} \sum_{i=1}^{N_{\mathrm{R}}} \frac{e_i^2}{\sqrt{A_i}}. \tag{8.39}$$

Here, I is the segmented image of size $N \times M$, N_{R} is the number of resulted regions, A_i is the area of region R_i and e_i is defined as the sum of the Euclidian distances between the RGB color vectors of the pixels of region R_i and the color vector designated as representative to that region. A small value of $F(I)$ indicates a *good* segmentation. However, the authors of [14] make several observations regarding this quality measure: (2) a large number of regions in the segmented image is penalized only by the factor $\sqrt{N_{\mathrm{R}}}$ and (2) the average color error of small regions is usually close to zero; therefore, the function F tends to evaluate in a favorable manner the very noisy segmentation results. Consequently, Borsotti et al. propose two improved versions of the original F metric proposed by Liu:

$$\begin{cases} F'(I) = \frac{1}{10000(N \cdot M)} \sqrt{\sum\limits_{A=1}^{\mathrm{Max}} [N_{\mathrm{A}}(A)]^{1+\frac{1}{A}}} \cdot \sum\limits_{i=1}^{N_{\mathrm{R}}} \frac{e_i^2}{\sqrt{A_i}}, \\ Q(I) = \frac{1}{10000(N \cdot M)} \sqrt{N_{\mathrm{R}}} \cdot \sum\limits_{i=1}^{N_{\mathrm{R}}} \left[\frac{e_i^2}{1+\log A_i} + \left(\frac{N_{\mathrm{A}}(A_i)}{A_i} \right)^2 \right], \end{cases} \tag{8.40}$$

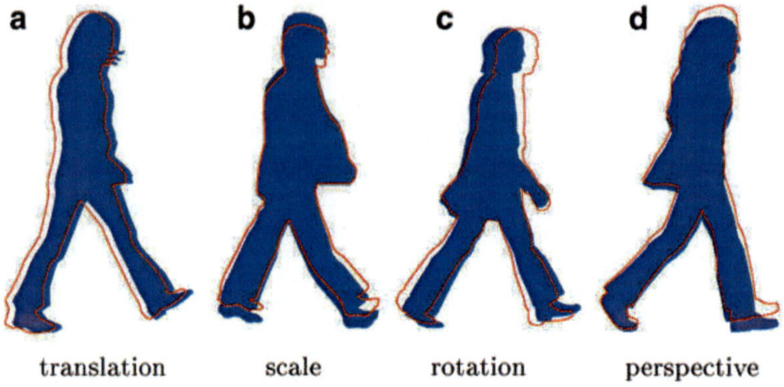

Fig. 8.37 Four types of alterations according to [55]

where $N_A(A)$ is the number of regions having exactly area A and Max is the area of the largest region in the segmented image. The exponent $(1+\frac{1}{A})$ has the role to enhance the small regions' contribution. The term $\sqrt{A_i}$ was replaced with $(1+\log A_i)$ in order to increase the penalization of non-homogeneous regions. The two metrics were scaled by a factor of 10 in order to obtain a range of values similar to the one of F.

A pertinent observation is the one in [103]: for a given image, the resulting segmentation is not unique, depending on the segmentation approach that is used, or the choice of input parameters. The region boundaries slightly differ as a result of very minor changes of the values of the input parameters or minor changes in the image: the most common change being a *shift* of a few pixels. The authors of [103] stated that the shift can be a translation, a rotation, or even a dilation, but they propose the metric called *shift variance* restricted to the case of translations. In [55], four types of *alterations* are identified: translation, scale change, rotation, and perspective (change) (see Fig. 8.37).

Hemery et al. [55] consider that a quality metric, first of all, should fulfill several properties: (1) *symmetry*, since the metric should penalize in the same way two results exhibiting the same alteration, but in opposite directions; (2) *strict monotony*, because the metric should penalize the results the more they are altered; (3) *uniform continuity*, since the metric should not have an important gap between two close results; and finally, (4) *topological dependency* given that the metric result should depend on the size or the shape of the localized object.

The metric proposed in [55] is region based and the methodology relies on finding the correspondence between the objects from the manual (ground truth) segmentation and the resulting regions of the segmentation. This first step of matching allows also for the detection of missed objects (under-detection) and the detection of too many objects (oversegmentation). For the implementation of the matching phase, the authors used the matching score matrix proposed in [100]. Then they compute the recovery of objects using the so-called *PAS* metric [41] or some other metric as defined in [83].

Further on, Hemery et al. proposed in [56] the metric called the cumulative similarity of correct comparison (*SCC*) and a variant of it, referenced to the maximum value of *SCC*, and proved it to respect all four properties mentioned above. However, according to [90], all quality metrics should reflect the human assessment, and the quality measures should be compared against the subjective opinion and measure the agreement between the two; therefore, the authors proposed the attribute-specific severity evaluation tool (ASSET) and rank agreement measure (RAM).

8.6 Conclusions

Recently *TurboPixel* segmentation or *SuperPixels* [75] imposed, as an approach for image oversegmentation, whose result is a lattice-like structure of superpixel regions of relatively uniform size. An approach based on learning eigen-images from the image to be segmented is presented in [145], the underlying idea being the one of pixel clustering: for the pixels in each local window, a linear transformation is introduced to map their color vectors to be the cluster indicator vectors. Based on the eigen-images constructed in an error minimization process, a multidimensional image gradient operator is defined to evaluate the gradient, which is supplied to the TurboPixel algorithm to obtain the final superpixel segmentations.

Even we presented the color segmentation frameworks separately, there is a fine frontier between them and quite often hybrid techniques emerge, that combine, for instance, pyramids and watersheds [4] and the approach proposed by Serra [119]. However, the segmentation approaches evolved toward unanimously accepted frameworks with relatively well-defined characteristics. In addition, the quality segmentation metric offer the way to refine the segmentation process, thus leading to very effective segmentation approaches and, consequently, to the best results.

The segmentation process requires addressing three kinds of issues: firstly the features capturing the homogeneity of regions, secondly, the similarity measures or distance functions between features content, and finally,f the segmentation framework which optimizes the segmentation map in function of the feature/metrics tandem. Given that for each problem there exists a plethora of approaches, in the last decades we are spectators to the development of many combinations of those, in single or multiple-scale ways. Surprisingly, if several segmentation frameworks imposed themselves as *standard* techniques: pyramidal approaches, watershed, JSEG, graph cuts, normalized cuts, active contours, or more recently, TurboPixels, very few advances have been made to address the first two issues of features and metrics. Even if, given the increasing computation capabilities, new frameworks based on graph theory have appeared and offer direct links to the semantic level of image processing. All these recent algorithms are based on the fundamental idea that *cuts* in images is intimately connected to maximizing a heterogeneity or homogeneity function. So we could not expect that such frameworks, as perfect as they are, will solve issues related to choice of the pair attribute-metric.

Sad enough, even today the question still remains since Haralick: which is the best value for the parameters of the homogeneity criteria? There are no recommended recipes! One conclusion would be the fact that the segmentation process remains the privilege of experts. However, there is a growing demand for the development of segmentation quality metrics that allow for the quantitative and objective evaluation, which consequently will lead to the automatic choice of values for the input parameters of the segmentation approaches. To conclude, the question is what is the quality of all these quality metrics? In the end, it's the human observer to give the answer [90].

Fortunately, new perspectives come from psychophysics with perceptual theory, in particular, Gestalt theory. As the homogeneity/heterogeneity definition have been expressed as complexity of a feature distribution, these perceptual theories search to explain what are the physical parameters that are taken into account by the human visual system. Such developments are not new, and were initiated in 1923 by Max Wertheimer, under the assumption that "a small sets of geometric laws governing the perceptual synthesis of phenomenal objects, or 'Gestalt' from the atomic retina input" [35]. Nevertheless, this theory is not straightforward in image processing, because it addresses the human vision and not image acquisition, processing, and rendering. Introducing such developments in image segmentation induces the transformation of visual properties in quantitative features. According to [118], there are seven Gestalt-grouping principles that assist in arranging forms: *proximity/nearness*, *similarity*, *uniform connectedness*, *good continuation*, *common fate*, *symmetry*, and *closure*. Most of them are easy to adapt in image processing, as they talk about shape, topological organization, or neighborhood. Under another point of view, these laws are the correct framework for the right quality metrics for segmentation based on validated human vision properties.

But, what is the link between the similarity law from Gestalt theory and the homogeneity property in segmentation? Randall in [105] links the similarity law *to grouping into homogeneus regions of color or texture*. Several works in physiology and in human vision have explored this process with stochastic or regular patterns, but nowadays this work is in progress for color patterns as well. Nevertheless, the definition of homogeneity is still imperfect, often reduced to basic moments available for particular scales of the image. The future trends should be around these questions.

To end this chapter, it is interesting to see that Randall apologizes for a recursive grouping through edge extraction and region clustering, finally as Paillоncy suggested in his original paper.

Acknowledgment We would like to thank Martin Druon, Audrey Ledoux, and Julien Dombre (XLIM-SIC UMR CNRS 6172, Université de Poitiers, France), Diana Stoica and Alexandru Căliman (MIV Imaging Venture, Transilvania University, Braşov, România) for the results they provided and for the fruitful discussions. Image "angel" is courtesy of Centre d'Etudes Supérieurs de Civilisation Médiévale (CESCM), UMR 6223, Poitiers, France, while the melanoma image is courtesy of Dermnet Skin Disease Image Atlas, http://www.dermnet.com.

References

1. Seo N (2008) Tutorial: OpenCV haartraining, Rapid object detection with a cascade of boosted classifiers based on haar-like features, http://note.sonots.com/SciSoftware/haartraining.html
2. Bradski G, Kaehler A, Pisarevsky V (2005) Learning-based computer vision with Intel's open source computer vision library. Intel Technology Journal, vol. 09, issue 02, May 2005
3. Ameling S, Wirth S, Shevchenko N, Wittenberg T, Paulus D, Münzenmayer C (2009) Detection of lesions in colonoscopic images: a review. In: Dössel O, Schlegel WC (eds) World congress on medical physics and biomedical engineering, vol 25/IV. Springer, Heidelberg, pp 995–998
4. Angulo J, Serra J (2003) Color segmentation by ordered mergings. In: Image Processing, 2003. ICIP 2003. Proceedings. 2003 International Conference on, vol 2, pp II – 125–8 vol 3, DOI:10.1109/ICIP.2003.1246632
5. Antonisse HJ (1982) Image segmentation in pyramids. Comput Graph Image Process 19(4):367–383, DOI:10.1016/0146-664X(82)90022-3
6. Arbelaez PA, Cohen LD (2004) Segmentation d'images couleur par partitions de voronoi - color image segmentation by voronoi partitions. Traitement du signal 21(5):407–421
7. Bardet JM (1998) Dimension de corrlation locale et dimension de hausdorff des processus vectoriels continus - local correlation dimension and hausdorff dimension of continuous random fields. Comptes Rendus de l'Acadmie des Sciences - Series I - Mathematics 326(5):589–594
8. Barnard K, Duygulu P, Freitas OD, Forsyth D (2002) Object recognition as machine translation - part 2: exploiting image data-base clustering models. In: European Conference on Computer Vision
9. Barnard K, Duygulu P, Forsyth D, de Freitas N, Blei DM, Jordan MI (2003) Matching words and pictures. J Mach Learn Res 3:1107–1135
10. Beucher S (1982) Watersheds of functions and picture segmentation. Acoustics, Speech, and Signal Processing, IEEE International Conference on ICASSP '82 7:1928–1931
11. Beucher S (1994) Watershed, hierarchical segmentation and waterfall algorithm. In: Serra J, Soille P (eds) Mathematical morphology and its applications to image processing, computational imaging and vision, Kluwer Academic Publishers, Fontainebleau, France, vol 2. pp 69–76
12. Bhanu B, Lee S, Ming J (1991) Closed-loop adaptive image segmentation. In: Computer vision and pattern recognition 1991, Maui, Hawaii, pp 734–735
13. Bister M, Cornelis J, Rosenfeld A (1990) A critical view of pyramid segmentation algorithms. Pattern Recogn Lett 11:605–617, DOI:10.1016/0167-8655(90)90013-R
14. Borsotti M, Campadelli P, Schettini R (1998) Quantitative evaluation of color image segmentation results. Pattern Recogn Lett 19:741–747
15. Boykov Y, Jolly M (2001) Interactive graph cuts for optimal boundary and region segmentation of objects in n-d images. In: International conference on computer vision, vol 1, pp 105–112
16. Canny J (1986) A computational approach to edge detection. IEEE Trans Pattern Anal Mach Intell 8(6):679–698
17. Caputo B, Vincze M (eds) (2008) Cognitive Vision, 4th International Workshop - Revised Selected Papers, Santorini, Greece, May 12, 2008
18. Cesar Jr RM, Bengoetxea E, Bloch I, Larrañaga P (2005) Inexact graph matching for model-based recognition: evaluation and comparison of optimization algorithms. Pattern Recognition, Volume 38, Issue 11
19. Chanussot J, Lambert P (1998) Total ordering based on space filling curves for multivalued morphology. In: Proceedings of the fourth international symposium on Mathematical morphology and its applications to image and signal processing, Kluwer Academic Publishers, Norwell, MA, USA, ISMM '98, pp 51–58

20. Chanussot J, Lambert P (1999) Watershed approaches for color image segmentation. In: NSIP'99, pp 129–133
21. Chaudhuri B, Sarkar N (1995) Texture segmentation using fractal dimension. IEEE Trans Pattern Anal Mach Intell 17(1):72–77
22. Chen J, Pappas T, Mojsilovic A, Rogowitz B (2005) Adaptive perceptual color-texture image segmentation. IEEE Trans Image Process 14(10):1524–1536
23. Chi CY, Tai SC (2006) Perceptual color contrast based watershed for color image segmentation. In: Systems, man and cybernetics, 2006. SMC '06. IEEE international conference on, vol 4, pp 3548–3553
24. Commission Internationale de l'Eclairage (CIE) (2008) Colorimetry - part 4: Cie 1976 l*a*b* colour spaces. Tech. rep., CIE
25. Commission Internationale de l'Eclairage (CIE) (1995) Industrial colour-difference evaluation. CIE Publication 116
26. Commission Internationale de l'Eclairage (CIE) (2001) Technical report: improvement to industrial colordifference evaluation. CIE Publication 142
27. Clarke FJJ, McDonald R, Rigg B (1984) Modification to the JPC79 Colour–difference Formula. J Soc Dyers Colourists 100(4):128–132
28. Cohen LD (1991) On active contour models and balloons. CVGIP: Image Underst 53:211–218, DOI:10.1016/1049-9660(91)90028-N
29. Comaniciu D, Meer P (2002) Mean shift: a robust approach toward feature space analysis. IEEE Trans Pattern Anal Mach Intell 24:603–619, DOI:10.1109/34.1000236
30. Cordella LP, Foggia P, Sansone C, Vento M (2001) An improved algorithm for matching large graphs. In: 3rd IAPR-TC15 Workshop on Graph-based representations in pattern recognition, Cuen, pp 149–159
31. Couprie M, Bertrand G (1997) Topological gray-scale watershed transform. In: Proceedings of SPIE vision geometry V, vol 3168, pp 136–146
32. Cousty J, Bertrand G, Najman L, Couprie M (2009) Watershed cuts: minimum spanning forests and the drop of water principle. IEEE Trans Pattern Anal Mach Intell 31(8):1362–1374
33. Deng Y, Manjunath BS (2001) Unsupervised segmentation of color-texture regions in images and video. IEEE Trans Pattern Anal Mach Intell (PAMI '01) 23(8):800–810
34. Deng Y, Manjunath BS, Shin H (1999) Color image segmentation. In: Proc. IEEE computer society conference on computer vision and pattern recognition CVPR'99, Fort Collins, CO, vol 2, pp 446–51
35. Desolneux A, Moisan L, Morel JM (2003) Computational gestalts and perception thresholds. J Physiol 97:311–324
36. DIN 6176. Farbmetrische Bestimmung von Farbabständen bei Körperfarben nach der DIN99-Formel (Colorimetric evaluation of colour differences of surface colours according to DIN99 formula), DIN Deutsches Institut für Normung e. V., Burggrafenstraße 6, 10787 Berlin, Germany
37. Dombre J (2003) Multi-scale representation systems for indexing and restoring color medieval archives, PhD thesis, University of Poitiers, France, http://tel.archives-ouvertes.fr/tel-00006234/
38. Domon M, Honda E (1999) Correlation of measured fractal dimensions with lacunarities in computer-generated three-dimensional images of cantor sets and those of fractal brownian motion. In: FORMA, vol 14, pp 249–263
39. Duygulu P, Barnard K, Freitas JFG de, Forsyth DA (2002) Object recognition as machine translation: learning a lexicon for a fixed image vocabulary. In: Proceedings of the 7th European conference on computer vision-part IV, Springer-Verlag, London, UK, UK, ECCV 02, pp 97–112
40. Edgar G (1990) Measure, topology and fractal geometry. Springer, New York
41. Everingham M, Zisserman A, Williams C, Van Gool L, Allan M, Bishop C, Chapelle O, Dalal N, Deselaers T, Dork G, Duffner S, Eichhorn J, Farquhar J, Fritz M, Garcia C, Griffiths T, Jurie F, Keysers D, Koskela M, Laaksonen J, Larlus D, Leibe B, Meng H, Ney H, Schiele B, Schmid C, Seemann E, Shawe-Taylor J, Storkey A, Szedmak S, Triggs B,

Ulusoy I, Viitaniemi V, Zhang J (2006) The 2005 pascal visual object classes challenge. In: Machine learning challenges. Evaluating predictive uncertainty, visual object classification, and recognising tectual entailment, Lecture notes in computer science, vol 3944. Springer, Berlin, pp 117–176
42. Falconer K (1990) Fractal Geometry, mathematical foundations and applications. Wiley, New York
43. Feagin R (2005) Heterogeneity versus homogeneity: a conceptual and mathematical theory in terms of scale-invariant and scale-covariant distributions. Ecol Complex 2:339–356
44. Ford L, Fulkerson D (1962) Flows in networks. Princeton University Press, Princeton
45. Fu K, Mui J (1981) A survey on image segmentation. Pattern Recogn 13(1):3–16
46. Funt BV, Finlayson GD (1995) Color constant color indexing. IEEE Trans Pattern Anal Mach Intell 17:522–529
47. Galloway MM (1975) Texture analysis using gray level run lengths. Comput Graph Image Process 4(2):172–179, DOI:10.1016/S0146-664X(75)80008-6
48. Gil D, Radeva P (2003) Curvature vector flow to assure convergent deformable models for shape modelling. In: EMMCVPR, pp 357–372
49. Gonzalez RC, Woods RE (2006) Digital image processing, 3rd edn. Prentice-Hall, Inc., NJ
50. Hanbury A (2003) A 3d-polar coordinate colour representation well adapted to image analysis. In: Proceedings of the 13th Scandinavian conference on image analysis, Springer, Berlin, Heidelberg, SCIA'03, pp 804–811
51. Hanson A, Riseman E (1978) Visions: a computer system for interpreting scenes. In: Hanson A, Riseman E (eds) Computer vision systems. Academic, New York, pp 303–333
52. Haralick R, Shapiro L (1985) Image segmentation techniques. Comput Vis Graph Image Process 29(1):100–132
53. Harris C, Stephens M (1988) A Combined Corner and Edge Detection, in Proceedings of the 4th Alvey Vision Conference, volume 15, pp 147–151
54. He L, Han CY, Everding B, Wee WG (2004) Graph matching for object recognition and recovery. Pattern recogn 37:1557–1560
55. Hemery B, Laurent H, Rosenberger C (2009) Evaluation metric for image understanding. In: ICIP, pp 4381–4384
56. Hemery B, Laurent H, Rosenberger C (2010) Subjective evaluation of image understanding results. In: European Signal Processing Conference (EUSIPCO), August 23–27, Aalborg, Denmark
57. Huang J, Kumar SR, Mitra M, Zhu WJ, Zabih R (1997) Image indexing using color correlograms. In: Proceedings of the 1997 conference on computer vision and pattern recognition (CVPR '97), IEEE Computer Society, Washington, CVPR '97, pp 762–768
58. Huang ZK, Liu DH (2007) Segmentation of color image using em algorithm in hsv color space. In: Information acquisition, 2007. ICIA '07. International conference on, pp 316–319, DOI:10.1109/ICIA.2007.4295749
59. Ionescu M, Ralescu A (2004) Fuzzy hamming distance in a content-based image retrieval system. In: Fuzzy systems, 2004. Proceedings. 2004 IEEE international conference on, vol 3, pp 1721–1726
60. Ivanovici M, Richard N (2009) Fractal dimension of colour fractal images. IEEE TransImage Process 20(1):227–235
61. Ivanovici M, Richard N (2009) The lacunarity of colour fractal images. In: ICIP'09 - IEEE international conference on image processing, Cairo, Egypt, pp 453–456
62. Jain AK (1989) Fundamentals of digital image processing. Prentice-Hall, Inc., NJ, USA
63. Jing X, Jian W, Feng Y, Zhi-ming C (2008) A level set method for color image segmentation based on bayesian classifier. In: Computer science and software engineering, 2008 International conference on, vol 2, pp 886–890, DOI:10.1109/CSSE.2008.1193
64. Jolion JM, Montanvert A (1991) The adaptive pyramid: a framework for 2d image analysis. CVGIP: Image underst 55:339–348
65. Jones M, Viola P (2003) Fast multi-view face detection, Technical Report, Mitsubishi Electric Research Laboratories

66. Kapur T, Grimson WEL, Kikinis R (1995) Segmentation of brain tissue from mr images. In: Proceedings of the first international conference on computer vision, virtual reality and robotics in medicine, Springer, London, UK, CVRMed '95, pp 429–433
67. Kass M, Witkin A, Terzopoulos D (1988) Snakes: active contour models. Int J Comput Vision 1(4):321–331
68. Keller J, Chen S (1989) Texture description and segmentation through fractal geometry. Comput Vis Graph Image process 45:150–166
69. Kiser C, Musial C, Sen P (2008) Accelerating Active Contour Algorithms with the Gradient Diffusion Field. In: Proceedings of international conference on pattern recognition (ICPR) 2008
70. Kolasa J, Rollo C (1991) chap The heterogeneity of heterogeneity: a glossary. Ecological heterogeneity (Ecological studies), 1st edn. Springer, New-York, pp 1–23
71. Komati KS, Salles EO, Filho MS (2009) Fractal-jseg: jseg using an homogeneity measurement based on local fractal descriptor. Graphics, patterns and images, SIBGRAPI Conference on 0:253–260
72. Kropatsch W (1995) Building irregular pyramids by dual-graph contraction. Vision Image Signal Process, IEE Proc - 142(6):366–374, DOI:10.1049/ip-vis:19952115
73. Kwatra V, Schödl A, Essa I, Turk G, Bobick A (2003) Graphcut textures: image and video synthesis using graph cuts. In: ACM SIGGRAPH 2003 Papers, ACM, New York, SIGGRAPH '03, pp 277–286
74. Lay J, Guan L (2004) Retrieval for color artistry concepts. IEEE Trans Image Process 13(3):326–339
75. Levinshtein A, Stere A, Kutulakos KN, Fleet DJ, Dickinson SJ, Siddiqi K (2009) Turbopixels: fast superpixels using geometric flows. IEEE Trans Pattern Anal Mach Intell: 31(12):2290–2297
76. Li B, Loehle C (1995) Wavelet analysis of multiscale permeabilities in the subsurface. Geophys Res Lett 22(23):3123–3126
77. Li H, Reynolds JF (1995) On definition and quantification of heterogeneity. Oikos 73(2):280–284
78. Liu J, Yang YH (1994) Multiresolution color image segmentation. IEEE Trans Pattern Anal Mach Intell 16:689–700, DOI:10.1109/34.297949
79. MacAdam D (1942) Visual sensitivities to color differences in daylight. JOSA 32(5):247–273
80. Mandelbrot B (1982) The fractal geometry of nature. W.H. Freeman and Co, New-York
81. Marfil R, Molina-Tanco L, Bandera A, Rodríguez J, Sandoval F (2006) Pyramid segmentation algorithms revisited. Pattern Recogn 39:1430–1451
82. Marfil R, Rodrguez JA, Bandera A, Sandoval F (2004) Bounded irregular pyramid: a new structure for color image segmentation. Pattern Recogn 37(3):623–626, DOI:10.1016/j.patcog.2003.08.012
83. Martin D, Fowlkes C, Tal D, Malik J (2001) A database of human segmented natural images and its application to evaluating segmentation algorithms and measuring ecological statistics. In: Computer vision, 2001. ICCV 2001. Proceedings. Eighth IEEE international conference on, vol 2, pp 416–423
84. Martin V, Thonnat M, Maillot N (2006) A learning approach for adaptive image segmentation. In: Proceedings of the fourth IEEE international conference on computer vision systems, IEEE Computer Society, Washington, pp 40–48
85. Meyer F (1992) Color image segmentation. In: Image processing and its applications, International conference on, pp 303–306
86. Micusik B, Hanbury A (2005) Supervised texture detection in images. In: Conference on computer analysis of images and patterns (CAIP), pp. 441–448, Versailles, France
87. Mojsilovic A, Hu H, Soljanin E (2002) Extraction of perceptually important colors and similarity measurement for image matching, retrieval and analysis. IEEE Trans Image Process 11(11):1238–1248

88. Moravec H (1980) Obstacle avoidance and navigation in the real world by a seeing robot rover. In: tech. report CMU-RI-TR-80-03, Robotics Institute, Carnegie Mellon University & doctoral dissertation, Stanford University, CMU-RI-TR-80-03
89. Mortensen EN, Barrett WA (1998) Interactive segmentation with intelligent scissors. In: Graphical models and image processing, pp 349–384
90. Nachlieli H, Shaked D (2011) Measuring the quality of quality measures. IEEE Trans Image Process 20(1):76–87
91. Nadenau M (2000) Integration of human color vision models into high quality image compression, PhD thesis, École Polytechnique Fédérale de Lausanne (EPFL), http://infoscience.epfl.ch/record/32772
92. Ozkan D, Duygulu P (2006) Finding people frequently appearing in news. In: Sundaram H, Naphade M, Smith J, Rui Y (eds) Image and video retrieval, lecture notes in computer science, vol 4071. Springer, Berlin, pp 173–182
93. Pailloncy JG, Deruyver A, Jolion JM (1999) From pixels to predicates revisited in the graphs framework. In: 2nd international workshop on graph based representations,GbR99
94. Pal NR, Pal SK (1993) A review on image segmentation techniques. Pattern Recogn 26(9):1277–1294
95. Papoulis A (1991) Probability, random variables, and stochastic processes, 3rd edn. McGraw-Hill, New York
96. Park HK, Chung MJ (2002) Exernal force of snakes: virtual electric field. Electron Lett 38(24):1500–1502
97. Pass G, Zabih R, Miller J (1996) Comparing images using color coherence vectors. In: ACM multimedia, pp 65–73
98. Pauli H (1976) Proposed extension of the CIE recommendation on Uniform color spaces, color difference equations, and metric color terms. JOSA 66(8):866–867
99. Paulus D, Hornegger J, Niemann H (1999) Software engineering for image processing and analysis. In: Jähne B, Gei"sler P, Hau"secker H (eds) Handbook of computer vision and applications, Academic, San Diego, pp 77–103
100. Phillips I, Chhabra A (1999) Empirical performance evaluation of graphics recognition systems. IEEE Trans Pattern Anal Mach Intell 21(9):849–870, DOI:10.1109/34.790427
101. Plotnick R, Gardner R, O'Neill R (1993) Lacunarity indices as measures of landscape texture. Lanscape Ecol 8(3):201–211
102. Pratt WK (2001) Digital image processing: PIKS Inside, 3rd edn. Wiley, New York
103. Prewer D, Kitchen L (2001) Soft image segmentation by weighted linked pyramid. Pattern Recogn Lett 22:123–132
104. Priese L, Rehrmann V (1993) On hierarchical color segmentation and applications. In: Proceedings, Proceedings of the conference on computer vision and pattern recognition, pp 633–634
105. Randall J, Guan L, Li W, XZhang (2008) The hcm for perceptual image segmentation. Neurocomputing 71(10-12):1966–1979
106. Renyi A (1955) On a new axiomatic theory of probability. Acta Mathematica Hungarica 6(3-4):285–335
107. Rezaee M, van der Zwet P, Lelieveldt B, van der Geest R, Reiber J (2000) A multiresolution image segmentation technique based on pyramidal segmentation and fuzzy clustering. IEEE Trans Image Process 9(7):1238–1248, DOI:10.1109/83.847836
108. Richard N, Bringier B, Rollo E (2005) Integration of human perception for color texture management. In: Signals, circuits and systems, 2005. ISSCS 2005. International symposium on, vol 1, pp 207–210
109. Roerdink JB, Meijster A (2001) The wastershed transform: definitions, algorithms and parallelization strategies. Fundamenta Informaticae 41:187–228
110. Rosenfeld A (1970) Connectivity in digital pictures. J ACM 17(1):146–160
111. Rosenfeld A (1974) Adjacency in digital pictures. Inform Contr 26(1):24–33
112. Rosenfeld A (1979) Digital topology. Am Math Mon 86(8):621–630

113. Rosenfeld A (1986) Some pyramid techniques for image segmentation. Springer, London, pp 261–271
114. Rubner Y, Guibas L, Tomasi C (1997) The earth mover's distance, multi-dimensional scaling, and color-based image retrieval. In: DARPA97, pp 661–668
115. Rubner Y, Tomasi C, Guibas LJ (1998) A metric for distributions with applications to image databases. In: Proceedings of the 1998 IEEE international conference on computer vision, Bombay, India, pp 59–66
116. Saarinen K (1994) Color image segmentation by a watershed algorithm and region adjacency graph processing. In: Image processing, 1994. Proceedings. ICIP-94., IEEE international conference, vol 3, pp 1021–1025, DOI:10.1109/ICIP.1994.413690
117. Saunders S, Chen J, Drummer T, Gustafson E, Brosofske K (2005) Identifying scales of pattern in ecological data: a comparison of lacunarity, spectral and wavelet analyses. Ecol Complex 2:87–105
118. Schiffman HR (1996) Sensation and perception: an integrated approach, 4th edn. Wiley, New York
119. Serra J (2006) A lattice approach to image segmentation. J Math Imaging Vis 24:83–130, DOI:10.1007/s10851-005-3616-0
120. Seve R (1991) New formula for the computation of CIE 1976 hue difference. Color Res Appl 16(3):217–218
121. Seve R (1996) Practical formula for the computation of CIE 1976 hue difference. Color Res Appl21(4):314–314
122. Shi J, Malik J (2000) Normalized cuts and image segmentation. IEEE Trans Pattern Anal Mach Intell (PAMI) 22(8):888–905
123. Singh A, Terzopoulos D, Goldgof DB (1998) Deformable models in medical image analysis, 1st edn. IEEE Computer Society Press, Los Alamitos
124. Sinop AK, Grady L (2007) A seeded image segmentation framework unifying graph cuts and random walker which yields a new algorithm. In: Computer vision, IEEE international conference on, IEEE Computer Society, Los Alamitos, pp 1–8
125. Smith J, Chang SF (1995) Single color extraction and image query. In: Image processing, 1995. Proceedings., International conference on, vol 3, pp 528–531, DOI:10.1109/ICIP.1995.537688
126. Stricker M, Orengo M (1995) Similarity of color images. In: Storage and retrieval for image and video databases, pp 381–392
127. Sum KW, Cheung PYS (2007) Boundary vector field for parametric active contours. Pattern Recogn 40:1635–1645, DOI:10.1016/j.patcog.2006.11.006
128. Swain MJ, Ballard DH (1991) Color indexing. Int J Comput Vis 7(1):11–32
129. Tang X (1998) Texture information in run-length matrices. IEEE Trans Image Process 7(11):1602–1609, DOI:10.1109/83.725367
130. Tao Wang IC, Basu A (2009) Fluid vector flow and applications in brain tumor segmentation. IEEE Trans Biomed Eng 56(3):781–789
131. Terzopoulos D (2003) Deformable models: classic, topology-adaptive and generalized formulations. In: Osher S, Paragios N (eds) Geometric level set methods in imaging, vision, and graphics, chap 2. Springer, New York, pp 21–40
132. Tremeau A, Colantoni P (2000) Regions adjacency graph applied to color image segmentation. IEEE Trans Image Process 9(4):735–744
133. Turiac M, Ivanovici M, Radulescu T, Buzuloiu V (2010) Variance-driven active contours. In: IPCV, pp 83–86
134. Turner M, Gardner R, ONeill R (2001) Landscape ecology in theory and practice: Pattern and process. Springer, New York
135. Urdiales C, Dominguez M, de Trazegnies C, Sandoval F (2010) A new pyramid-based color image representation for visual localization. Image Vis Comput 28(1):78–91, DOI:10.1016/j.imavis.2009.04.014
136. Vincent L, Soille P (1991) Watersheds in digital spaces: an efficient algorithm based on immersion simulations. IEEE PAMI 13(6):583–598

137. Viola P, Jones M (2001) Robust real-time object detection. In: 2nd International workshop on statistical and computational theories of vision – Modeling, learning, computing, and sampling. Vancouver, Canada
138. Voss R (1986) Random fractals: characterization and measurement. Scaling phenomena in disordered systems 10(1):51–61
139. Willersinn D, Kropatsch W (94) Dual graph contraction for irregular pyramids. In: International conference on pattern recognition, Jerusalem, pp 251–256
140. Wirtz S, Paulus D (2010) Model-based recognition of 2d objects in perspective images. In: Proceedings of the 10th international conference on pattern recognition and image analysis: new information technologies (PRIA-10-2010), St. Petersburg, Russia, 978-5-7325-0972-4, pp 259–261
141. Witkin A, Terzopoulos D, Kass M (1987) Signal matching through scale space. Int J Comput Vis 1:133–144
142. Witkin AP (1983) Scale-space filtering. In: International joint conference on artificial intelligence, pp 1019–1022
143. Wu Q, Castleman KR (2008) Image segmentation. In: Microscope image processing. Academic, Burlington, pp 159–194, DOI:10.1016/B978-0-12-372578-3.00009-X
144. Xia Y, Feng D, Zhao R (2006) Morphology-based multifractal estimation for texture segmentation. IEEE Trans Image Process 15(3):614–623, DOI:10.1109/TIP.2005.863029
145. Xiang S, Pan C, Nie F, Zhang C (2010) Turbopixel segmentation using eigen-images.IEEE Trans Image Process 19(11):3024–3034, DOI:10.1109/TIP.2010.2052268
146. Xu C, Prince JL (1997) Gradient vector flow: a new external force for snakes. In: Proceedings of the 1997 Conference on Computer Vision and Pattern Recognition (CVPR '97), IEEE Computer Society, Washington, DC, USA, pp 66–71
147. Xu C, Prince JL (1998) Generalized gradient vector flow external forces for active contours. Signal Process 71:131–139
148. Xu C, Prince JL (1998) Snakes, shapes, and gradient vector flow. IEEE Trans Image Process 7(3):359–369
149. Xu Y, Duygulu P, Saber E, Tekalp AM, Yarman-Vural FT (2003) Object-based image labeling through learning by example and multi-level segmentation. Pattern Recogn 36(6):1407–1423, DOI:10.1016/S0031-3203(02)00250-9
150. Yu Sy, Zhang Y, Wang Yg, Yang J (2008) Unsupervised color-texture image segmentation. J Shanghai Jiaotong University (Science) 13:71–75
151. Zahn CT (1971) Graph-theoretical methods for detecting and describing gestalt clusters. IEEE Trans Comput 20:68–86
152. Zhang YJ (2006) Advances in image and video segmentation. IRM Press, USA
153. Zhao R, Grosky WI (2001) Bridging the semantic gap in image retrieval, in Distributed multimedia databases: Techniques and applications, IGI Global, pp 14–36

Chapter 9
Parametric Stochastic Modeling for Color Image Segmentation and Texture Characterization

Imtnan-Ul-Haque Qazi, Olivier Alata, and Zoltan Kato

Black should be made a color of light

Clemence Boulouque

Abstract Parametric stochastic models offer the definition of color and/or texture features based on model parameters, which is of interest for color texture classification, segmentation and synthesis.

In this chapter, distribution of colors in the images through various parametric approximations including multivariate Gaussian distribution, multivariate Gaussian mixture models (MGMM) and Wishart distribution, is discussed. In the context of Bayesian color image segmentation, various aspects of sampling from the posterior distributions to estimate the color distribution from MGMM and the label field, using different move types are also discussed. These include reversible jump mechanism from MCMC methodology. Experimental results on color images are presented and discussed.

Then, we give some materials for the description of color spatial structure using Markov Random Fields (MRF), and more particularly multichannel GMRF, and multichannel linear prediction models. In this last approach, two dimensional complex multichannel versions of both causal and non-causal models are discussed to perform the simultaneous parametric power spectrum estimation of the luminance

I.-U.-H. Qazi (✉)
SPARCENT Islamabad, Pakistan Space & Upper Atmosphere Research Commission, Pakistan
e-mail: manager.nsdi@suparco.gov.pk

O. Alata
Laboratory Hubert Curien, UMR CNRS 5516, Jean Monnet University, Saint-Etienne, France
e-mail: olivier.alata@univ-st-etienne.fr

Z. Kato
Department of Image Processing and Computer Graphics, Institute of Informatics, University of Szeged, Hungary

C. Fernandez-Maloigne (ed.), *Advanced Color Image Processing and Analysis*,
DOI 10.1007/978-1-4419-6190-7_9,

and the chrominance channels of the color image. Application of these models to the classification and segmentation of color texture images is also illustrated.

Keywords Stochastic models • Multivariate Gaussian mixture models • Wishart distribution • Multichannel complex linear prediction models • Gaussian Markov Random Field • Parametric spectrum estimation • Color image segmentation • Color texture classification • Color texture segmentation • Reversible jump Markov chain Monte Carlo

In this chapter, the support of an image will be called $E \subset \mathbb{Z}^2$ and a site, or a pixel, will be denoted $\mathbf{x} = (x_1, x_2) \in E$. A color image can be either a function $\boldsymbol{f} : E \mapsto V^{\mathrm{rgb}}$ with $\boldsymbol{f}(\mathbf{x}) = (f_{\mathrm{R}}(\mathbf{x}), f_{\mathrm{G}}(\mathbf{x}), f_{\mathrm{B}}(\mathbf{x}))$, or a function $\boldsymbol{f} : E \mapsto V^{\mathrm{hls}}$ with $\boldsymbol{f}(\mathbf{x}) = (f_{\mathrm{L}}(\mathbf{x}), f_{\mathrm{H}}(\mathbf{x}), f_{\mathrm{S}}(\mathbf{x}))$, or a function $\boldsymbol{f} : E \mapsto V^{l c_1 c_2}$ with $\boldsymbol{f}(\mathbf{x}) = (f_{\mathrm{L}}(\mathbf{x}), f_{\mathrm{C_1}}(\mathbf{x}), f_{\mathrm{C_2}}(\mathbf{x}))$, in a "Red Green Blue" (RGB) space, in a color space of type "Hue Luminance Saturation" (HLS), and in a color space with one achromatic component and two chromatic components like CIE L*a*b* [11], respectively.

When using stochastic models, the observation of a color image which can be also denoted $\boldsymbol{f} = \{\boldsymbol{f}(\mathbf{x})\}_{\mathbf{x} \in E}$ following previous definitions may be considered as a realization of a random field $\boldsymbol{F} = \{\boldsymbol{F}(\mathbf{x})\}_{\mathbf{x} \in E}$. The $\boldsymbol{F}(\mathbf{x})$ are multidimensional (n-d) random vectors whose sample spaces of random variables inside the vectors may be different, depending on the used color space. The mean vector at $\mathbf{x}$ will then be denoted $\mathbf{m}_{\boldsymbol{f}}(\mathbf{x}) = \mathrm{E}\{\boldsymbol{F}(\mathbf{x})\}$, $\forall \mathbf{x} \in E$, with $E\{.\}$ the vector containing the expectation of each random variables.

Two main domains in parametric stochastic modeling of the observation have been studied as for gray-level images, which may be described by random fields of one dimension (1-d) random variables:

- The description (or approximation) of the multidimensional distribution[1] of the colors (see Sects. 9.1.1 and 9.2).
- The description of the spatial structures between the random variables, i.e., the spatial correlations or the spatial dependencies between the random vectors.

Parametric stochastic models for color texture characterization are mostly extensions of ones proposed for gray-level texture characterization [1, 6]. One of the major advantages of these models is that they both offer the definition of texture features based on model parameters, which is of interest for texture classification [24, 35], tools for color texture segmentation [21] and the possibility to synthesize textures [36]. Let us notice that there are numerous works about stochastic modeling for color textures (classification and/or segmentation) that only consider the spatial structure in luminance channel [32,51]. The chromatic information is only described by color distribution which neglects spatial information of chromatic information. This approach is not further discussed in this chapter.

[1] Or distributions when there is several homogeneous regions in the image, each region having its own distribution.

As for gray-level images, parametric stochastic models for color texture characterization has been developed following two axes:

- The description of the spatial dependencies of random vectors from a probabilistic point of view, classically thanks to the theory of Markov random fields (MRF) (see Sect. 9.1.2). This approach has also been extended to multispectral or hyperspectral images [56].
- The development of multichannel spectral analysis tools [52] based on prediction linear models (see Sects. 9.1.3 and 9.3) which can be more defined as a "signal processing" approach.[2]

These two approaches become almost similar when the spatial dependencies may be considered as linear: each random vector is a weighted sum of neighboring random vectors to which a random vector (the excitation) is added. The associated MRF is then called "Gauss Markov Random Field" (GMRF). Another interesting aspect of these developments is the recent use of various color spaces [9, 36] as the first extensions was only done with RGB color space like in [24]. In Sect. 9.3, the works presented in [52,54,55] will be summarized for its comparative study of three color spaces, RGB, IHLS (improved HLS color space) [22], and L*a*b* [11] in the context of multichannel linear prediction for texture characterization, classification of textures, and segmentation of textured images.

In the next section, we present the models which will be used in the two next sections: for color image segmentation, firstly, and for texture characterization, classification, and segmentation, secondly.

9.1 Stochastic Parametric Data Description

9.1.1 Distribution Approximations

In 1-d case, many parametric models exist to approach the distribution of the observation when considering the family of discrete and continuous probability laws.[3] There also exists measures in order to evaluate the distance between two probability distributions [4]. Sometimes, these measures can be directly computed from the parameters of the probability distribution as the Kullback–Leibler divergence in a Gaussian case, for example.

The family of n-d parametric laws is smaller than in the 1-d case. Moreover, measures between two n-d probability distribution are often hard to compute and an approximation is sometimes needed (see [25], for example). For color

[2]The origin of this approach may be the linear prediction-based spectral analysis of signals like speech signals [42], for example.

[3]See http://en.wikipedia.org/wiki/List_of_probability_distributions for 1-d and n-d cases, for example.

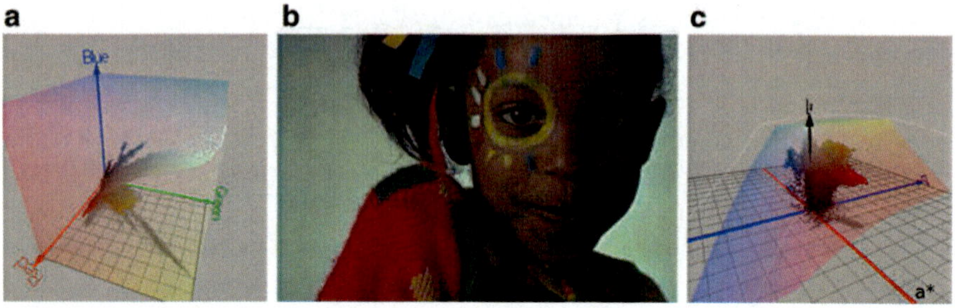

Fig. 9.1 A color image (**b**) and its distributions in two color spaces: (**a**) RGB color space and (**c**) L*a*b* color space. Figures (a) and (c) have been obtained with "colorspace": http://www.couleur.org/index.php?page=colorspace

images, another classical way to approach the distribution uses nonparametric 3-d histograms [54]. Even if a histogram is easy to compute, many questions remains open as how to choose the number of the bins? or the support (the width in 1-d case) of the bins should be always the same? In the following, we give the main parametric models used to describe the distribution of colors in an image.

The most classical approximation is the multivariate gaussian distribution (MGD) defined by its probability density function (pdf) parameterized by $\Theta = \{\mathbf{m}_f, \Sigma_F\}$:

$$p(f(\mathbf{x})|\Theta) = \frac{(2\pi)^{-\omega/2}}{\sqrt{\det(\Sigma_F)}} \exp\left[-\frac{(f(\mathbf{x}) - \mathbf{m}_f)^T (\Sigma_F)^{-1} (f(\mathbf{x}) - \mathbf{m}_f)}{2}\right], \tag{9.1}$$

where ω is the dimension of the real vector $f(\mathbf{x})$ and Σ_F the variance–covariance matrix. Of course, this approximation can only be used when the values associated to the different axis may be considered as real values like for RGB color space. In this space, $\omega = 3$. Although this approximation is simple and mostly used, it may not be an accurate approximation when the distribution of colors is neither gaussian nor unimodal (see Fig. 9.1). Multivariate Gaussian mixture model (MGMM) is one of the most used parametric models to approximate a multimodal probability density function (see [2], for example, and for a study of MGMM applied to color images) especially when no information is available from the distribution of data. The MGMM is defined as:

$$p(f(\mathbf{x})|\Theta) = \sum_{k=1}^{K} p_k p(f(\mathbf{x})|\Theta_k), \tag{9.2}$$

where $p_1, \ldots, p_K$ are the prior probabilities of each Gaussian component of the mixture, and $K > 1$ is the number of components of MGMM. Each $\Theta_k = \{\mathbf{m}_k, \Sigma_k\}$, $k = 1, \ldots, K$, is the set of model parameters defining the kth Gaussian component of the mixture model (see (9.1)). The prior probability values must satisfy following conditions:

$$p_k > 0,\ k = 1,\ldots,K, \text{ and } \sum_{k=1}^{K} p_k = 1. \tag{9.3}$$

Thus, the complete set of mixture model parameters is $\Theta = \{p_1,\ldots,p_K,\Theta_1,\ldots,\Theta_K\}$.

The parameters of the MGD are classically estimated with the formula of the empirical mean and the empirical estimators of second-order statistics. The estimation of the MGMM parameter set from a given data set is generally done by maximizing the likelihood function, $f(\Theta) = p(\mathbf{f}|\Theta)$:

$$\hat{\Theta} = \underset{\Theta}{\operatorname{argmax}}\, p(\mathbf{f}|\Theta). \tag{9.4}$$

To estimate Θ, it is generally assumed that the random vectors of the random field are independent. Thus, $f(\Theta) = \prod_{\mathbf{x}\in E} p(\mathbf{f}(\mathbf{x})|\Theta)$. In such a context, the expectation-maximization (EM) algorithm is a general iterative technique for computing maximum likelihood estimation (MLE) widely used when observed data can be considered as incomplete [13]. The algorithm consists of two steps, an E-step and an M-step, which produce a sequence of estimates $\Theta^{(t)}$, $t = 0,1,2,\ldots$ by repeating these two steps. The last estimate gives $\hat{\Theta}$ (see (9.4)). For more details about EM algorithm, see [2, 13], for example. Estimation of the number of components in an MGMM is a model selection problem. In [2], on a benchmark of color images, the distributions of the images are well approximated using more than twenty components. Different information criteria for selecting the number of components are also compared. Nevertheless, the estimate obtained from EM algorithm is dependent from the initial estimate. To our knowledge, the most accurate alternative to EM algorithm is a stochastic algorithm which both estimates the parameters of the mixture and its number of components: the reversible jump Markov chain Monte Carlo (RJMCMC) algorithm [17, 57]. In Sect. 9.2, we provide an extension of the works of Richardson and Green [57] to the segmentation of color images [33].

To end this section, we also recall the Wishart distribution. The Wishart distribution is a generalization to multiple dimensions of the chi-square distribution. To achieve robustness and stability of approximation, this model takes into account multiple observations to define the probability instead of using a single observation. Let us define $\mathbf{J}(\mathbf{x})$, $\mathbf{x} \in E$, a matrix of α vectors with dimension ω ($\alpha \geq \omega$) issued from a finite set of vectors of the image including $\mathbf{f}(\mathbf{x})$. The density of the Wishart distribution is:

$$p(\mathbf{J}(\mathbf{x})|\Theta) = \frac{|\mathbf{M}(\mathbf{x})|^{\frac{1}{2}(\alpha-\omega-1)} \exp\left(-\frac{1}{2}\mathrm{Tr}\left(\Sigma_F^{-1}\mathbf{M}(\mathbf{x})\right)\right)}{2^{\alpha(\omega/2)}\pi^{\omega(\omega-1)/4}\,|\Sigma_F|^{\alpha/2}\prod_{i=1}^{\omega}\Gamma\left(\frac{1}{2}(\alpha+1-i)\right)} \tag{9.5}$$

with $\mathbf{M}(\mathbf{x}) = \mathbf{J}(\mathbf{x})\mathbf{J}(\mathbf{x})^T$, a half-positive definite matrix with size $\omega \times \omega$ and Γ, the gamma function. Σ_F is the variance–covariance matrix computed from the support used to define the finite set of vectors in $\mathbf{J}(\mathbf{x})$. In Sect. 9.3, the Wishart distribution will be used for color texture segmentation not directly on the color vectors of the image but on the linear prediction errors computed from a linear prediction model [55].

In the two next sections, we present stochastic models for the description of color spatial structures.

9.1.2 MRF and GMRF

The two main properties of an MRF, associated to a reflexive and symmetric graph [20] issued from the definition of a neighborhood for each site, $V(\mathbf{x})$, $\mathbf{x} \notin V(\mathbf{x})$, are:

- The local conditional probability of $\boldsymbol{f}(\mathbf{x})$ considering all other realizations depends only from the realizations on the neighboring sites:

$$p\left(\boldsymbol{f}(\mathbf{x}) \middle| \{\boldsymbol{f}(\mathbf{y}\}_{\mathbf{y}\in E\setminus\{\mathbf{x}\}}\right) = p\left(\boldsymbol{f}(\mathbf{x}) \middle| \{\boldsymbol{f}(\mathbf{y})\}_{\mathbf{y}\in V(\mathbf{x})}\right). \tag{9.6}$$

- If $p(\boldsymbol{f}) > 0$, $\forall \boldsymbol{f} \in \Omega_f$, Ω_f the sampling space of $\boldsymbol{F}$, the joint probability can be written as a Gibbs distribution (theorem of Hammersley–Clifford):

$$p(\boldsymbol{f}) \propto \exp(-U(\boldsymbol{f})) = \exp\left(-\sum_{c\in Cl} f_c(\boldsymbol{f})\right), \tag{9.7}$$

and "$\propto$" symbol means that the probability density function of (9.7) is unormalized. Cl is the set of cliques[4] defined from the neighboring system, $U(\boldsymbol{f})$ the energy of the realization $\boldsymbol{f}$, and f_c a potential whose value only depends from the realization of the sites of the clique c.

This model offers the possibility to use potentials adapted to the properties which one aims to study in an image or a label field. This is the main reason why many image segmentation algorithms have been developed based on MRF. For image segmentation, the energy is the sum of a data driven energy which may contain color and/or texture information and an "internal" energy which potentials model the properties of the label field. The segmentation of the image is then obtained by finding the realization of the label field that minimizes the energy. These aspects will be more developed in Sects. 9.2 and 9.3.

9.1.2.1 Multichannel GMRF

For color texture characterization, the GMRF in its vectorial form has been mainly used. Let us suppose that the random vectors are centered. For this model, the formula of (9.6) becomes:

[4] A clique can contain neighboring sites against each other or just a singleton.

$$p\left(\boldsymbol{f}(\mathbf{x}) \,\middle|\, \{\boldsymbol{f}(\mathbf{y})\}_{\mathbf{y}\in V(\mathbf{x})}\right) \propto \exp\left\{-\frac{1}{2}\left\|\boldsymbol{e_f}(\mathbf{x})\right\|^2_{\Sigma_{\boldsymbol{E_F}}}\right\} \tag{9.8}$$

with $\boldsymbol{e_f}(\mathbf{x}) = \boldsymbol{f}(\mathbf{x}) + \sum_{\mathbf{y}\in V(\mathbf{x})} \mathbf{A}_{\mathbf{y}-\mathbf{x}}\boldsymbol{f}(\mathbf{y})$ and $\left\|\boldsymbol{e_f}(\mathbf{x})\right\|^2_{\Sigma_{\boldsymbol{E_F}}} = \boldsymbol{e_f}(\mathbf{x})^T \Sigma_{\boldsymbol{E_F}}^{-1} \boldsymbol{e_f}(\mathbf{x})$, $\Sigma_{\boldsymbol{E_F}}$ the conditional variance matrix associated to the random vectors $\boldsymbol{E_F} = \{\boldsymbol{E_F}(\mathbf{x})\}_{\mathbf{x}\in E}$. Following (9.8), the model can also be defined by a linear relation between random vectors:

$$\boldsymbol{F}(\mathbf{x}) = -\sum_{\mathbf{y}\in V(\mathbf{x})} \mathbf{A}_{\mathbf{y}-\mathbf{x}}\boldsymbol{F}(\mathbf{y}) + \boldsymbol{E_F}(\mathbf{x}) = -\sum_{\mathbf{y}\in D} \mathbf{A}_{\mathbf{y}}\boldsymbol{F}(\mathbf{x}-\mathbf{y}) + \boldsymbol{E_F}(\mathbf{x}) \tag{9.9}$$

with D the neighboring support used around each site. D is conventionally a finite non-causal support of order o, called D_o^{NC} in this chapter (see Fig. 9.3c):

$$D_1 = \left\{\mathbf{y}\in\mathbb{Z}^2, \arg\min_{\mathbf{y}} \|\mathbf{y}\|_2, \mathbf{y} \neq (0,0)\right\}$$

$$D_k = \left\{\mathbf{y}\in\mathbb{Z}^2, \arg\min_{\mathbf{y}\notin \bigcup_{1\le l\le k-1} D_l} \|\mathbf{y}\|_2, \mathbf{y} \neq (0,0)\right\}, k > 1$$

$$D_o^{\mathrm{NC}} = \bigcup_{1\le k\le o} D_k. \tag{9.10}$$

From (9.8) and (9.9), the conditional law of $\boldsymbol{F}(\mathbf{x})$ is a Gaussian law with the same covariance matrix for all sites but with a mean vector which depends on the neighborhood: $\mathbf{m}_{\boldsymbol{f}}(\mathbf{x}) = -\sum_{\mathbf{y}\in V(\mathbf{x})} \mathbf{A}_{\mathbf{y}-\mathbf{x}}\boldsymbol{f}(\mathbf{y})$. As for the scalar case [20], the random vectors of the family $\boldsymbol{E_F}$ are correlated (see Sect. 9.1.3.1). The parameter set of the model is $\Theta_{\mathrm{MGMRF}} = \left\{\{\mathbf{A}_{\mathbf{y}}\}_{\mathbf{y}\in D_{o,2}^{\mathrm{NC}}}, \Sigma_{\boldsymbol{E_F}}\right\}$, $D_{o,2}^{\mathrm{NC}}$ the upper half of support D_o^{NC} as $\mathbf{A}_{\mathbf{y}} = \mathbf{A}_{-\mathbf{y}}$ in the real case. The parameters in the matrices $\mathbf{A}_{\mathbf{y}}$, $\mathbf{y}\in D_{o,2}^{\mathrm{NC}}$ can be used as color texture features describing the spatial structure of each plane and the spatial interaction between planes [24]. Different algorithms for the estimation of the parameters of this model have been detailed in [56]: estimation in the maximum likelihood sense, estimation in the maximum pseudo-likelihood sense,[5] and estimation in the minimum mean squared error (MMSE) sense.

[5]The parameters are estimated by maximizing the probability $\prod_{\mathbf{x}\in E} p\left(\boldsymbol{f}(\mathbf{x}) \,\middle|\, \{\boldsymbol{f}(\mathbf{y})\}_{\mathbf{y}\in V(\mathbf{x})}\right)$ despite the fact that the $\boldsymbol{F}(\mathbf{x})$ are not independent relatively to each other.

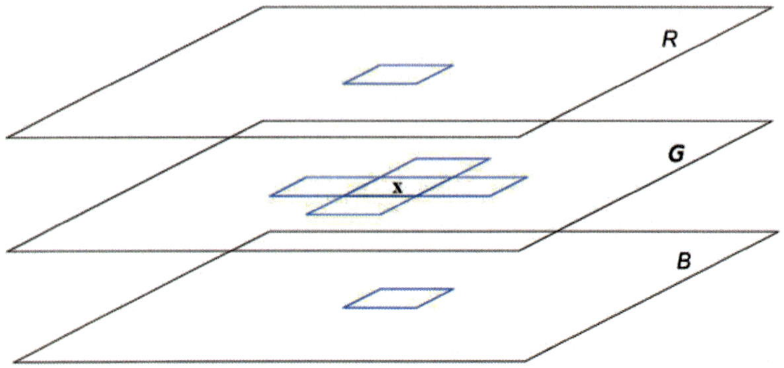

Fig. 9.2 3-d neighborhood based on the nearest neighbors, for the model associated to the "G" channel in an RGB color space

9.1.2.2 3-d GMRF

Rather than using a 2-d multichannel model for the color image, a 3-d scalar model for each plane, $i = 1,2,3$, may be preferred:

$$p\left(f_i(\mathbf{x})\,\middle|\,\{f_j(\mathbf{y})\}_{\mathbf{y}\in V_{i,j}(\mathbf{x})}, j = 1,2,3\right) \propto \exp\left\{-\frac{1}{2\sigma^2_{e_{f,i}}} e_{f,i}(\mathbf{x})^2\right\} \tag{9.11}$$

with $e_{f,i}(\mathbf{x}) = f_i(\mathbf{x}) + \sum_{j=1,2,3} \sum_{\mathbf{y}\in V_{i,j}(\mathbf{x})} a_{i,j}(\mathbf{y}-\mathbf{x}) f_j(\mathbf{y})$ and $\sigma^2_{e_{f,i}}$ is the conditional variance associated to the random variables $E_{f,i} = \{E_{f,i}(\mathbf{x})\}$. Unlike the MGMRF, this model offers the possibility of having different supports from one plane to another. Figure 9.2 provides an example of 3-d neighboring system based on the nearest neighbors in the three planes. Estimation in the maximum likelihood sense is still given in [56].

9.1.3 *Linear Prediction Models*

As for MRF, multichannel (or vectorial) models and 3-d scalar models, each by plane, have been proposed in literature.

9.1.3.1 Multichannel (or Vectorial) Linear Prediction Models

For a general approach of 2-d multichannel linear prediction, the model is supposed to be complex. Complex vectors allow to describe the color image as a two-channel process: one channel for the achromatic values and one channel for the chromatic values. when considering a HLS color space, it gives:

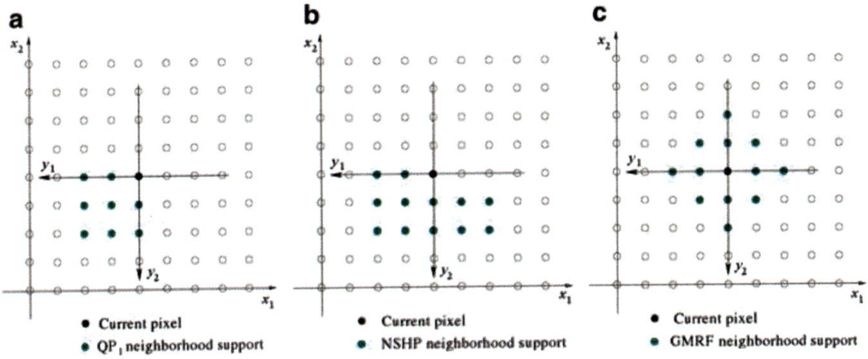

Fig. 9.3 Neighborhood support regions for QP1 (**a**) and NSHP (**b**) causal models of order (o_1, o_2) with $o_1 = 2$ and $o_2 = 2$ and for GMRF model (**c**) of order $o = 3$

$$f(\mathbf{x}) = \begin{bmatrix} f_1(\mathbf{x}) = l \\ f_2(\mathbf{x}) = se^{ih} \end{bmatrix} \tag{9.12}$$

with h in radians, and for a color space with one achromatic component and two chromatics components:

$$f(\mathbf{x}) = \begin{bmatrix} f_1(\mathbf{x}) = l \\ f_2(\mathbf{x}) = c_1 + i \times c_2 \end{bmatrix}. \tag{9.13}$$

This approach has been extensively studied in [53]. The 2-d multichannel complex linear prediction model is defined by the following relationship between complex random vectors:

$$\boldsymbol{F}(\mathbf{x}) = \hat{\boldsymbol{F}}(\mathbf{x}) + \mathbf{m} + \boldsymbol{E_F}(\mathbf{x}) \tag{9.14}$$

with $\hat{\boldsymbol{F}}(\mathbf{x}) = -\sum_{\mathbf{y} \in D} \mathbf{A_y} (\boldsymbol{F}(\mathbf{x} - \mathbf{y}) - \mathbf{m})$ the linear prediction of $\boldsymbol{F}(\mathbf{x})$ using the finite prediction support $D \subset \mathbb{Z}^2$ and the set of complex matrices $\{\mathbf{A_y}\}_{\mathbf{y} \in D}$. The vector random field $\boldsymbol{E_F} = \{\boldsymbol{E_F}(\mathbf{x})\}_{\mathbf{x} \in E}$ is called the excitation, or the linear prediction error (LPE), of the model whose statistical properties may be different depending on the prediction support as we shall see in the following. The prediction supports conventionally used in the literature are the causal supports, quarter plane (QP) or nonsymmetric half plane (NSHP), whose size is defined by a couple of integers called the order, $\mathbf{o} = (o_1, o_2) \in \mathbb{N}^2$ (see Figs. 9.3a and b):

$$D_{\mathbf{o}}^{\mathrm{QP1}} = \{\mathbf{y} \in \mathbb{Z}^2, 0 \leq y_1 \leq o_1, 0 \leq y_2 \leq o_2, \mathbf{y} \neq (0,0)\} \tag{9.15}$$

$$D_{\mathbf{o}}^{\mathrm{DPNS}} = \{\mathbf{y} \in \mathbb{Z}^2, 0 < y_1 \leq o_1 \text{ pour } y_2 = 0, \\ -o_1 < y_1 \leq o_1 \text{ pour } 0 < y_2 \leq o_2\} \tag{9.16}$$

or the noncausal (NC) support of order $o \in \mathbb{N}$ already defined (see (9.10) and Fig. 9.3c). These models allow a multidimensional spectral analysis of a color process. For HLS or LC_1C_2 color spaces (see (9.12) and (9.13)), the power spectral density function (PSD) of the process may be defined from the PSD of $\boldsymbol{E_F}$ and the set of matrices $\{\mathbf{A_y}\}_{\mathbf{y}\in D}$:

$$\mathbf{S}_{\boldsymbol{F}}(\nu) = \mathbf{A}(\nu)^{-1}\mathbf{S}_{\boldsymbol{E_F}}(\nu)\mathbf{A}^H(\nu)^{-1} = \begin{bmatrix} S_{LL}(\nu) & S_{LC}(\nu) \\ S_{CL}(\nu) & S_{CC}(\nu) \end{bmatrix} \tag{9.17}$$

with $\nu = \{\nu_1, \nu_2\} \in \mathbb{R}^2$, the 2-d normalized frequency, and $\mathbf{S}_{\boldsymbol{E_F}}(\nu)$, the PSD of the excitation and $\mathbf{A}^H(\nu)$, the Hermitian matrix of $\mathbf{A}(\nu)$:

$$\mathbf{A}(\nu) = 1 + \sum_{\mathbf{y}\in D} \mathbf{A_y} \exp(-i2\pi\langle \nu, \mathbf{y}\rangle). \tag{9.18}$$

In (9.17), $S_{LL}(\nu)$ is the PSD of the "achromatic" (or luminance) channel of the image, $S_{CC}(\nu)$, the PSD of the "chromatic" channel, and $S_{LC}(\nu) = S^*_{CL}(\nu)$ the inter-spectrum of the two channels. Let us notice that the image has three real channels in RGB color space and the PSD is written:

$$\mathbf{S}_{\boldsymbol{F}}(\nu) = \begin{bmatrix} S_{\mathrm{RR}}(\nu) & S_{\mathrm{RG}}(\nu) & S_{\mathrm{RB}}(\nu) \\ S_{\mathrm{GR}}(\nu) & S_{\mathrm{GG}}(\nu) & S_{\mathrm{GB}}(\nu) \\ S_{\mathrm{BR}}(\nu) & S_{\mathrm{BG}}(\nu) & S_{\mathrm{BB}}(\nu) \end{bmatrix}. \tag{9.19}$$

When the support is causal (QP or NSHP), the model is an extension of the classic autoregressive (AR) model. Its excitation is supposed to be a white noise and its PSD is constant and equal to the variance–covariance matrix of $\boldsymbol{E_F}$: $\mathbf{S}_{\boldsymbol{E_F}}(\nu) = \Sigma_{\boldsymbol{E_F}}$ and $\mathbf{S}_{\boldsymbol{F}}(\nu) = \mathbf{A}(\nu)^{-1}\Sigma_{\boldsymbol{E_F}}\mathbf{A}^H(\nu)^{-1}$. In the case of QP1 support (see (9.15)), the PSD has an anisotropy which can be corrected by the mean of an estimator based on an harmonic mean (HM) of the PSD obtained from models with QP1 and QP2 supports of order $\mathbf{o} = (o_1, o_2) \in \mathbb{N}^2$ [28] (see (9.17)):

$$\mathbf{S}_{\boldsymbol{F}}^{\mathrm{HM}}(\nu) = 2\left(\mathbf{S}_{\boldsymbol{F}}^{\mathrm{QP1}}(\nu)^{-1} + \mathbf{S}_{\boldsymbol{F}}^{\mathrm{QP2}}(\nu)^{-1}\right)^{-1} \tag{9.20}$$

with

$$\mathbf{D}_{\mathbf{o}}^{QP2} = \left\{\mathbf{y}\in\mathbb{Z}^2, -o_1 \le y_1 \le 0, 0 \le y_2 \le o_2, \mathbf{y} \neq (0,0)\right\}. \tag{9.21}$$

When using an NC support, the model is the MGMRF model (see Sect. 9.1.2.1). For the GMRF, due to the orthogonality condition on random variables [56] between the components of $\boldsymbol{E_F}(\mathbf{x})$ and the components of $\boldsymbol{F}(\mathbf{y})$, $\mathbf{y}\in E\backslash\{\mathbf{x}\}$, the PSD of $\boldsymbol{E_F}$ becomes:

$$\mathbf{S}_{\boldsymbol{E_F}}(\nu) = \mathbf{A}(\nu)\Sigma_{\boldsymbol{E_F}} \tag{9.22}$$

and

$$\mathbf{S}_F^{\mathrm{GMRF}}(\nu) = \Sigma_{E_F} \mathbf{A}^H(\nu)^{-1}. \tag{9.23}$$

The different estimations of the PSD using the HM method (called in the following PSD_HM method), the NSHP support (PSD_NSHP method) or the NC support (PSD_GMRF method) are obtained from the estimation of set of matrices $\{\mathbf{A}_\mathbf{y}\}_{\mathbf{y}\in D}$. To this aim, an MMSE or an ML estimation can be done for each model [56] (see Sect. 9.1.2.1). For the causal models and under a Gaussian assumption for the excitation process, the MMSE estimation provides the same estimate as the ML estimation.

In Sect. 9.3, this approach will be detailed and compared for three color spaces, IHLS, L*a*b* and RGB, for spectral analysis, texture characterization and classification, and textured image segmentation.

9.1.3.2 MSAR Model

The multispectral simultaneous autoregressive (MSAR) model [21, 24] is defined in a similar way to the 3-d GMRF (see Sect. 9.1.2.2) apart the excitation process is supposed to be a white noise. In [24,35], an MMSE method of parameter estimation is given.

In this section, we only presented stochastic models for stationary processes or at least for stationary processes by windows. Extensions of these models exist in order to describe textures whose spatial structure evolves spatially [62]. In a theoretical point of view, it is relatively simple, just to vary spatially model parameters. But, in a practical implementation, it is quite difficult. To our knowledge, a few researches has been done on this topic contrary to the domain of signal processing.[6] There remains a vast field of investigation to do on the subject.

In the next section, we provide a way to use MGMM (see (9.2)) for making color image segmentation.

9.2 Mixture Models and Color Image Segmentation

MRF modeling and MCMC methods are successfully used for supervised [16, 19, 26, 50] and unsupervised [33, 38, 39, 65] color image segmentation. In this section, we present a method proposed in [30, 33] for automatic color image segmentation, which adopts a Bayesian model using a first-order MRF. The observed image is represented by a mixture of multivariate Gaussian distributions while inter-pixel interaction favors similar labels at neighboring sites. In a Bayesian framework [15],

---

[6] See the studies about time-varying AR (TVAR), for example.

we are interested in the *posterior distribution* of the unknowns given the observed image. Herein, the unknowns comprise the hidden label field configuration, the Gaussian mixture parameters, the MRF hyperparameter, and the number of mixture components (or classes). Then a RJMCMC algorithm is used to sample from the whole posterior distribution in order to obtain a MAP estimate via simulated annealing [15]. RJMCMC has been applied to various problems, such as univariate Gaussian mixture identification [57] and its applications for inference in hidden Markov models [58], intensity-based image segmentation [3], and computing medial axes of 2D shapes [66]. Following [33], we will develop a RJMCMC sampler for identifying multivariate Gaussian mixtures and apply it to unsupervised color image segmentation. RJMCMC allows us the direct sampling of the whole posterior distribution defined over the combined model space thus reducing the optimization process to a single simulated annealing run. Another advantage is that no coarse segmentation neither exhaustive search over a parameter subspace is required. Although for clarity of presentation we will concentrate on the case of three-variate Gaussians, it is straightforward to extend the equations to higher dimensions.

9.2.1 Color Image Segmentation Model

The model assumes that the real-world scene consists of a set of regions whose observed color changes slowly, but across the boundary between them, they change abruptly. What we want to infer is a *labeling* l consisting of a simplified, abstract version of the input image: regions has a constant value (called a *label* in our context) and the discontinuities between them form a curve—the contour. Such a labeling l specifies a *segmentation*. Taking the probabilistic approach, one usually wants to come up with a *probability measure* on the set Ω_l of all possible segmentations of the input image and then select the one with the highest probability. Note that Ω_l is finite, although huge. A widely accepted standard, also motivated by the human visual system [34,47], is to construct this probability measure in a Bayesian framework [8,46,64]. We will assume that we have a set of observed ($\boldsymbol{F}$) and hidden (L) random variables. In our context, the *observation* $\boldsymbol{f} = \{\boldsymbol{f}(\mathbf{x})\}_{\mathbf{x} \in E}$ represents the color values used for partitioning the image, and the hidden entity $l \in L$ represents the segmentation itself. Note that color components are normalized, i.e., if the color space is RGB, $0 < f_i(\mathbf{x}) < 1$, i =R; G; B. Furthermore, a segmentation l assigns a label $l(\mathbf{x})$ from the set of labels $\Lambda = \{1,2,\ldots,K\}$ to each site $\mathbf{x}$.

First, we have to quantify how well any occurrence of l fits $\boldsymbol{f}$. This is expressed by the probability distribution $P(\boldsymbol{f}|l)$—the *imaging model*. Second, we define a set of properties that any segmentation l must possess regardless the image data. These are described by $P(l)$, the prior, which tells us how well any occurrence l satisfies these properties. For that purpose, $l(\mathbf{x})$ is modeled as a discrete random variable taking values in Λ. The set of these labels $l = \{l(\mathbf{x})\}_{\mathbf{x} \in E}$ is a random field, called the *label process*. Furthermore, the observed color features are supposed to be a realization f from another random field, which is a function of the label process l. Basically, the *image process* f represents the manifestation of the underlying label process. The multivariate Normal density is typically an appropriate model for such classification problems where the feature vectors $f(\mathbf{x})$ for a given class λ are mildly corrupted versions of a single mean vector $\mathbf{m}_\lambda$ [45, 51]. Applying these ideas, the *image process* f can be formalized as follows: $P(f(\mathbf{x}) \mid l(\mathbf{x}))$ follows a three-variate Gaussian distribution $N(\mathbf{m}, \Sigma)$, each pixel class $\lambda \in \Lambda = \{1, 2, \ldots, K\}$ is represented by its mean vector $\mathbf{m}_\lambda$ and covariance matrix Σ_λ. As for the *label process* l, a MRF model is adopted [31] over a nearest neighborhood system. According to the *Hammersley–Clifford theorem* [15], $P(l)$ follows a Gibbs distribution:

$$P(l) = \frac{1}{Z}\exp(-U(l)) = \frac{1}{Z}\exp\left(-\sum_{C\in\mathscr{C}} V_C(l_C)\right), \tag{9.24}$$

where $U(l)$ is called an *energy function*, $Z = \sum_{l\in\Omega_l}\exp(-U(l))$ is the normalizing constant (or *partition function*) and V_C denotes the *clique potentials* of cliques $C \in \mathscr{C}$ having the label configuration l_C. The prior $P(l)$ will represent the simple fact that segmentations should be locally homogeneous. Therefore we will define clique potentials V_C over pairs of neighboring pixels (*doubletons*) such that similar classes in neighboring pixels are favored:

$$V_C = \beta \cdot \delta(l(\mathbf{x}), l(\mathbf{y})) = \begin{cases} +\beta \text{ if } l(\mathbf{x}) \neq l(\mathbf{y}) \\ -\beta \text{ otherwise,} \end{cases} \tag{9.25}$$

where β is a hyper-parameter controlling the interaction strength. As β increases, regions become more homogeneous. The energy is proportional to the length of the region boundaries. Thus homogeneous segmentations will get a higher probability, as expected.

Factoring the above distributions and applying the Bayes theorem gives us the posterior distribution $P(l|f) \propto P(f|l)P(l)$. Note that the constant factor $1/P(f)$ has been dropped as we are only interested in $\widehat{l}$ which *maximizes* the posterior, i.e. the maximum A posteriori (MAP) estimate of the hidden field L:

$$\widehat{l} = \arg\max_{l\in\Omega_l} P(f \mid l)P(l). \tag{9.26}$$

The models of the above distributions depend also on certain parameters. Since neither these parameters nor l is known, both has to be inferred from the only observable entity f. This is known in statistics as the *incomplete data* problem and a fairly standard tool to solve it is *expectation maximization* [13] and its variants. However, our problem becomes much harder when the number of labels K is unknown. When this parameter is also being estimated, the unsupervised segmentation problem may be treated as a *model selection* problem over a combined model space. From this point of view, K becomes a *model indicator* and the observation f is regarded as a three-variate Normal *mixture* with K components corresponding to clusters of pixels which are homogeneous in color.

The goal of our analysis is inference about the number K of Gaussian mixture components (each one corresponds to a label), the component parameters $\Theta = \{\Theta_\lambda = (\mathbf{m}_\lambda, \Sigma_\lambda) \mid \lambda \in \Lambda\}$, the component weights p_λ summing to 1, the inter-pixel interaction strength β, and the segmentation l. The only observable entity is f, thus the posterior distribution becomes:

$$P(K,p,\beta,l,\Theta \mid f) = P(K,p,\beta,l,\Theta,f)/P(f) \tag{9.27}$$

with $p = \{p_1, \cdots, p_K\}$. Note that $P(f)$ is constant; hence, we are only interested in the joint distribution of the variables K,p,β,l,Θ,f:

$$P(K,p,\beta,l,\Theta,f) = P(l,f \mid \Theta,\beta,p,K)P(\Theta,\beta,p,K). \tag{9.28}$$

In our context, it is natural to impose conditional independences on (Θ,β,p,K) so that their joint probability reduces to the product of priors:

$$P(\Theta,\beta,p,K) = P(\Theta)P(\beta)P(p)P(K). \tag{9.29}$$

Let us concentrate now on the posterior of (l,f):

$$P(l,f \mid \Theta,\beta,p,K) = P(f \mid l,\Theta,\beta,p,K)P(l \mid \Theta,\beta,p,K). \tag{9.30}$$

Before further proceeding, we can impose additional conditional independences. Since each pixel class (or label) is represented by a Gaussian, we obtain

$$\begin{aligned} P(f \mid l,\Theta,\beta,p,K) &= P(f \mid l,\Theta) \\ &= \prod_{\mathbf{x}\in E} \frac{1}{\sqrt{(2\pi)^3 |\Sigma_{l(\mathbf{x})}|}} \exp\left(-\tfrac{1}{2}(f(\mathbf{x}) - \mathbf{m}_{l(\mathbf{x})})\Sigma_{l(\mathbf{x})}^{-1}(f(\mathbf{x}) - \mathbf{m}_{l(\mathbf{x})})^T\right) \end{aligned} \tag{9.31}$$

and $P(l \mid \Theta,\beta,p,K) = P(l \mid \beta,p,K)$.

Furthermore, the component weights p_λ, $\lambda \in \Lambda$, can be incorporated into the underlying MRF label process as an external field strength. Formally, this is done via the *singleton* potential (probability of individual pixel labels):

$$P(l \mid \beta,p,K) = P(l \mid \beta,K) \prod_{\mathbf{x}\in E} p_{l(\mathbf{x})}. \tag{9.32}$$

Since the label process follows a Gibbs distribution [15], we can also express the above probability in terms of an energy:

$$P(l \mid \beta,p,K) = \frac{1}{Z(\beta,p,K)} \exp(-U(l \mid \beta,p,K)) \text{ , where} \tag{9.33}$$

$$U(l \mid \beta, p, K) = \beta \sum_{\{\mathbf{x},\mathbf{y}\} \in \mathscr{C}} \delta(l(\mathbf{x}), l(\mathbf{y})) - \sum_{\mathbf{x} \in E} \log(p_{l(\mathbf{x})}). \tag{9.34}$$

$\{\mathbf{x},\mathbf{y}\}$ denotes a *doubleton* containing the neighboring pixel sites $\mathbf{x}$ and $\mathbf{y}$. The basic idea is that segmentations has to be homogeneous *and* only those labels are valid in the model for which we can associate fairly big regions. The former constraint is ensured by the doubletons while the latter one is implemented via the component weights. Indeed, invalid pixel classes typically get only a few pixels assigned; hence, no matter how homogeneous are the corresponding regions, the above probability will be low. Unfortunately, the partition function $Z(\beta, p, K)$ is not tractable [37]; thus, the comparison of the likelihood of two differing MRF realizations from (9.33) is infeasible. Instead, we can compare their pseudo-likelihood [33, 37]:

$$P(l \mid \beta, p, K) \approx \prod_{\mathbf{x} \in E} \frac{p_{l(\mathbf{x})} \exp\left(-\beta \sum_{\forall \mathbf{y}:\{\mathbf{x},\mathbf{y}\} \in \mathscr{C}} \delta(l(\mathbf{x}), l(\mathbf{y}))\right)}{\sum_{\lambda \in \Lambda} p_\lambda \exp\left(-\beta \sum_{\forall \mathbf{y}:\{\mathbf{x},\mathbf{y}\} \in \mathscr{C}} \delta(\lambda, l(\mathbf{y}))\right)}. \tag{9.35}$$

Finally, we get the following approximation for the whole posterior distribution [33]:

$$P(K, p, \beta, l, \Theta \mid f) \propto P(f \mid l, \Theta) \prod_{\mathbf{x} \in E} \frac{p_{l(\mathbf{x})} \exp\left(-\beta \sum_{\forall \mathbf{y}:\{\mathbf{x},\mathbf{y}\} \in \mathscr{C}} \delta(l(\mathbf{x}), l(\mathbf{y}))\right)}{\sum_{\lambda \in \Lambda} p_\lambda \exp\left(-\beta \sum_{\forall \mathbf{y}:\{\mathbf{x},\mathbf{y}\} \in \mathscr{C}} \delta(\lambda, l(\mathbf{y}))\right)}$$

$$\times P(\beta) P(K) \prod_{\lambda \in \Lambda} P(\mathbf{m}_\lambda) P(\Sigma_\lambda) P(p_\lambda). \tag{9.36}$$

In order to simplify our presentation, we will follow [33, 57] and chose uniform reference priors for K, $\mathbf{m}_\lambda$, Σ_λ, p_λ ($\lambda \in \Lambda$). However, we note that informative priors could improve the quality of estimates, especially in the case of the number of classes. Although it is theoretically possible to sample β from the posterior, we will set its value a priori. The reasons are as follows:

- Due to the approximation by the pseudo-likelihood, the posterior density for β may not be proper [3].
- Being a hyper-parameter, β is largely independent of the input image. As long as it is large enough, the quality of segmentations are quite similar [31]. In addition, it is also independent of the number of classes since doubleton potentials will only check whether two neighboring labels are equal.

As a consequence, $P(\beta)$ is constant.

9.2.2 Sampling from the Posterior Distribution

A broadly used tool to sample from the posterior distribution in (9.36) is the Metropolis–Hastings method [23]. Classical methods, however, cannot be used due to the changing dimensionality of the parameter space. To overcome this limitation, a promising approach, called reversible jump MCMC (RJMCMC), has been proposed in [17]. When we have multiple parameter subspaces of different dimensionality, it is necessary to devise different *move types* between the subspaces [17]. These will be combined in a so-called *hybrid sampler*. For the color image segmentation model, the following move types are needed [33]:

1. Sampling the labels l (i.e., re-segment the image)
2. Sampling Gaussian parameters $\Theta = \{(\mathbf{m}_\lambda, \Sigma_\lambda)\}$
3. Sampling the mixture weights $p_\lambda\,(\lambda \in \Lambda)$
4. Sampling the MRF hyperparameter β
5. Sampling the number of classes K (splitting one mixture component into two, or combining two into one)

The only randomness in scanning these move types is the random choice between splitting and merging in move (5). One iteration of the hybrid sampler, also called a *sweep*, consists of a complete pass over these moves. The first four move types are conventional in the sense that they do not alter the dimension of the parameter space. In each of these move types, the posterior distribution can be easily derived from (9.36) by setting unaffected parameters to their current estimate. For example, in move (1), the parameters K, p, β, Θ are set to their estimates $\widehat{K}$, $\widehat{p}$, $\widehat{\beta}$, $\widehat{\Theta}$. Thus, the posterior in (9.36) reduces to the following form:

$$\begin{aligned} P(K,p,\beta,l,\Theta \mid f) &\propto P(f \mid l,\widehat{\Theta})P(l \mid \widehat{\beta},\widehat{p},\widehat{K}) \\ &\propto \prod_{\mathbf{x}\in E}\left(\frac{1}{\sqrt{(2\pi)^3|\widehat{\Sigma}_{l(\mathbf{x})}|}}\exp\left(-\frac{1}{2}\left(f(\mathbf{x})-\widehat{\mathbf{m}}_{l(\mathbf{x})}\right)\widehat{\Sigma}_{l(\mathbf{x})}^{-1}(f(\mathbf{x})-\widehat{\mathbf{m}}_{l(\mathbf{x})})^T\right)\right) \\ &\quad \times \prod_{\mathbf{x}\in E}\widehat{p}_{l(\mathbf{x})}\exp\left(-\widehat{\beta}\sum_{\forall \mathbf{y}:\{\mathbf{x},\mathbf{y}\}\in\mathscr{C}}\delta(l(\mathbf{x}),l(\mathbf{y}))\right). \end{aligned} \tag{9.37}$$

Basically, the above equation corresponds to a segmentation with *known parameters*.

In our experiments, move (4) is never executed since β is fixed a priori. As for moves (2) and (3), a closed form solution also exists: Using the current label field $\widehat{l}$ as a *training set*, an unbiased estimate of p_λ, $\mathbf{m}_\lambda$, and Σ_λ can be obtained as the zeroth, first, and second moments of the labeled data [31, 45].

Hereafter, we will focus on move (5), which requires the use of the reversible jump mechanism. This move type involves changing K by 1 and making necessary corresponding changes to l, Θ, and p.

9.2.2.1 Reversible Jump Mechanism

First, let us briefly review the reversible jump technique. A comprehensive introduction by Green can be found in [18]. For ease of notation, we will denote the set of unknowns $\{K, p, \beta, l, \Theta\}$ by χ and let $\pi(\chi)$ be the target probability measure (the posterior distribution from (9.36), in our context). A standard tool to sample from $\pi(\chi)$ is the Metropolis–Hastings method [23, 44]: Assuming the current state is χ,

1. First a candidate new state is drawn from the *proposal measure* $q(\chi, \chi')$, which is an essentially arbitrary joint distribution. Often a uniform distribution is adopted in practice.
2. Then χ' is accepted with probability $\mathscr{A}(\chi, \chi')$—the so-called *acceptance probability*.

If χ' is rejected, then we stay in the current state χ. Otherwise, a transition $\chi \to \chi'$ is made. The sequence of accepted states is a Markov chain. As usual in MCMC [18], this chain has to be reversible which implies that the transition kernel P of the chain satisfies the *detailed balance* condition:

$$\int \pi(\mathrm{d}\chi) P(\chi, \mathrm{d}\chi') = \int \pi(\mathrm{d}\chi') P(\chi', \mathrm{d}\chi). \tag{9.38}$$

From the above equation, $\mathscr{A}(\chi, \chi')$ can be formally derived [18, 23]:

$$\mathscr{A}(\chi, \chi') = \min\left(1, \frac{\pi(\chi') q(\chi, \chi')}{\pi(\chi) q(\chi', \chi)}\right). \tag{9.39}$$

The implementation of these transitions are quite straightforward. Following Green [18], we can easily separate the random and deterministic parts of such a transition in the following manner:

- At the current state χ, we generate a random vector u of dimension r from a known density p. Then the candidate new state is formed as a deterministic function of the current state χ and the random numbers in u: $\chi' = h(\chi, u)$.
- Similarly, the reverse transition $\chi' \to \chi$ would be accomplished with the aid of r' random numbers u' drawn from p', yielding $\chi = h'(\chi', u')$.

If the transformation from (χ, u) to (χ', u') is a *diffeomorphism* (i.e., both the transformation and its inverse are differentiable), then the *detailed balance* condition is satisfied when [18]

$$\pi(\chi) p(u) \mathscr{A}(\chi, \chi') = \pi(\chi') p'(u') \mathscr{A}(\chi', \chi) \left| \frac{\partial(\chi', u')}{\partial(\chi, u)} \right|, \tag{9.40}$$

where the last factor is the Jacobian of the diffeomorphism. Note that it appears in the equality only because the proposal destination $\chi' = h(\chi, u)$ is specified

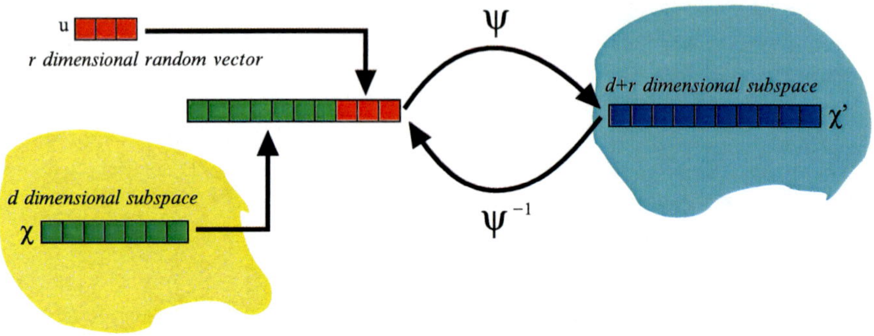

Fig. 9.4 ψ is a *diffeomorphism* which transforms back and forth between parameter subspaces of different dimensionality. *Dimension matching* can be implemented by generating a random vector u such that the dimensions of (χ, u) and χ' are equal

indirectly. The *acceptance probability* is derived again from the *detailed balance* equation [17, 18]:

$$\mathscr{A}(\chi, \chi') = \min\left(1, \frac{\pi(\chi')p'(u')}{\pi(\chi)p(u)} \left|\frac{\partial(\chi', u')}{\partial(\chi, u)}\right|\right). \tag{9.41}$$

The main advantage of the above formulation is that it remains valid in a variable dimension context. As long as the transformation $(\chi, u) \mapsto (\chi', u')$ remains a *diffeomorphism*, the dimensions of χ and χ' (denoted by d and d') can be different. One necessary condition for that is the so-called *dimension matching* (see Fig. 9.4). Indeed, if the $d + r = d' + r'$ equality failed then the mapping and its inverse could not both be differentiable.

In spite of the relatively straightforward theory of reversible jumps, it is by far not evident how to construct efficient jump proposals in practice. This is particularly true in image-processing problems, where the dimension of certain inferred variables (like the labeling l) is quite big. Although there have been some attempt [7, 18] to come up with general recipes on how to construct efficient proposals, there is still no good solution to this problem.

In the remaining part of this section, we will apply the reversible jump technique for sampling from the posterior in (9.36). In particular, we will construct a diffeomorphism ψ along with the necessary probability distributions of the random variables u such that a reasonable acceptance rate of jump proposals is achieved. In our case, a jump proposal may either be a *split* or *merge* of classes. In order to implement these proposals, we will extend the moment matching concept of Green [17, 57] to three-variate Gaussians. However, our construction is admittedly ad hoc and fine-tuned to the color image segmentation problem. For a theoretical treatment of the multivariate Gaussian case, see the works of Stephens [60, 61].

9.2.2.2 Splitting One Class into Two

The *split proposal* begins by randomly choosing a class λ with a uniform probability $P^{\text{split}}_{\text{select}}(\lambda) = 1/K$. Then K is increased by 1 and λ is split into λ_1 and λ_2. In doing so, a new set of parameters need to be generated. Altering K changes the dimensionality of the variables Θ and p. Thus, we shall define a deterministic function ψ as a function of these Gaussian mixture parameters:

$$(\Theta^+, p^+) = \psi(\Theta, p, u), \tag{9.42}$$

where the superscript $+$ denotes parameter vectors after incrementing K. u is a set of random variables having as many elements as the degree of freedom of joint variation of the current parameters (Θ, p) and the proposal (Θ^+, p^+). Note that this definition satisfies the *dimension-matching* constraint [17] (see Fig. 9.4), which guarantees that one can jump back and forth between different parameter subspaces. The new parameters of λ_1 and λ_2 are assigned by matching the 0th, 1th, 2th moments of the component being split to those of a combination of the two new components [33, 57]:

$$p_\lambda = p^+_{\lambda_1} + p^+_{\lambda_2}, \tag{9.43}$$

$$p_\lambda \mathbf{m}_\lambda = p^+_{\lambda_1} \mathbf{m}^+_{\lambda_1} + p^+_{\lambda_2} \mathbf{m}^+_{\lambda_2}, \tag{9.44}$$

$$p_\lambda \left(\mathbf{m}_\lambda \mathbf{m}^T_\lambda + \Sigma_\lambda\right) = p^+_{\lambda_1} \left(\mathbf{m}^+_{\lambda_1} \mathbf{m}^{+T}_{\lambda_1} + \Sigma^+_{\lambda_1}\right) + p^+_{\lambda_2} \left(\mathbf{m}^+_{\lambda_2} \mathbf{m}^{+T}_{\lambda_2} + \Sigma^+_{\lambda_2}\right). \tag{9.45}$$

There are 10 degrees of freedom in splitting λ since covariance matrices are symmetric. Therefore, we need to generate a random variable $u1$, a random vector $u2$, and a symmetric random matrix $u3$. We can now define the diffeomorphism ψ which transforms the old parameters (Θ, p) to the new (Θ^+, p^+) using the above moment equations and the random numbers $u1$, $u2$, and $u3$ [33]:

$$p^+_{\lambda_1} = p_\lambda u1, \tag{9.46}$$

$$p^+_{\lambda_2} = p_\lambda (1 - u1), \tag{9.47}$$

$$\mathbf{m}^+_{\lambda_1,i} = \mathbf{m}_{\lambda,i} + u2_i \sqrt{\Sigma_{\lambda,i,i} \frac{1-u1}{u1}}, \tag{9.48}$$

$$\mathbf{m}^+_{\lambda_2,i} = \mathbf{m}_{\lambda,i} - u2_i \sqrt{\Sigma_{\lambda,i,i} \frac{u1}{1-u1}}, \tag{9.49}$$

$$\Sigma^+_{\lambda_1,i,j} = \begin{cases} u3_{i,i} \left(1 - u2_i^{\,2}\right) \Sigma_{\lambda,i,i} \dfrac{1}{u1} & \text{if } i = j \\ u3_{i,j} \Sigma_{\lambda,i,j} \sqrt{\left(1 - u2_i^{\,2}\right)} \sqrt{\left(1 - u2_j^{\,2}\right) u3_{i,i} u3_{j,j}} & \text{if } i \neq j, \end{cases} \tag{9.50}$$

$$\Sigma^{+}_{\lambda_2,i,j} = \begin{cases} \left(1 - u3_{i,i}\right)\left(1 - u2_i{}^2\right)\Sigma_{\lambda,i,i}\dfrac{1}{u1} & \text{if } i = j \\ \left(1 - u3_{i,j}\right)\Sigma_{\lambda,i,j} & \\ \quad\times\sqrt{\left(1 - u2_i{}^2\right)\left(1 - u2_j{}^2\right)}\sqrt{\left(1 - u3_{i,i}\right)\left(1 - u3_{j,j}\right)} & \text{if } i \neq j. \end{cases} \tag{9.51}$$

The random variables u are chosen from the interval $(0,1]$. In order to favor splitting a class into roughly equal portions, beta$(1.1,1.1)$ distributions are used. To guarantee numerical stability in inverting $\Sigma^{+}_{\lambda_1}$ and $\Sigma^{+}_{\lambda_2}$, one can use some regularization like in [12], or one can use the well-known Wishart distribution [41]. However, we did not experience such problems, mainly because the obtained covariance matrices are also reestimated from the image data in subsequent move types. Therefore, as long as our input image can be described by a mixture of Gaussians, we can expect that the estimated covariance matrices are correct.

The next step is the reallocation of those sites $\mathbf{x} \in E_\lambda$ where $l(\mathbf{x}) = \lambda$. This reallocation is based on the new parameters and has to be completed in such a way as to ensure the resulting labeling l^+ is drawn from the posterior distribution with $\Theta = \Theta^+$, $p = p^+$, and $K = K + 1$. At the moment of splitting, however, the neighborhood configuration at a given site $\mathbf{x} \in E_\lambda$ is unknown. Thus, the calculation of the term $P(l^+ \mid \widehat{\beta}, p^+, K+1)$ is not possible. First, we have to provide a tentative labeling of the sites in E_λ. Then we can sample the posterior distribution using a Gibbs sampler. Of course, a tentative labeling might be obtained by allocating λ_1 and λ_2 at random. In practice, however, we need a labeling l^+ which has a relatively high posterior probability in order to maintain a reasonable acceptance probability. To achieve this goal, we use a few step (around 5 iterations) of ICM [5] algorithm to obtain a suboptimal initial segmentation of E_λ. The resulting label map can then be used to draw a sample from the posterior distribution using a one-step Gibbs sampler [15]. The obtained l^+ has a relatively high posterior probability since the tentative labeling was close to the optimal one.

9.2.2.3 Merging Two Classes

A pair (λ_1, λ_2) is chosen with a probability inversely proportional to their distance:

$$P^{\text{merge}}_{\text{select}}(\lambda_1, \lambda_2) = \frac{1/d(\lambda_1, \lambda_2)}{\sum\limits_{\lambda \in \Lambda} \sum\limits_{\kappa \in \Lambda} 1/d(\lambda, \kappa)}, \tag{9.52}$$

where $d(\lambda_1, \lambda_2)$ is the symmetric Mahalanobis distance between the classes λ_1 and λ_2 defined as:

$$d(\lambda_1, \lambda_2) = (\mathbf{m}_{\lambda_1} - \mathbf{m}_{\lambda_2})\Sigma^{-1}_{\lambda_1}(\mathbf{m}_{\lambda_1} - \mathbf{m}_{\lambda_2}) + (\mathbf{m}_{\lambda_2} - \mathbf{m}_{\lambda_1})\Sigma^{-1}_{\lambda_2}(\mathbf{m}_{\lambda_2} - \mathbf{m}_{\lambda_1}). \tag{9.53}$$

In this way, we favor merging classes that are close to each other thus increasing acceptance probability. The merge proposal is deterministic once the choices of λ_1 and λ_2 have been made. These two components are merged, reducing K by 1. As in the case of splitting, altering K changes the dimensionality of the variables Θ and p. The new parameter values (Θ^-, p^-) are obtained from (9.43)–(9.45). The reallocation is simply done by setting the label at sites $\mathbf{x} \in E_{\{\lambda_1,\lambda_2\}}$ to the new label λ. The random variables u are obtained by back substitution into (9.46)–(9.51).

9.2.2.4 Acceptance Probability

As discussed in Sect. 9.2.2.1, the split or merge proposal is accepted with a probability relative to the probability ratio of the current and the proposed states. Let us first consider the acceptance probability $\mathscr{A}_{\text{split}}$ for the split move. For the corresponding merge move, the acceptance probability is obtained as the inverse of the same expression with some obvious differences in the substitutions.

$$\mathscr{A}_{\text{split}}(K,\widehat{p},\widehat{\beta},\widehat{l},\widehat{\Theta},K+1,p^+,\widehat{\beta},l^+,\Theta^+) = \min(1,A), \text{ where} \tag{9.54}$$

$$A = \frac{P(K+1,p^+,\widehat{\beta},l^+,\Theta^+ \mid f)}{P(K,\widehat{p},\widehat{\beta},\widehat{l},\widehat{\Theta} \mid f)} \frac{P_{\text{merge}}(K+1)P^{\text{merge}}_{\text{select}}(\lambda_1,\lambda_2)}{P_{\text{split}}(K)P^{\text{split}}_{\text{select}}(\lambda)P_{\text{realloc}}} \times \frac{1}{P(u1)\prod_{i=1}^{3}\left(P(u2_i)\prod_{j=i}^{3}P(u3_{i,j})\right)} \left|\frac{\partial\psi}{\partial(\Theta_\lambda,p_\lambda,u)}\right|. \tag{9.55}$$

P_{realloc} denotes the probability of reallocating pixels labeled by λ into regions labeled by λ_1 and λ_2. It can be derived from (9.37) by restricting the set of labels Λ^+ to the subset $\{\lambda_1,\lambda_2\}$ and taking into account only those sites $\mathbf{x}$ for which $l(\mathbf{x})^+ \in \{\lambda_1,\lambda_2\}$:

$$\begin{aligned} P_{\text{realloc}} &\approx \prod_{\forall \mathbf{x}: l(\mathbf{x})^+ \in \{\lambda_1,\lambda_2\}} \frac{1}{\sqrt{(2\pi)^3 |\Sigma^+_{l(\mathbf{x})^+}|}} \exp\left(-\tfrac{1}{2}\left(f(\mathbf{x})-\mathbf{m}^+_{l(\mathbf{x})^+}\right)\Sigma^{+-1}_{l(\mathbf{x})^+}\left(f(\mathbf{x})-\mathbf{m}^+_{l(\mathbf{x})^+}\right)^T\right) \\ &\times \prod_{\forall \mathbf{x}: l(\mathbf{x})^+ \in \{\lambda_1,\lambda_2\}} p^+_{l(\mathbf{x})^+} \exp\left(-\widehat{\beta}\textstyle\sum_{\forall \mathbf{y}:\{\mathbf{x},\mathbf{y}\}\in\mathscr{C}} \delta(l(\mathbf{x})^+, l(\mathbf{y})^+)\right). \end{aligned} \tag{9.56}$$

The last factor is the Jacobian determinant of the transformation ψ:

$$\left|\frac{\partial\psi}{\partial(\Theta_\lambda,p_\lambda,u)}\right| = -p_\lambda \prod_{i=1}^{3}\left(\frac{\Sigma^2_{i,i}}{u1\,(u1-1)}\,(1-u2_i{}^2)\,(1-u3_{i,i})\;u3_{i,i}\prod_{j=i}^{3}\frac{\Sigma_{i,j}}{u1\,(u1-1)}\right). \tag{9.57}$$

The acceptance probability for the merge move can be easily obtained with some obvious differences in the substitutions as

$$\mathscr{A}_{\text{merge}}(K,\widehat{p},\widehat{\beta},\widehat{l},\widehat{\Theta};K-1,p^{-},\widehat{\beta},l^{-},\Theta^{-}) = \min\left(1,\frac{1}{A}\right). \tag{9.58}$$

9.2.3 *Optimization According to the MAP Criterion*

The following MAP estimator is used to obtain an optimal segmentation $\widehat{l}$ and model parameters $\widehat{K},\widehat{p},\widehat{\beta},\widehat{\Theta}$:

$$(\widehat{l},\widehat{K},\widehat{p},\widehat{\beta},\widehat{\Theta}) = \arg\max_{K,p,\beta,l,\Theta} P(K,p,\beta,l,\Theta \mid f) \tag{9.59}$$

with the following constraints: $l \in \Omega, K_{\min} \leq K \leq K_{\max}, \Sigma_{\lambda\in\Lambda} p_\lambda = 1, \forall\lambda \in \Lambda : 0 \leq \mathbf{m}_{\lambda,i} \leq 1, 0 \leq \Sigma_{\lambda,i,i} \leq 1$, and $-1 \leq \Sigma_{\lambda,i,j} \leq 1$. Equation (9.59) is a combinatorial optimization problem which can be solved using simulated annealing [15, 33]:

Algorithm 1 (RJMCMC Segmentation)

① *Set $k=0$. Initialize $\widehat{\beta}^0$, $\widehat{K}^0$, $\widehat{p}^0$, $\widehat{\Theta}^0$, and the initial temperature T_0.*

② *A sample $(\widehat{l}^k,\widehat{K}^k,\widehat{p}^k,\widehat{\beta}^k,\widehat{\Theta}^k)$ is drawn from the posterior distribution using the* hybrid sampler *outlined in Sect. 9.2.2. Each sub-chain is sampled via the corresponding move type while all the other parameter values are set to their current estimate.*

③ *Goto Step ② with $k=k+1$ and T_{k+1} until $k<\mathscr{K}$.*

As usual, an exponential annealing schedule ($T_{k+1} = 0.98T_k$, $T_0 = 6.0$) was chosen so that the algorithm would converge after a reasonable number of iterations. In our experiments, the algorithm was stopped after 200 iterations ($T_{200} \approx 0.1$).

9.2.4 *Experimental Results*

The evaluation of segmentation algorithms is inherently subjective. Nevertheless, there have been some recent works on defining an objective quality measure. Such a boundary benchmarking system is reported in [43] that we will use herein to quantify our results. The ground truth is provided as human-segmented images (each image is processed by several subjects). The output of the benchmarked segmentation algorithm is presented to the system as a soft boundary map where higher values mean greater confidence in the existence of a boundary. Then, two quantities are computed:

Precision is the probability that a machine-generated boundary pixel is a true boundary pixel. It measures the noisiness of the machine segmentation with respect to the human ones.

Recall is the probability that a true boundary pixel is detected. It tells us how much the ground truth is detected.

From these values, a precision–recall curve is produced which shows the trade-off between the two quantities (see Fig. 9.6). We will also summarize the performance in a single number: the maximum *F-measure* value across an algorithm's precision–recall curve. The *F-measure* characterizes the distance of a curve from the origin which is computed as the harmonic mean of precision and recall [43]. Clearly, for nonintersecting precision–recall curves, the one with a higher maximum *F-measure* will dominate.

The presented algorithm has been tested on a variety of real color images. First, the original images were converted from RGB to LHS color space [59] in which chroma and intensity informations are separated. Results in other color spaces can be found in [30]. The dynamic range of color components was then normalized to $(0,1)$. The number of classes K was restricted to the interval $[1,50]$ and β has been set to 2.5. This value gave us good results in all test cases. This is demonstrated in Fig. 9.6, where we plot precision–recall curves for $\beta = 2.5$, $\beta = 0.5$, and $\beta = 10.0$.

Independently of the input image, we start the algorithm with two classes ($\widehat{K}^0 = 2$), each of them having equal weights ($\widehat{p}_0^0 = \widehat{p}_1^0 = 0.5$). The initial mean vectors were set to $[0.2, 0.2, 0.2]$ and $[0.7, 0.7, 0.7]$, and both covariance matrices were initialized as

$$\widehat{\Sigma}_0^0 = \widehat{\Sigma}_1^0 = \begin{pmatrix} 0.05 & 0.00001 & 0.00001 \\ 0.00001 & 0.05 & 0.00001 \\ 0.00001 & 0.00001 & 0.05 \end{pmatrix}.$$

As an example, we show in Fig. 9.5 these initial Gaussians as well as the final estimates. In spite of the rough initialization, the algorithm finds the three meaningful classes and an accurate segmentation is obtained.

In subsequent figures, we will compare the presented method to JSEG [14], which is a recent unsupervised color image segmentation algorithm. It consists of two independent steps:

1. Colors in the image are quantized to several representative classes. The output is a class map where pixels are replaced by their corresponding color class labels.
2. A region-growing method is then used to segment the image based on the multi-scale *J-images*. A *J-image* is produced by applying a criterion to local windows in the class-map (see [14] for details on that).

JSEG is also region based, uses similar cues (color similarity and spatial proximity) to RJMCMC, and it is fully automatic. We have used the program provided by the authors [14] and kept its default settings throughout our test: automatic color quantization threshold and number of scales, the region merge threshold was also set to its default value (0.4). Note that JSEG is not model based; therefore, there are no pixel classes. Regions are identified based on the underlying color properties of the input image. Although we also show the number of labels for JSEG in our test results, these numbers reflect the number of detected *regions*. In RJMCMC,

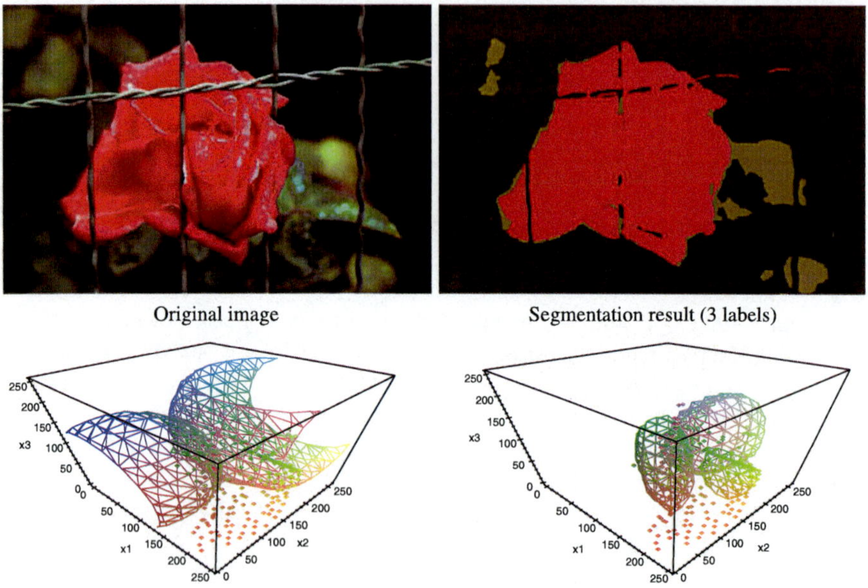

Original image — Segmentation result (3 labels)

Initial Gaussians — Final estimation (3 classes)

Fig. 9.5 Segmentation of image *rose41*

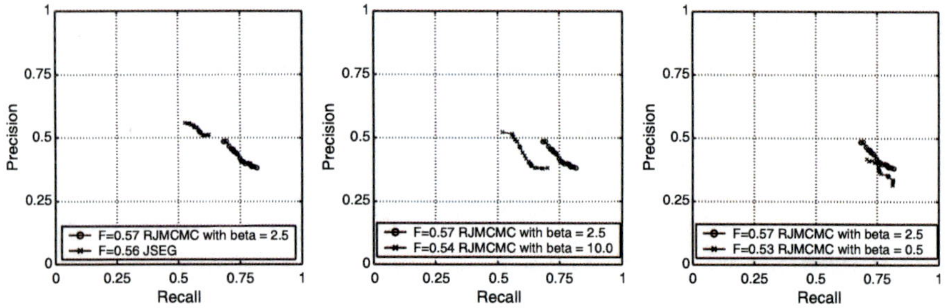

RJMCMC vs. JSEG — RJMCMC $\beta = 2.5$ vs. $\beta = 10.0$ — RJMCMC $\beta = 2.5$ vs. $\beta = 0.5$

Fig. 9.6 Precision–recall curves for JSEG and RJMCMC

however, the *same* label is assigned to spatially distant regions if they are modeled by the same Gaussian component. Segmentation results are displayed as a *cartoon image* where pixel values are replaced by their label's average color in order to help visual evaluation of the segmentation quality.

In Fig. 9.7, we show a couple of results obtained on the Berkeley segmentation data set [43], and in Fig. 9.6, we plot the corresponding precision–recall curves. Note that RJMCMC has a slightly higher *F-measure* which ranks it over JSEG. However, it is fair to say that both method perform equally well but behave differently: while JSEG tends to smooth out fine details (hence, it has a higher

Fig. 9.7 Benchmark results on images from the Berkeley segmentation data set

precision but lower recall value), RJMCMC prefers to keep fine details at the price of producing more edges (i.e., its recall values are higher at a lower precision value).

After showing the interest of an MGMM approximation of the color distribution, we now present how to use the linear prediction models for the characterization of the spatial structure of color textures.

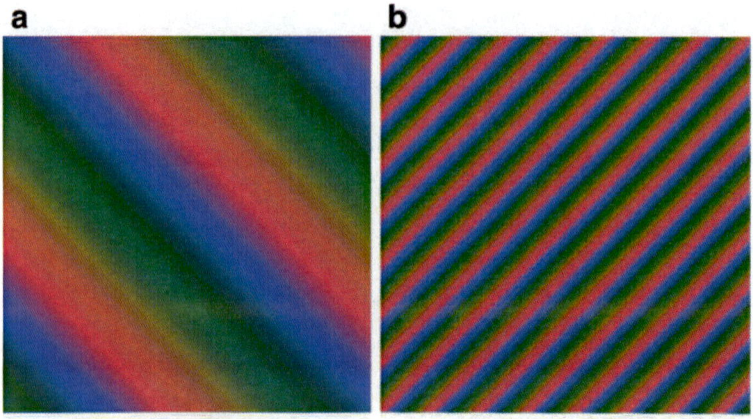

Fig. 9.8 Chromatic sinusoids in IHLS color space

9.3 Linear Prediction Models and Spectral Analysis

*9.3.1 Spectral Analysis in IHLS and L*a*b**

In order to compare the different PSD estimation methods, (see Sect. 9.1.3.1) as it was done in the case of gray-level images [10], we have generated synthetic images containing noisy sinusoids. In [53], this comparison is presented for the two color spaces IHLS and L*a*b*. As an initial supposition, we considered IHLS color space more appropriate for this kind of spectral estimation as it has an achromatic axis which is perpendicular to the chromatic plane. It is for this reason that we did not consider other spaces in [52]. However, other color spaces of the type IHS could also be used for this type of analysis.

9.3.1.1 Comparison of PSD Estimation Methods

The noisy sinusoidal images contain in them a simulated 2-d real sinusoid for the luminance channel and a simulated complex 2-d sinusoid for the chrominance channel. A single realization of a multichannel Gaussian white noise vector is added:

$$f(\mathbf{x}) = \begin{bmatrix} A_l \cos\left(2\pi \langle \mathbf{x}, \nu_l \rangle + \phi_l\right) \\ A_c \times \exp\left(j(2\pi \langle \mathbf{x}, \nu_c \rangle + \phi_c)\right) \end{bmatrix} + \boldsymbol{b}(\mathbf{x}) \tag{9.60}$$

with A_i, ϕ_i, and ν_i, $i = l$ or c, the amplitudes, phases, and 2-d normalized frequencies of the sinusoids in the two channels, respectively. The Fig. 9.8 shows a chromatic sinusoid (with constant luminance value and additive white noise whose covariance

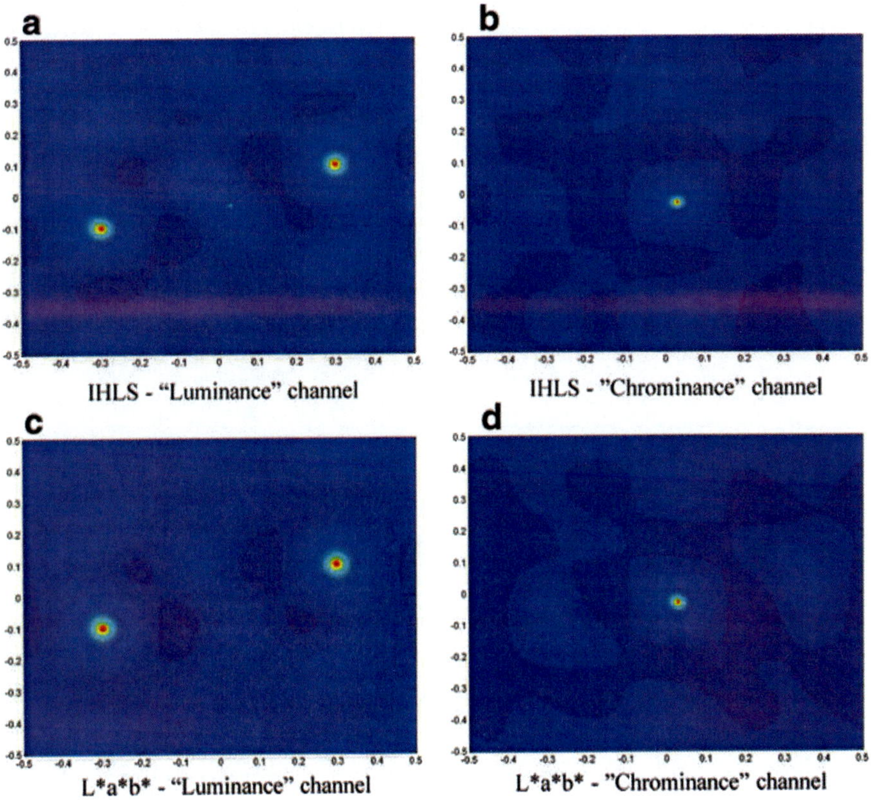

Fig. 9.9 Spectral estimates ($\nu \in [-0.5, 0.5]^2$) computed through PSD_HM method in IHLS and L*a*b* color spaces. The images have gone through a double transformation, i.e., from IHLS or L*a*b* to RGB then RGB to IHLS or L*a*b*

matrix is null for (9.60)) generated in IHLS color space. It is the description of a circular function in the plane which is perpendicular to the achromatic axis. Due to this fact, all of colors appear for a given saturation value and create a wave whose orientation and variations depend upon the values of ν_c, $\nu_c = (0.05, 0.05)$ for Fig. 9.8a and $\nu_c = (-0.3, 0.3)$ for Fig. 9.8b.

Figure 9.9 shows the spectral estimation examples using PSD_HM method (see Sect. 9.1.3.1) on the noisy sinusoids, $\nu_l = (0.3, 0.1)$ et $\nu_c = (0.03, -0.03)$ in the IHLS and L*a*b* color spaces. It is to note that the two symmetric lobes in S_{LL}^{HM} are well localized (with respect to the estimation error) around ν_l and $-\nu_l$ (see Figs. 9.9a and c). Also, the lobe in S_{CC}^{HM} is well localized around ν_c (see Figs. 9.9b and d).

In order to have precise comparative information of different spectral estimation methods, we estimated the mean (accuracy) and the variance (precision) of the estimations from multiple frequencies ($\nu_l = (0.3, 0.3)$ et $\nu_c = (0.05, 0.05)$ - $\nu_l = (0.05, 0.3)$ et $\nu_c = (-0.3, 0.3)$), multiple white noise sequences with SNR = 0 dB,

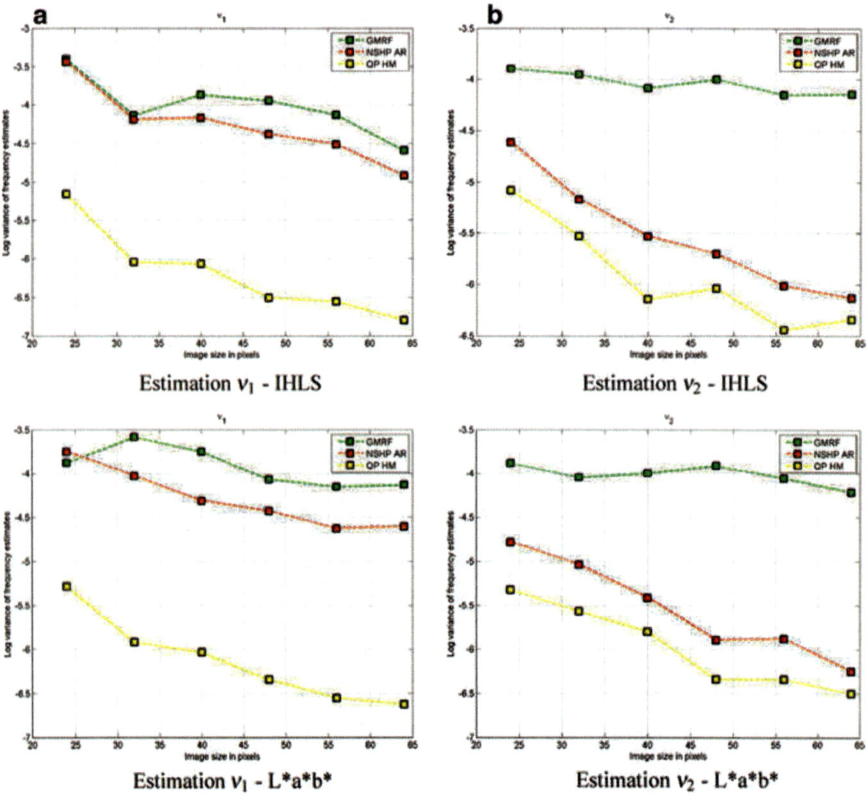

Fig. 9.10 Comparison of precisions of spectral analysis methods in IHLS and L*a*b* color spaces for luminance channel, through log variance of the estimated frequencies plotted against the image size

multiple image sizes ranging from 24×24 sites to 64×64 sites (see Figs. 9.10 and 9.11 for the precision of the spectral estimation in the luminance and chrominance channels, respectively) and taking same size of prediction support region for the three models. Globally, we obtained similar results as for the gray-level images in terms of mean error and variance. For causal models, isotropy of the PSD_HM estimates is better than isotropy for PSD_NSHP estimates. In these curves, there is no significant difference in the estimates with respect to the two considered color spaces.

9.3.1.2 Study of Luminance-Chrominance Interference

In [53] and [54], we have presented an experiment to study the interchannel interference between luminance and chrominance spectra associated to the color space transformations between RGB and IHLS or L*a*b*.

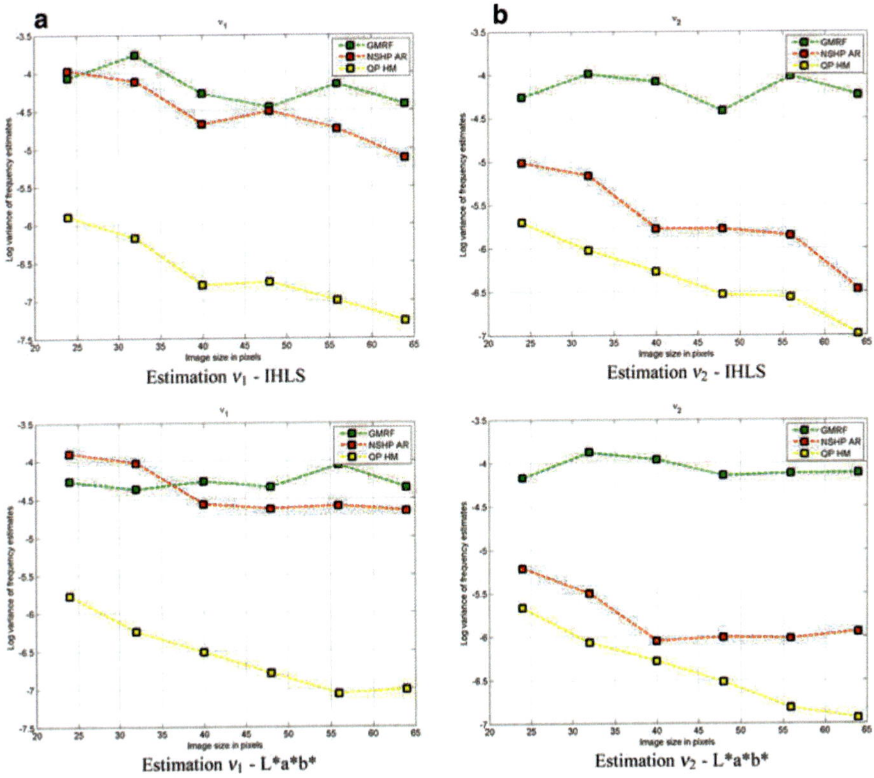

Fig. 9.11 Comparison of precisions of spectral analysis methods in IHLS and L*a*b* color spaces for chrominance channel, through log variance of the estimated frequencies plotted against the image size

The two channel complex sinusoidal images used in these tests are generated in perceptual color spaces, i.e., IHLS or L*a*b* (see (9.60)), transformed to RGB and again re-transformed to the perceptual color spaces before spectrum estimation. The spectral analysis using these models reveals the lobes associated to the successive transformations.

If we observe $S_{\mathrm{LL}}^{\mathrm{HM}}$ obtained from spectral analysis in IHLS color space (see Fig. 9.9a), we observe an extra frequency peak. This frequency peak is localized at the normalized frequency position of the chrominance channel. However, the spectral estimate of the image generated with same parameters in L*a*b* color space shows negligible interference of this chrominance in the luminance channel (see Fig. 9.9c). However, the interference of the luminance channel frequencies in the chrominance channel are less significant and, hence, could not be visualized in $S_{\mathrm{CC}}^{\mathrm{HM}}$ for both the used color spaces (see Figs. 9.9b and d). The degree of separation between the two channels, offered by each of these two color spaces could be characterized through a quantitative analysis of these interferences.

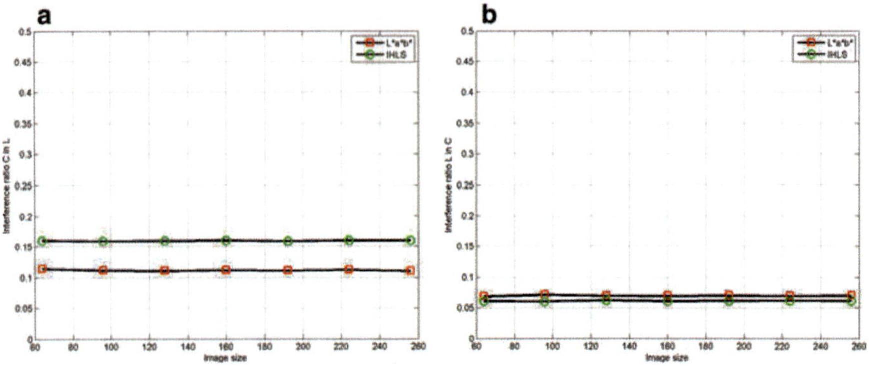

Fig. 9.12 Comparison of the interference between luminance and chrominance information i.e. IR_{CL} *(left) and* IR_{LC} *(right) in the two color spaces*

In order to measure these interferences, we generated 20 images for each color space (IHLS and L*a*b*), of size $n \times n$ with $n \in \{64, 96, 128, 160, 192, 224, 256\}$ containing the sinusoids having similar amplitude and phase ($A_{|} = 0.25$, $A_c = 0.25$, $\phi_r = 30°$, and $\phi_c = 30°$), and three different frequency sets: each set consisted of the same real frequency component $\nu_{|} = (0.3, 0.3)$, whereas chrominance channel frequencies were varied $\nu_c \in \{(-0.3, 0.3), (0.3, -0.3), (-0.3, -0.3)\}$. These images were created with a zero mean value. In [53], the PSDs were calculated for three different levels of SNR, SNR $\in \{-3, 0, 3\}$ dB. Here, we present the results for SNR = 0 dB. These PSDs were estimated using only harmonic mean method, PSD_HM with QP model of order $(2, 2)$.

We measure the level of interference of chrominance channel in luminance channel by using the ratio IR_{CL}, defined as:

$$IR_{\mathrm{CL}} = \frac{A_{c|}}{A_{|}} \tag{9.61}$$

with $A_{|c}$, the mean value (over 20 images, of size n and a given set of frequencies $(\nu_{|}, \nu_c)$) of the amplitude of the lobe associated to the chromatic sinusoid appearing in the luminance channel. Similarly, interference of the luminance channel in the chrominance channel is measured by the ratio IR_{LC}:

$$IR_{\mathrm{LC}} = \frac{A_{|c}}{A_c} \tag{9.62}$$

with $A_{|c}$, the mean value (over 20 images, of size $n \times n$ and a given set of frequencies $(\nu_{|}, \nu_c)$) of the amplitude of the lobe associated to the luminance sinusoid appearing in the chrominance channel.

Plots of these interferences for different image sizes, in IHLS and L*a*b* color spaces are given in Fig. 9.12. These ratios have been calculated for the frequency sets $\{(0.3, 0.3), (0.3, -0.3)\}$, and an SNR = 0 dB. These curves are presented on the same scales for a better comparison. From these results, globally we have concluded:

- The interference ratio values calculated for the luminance frequencies appearing in the chrominance channel IR_{LC} are approximately one half of those calculated for the chrominance frequencies appearing in the luminance channels IR_{CL} (see Fig. 9.12).
- The values of IR_{LC} are approximately the same in both the color spaces.
- The values of IR_{CL} are much significant in the IHLS color space than in L*a*b* color space.

In the Sect. 9.3.2, we present, how spectral analysis exploiting the luminance-chrominance decorrelation could be useful for characterization of the spatial structures in color textures. Figure 9.13 shows the spectral analysis of a color texture. The PSDs of the two channels obtained using different models could be compared with corresponding magnitude spectra of these channels obtained through Discrete Fourier Transform. The cross spectra show the correlations which exist between the two channels.

9.3.2 *Color Texture Classification*

In this section, we present a color texture classification method through the spectral estimates computed using the 2-D multichannel complex linear prediction models (see Sect. 9.1.3.1). We used three different data sets (DS) containing the images taken from Vistex and Outex databases.

In DS_1, each 512×512 image from Vistex was considered as an individual class. For each textured color image, i.e., for each class, the image feature cues were computed on the subimage blocks of size 32×32, hence forming 256 sub images for each image. Training data set for each color texture consisted of 96 sub images, while the remaining 160 sub images were used as the test data set, for each textured color image. By this configuration, we had a total of 2,304 training and 3,840 test subimages in total.

In the second data set DS_2, 54 images from Vistex database are used. The 54 original Vistex images of dimensions 512×512 were split into 16 samples of 128×128. DS_2 is available on the Outex web site[7] as test suite *Contrib_TC_00006*. For each texture, half of the samples were used in the training set and the other half served as testing data.

The third data set DS_3 included 68 images of the Outex database [48]. From the 68 Outex images of size 746×538 originally, 20 samples of size 128×128 were obtained. The training and test sets were chosen in the same way as in DS_2, thus giving a total of 680 samples in each of training and test set. At the Outex site, this is the test suite *Outex_TC_00013*.

[7] http://www.outex.oulu.fi/

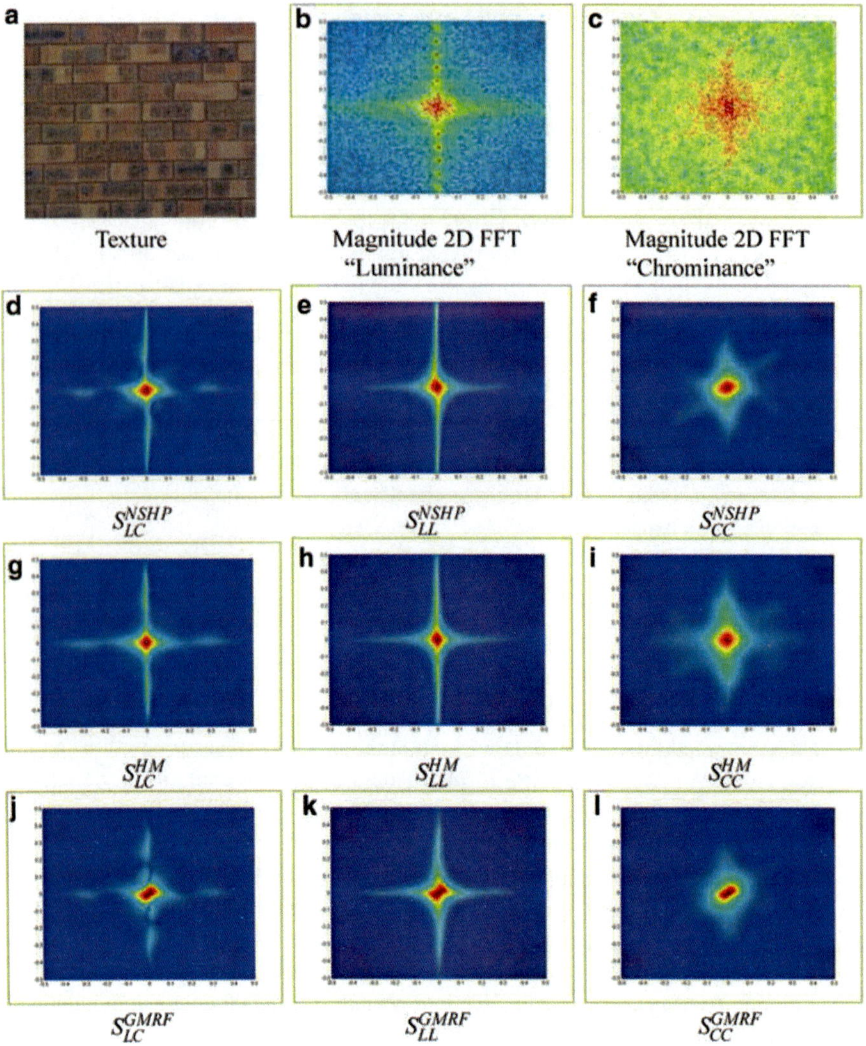

Fig. 9.13 Spectral analysis of a color texture, FFT stands for Fast Fourier Transform

9.3.2.1 Distance Measures

To measure overall closeness of luminance and chrominance spectra at all frequencies, spectral distance measures are used. In [4], the author has presented a discretized symmetric extension of Kullback–Leibler (KL) divergence for spectral distance between two spectra. We use the same distance to measure the closeness of luminance and chrominance spectra. The spectral distance measure is given as:

$$\mathrm{KL}_\beta\left(S_{1,\beta},S_{2,\beta}\right)=\frac{1}{2}\times\sum_{\nu_1,\nu_2}\left|\sqrt{\frac{S_{1,\beta}\left(\nu_1,\nu_2\right)}{S_{2,\beta}\left(\nu_1,\nu_2\right)}}-\sqrt{\frac{S_{2,\beta}\left(\nu_1,\nu_2\right)}{S_{1,\beta}\left(\nu_1,\nu_2\right)}}\right|^2, \tag{9.63}$$

where $\beta \in \{\mathrm{LL},\mathrm{CC}\}$ (see (9.17)). The spectral distance measure given in (9.63) gives the closeness of each channel individually.

3-D color histogram cubes are used as pure color feature cues in the discussed method. In order to measure the closeness of 3-D color histogram cubes, symmetrized KL divergence (KLS), given in [29] is used:

$$\mathrm{KLS}\left(\mathscr{H}_1,\mathscr{H}_2\right)=\frac{\mathrm{KL}(\mathscr{H}_1,\mathscr{H}_2)+\mathrm{KL}(\mathscr{H}_2,\mathscr{H}_1)}{2}, \tag{9.64}$$

where $\mathrm{KL}\left(\mathscr{H}_1,\mathscr{H}_2\right)$ is KL divergence between two histograms $\mathscr{H}_1$ and $\mathscr{H}_2$, given as:

$$\mathrm{KL}\left(\mathscr{H}_1,\mathscr{H}_2\right)=\sum_{i,j,k=(1,1,1)}^{B,B,B}\frac{N_{1,ijk}}{|\Lambda|\times\gamma^3}\times\log\frac{N_{1,ijk}}{N_{2,ijk}}, \tag{9.65}$$

where Δ is the number of pixels, Γ is the number of cubic bins, probabilities $\mathscr{H}_1$ and $\mathscr{H}_2$ represent the probability distributions of the pure color information of the images computed through 3-D color histograms and B is the number of bins per axes.

9.3.2.2 The Probabilistic Cue Fusion

In this approach, a posteriori class probabilities are computed using each of these three feature cues independently. The different a posteriori class probabilities obtained through each of these three cues are combined by multiplying these individual a posteriori class probabilities. A pattern x is assigned the label $\hat{\omega}$ which maximizes the product of the a posteriori probabilities provided by each of the independent feature cues (in our case, $K=3$):

$$\hat{\omega}=\underset{\omega_i,i\in\{1,\ldots,n\}}{\arg\max}\left(\prod_{k=1}^{K}P_k\left(w_i|x\right)\right), \tag{9.66}$$

where n is the number of texture classes. In order to quantify these probabilities, we used a distance-based normalized similarity measure which is given as:

$$P_k\left(w_i|x\right)=\frac{\frac{1}{1+d_k(x,x_i)}}{\sum_{j=1}^{n}\frac{1}{1+d_k\left(x,x_j\right)}}, \tag{9.67}$$

Table 9.1 Average percentage classification results obtained for DS_1, DS_2 and DS_3 using 3D color histograms

	H_{RGB}	H_{IHLS}	H_{Lab}
DS_1 ($B = 10$)	96.4	96.4	91.8
DS_2 ($B = 16$)	99.5	100.0	99.1
DS_3 ($B = 16$)	94.0	94.5	92.2
Average	96.6	97.0	94.4

where $d_k(x, x_i)$ is the Kullback–Leibler distance measure for respective feature cue. In (9.66), we utilize the degree of independence between the different feature cues to obtain a better result when the cues are fused together. More the individual feature cues are decorrelated, better will be the results computed after their fusion through (9.66).

9.3.2.3 Experiment Results on Texture Classification

We conducted experiments to evaluate the color texture characterization based on the pure color distribution using 3-D histogram information. For the three data sets, these 3-D histograms were computed for different number of bin cubes $B \times B \times B$. For DS_1, $B \in \{4, 6, 9, 10\}$ and for DS_2 and DS_3 $B \in \{8, 12, 16\}$.

The choice of the number of bin cubes for the 3-D color histograms was made keeping in view, the sizes of the test and training subimages in each data set. For the small subimage sizes, i.e., the color textures in DS_1, small bin sizes are chosen. Whereas, for the large sub image sizes, i.e., the color textures in DS_2 and DS_3, larger bin sizes are chosen. For all the three color spaces, i.e., RGB, IHLS, and L*a*b* the tests were performed on the three data sets. For each test texture subimage, 3-D histogram was computed. Then symmetrized Kullback–Leibler divergence was computed using (9.64). Finally, a class label was assigned to the test texture subimage using nearest neighbor method. Average percentage classification results obtained for DS_1, DS_2, and DS_3 are shown in Table 9.1. For the data set DS_1, maximum percentage classification is achieved in the RGB color space with $B = 10$. While for the data sets DS_2 and DS_3, maximum percentage classification achieved is in IHLS color space with $B = 16$. As we have larger subimage sizes in DS_2 and DS_3; therefore, we have higher percentage classification values in these data sets than that of the values obtained for the DS_1 data set. For a given data set and fixed number of 3D histogram bins, percentage classification obtained in different color spaces varies significantly. This indicates that all the color spaces do not estimate the global distribution of the color content in the same manner and this estimate depend upon the shape of the color space gamut. The bins considered for 3D histograms are of regular cubical shape. The color spaces with regular-shaped color gamut, i.e., RGB and IHLS are more appropriate for this kind of bin shape and therefore show slightly better results than those of the L*a*b* color space.

It is important to note that L*a*b* color space has proven to give better results for the estimation of global distribution of pure color content of an image when used with a parametric approximation with MGMM [2] (see (9.2)). This parametric

Table 9.2 Average percentage classification results obtained for DS_1, DS_2 and DS_3 using structure feature cues with IHLS and L*a*b* color spaces

	L		C		LC	
	IHLS	L*a*b*	IHLS	L*a*b*	IHLS	L*a*b*
DS_1	87.4	87.7	85.8	92.1	95.4	97.2
DS_2	91.4	90.3	87.5	91.2	97.4	96.5
DS_3	75.1	79.4	73.2	78.5	84.1	88.0
Average	84.6	85.8	82.1	87.3	91.3	93.9

Table 9.3 Average percentage classification of DS_1, DS_2 and DS_3 with RGB color space

	R	G	B	RGB
DS_1	78.3	80.0	83.1	85.9
DS_2	89.6	88.4	90.3	92.1
DS_3	75.3	76.6	72.2	82.8
Average	81.1	81.7	81.9	86.9

approximation is *well suited* to irregular gamut shape of the L*a*b* color space and, hence, authors in [2] have indicated L*a*b* as the better performing color space for parametric multimodal color distribution approximation. Here we do not use such a parametric approximation for the spatial distribution of color content in textured images as:

- Image size in the DS_1 is 32×32. For such a small image size, it is probable that one will have to face a significant problem of numerical instabilities while calculating the model parameters for MGMM.
- Similarity metrics used for the distance measures between two MGMM distributions are not very well suited to the problem and have a tendency to produce suboptimal results.

Let us discuss now about the structure feature cues.

To compute luminance and chrominance spatial structure feature cues, auto spectra were estimated using the approach given Sect. 9.1.3.1. The auto spectra are computed in Cartesian coordinates for normalized frequency range $\nu = (\nu_1, \nu_2) \in [-0.5, 0.5]^2$. Then in order to compute the overall closeness of luminance (L) and chrominance (C) spectra at all frequencies, spectral distance measure, given in (9.63) is used.

Again, a class label was assigned to the test texture subimage using the nearest neighbor method, based on the information from both luminance and chrominance structure feature cues individually. The individual and independent information obtained through these two spatial structure cues is then combined using (9.66) and (9.67). This gives us the class label assignment based on both the luminance and chrominance structure information. These results for the two perceptual color spaces, i.e., IHLS and L*a*b* are shown in Table 9.2. Results for RGB are given in Table 9.3.

It is clear from these results that the percentage classification results obtained by individual channels in RGB color space are inferior than those obtained by the luminance and chrominance spectra using our approach in both the perceptual color

Table 9.4 Comparison of best average percentage classification results for DS_1 with state of the art results presented in [51]. Best results are indicated in italic faces

DS_1	Our method	[51]	
Structure	*97.2*	91.2	(with DCT scale 4)
Colour	*96.4*	90.6	(with RGB mean and cov.)
Structure + colour	*99.3*	96.6	(with both of above)

spaces. One can also see that for the same test conditions and same information fusion approach with proposed method, combined overall results in IHLS and L*a*b* color spaces are approximately 6% to 7% higher (for the used data sets) than those computed in the RGB color space. This provides an experimental evidence to the hypothesis over the choice of a perceptual color space for color texture classification instead of the standard RGB color space. Complete results can be obtained in [53, 54].

Average percentage classification of color textures obtained in different color spaces can easily be compared to the average percentage classification results of color textures computed through other existing approaches. In the case of DS_1, best known results are presented in [51].

Comparison of the results achieved by our approach with the results presented in [51] is given in Table 9.4. Best value for average percentage classification achieved using only the structure information in [51], is 91.2% which is obtained using wavelet-like DCT coefficients as structure descriptors. Compared to this value of average percentage classification, we observe a significant increase in average percentage classification value of 24 color textures with our method. In our work, the best average percentage classification achieved using only structure feature cues is 97.24%.

In the case of DS_2 and DS_3 test data sets, best average percentage classification results are presented in [40]. The authors have compared the results of a large number of existing texture descriptors for both DS_2 and DS_3 without concentrating on the performance of a given algorithm. Comparison of results with our approach for these two test data sets with best results presented in [40] are given in Tables 9.5 and 9.6 respectively. In [40], the best reported results using different feature cues for each case, are not obtained using the same descriptors.

For example, for DS_2, the best results presented using only color feature cues are obtained using 3-D histograms in $I_1I_2I_3$ color space with $B = 32$ and the best results presented for structure information are computed using local binary patterns (LBP) [49], i.e., $LBP_{16,2}$ in L*a*b*. Then the best results by fusing both the feature cues are presented for 3-D histograms in RGB with $B = 16$ used as color feature cue and $LBP^{u2}_{16,2}$ as structure feature cues. The decision rule used for fusion is the Borda count.

In [40], for DS_3, the best results presented using only color feature cues are obtained by 3-D histograms in HSV color space with $B = 16$ and the best results presented for structure information are computed through $LBP_{8,1}$ in RGB color

Table 9.5 Comparison of best average percentage classification results for DS_2 with state of the art results presented in [40]

DS_2	Our method	[40]	
Structure	96.5	100.0	($LBP_{16,2}$ L*a*b*)
Colour	*100.0*	100.0	(3D hist. $I_1I_2I_3$)
Structure + colour	99.1	99.8	($LBP^{u2}_{16,2}$ + 3D hist. RGB)

Table 9.6 Comparison of best average percentage classification results for DS_3 with state-of-the-art results presented in [40]

DS_3	Our method	[40]	
Structure	*88.0*	87.8	($LBP_{8,1} +^{u2}_{16,3} +^{u2}_{24,5}$ L*a*b*)
Colour	94.5	95.4	(3D hist. HSV)
Structure + colour	89.0	94.6	($Gabor_{3,4}$ + 3D hist. RGB)

space. Then the best results by fusing both the feature cues are presented for 3-D histograms in RGB with $B = 16$ used as color feature cue and $Gabor_{3,4}$ as structure feature cues. The decision rule used for fusion is the maximum dissimilarity.

As we use the same color and texture features for all the data sets along with the same fusion method; therefore, the comparison of the results through our approach for the different data sets is more judicious. It can be noted that for the two test data sets DS_2 and DS_3, our method and the best results reported so far, i.e., as in [40] are approximately of the same order when individual color and texture feature cues are considered. For DS_2, the best results with our approach and the ones in [40] are approximately the same even when the two feature cues are fused. For DS_3, authors in [40] report an average percentage classification of 94.6%. The corresponding percentage computed with our approach is 88.97%. It is to note here that in [40], the main objective was to produce maximum percentage classification using different combinations of color and texture features and different fusion methods. Contrarily in our work, the main goal is to analyze the effect of luminance chrominance spectral decorrelation in the perceptual color spaces and its implications on the color texture classification. Even under this consideration, the presented approach outperforms the state of art in certain cases while compete well in other cases in terms of average percentage classification results.

We now discuss how complex 2-D vectorial linear prediction could be used in the context of supervised segmentation of color textures.

9.3.3 Segmentation of Color Textures

In a supervised context, given a training sample of a color texture, it is possible to determine the set of parameters (see (9.14)), $\Theta = \left\{\mathbf{m}, \{\mathbf{A_y}\}_{\mathbf{y}\in D}, \Sigma_{E_F}\right\}$ for each texture appearing in the image. With these parameters, the Linear Prediction Error (LPE) sequence $\boldsymbol{e} = \{\boldsymbol{e}(\mathbf{x})\}_{\mathbf{x}\in E}$ associated to each texture, for the whole image can be calculated. Then, using Bayesian approach in segmentation (see Sect. 9.2.1 and (9.69)), it is possible to derive the LPE distributions using parametric models.

9.3.3.1 Label Field Estimation

First phase of the color texture segmentation method assigns the class labels, $l = \{l(\mathbf{x})\}_{\mathbf{x}\in E}$, without taking into account any spatial regularization. This assignment is done following a maximum likelihood criterion that maximizes the product of probabilities with an independence supposition over LPE:

$$\hat{l}(\mathbf{x}) = \underset{\lambda=1,\ldots,K}{\arg\max} \left(p\left(\boldsymbol{e}(\mathbf{x})|\hat{\Theta}_\lambda\right)\right), \tag{9.68}$$

where K is the total number of classes in the image and the set of parameters used in (9.68) are the ones estimated using different approximations of the LPE distribution: multivariate Gaussian distribution (MGD) (see (9.1)), multivariate Gaussian mixture model (MGMM) (see (9.2)), and Wishart distribution (see (9.5)). In this last case, we define $\mathbf{J}(\mathbf{x}) = [\boldsymbol{e}(\mathbf{x}-\mathbf{1}_v), \boldsymbol{e}(\mathbf{x}-\mathbf{1}_h), \boldsymbol{e}(\mathbf{x}), \boldsymbol{e}(\mathbf{x}+\mathbf{1}_v), \boldsymbol{e}(\mathbf{x}+\mathbf{1}_h)]$ with $\mathbf{1}_h = (1, \emptyset)$ et $\mathbf{1}_v = (\emptyset, 1)$, which contains the LPE vector at site $\mathbf{x}$ and its four nearest-neighbor vectors. For more details, see [55].

During the second phase, a maximum A posteriori (MAP) type estimation is carried out in order to determine the final class labels for the pixels [1] (see Sect. 9.2.1). A Markovian hypothesis is made over $P(l|\boldsymbol{f})$, derived through a Gibbs distribution:

$$P(l|\boldsymbol{f}) \propto \exp(-U_D(\boldsymbol{f}, l) - U_i(l)), \tag{9.69}$$

where U_D is the energy of the given observation field $\boldsymbol{f}$ and of class label field l, whereas U_i is an energy related only to the label field. U_D is calculated as:

$$U_D(\boldsymbol{f}, l) = \sum_{\mathbf{x}} \left(-\log\left(p\left(\boldsymbol{e}(\mathbf{x})|\Theta_{l(\mathbf{x})}\right)\right)\right), \tag{9.70}$$

where $p\left(\boldsymbol{e}(\mathbf{x})|\Theta_{l(\mathbf{x})}\right)$ is the conditional probability of the LPE given the texture class in $\mathbf{x}$, i.e., $l(\mathbf{x})$.

We propose to use an internal energy associated to the label field consisting of two terms: $U_i(l) = U_{i,1}(l) + U_{i,2}(l)$. $U_{i,1}(l)$ corresponds to the Gibbs energy term associated to a Potts model [1]:

$$U_{i,1}(l) = \beta \left(\sum_{\langle \mathbf{x},\mathbf{y}\rangle_1} 1_{(l(\mathbf{x})\neq l(\mathbf{y}))} + \sum_{\langle \mathbf{x},\mathbf{y}\rangle_2} 1_{(l(\mathbf{x})\neq l(\mathbf{y}))} \right) \tag{9.71}$$

with β the weight term, or the hyperparameter, of Potts model, and $\langle \mathbf{x},\mathbf{y}\rangle_p$, $p = 1, 2$, describes $\|\mathbf{x}-\mathbf{y}\|_2 = \sqrt{p}$, $(\mathbf{x},\mathbf{y}) \in E^2$, $\mathbf{x} \neq \mathbf{y}$. Let us notice that this model

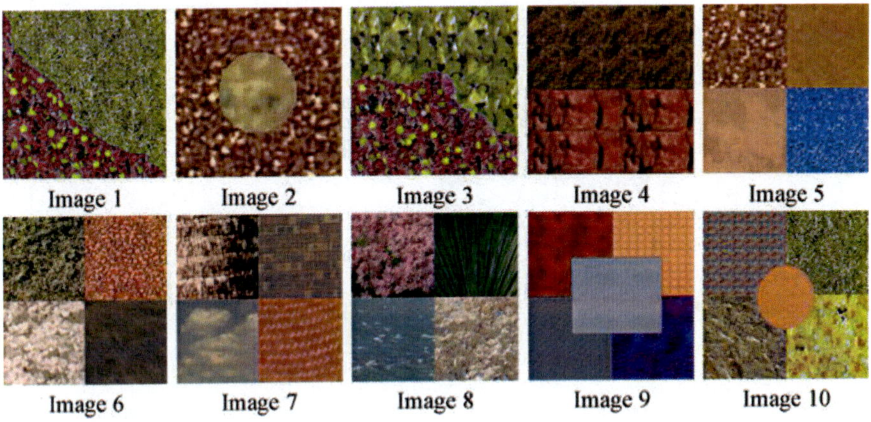

Fig. 9.14 Data base of ten color images used for simulations

is almost the same used in Sect. 9.2 but with an eight-connected neighborhood. In many works, as in [1], for example, the second sum is weighted by $\frac{1}{\sqrt{2}}$.

To calculate $U_{i,2}(l)$, we use an energy term which depends on the size of the region pertaining to a single class label. The size of the region $|R_i|$, $i = 1,\ldots,n_R$, n_R being the total number of the regions in the label field, follows a probability distribution which favors the formation of larger regions [63]. This term is defined as:

$$U_{i,2}(l) = \gamma \left(\sum_{i=1}^{n_R} |R_i|^{\kappa} \right), \tag{9.72}$$

where κ is a constant [63]. γ is again a hyperparameter whose value will be given in the next section.

9.3.3.2 Experimental Results

The ground truth data associated with complex natural images is difficult to estimate and its extraction is highly influenced by the subjectivity of the human operator. Thus, the evaluation of the proposed parametric texture segmentation framework was performed on natural as well as synthetic color textures which possess unambiguous ground truth data. The test images were taken from the color texture database used in [27]. The database was constructed using color textures from Vistex and Photoshop databases. In the first phase of the supervised color texture segmentation, a single subimage of size 32×32 was used as the training image for each class. Image observation model parameters, multichannel prediction error, and parameter sets for the used parametric approximations were computed for this sub image. Now these parameters were used to compute the initial class label field for each of the ten test textured color images shown in Fig. 9.14.

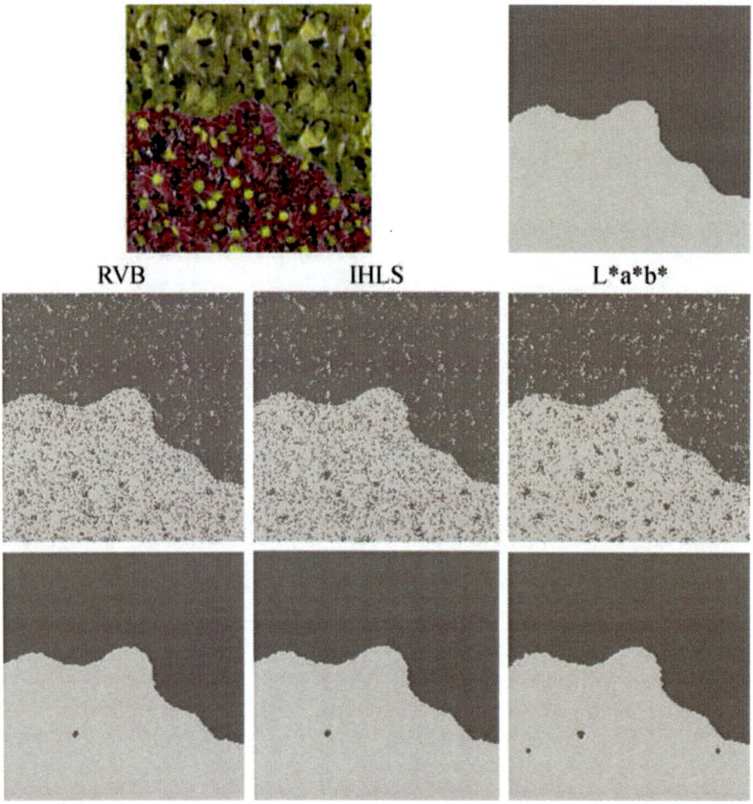

Fig. 9.15 Segmentation results without spatial regularization ($\beta = \gamma = 0$, 2nd row) and with regularization (3rd row) for the textured image 3 (first row on the left)—2D QP AR model and MGMM—RGB, IHLS and L*a*b* color spaces. Ground truth is presented in first row on the right

For the test images 3 and 10, the initial segmentation results are presented in the second row of Figs. 9.15 and 9.16.

In the second phase of the algorithm, initial segmentation is refined by spatial regularization using Potts model and the region size energy term. An iterative method (ICM—Iterated Conditional Mode) is used to compute the convergence of the class label field. In these experiments we used β, i.e., the hyperparameter of the Potts model, as a *progressively varying parameter*. We used the regularized segmentation result obtained through one value of β as an initial class label field for the next value of β. The value of the hyperparameter β was varied from 0.1 to 4.0 with an exponential interval. For the region energy term, we fixed the hyperparameter $\gamma = 2$ and the coefficient $\kappa = 0.9$ [63]. Third rows of Figs. 9.15 and 9.16 show the segmentation results with spatial regularization and with MGMM for LPE distribution using a number of components equal to five for each texture.

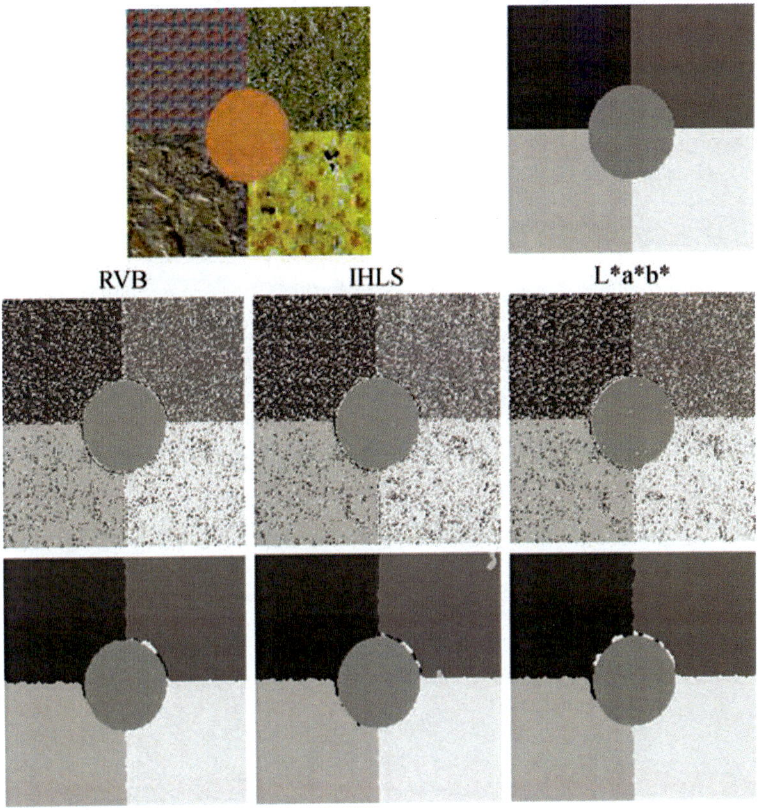

Fig. 9.16 Segmentation results without spatial regularization ($\beta = \gamma = 0$, 2nd row) and with regularization (3rd row) for the textured image 10 (1st row on the left)—2D QP AR model and MGMM—RGB, IHLS, and L*a*b* color spaces. Ground truth is presented in first row on the right

Average percentage pixel classification error of 10 color-textured images without spatial regularization, i.e., ($\beta = \gamma = 0$) are given in Table 9.7 and their curves against the values of β are given in Fig. 9.17.

Overall performance of MGMM and Wishart approximations before spatial regularization are better than the Gaussian approximation for the three color spaces Table 9.7. This result enforces the initial hypothesis that the Gaussian approximation can not optimally approach the LPE distribution. Overall performance of the MGMM approximation after the spatial regularization is better than the other two approximations for all the three color spaces. the Wishart distribution does not show the global minimum values of the average percentage error, yet the initial value of the percentage error in this case is much lower than the other two error models. This is attributed to the robust and stable prior term computed in the case of Wishart

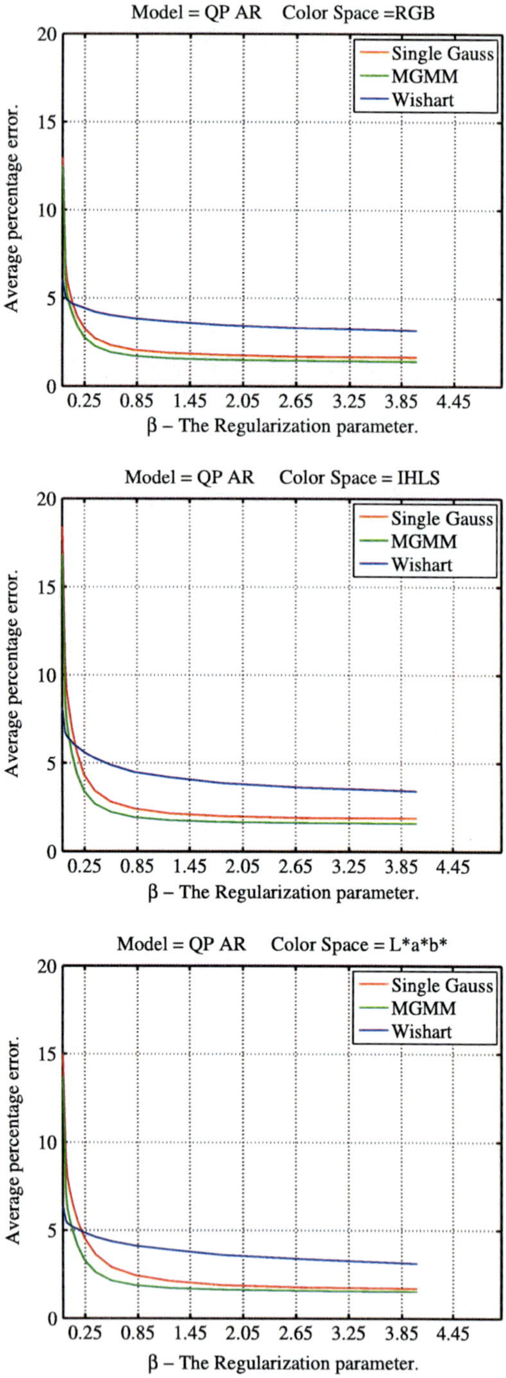

Fig. 9.17 Comparison of the computed segmentation results with three parametric models of LPE distribution, in RGB, IHLS, and L*a*b* color spaces

Table 9.7 Mean percentages of pixel classification errors for the ten color-textured images (see Fig. 9.14) without spatial regularization and with spatial regularization. Best results are indicated in italic faces

	Without regularization			With regularization		
	RGB	IHLS	L*a*b*	RGB	IHLS	L*a*b*
Single Gauss	12.98	18.41	14.97	1.62	1.85	1.68
MGMM	12.47	16.83	13.67	*1.41*	*1.58*	*1.52*
Wishart Distribution	*6.04*	*8.14*	*6.35*	3.15	3.37	3.09

distribution as it considers multiple observations, i.e., LPE vectors to compute the probability of a given observation (LPE vector).

The better performance of the RGB color space in the case of synthetic color texture images may appear as a contradiction to our prior findings on color texture characterization [53, 54] (see Sect. 9.3.2.3). In [53, 54], the mathematical models used were based on the decorrelation of the different channels in color images; therefore, RGB (having a higher interchannel correlation characteristic) showed inferior results than the two perceptual color spaces. Whereas in this chapter, the parametric models used for the approximation of multichannel LPE distribution (for example, MGMM) exploit the interchannel correlation of the color planes through modeling of the joint probability distributions of the LPE.

In [53, 55], complete results on these approaches are presented, particularly for satellite images which prove the suitability of using L*a*b* color space rather than using RGB or IHLS color space. These results reinforce the results obtained for classification of color textures.

9.4 Conclusion

In this chapter, we provided the definition of parametric stochastic models which can be used for color image description and color image processing.

Multivariate gaussian mixture model is a widely used model for various data analysis applications. We presented recent works done with this model for unsupervised color image segmentation. During the proposed reversible jump Markov chain Monte Carlo algorithm, all the parameters of the mixture are estimated including its number of components which is the unknown number of classes in the image. Segmentation results are compared to JSEG method showing the accuracy of the approach.

2-D complex multichannel linear prediction not only allows the separation of luminance and chrominance information but also the simultaneous derivation of second-order statistics for the two channels individually as well as the characterization of interchannel correlations. With the help of the computations over RGB to IHLS/L*a*b* and IHLS/L*a*b* to RGB transformations, we have compared the luminance–chrominance interference introduced through these transformations. this experiment has shown more luminance–chrominance interference in IHLS color space as compared to L*a*b* color space.

During the PhD of Imtnan Qazi [53], these parametric models have been used for classication of color texture databases and segmentation of color textures. Globally, we have obtained the best results with 2-D QP AR model in L*a*b* color space, knowing that we have not optimized the parameter estimation in the case of GMRF. Other color spaces could also be used in the future.

A potential perspective is to exploit the combination of linear prediction models with decomposition methods. Such an approach shall allow the characterization of the color textures following their deterministic and random parts separately.

References

1. Alata O, Ramananjarasoa C (2005) Unsupervised textured image segmentation using 2-D quarter plane autoregressive model with four prediction supports. Pattern Recognit Lett 26:1069–1081
2. Alata O, Quintard L (2009) Is there a best color space for color image characterization or representation based on multivariate gaussian mixture model? Comput Vision Image Underst 113:867–877
3. Barker SA, Rayner PJW (2000) Unsupervised image segmentation using Markov random field models. Pattern Recognit 33(4):587–602
4. Basseville M (1989) Distance measures for signal processing and pattern recognition. Signal Process 4(18):349–369
5. Besag J (1986) On the statistical analysis of dirty pictures. J R Stat Soc Ser B 48(3):259–302
6. Bouman C, Liu B (1991) Multiple resolutions segmentation of textured images. IEEE Trans Pattern Anal Mach Intell 13(2):99–113
7. Brooks SP, Giudici P, Roberts GO (2003) Efficient construction of reversible jump Markov chain Monte Carlo proposal distributions. J R Stat Soc Ser B 65:3–55
8. Chalmond B (2003) Modeling and inverse problems in image analysis. Springer, New York
9. Chindaro S, Sirlantzis K, Fairhurst M (2005) Analysis and modelling of diversity contribution to ensemble-based texture recognition performance. In: Proceedings of MCS. Lecture notes in computer science (LNCS), vol 3541. Springer, Berlin, pp 387–396
10. Cariou C, Rouquette S, Alata O (2008) Two-dimensional signal analysis—Chapter 3, 2-D spectral analysis. Wiley, ISTE
11. Commission Internationale de l'Eclairage (1986) Colorimetry. CIE 15.2, Vienna
12. Cremers D, Tischhauser F, Weickert J, Schnorr C (2002) Diffusion snakes: Introducing statistical shape knowledge into the Mumford-Shah functional. Int J Comput Vision 50(3):295–313
13. Dempster AP, Laird NM, Rubin DB (1977) Maximum likelihood from incomplete data via the EM algorithm, J R Stat Soc Ser B 39(1):1–38
14. Deng Y, Manjunath BS (2001) Unsupervised segmentation of color-texture regions in images and video. IEEE Trans Pattern Anal Mach Intell 23(8):800–810. http://vision.ece.ucsb.edu/segmentation/jseg/
15. Geman S, Geman D (1984) Stochastic relaxation, Gibbs distributions and the Bayesian restoration of images. IEEE Trans Pattern Anal Mach Intell 6:721–741
16. Giordana N, Pieczynski W (1997) Estimation of generalized multisensor hidden Markov chains and unsupervised image segmentation. IEEE Trans Pattern Anal Mach Intell 19(5):465–475
17. Green PJ (1995) Reversible jump Markov chain Monte Carlo computation and Bayesian model determination. Biometrika 82(4):711–732

18. Green PJ (2003) Trans-dimensional Markov chain Monte Carlo. In: Green PJ, Hjort NL, Richardson S (eds) Highly structured stochastic systems. OUP, Oxford
19. Gupta L, Sortrakul T (1998) A Gaussian-mixture-based image segmentation algorithm. Pattern Recognit 31(3):315–325
20. Guyon X (1995) Random fields on a network—Modeling, statistics and application. Probability and its applications series. Springer, New York
21. Haindl M, Mikes S (2006) Unsupervised texture segmentation using multispectral modelling approach. In: Proceedings of international conference on pattern recognition (ICPR), II-203–II-206. http://ieeexplore.ieee.org/xpls/abs_all.jsp?arnumber=1699182&tag=1
22. Hanbury A, Serra J (2002) A 3D-polar coordinate colour representation suitable for image analysis. TR-77, PRIP, Vienna University of Technology, Vienna
23. Hastings WK (1970) Monte Carlo sampling methods using Markov chains and their application. Biometrika 57:97–109
24. Hernandez OJ, Cook J, Griffin M, Rama CD, McGovern M (2005) Classification of color textures with random field models and neural networks. J Comput Sci Technol 5(3):150–157
25. Hershey JR, Olsen PA (2007) Approximating the Kullback Leibler divergence between Gaussian mixture models. In: Proceedings of international conference on acoustics, speech and signal processing (IEEE ICASSP), IV-317–IV-320. http://ieeexplore.ieee.org/xpl/articleDetails.jsp?arnumber=4218101
26. Huang CL, Cheng TY, Chen CC (1992) Color images segmentation using scale space filter and Markov random field. Pattern Recognit 25(10):1217–1229
27. Ilea DE, Whelan PF (2008) CTex—An adaptive unsupervised segmentation algorithm based on color-texture coherence. IEEE Trans Image Process 17(10):1926–1939
28. Jackson LB, Chien HC (1979) Frequency and bearing estimation by two-dimensional linear prediction. In: Proceedings of international conference on acoustics, speech and signal processing (IEEE ICASSP), pp 665–668. http://ieeexplore.ieee.org/xpl/articleDetails.jsp?arnumber=1170793
29. Johnson D, Sinanovic S (2001) Symmetrizing the Kullback–Leibler distance. IEEE Trans Inform Theory. http://citeseerx.ist.psu.edu/viewdoc/summary?doi=0.1.1.26.2327
30. Kato Z (1999) Bayesian color image segmentation using reversible jump Markov chain Monte Carlo, Research Report 01/99-R055. ERCIM/CWI, Amsterdam. Available as a CWI Research Report PNA-R9902, ISSN 1386–3711. http://www.ercim.org/publication/technical_reports/055-abstract.html
31. Kato Z, Pong TC, Lee JCM (2001) Color image segmentation and parameter estimation in a Markovian framework. Pattern Recognit Lett 22(3–4):309–321
32. Kato Z, Pong TC (2006) A Markov random field image segmentation model for color textured images. Image Vision Comput 24(10):1103–1114
33. Kato Z (2008) Segmentation of color images via reversible jump MCMC sampling. Image Vision Comput 26(3):361–371
34. Kersten D, Mamassian P, Yuille A (2004) Object perception as Bayesian inference. Ann Rev Psychol 55:271–304
35. Khotanzad A, Hernandez OJ (2006) A classification methodology for color textures using multispectral random field mathematical models. Math Comput Appl 11(2):111–120
36. Kokaram A (2002) Parametric texture synthesis for filling holes un pictures. In: Proceedings of international conference on image processing (IEEE ICIP), pp 325–328. http://ieeexplore.ieee.org/xpl/articleDetails.jsp?arnumber=1038026
37. Lakshmanan S, Derin H (1989) Simultaneous parameter estimation and segmentation of Gibbs random fields using simulated annealing. IEEE–PAMI 11(8). 799–813
38. Langan DA, Modestino JW, Zhang J (1998) Cluster validation for unsupervised stochastic model-based image segmentation. IEEE Trans Image Process 7(2):180–195
39. Liu J, Yang YH (1994) Multiresolution color image segmentation. IEEE Trans Pattern Anal Mach Intell 16(7):689–700
40. Maenpaa T, Pietikainen M (2004) Classification with color and texture: jointly or separately? Pattern Recognit 37(8):1629–1640

41. Mardia KV, Kent JT, Bibby JM (1979) Multivariate Analysis. Academic, Duluth
42. Markel JD, Gray AH Jr (1976) Linear prediction of speech. Communication and cybernetics series. Springer, New York
43. Martin D, Fowlkes C, Tal D, Malik J (2001) A database of human segmented natural images and its application to evaluating segmentation algorithms and measuring ecological statistics. In: Proceedings of IEEE international conference on computer vision, vol 2. University of California, Berkeley, pp 416–423.
44. Metropolis N, Rosenbluth A, Rosenbluth M, Teller A, Teller E (1953) Equation of state calculations by fast computing machines. J Chem Phys 21:1087–1092
45. Miao GJ, Clements MA (2002) Digital signal processing and statistical classification. Artech House, USA. ISBN 1580531350
46. Mumford D (1994) The Bayesian rationale for energy functionals. In: Romeny B (ed) Geometry-driven diffusion in computer vision. Kluwer Academic, Dordrecht, pp 141–153
47. Mumford D (1996) Pattern theory: a unifying perspective. In: Knill D, Richards W (eds) Perception as Bayesian inference. Cambridge University Press, UK, pp 25–62
48. Ojala T, Mäenpää T, Pietikäinen M, Viertola J, Kyllönen J, Huovinen S (2002) Outex—New framework for empirical evaluation of texture analysis algorithms. In: Proceedings of 16th international conference on pattern recognition. Qubec, pp 701–706
49. Ojala T, Pietikäinen M, Mäenpää T (2002) Multiresolution gray-scale and rotation invariant texture classification with local binary patterns. IEEE Trans Pattern Anal Mach Intell 24(7):971–987
50. Panjwani DK, Healey G (1995) Markov random field models for unsupervised segmentation of textured color images. IEEE Trans Pattern Anal Mach Intell 17(10):939–954
51. Permuter H, Francos J, Jermyn I (2006) A study of gaussian mixture models of color and texture features for image classification and segmentation. Pattern Recognit 39(4):695–706
52. Qazi I-U-H, Alata O, Burie J-C, Fernandez-Maloigne C (2010) Colour spectral analysis for spatial structure characterization of textures in ihls colour space. Pattern Recognit 43(3):663–675
53. Qazi I-U-H (2010) Luminance-chrominance linear prediction models for color textures: an application to satellite image segmentation. PhD Thesis, University of Poitiers, France
54. Qazi I-U-H, Alata O, Burie J-C, Moussa A, Fernandez-Maloigne C (2011) Choice of a pertinent color space for color texture characterization using parametric spectral analysis. Pattern Recognit 44(1):16–31
55. Qazi I-U-H, Alata O, Burie J-C, Abadi M, Moussa A, Fernandez-Maloigne C (2011) Parametric models of linear prediction error distribution for color texture and satellite image segmentation. Comput Vision Image Underst 115(8):1245–1262
56. Rellier G, Descombes X, Falzon F, Zerubia J (2004) Texture feature analysis using a gauss-markov model in hyperspectral image classification. IEEE Trans Geosci Remote Sens 42(7):1543–1551
57. Richardson S, Green PJ (1997) On Bayesian analysis of mixtures with an unknown number of components. J R Stat Soc Ser B 59(4):731–792
58. Robert C, Rydén T, Titterington DM (2000) Bayesian inference in hidden Markov models through the reversible jump Markov chain Monte Carlo method. J R Stat Soc Ser B 62(1): 57–75
59. Sangwine SJ, Horne REN (eds) (1998) The colour image processing handbook. Chapman & Hall, london
60. Stephens M (1997) Bayesian methods for mixtures of normal distributions. PhD Thesis, University of Oxford
61. Stephens M (2000) Bayesian analysis of mixture models with an unknown number of components—An alternative to reversible jump methods. Ann Stat 28(1):40–74
62. Suen P-H, Healey G (1999) Modeling and classifying color textures using random fields in a random environment. Pattern Recognit 32(6):1009–1017
63. Tu Z, Zhu S-C (2002) Image segmentation by data-driven Markov chain Monte Carlo. IEEE Trans Pattern Anal Mach Intell 24:657–673

64. Winkler G (2003) Image analysis, random fields and Markov chain Monte Carlo methods. 2nd edn, Springer, Berlin
65. Won CS, Derin H (1992) Unsupervised segmentation of noisy and textured images using Markov random fields. Comput Graphics Image Process: Graph Models Image Process 54(4):208–328
66. Zhu SC (1999) Stochastic jump-diffusion process for computing medial axes in Markov random fields. IEEE Trans Pattern Anal Mach Intell 21(11):1158–1169

Chapter 10
Color Invariants for Object Recognition

Damien Muselet and Brian Funt

What is without form is without color

Jacques Ferron

Abstract Color is a very important cue for object recognition, which can help increase the discriminative power of an object-recognition system and also make it more robust to variations in the lighting and imaging conditions. Nonetheless, even though most image acquisition devices provide color data, a lot of object-recognition systems rely solely on simple grayscale information. Part of the reason for this is that although color has advantages, it also introduces some complexities. In particular, the RGB values of a digital color image are only indirectly related to the surface "color" of an object, which depends not only on the object's surface reflectance but also on such factors as the spectrum of the incident illumination, surface gloss, and the viewing angle. As a result, there has been a great deal of research into color invariants that encode color information but at the same time are insensitive to these other factors. This chapter describes these color invariants, their derivation, and their application to color-based object recognition in detail. Recognizing objects using a simple global image matching strategy is generally not very effective since usually an image will contain multiple objects, involve occlusions, or be captured from a different viewpoint or under different lighting conditions than the model image. As a result, most object-recognition systems describe the image content in terms of a set of local descriptors—SIFT, for example—that describe the regions around a set of detected keypoints. This

D. Muselet (✉)
Laboratory Hubert Curien, UMR CNRS 5516, Jean Monnet University, Saint-Etienne, France
e-mail: damien.muselet@univ-st-etienne.fr

B. Funt
School of Computing Science, Simon Fraser University, Burnaby, Canada
e-mail: funt@sfu.ca

C. Fernandez-Maloigne (ed.), *Advanced Color Image Processing and Analysis*, DOI 10.1007/978-1-4419-6190-7_10, chapter includes a discussion of the three color-related choices that need to be made when designing an object-recognition system for a particular application: Color-invariance, keypoint detection, and local description. Different object-recognition situations call for different classes of color invariants depending on the particular surface reflectance and lighting conditions that will be encountered. The choice of color invariants is important because there is a trade-off between invariance and discriminative power. All unnecessary invariance is likely to decrease the discriminative power of the system. Consequently, one part of this chapter describes the assumptions underlying the various color invariants, the invariants themselves, and their invariance properties. Then with these color invariants in hand, we turn to the ways in which they can be exploited to find more salient keypoints and to provide richer local region descriptors. Generally but not universally, color has been shown to improve the recognition rate of most object-recognition systems. One reason color improves the performance is that including it in keypoint detection increases the likelihood that the region surrounding the keypoint will contain useful information, so descriptors built around these keypoints tend to be more discriminative. Another reason is that color-invariant-based keypoint detection is more robust to variations in the illumination than grayscale-based keypoint detection. Yet another reason is that local region descriptors based on color invariants more richly characterize the regions, and are more stable relative to the imaging conditions, than their grayscale counterparts.

Keywords Color-based object recognition • Color invariants • Keypoint detection • SIFT • Local region descriptors • Illumination invariance • Viewpoint invariance • Color ratios • Shadow invariance.

10.1 Introduction

10.1.1 Object Recognition

Given a target object as defined by a query image, the goal of an object-recognition system generally involves finding instances of the target in a database of images. As such, there are many similarities to content-based image retrieval. In some applications, the task is to retrieve images containing the target object; in others, it is to locate the target object within an image or images, and in others it is to identify the objects found in a given image. Whichever the goal, object recognition is made especially difficult by the fact that most of the time the imaging conditions cannot be completely controlled (see Fig. 10.1). For example, given two images representing the same object :

- The object may be rotated and/or translated in space.
- The acquisition devices may be different.
- The lighting may not be the same.

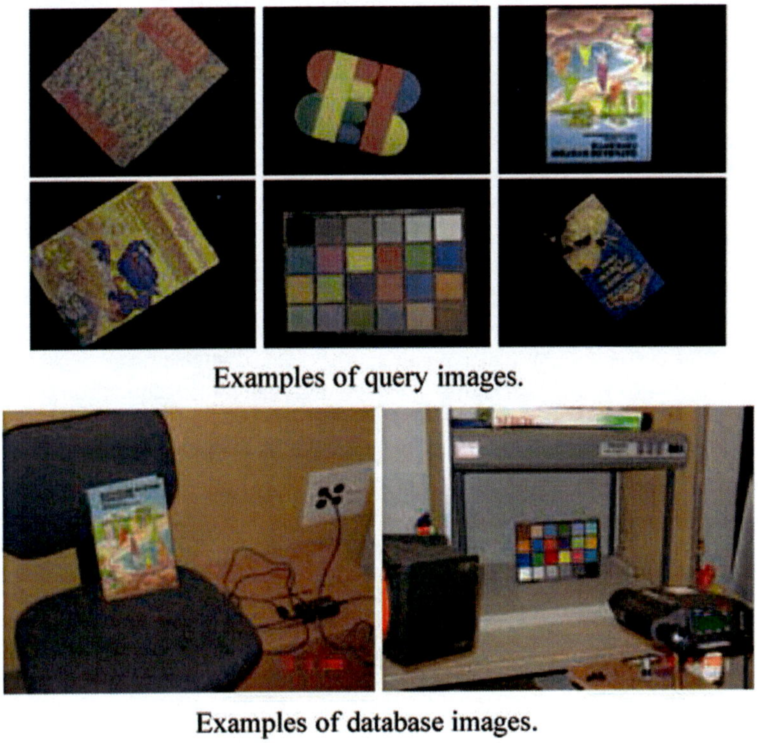

Fig. 10.1 These images are from the Simon Fraser University [5] which are available from http://www.cs.sfu.ca/~colour/data

Furthermore, an image may contain several objects (possibly partially occluded) against a cluttered background.

Since simple global image matching is not very effective when there are multiple objects, occlusions, viewpoint variations, uncontrolled lighting, and so on, the usual solution is to describe the image content in terms of a set of local keypoint descriptors centered around a set of keypoints. Ideally, a keypoint corresponds to a point on the object that remains stable across different views of the object. To find all instances of the target in a database images, the content of the query image $\mathbf{I_q}$ is compared to the content of each database image $\mathbf{I_d}$. The comparison requires three steps:

- Keypoint detection for both images $\mathbf{I_q}$ and $\mathbf{I_d}$.
- Keypoint description, i.e., describing the local regions around the keypoints to form a set of query descriptors and a set of database descriptors.
- Comparison of the sets of query and database descriptors. The similarity measure accounts for the fact that there could be several objects in one database image.

A threshold on the similarity measure is applied in order to determine the database images that contain the target object.

Object-recognition systems differ from one another in terms of their keypoint detection algorithms, the local keypoint descriptors used, and the similarity measure applied to the sets of keypoint descriptors. The choice for each of these components influences the recognition results.

Even though most acquisition devices provide color images, a lot of object-recognition systems are based, nonetheless, on grayscale information alone. Note that we will use the term "color" to refer to the camera's *RGB* sensor response to the incoming light at a given pixel. Hence, we are not referring to how humans perceive color, nor are we distinguishing between the colors of lights versus the colors of objects (see [71] for a thorough analysis of object color versus lighting color). Exploiting color for object recognition has two main advantages. First, two surfaces that differ in color can result in the same image grayscale value, and second finding discriminative features that are invariant to the acquisition conditions becomes easier when color is used [118]. With color, it is possible to increase the discriminative power of the system, while at the same time providing robustness to variations in the imaging and lighting conditions. However, there is a trade-off between invariance and discriminative power, so it is important to consider carefully the choice of invariants relative to the particular application.

10.1.2 Trade-off Between Discriminative Power and Invariance

The aim of the keypoint detection and keypoint description steps is to provide a good set of local descriptors for an image relative to the given image database. There are several criteria for assessing the effectiveness of the set of local keypoint descriptors; however, this section concentrates on the discriminative power of the keypoint descriptors, without devoting much attention to other criteria such as the time and memory required to compute them. For a given database, we consider the set of keypoint descriptors to be sufficiently discriminative if thresholding on the similarity measure distinguishes between the cases where a given image contains an instance of the target object from those where it does not.

The discriminative power of a set of keypoint descriptors depends on the image database. There is no single descriptor set that will yield the best results in all the cases. We categorize the types of databases according to the types of differences that may be found between two images of the same object. In particular, possible differences may occur due to changes in the:

- Lighting intensity
- Lighting color
- Lighting direction
- Viewpoint
- Ambient lighting effects

As well as other differences arise from changes in the:

- Highlights
- Shadows
- Shading

Ambient lighting, as described by Shafer [100], arises from diffuse light sources, interreflection of light between surfaces, unwanted infrared sensitivity of the camera, and lens flare.

The literature describes several keypoint detectors [68,78,107] and local keypoint descriptors [68, 77] that provide different levels of robustness across all these variations. For a particular application, the detectors and descriptors need to be insensitive only to the variations that are likely to occur in the images involved. For example, if the objects to be recognized consist only of matte surfaces, there is no need to use detectors and descriptors that are invariant to highlights. Any unnecessary invariance is likely to decrease the discriminative power of the system. Consequently, it is essential to understand color-image formation, and the impact any variations in the camera, lighting, or other imaging conditions may have on the resulting color image in order to be able to choose the most appropriate detectors and descriptors.

10.1.3 Overview

There are three color-related choices that need to be made when designing an object recognition system for a particular application. First is the choice of color invariants, which needs to be based on an understanding of how the image-acquisition conditions affect the resulting image. The color invariants are based on models of the possible variations that can occur. Consider two color images $\mathbf{I_q}$ and $\mathbf{I_d}$ representing the same object taken under different acquisition conditions. Let P_q and P_d be two pixels—one from each image—both imaging the same location on the object's surface. The pixel P_q has color $\mathbf{C}(P_q)$ in image $\mathbf{I_q}$ and the pixel P_d has color $\mathbf{C}(P_d)$ in image $\mathbf{I_d}$. Most color invariants assume that there exists a transform $\mathbf{F}$ between these colors such that :

$$\mathbf{C}(P_d) = \mathbf{F}(\mathbf{C}(P_q)). \tag{10.1}$$

However, the reader is warned that, in theory, metamerism means that no such function exists; invariance models ignore this difficulty and estimate a function that leads to the best invariance in practice.

The second and third color-related choices that have to be made concern the keypoint detectors and the local keypoint descriptors. Many keypoint detectors [68, 78, 107] and keypoint descriptors [68, 77] have been designed for grayscale data, but the question is how to modify these detectors and descriptors to include color, and will color make a difference? Section 10.3 presents color keypoint detectors, color key region detectors, and saliency-guided detection. It also discusses how machine learning can improve detection. Section 10.4 describes four approaches to introducing color into descriptors.

10.2 Color Invariance for Object Recognition

10.2.1 Classical Assumptions

The models of possible imaging variations are based on assumptions about color formation, the reflectance properties of the surfaces, the sensor sensitivities of the acquisition device, and the illumination incident at each point in the scene.

10.2.1.1 Models of Surface Reflection and Color-Image Formation

Most surface reflection models assume that the light hitting a surface is partially reflected by the air–surface interface (specular reflection), and that the remaining light enters the material body and is randomly reflected and absorbed within the material until the remaining light exits the material (diffuse body reflection). Let $S(\mathbf{x},\lambda)$ denote the spectral power distribution (SPD) of the light reflected from a surface patch and arriving at a camera having spectral sensitivities denoted $k(\lambda)$, $k = \{\mathrm{R},\mathrm{G},\mathrm{B}\}$. Following Wandell [113], we will call the incoming SPD at a pixel the "color signal." For a color camera, the R, G, and B components, $C^{\mathrm{R}}(P)$, $C^{\mathrm{G}}(P)$ and $C^{\mathrm{B}}(P)$, of its response are each obtained via the dot product of the color signal with the sensor sensitivity function of the corresponding color channel:

$$\begin{cases} C^{\mathrm{R}}(P) = \int_{\mathrm{vis}} R(\lambda)S(\mathbf{x},\lambda)\mathrm{d}\lambda \\ C^{\mathrm{G}}(P) = \int_{\mathrm{vis}} G(\lambda)S(\mathbf{x},\lambda)\mathrm{d}\lambda \\ C^{\mathrm{B}}(P) = \int_{\mathrm{vis}} B(\lambda)S(\mathbf{x},\lambda)\mathrm{d}\lambda \end{cases} \tag{10.2}$$

where the subscript vis means that the integral is over the visible range of wavelengths. Note that most cameras apply a tone curve correction or 'gamma' function to these linear outputs whenever the camera is not in "RAW" mode.

Three surface reflection models are widely used: the Kubelka–Munk model [39, 63], the dichromatic model as introduced by Shafer [100] and the Lambertian model [64, 122].

Assumption 1: Lambertian Reflectance

A Lambertian surface reflectance appears matte, has no specular component and has the property that the observed surface radiance is independent of the viewing location and direction. An ideal Lambertian surface also reflects 100% of the incident light, and so is pure white, but we will relax the usage here to include colored surfaces. If spatial location $\mathbf{x}$ on a surface with percent surface spectral reflectance $\beta(\mathbf{x},\lambda)$ is lit by light of SPD $E(\mathbf{x},\lambda)$ then the relative SPD of light $S_{\mathrm{Lambert}}(\mathbf{x},\lambda)$ reflected from this Lambertian surface is:

$$S_{\mathrm{Lambert}}(\mathbf{x},\lambda) = \beta(\mathbf{x},\lambda)\, E(\mathbf{x},\lambda). \tag{10.3}$$

Assumption 2: Dichromatic Reflectance

According to Shafer's dichromatic model [100] (see also Mollon's ([82]) account of how Monge described dichromatic reflection in 1789), the SPD $S_{\text{Dichromatic}}(\mathbf{x},\lambda)$ reflected by a non-matte surface is:

$$S_{\text{Dichromatic}}(\mathbf{x},\lambda) = m_{\text{bod}}(\theta)\,\beta(\mathbf{x},\lambda)\,E(\mathbf{x},\lambda) + m_{\text{int}}(\theta,\alpha)\,F_{\text{int}}(\mathbf{x},\lambda)\,E(\mathbf{x},\lambda), \quad (10.4)$$

where m_{bod} and m_{int} are the relative weightings of the body and interface components of the reflection and depend on the light θ and view α directions. $F_{\text{int}}(\mathbf{x},\lambda)$ represents the effect of Fresnel's laws at the interface.

Assumption 3: Kubelka–Munk Reflectance

Consider a material whose body reflectance at (spatial) position $\mathbf{x}$ is $\beta(\mathbf{x},\lambda)$ and whose Fresnel component is $F_{\text{int}}(\mathbf{x},\lambda)$. As Geusebroek [39] shows, the Kubelka–Munk model [63] predicts the SPD of the light $S_{\text{KM}}(\mathbf{x},\lambda)$ reflected from $\mathbf{x}$ lit by spectral power distribution $E(\mathbf{x},\lambda)$ to be given by :

$$S_{\text{KM}}(\mathbf{x},\lambda) = (1 - F_{\text{int}}(\mathbf{x},\lambda))^2\,\beta(\mathbf{x},\lambda)\,E(\mathbf{x},\lambda) + F_{int}(\mathbf{x},\lambda)\,E(\mathbf{x},\lambda). \quad (10.5)$$

The dichromatic and Kubelka–Munk models are similar in that they both describe the reflection in terms of specular and diffuse components. This kind of decomposition into two terms corresponding to two physical phenomena has been validated by Beckmann [10]. Both these models assume that the incident light has the same SPD from all directions. Shafer [100] also proposes extending the dichromatic model by adding a term for uniform ambient light $L_{\text{a}}(\lambda)$ of a different SPD. When this term is included, the reflected light is modeled as:

$$\begin{aligned} S_{\text{extended-Shafer}}(\mathbf{x},\lambda) &= m_{\text{bod}}(\theta)\,\beta(\mathbf{x},\lambda)\,E(\mathbf{x},\lambda) \\ &\quad + m_{\text{int}}(\theta,\alpha)\,F_{\text{int}}(\mathbf{x},\lambda)\,E(\mathbf{x},\lambda) + L_{\text{a}}(\lambda). \end{aligned} \quad (10.6)$$

10.2.1.2 Assumptions About Reflection Properties

Assumption 4: Neutral Interface Reflection

Reflection from the air–surface interface for many materials shows little variation across the visible wavelength range [100]. In other words, reflection from the interface is generally independent of wavelength:

$$F_{\text{int}}(\mathbf{x},\lambda) = F_{\text{int}}(\mathbf{x}). \quad (10.7)$$

Assumption 5: Matte Surface Reflection

For matte surface reflection often the Lambertian model suffices, but a more general matte model is to employ the weaker assumption that the specular component is zero within one of the more complex reflection models (e.g., Kubelka–Munk or dichromatic). Namely,

$$F_{\text{int}}(\mathbf{x},\lambda) = 0. \tag{10.8}$$

10.2.1.3 Assumptions About the Sensitivities of the Camera Sensors

Assumption 6: Normalized Camera Sensitivities

The spectral sensitivities $k(\lambda)$, $k = \text{R, G, B}$, of the camera can be normalized so that their integrals over the visible range are equal to a constant i_{RGB} :

$$\int_{\text{vis}} R(\lambda)\mathrm{d}\lambda = \int_{\text{vis}} G(\lambda)\mathrm{d}\lambda = \int_{\text{vis}} B(\lambda)\mathrm{d}\lambda = i_{\text{RGB}}. \tag{10.9}$$

Assumption 7: Narrowband Sensors

Some camera sensors are sensitive to a relatively narrow band of wavelengths, in which case it can be convenient to model their sensitivities as Dirac δ functions centered at wavelengths λ_k [35], $k = \text{R, G, B}$:

$$k(\lambda) = \delta(\lambda - \lambda_k), \;\; k = \text{R, G, B}. \tag{10.10}$$

This assumption holds only very approximately in practice; however, Finlayson et al. showed that narrower "sharpened" sensors often can be obtained as a linear combination of broader ones [26].

Assumption 8: Conversion from RGB to CIE 1964 XYZ

Geusebroek et al. [42] propose invariants with color defined in terms of CIE 1964 XYZ [122] color space. Ideally, the camera being used will have been color calibrated; however, when the camera characteristics are unknown, Geusebroek et al. assume they correspond to the ITU-R Rec.709 [58] or, equivalently, *sRGB* [106] standards. In this case, *linearized* RGB can be converted to CIE 1964 XYZ using the linear transformation:

$$\begin{bmatrix} C^X(P) \\ C^Y(P) \\ C^Z(P) \end{bmatrix} = \begin{pmatrix} 0.62 & 0.11 & 0.19 \\ 0.3 & 0.56 & 0.05 \\ -0.01 & 0.03 & 1.11 \end{pmatrix} \begin{bmatrix} C^R(P) \\ C^G(P) \\ C^B(P) \end{bmatrix}. \tag{10.11}$$

Note that in order to apply this transformation the RGB responses must be linear. Since most cameras output RGBs have a nonlinear "gamma" [88] or tone correction applied, it is essential to invert the gamma to ensure that the relationship between radiance and sensor response is linear before applying (10.11).

10.2.1.4 Assumptions About Illumination Properties

Assumption 9: Planckian Blackbody Illumination

Finlayson proposes an illumination model based on the Planckian model [122] of the blackbody radiator [23]. Planck's equation expresses the relative SPD of the light emitted by a blackbody radiator (e.g., a tungsten light bulb is approximately blackbody) as:

$$E(\lambda) = \frac{e\,c_1}{\lambda^5\left(\exp(\frac{c_2}{T\lambda}) - 1\right)}, \tag{10.12}$$

with

$$\begin{cases} e: \text{ illuminant intensity} \\ T: \text{ illuminant temperature} \\ c_1 = 3.74183 \times 10^{-16}\, Wm^2 \\ c_2 = 1.4388 \times 10^{-2}\, mK \end{cases} \tag{10.13}$$

Furthermore, since $\lambda \in [10^{-7}, 10^{-6}]$ for the visible range and $T \in [10^3, 10^4]$, Finlayson observes that $\exp(\frac{c_2}{T\lambda}) \gg 1$ and therefore simplifies the above equation to:

$$E(\lambda) = \frac{e\,c_1}{\lambda^5 \exp(\frac{c_2}{T\lambda})}. \tag{10.14}$$

Assumption 10: Constant Relative SPD of the Incident Light

Gevers assumes that the illuminant $E(\mathbf{x}, \lambda)$ can be expressed as a product of two terms. The first, $e(\mathbf{x})$, depends on position $\mathbf{x}$ and is proportional to the light intensity. The second is the relative SPD $E(\lambda)$ which is assumed to be constant throughout the scene [43]. Hence,

$$E(\mathbf{x}, \lambda) = e(\mathbf{x})\, E(\lambda). \tag{10.15}$$

Assumption 11: Locally Constant Illumination

Another common assumption is that the surface is identically illuminated at the locations corresponding to neighboring pixels [35, 113]. In other words, the incident illumination is assumed to be constant within a small neighborhood.

$$E(\mathbf{x}_1,\lambda) = E(\mathbf{x}_2,\lambda),\ \textit{for neighboring locations}\ \mathbf{x}_1\ \textit{and}\ \mathbf{x}_2. \tag{10.16}$$

There is no specific restriction on the size of the neighborhood, but the larger it is, the less likely the assumption is to hold. For a local descriptor, it needs to hold for neighborhoods corresponding in size to the descriptor's region of support. In most cases, the illumination at neighboring locations will only be constant if both the SPD of the light is spatially uniform and the surface is locally planar.

Assumption 12: Ideal White Illumination

The ideal white assumption is that the relative spectral distribution $E(\mathbf{x},\lambda)$ of the illuminant incident at location $\mathbf{x}$ is constant across all wavelengths. In other words,

$$E(\mathbf{x},\lambda) = E(\mathbf{x}). \tag{10.17}$$

Assumption 13: Known Illumination Chromaticity

It can be useful to know the chromaticity of the overall scene illumination when evaluating some color features that are invariant across shadows and highlights [118]. Chromaticity specifies color independent of any scaling. In some cases, the illumination chromaticity can be measured directly; however, when it is not known, it can be estimated from an analysis of the statistics of the colors appearing in the image [29, 36, 93, 123].

10.2.1.5 Mapping Colors Between Illuminants

Since the illumination may differ between the query and database images, it is necessary to have a way to map colors between illuminants. As mentioned earlier, although metamerism means that there is no unique mapping—two surface reflectances that yield the same RGB under one illuminant may in fact yield distinct RGBs under a second illuminant [71]—it is helpful to choose a mapping (function **F** in (10.1)) and hope that the error will not be too large.

Assumption 14: Diagonal Model

The diagonal model predicts the color $\mathbf{C}(P_d) = (C^R(P_d), C^G(P_d), C^B(P_d))^T$ of the pixel P_d from the color $\mathbf{C}(P_q) = (C^R(P_q), C^G(P_q), C^B(P_q))^T$ of the pixel P_q via a linear transformation **F** defined by the diagonal matrix [25, 62] :

$$\mathbf{C}(P_d) = \begin{pmatrix} a_R & 0 & 0 \\ 0 & a_G & 0 \\ 0 & 0 & a_B \end{pmatrix} \mathbf{C}(P_q). \tag{10.18}$$

for some scalings a_R, a_G, and a_B. The diagonal model holds, for example, for a Lambertian surface imaged by a camera with extremely narrowband sensors [35]. It also holds for the special case of illuminants and reflectances that are limited to low-dimensional linear models [25].

Assumption 15: Diagonal Model with Translation

Finlayson et al. extend the diagonal model by adding a translation in color space [31], in essence replacing the linear transformation with a limited affine one. Hence, the transformation **F** is defined by two matrices: one diagonal 3×3 matrix and one 3×1 matrix [31] :

$$\mathbf{C}(P_d) = \begin{pmatrix} a_R & 0 & 0 \\ 0 & a_G & 0 \\ 0 & 0 & a_B \end{pmatrix} \mathbf{C}(P_q) + \begin{pmatrix} b_R \\ b_G \\ b_B \end{pmatrix}. \tag{10.19}$$

The diagonal model combined with translation applies under the same conditions as the diagonal model, and takes into account any constant color offset such as a nonzero camera black level.

Assumption 16: 3 ×3 Linear Transformation

Another possible choice for mapping colors between illuminants is a full 3×3 linear transformation [67] :

$$\mathbf{C}(P_d) = \begin{pmatrix} a & b & c \\ d & e & f \\ g & h & i \end{pmatrix} \mathbf{C}(P_q). \tag{10.20}$$

Although the 3×3 model subsumes the diagonal model, more information about the imaging conditions is required in order to define the additional six parameters. For example, simply knowing the color of the incident illumination does not provide sufficient information since the RGB provides only 3 knowns for the 9 unknowns.

Assumption 17: Affine Transformation

A translational component can also be added to the 3×3 model [80]:

$$\mathbf{C}(P_d) = \begin{pmatrix} a & b & c \\ d & e & f \\ g & h & i \end{pmatrix} \mathbf{C}(P_q) + \begin{pmatrix} j \\ k \\ l \end{pmatrix}. \tag{10.21}$$

Assumption 18: Monotonically Increasing Functions

Finlayson [31] treats each color component independently and assumes that the value $C^k(P_d)$, $k = \text{R,G,B}$, of the pixel P_d can be computed from the value $C^k(P_q)$ of the pixel P_q via a strictly increasing function f^k:

$$C^k(P_d) = f^k(C^k(P_q)), k = \text{R,G,B}. \tag{10.22}$$

A function f^k is strictly increasing if $a > b \Rightarrow f^k(a) > f^k(b)$. The three monotonically increasing functions f^k, $k = \text{R,G,B}$, need not necessarily be linear.

For the above five models of illumination change, the greater the number of degrees of freedom (from diagonal model to affine transformation) potentially the better the result. However, metamerism means that no such function based on pixel color alone actually exists, so it is not a matter of simply creating a better approximation to this nonexistent function. Nevertheless, in the context of object recognition, these models are frequently used to normalize the images with the goal of making them at least partially invariant to the illumination. The above assumptions form the basis of most of the color-invariant features that follow.

10.2.2 The Color Invariants

There are many color-invariant features in use in the context of object recognition. Our aim is not to create an exhaustive list of color invariants but rather to understand how they follow from the assumptions described above. We classify the color invariants into three categories. The first category consists of those based on the ratio between the different color channels at a pixel or between the corresponding color channels from neighboring pixels. The second category consists of those based on the color distribution of the pixels from a local region. The third category consists of those based on spectral and/or spatial derivatives of the color components.

10.2.2.1 Intra- and Inter-color Channel Ratios

Ratios Between Corresponding Color Channels of Neighboring Pixels

- *Funt et al. approach [35]*

Following on the use of ratios in Retinex [65], Funt et al. propose a color invariant relying on the following three assumptions [35] :

- Lambertian model ((10.3)),
- Narrowband sensors ((10.10)),
- Constant illumination ((10.16)) $E(N_{3\times3}(P),\lambda)$ over the 3×3 neighborhood $N_{3\times3}(P)$ centered on pixel P.

We assume that to every pixel P corresponds a unique scene location $\mathbf{x}$ (i.e., there are no transparent surfaces) and for the rest of this chapter "pixel P" will refer explicitly to its image location and implicitly to the corresponding scene location $\mathbf{x}$. Likewise, we are assuming that spatial derivatives in the scene correspond to spatial derivatives in the image and vice versa. Given the Lambertian, narrowband, and constant illumination assumptions, the kth color component of pixel P for scene location $\mathbf{x}$ is given by:

$$\begin{aligned} C^k(P) &= \int_{\text{vis}} \beta(\mathbf{x},\lambda)\, E(N_{3\times3}(P),\lambda)\, k(\lambda)\mathrm{d}\lambda \\ &= \beta(\mathbf{x},\lambda_k)\, E(N_{3\times3}(P),\lambda_k)\, k(\lambda_k). \end{aligned} \tag{10.23}$$

Similarly, the kth color component $C^k(P_{\text{neigh}})$ of a neighboring pixel P_{neigh} is:

$$C^k(P_{\text{neigh}}) = \beta(\mathbf{x}_{\text{neigh}},\lambda_k)\, E(N_{3\times3}(P_{\text{neigh}}),\lambda_k)\, k(\lambda_k). \tag{10.24}$$

The ratio of two neighboring pixels

$$\begin{aligned} \frac{C^k(P)}{C^k(P_{\text{neigh}})} &= \frac{\beta(\mathbf{x},\lambda_k)\, E(N_{3\times3}(P),\lambda_k)\, k(\lambda_k)}{\beta(\mathbf{x}_{\text{neigh}},\lambda_k)\, E(N_{3\times3}(P_{\text{neigh}}),\lambda_k)\, k(\lambda_k)} \\ &= \frac{\beta(\mathbf{x},\lambda_k)}{\beta(\mathbf{x}_{\text{neigh}},\lambda_k)}, \end{aligned} \tag{10.25}$$

depends only on the spectral reflectances of the surfaces and the wavelength of the sensor's sensitivity λ_k.

Thus, if the surface is Lambertian, the sensors are narrowband and the light is locally uniform, the color channel ratios of neighboring pixels are insensitive to the color and intensity of the illumination, and to the viewing location. Hence, Funt et al. propose the color-invariant feature $(X^1(P),X^2(P),X^3(P))^T$ for pixel P as [35] :

$$\begin{cases} X^1(P) = \log(C^R(P_{\text{neigh}})) - \log(C^R(P)), \\ X^2(P) = \log(C^G(P_{\text{neigh}})) - \log(C^G(P)), \\ X^3(P) = \log(C^B(P_{\text{neigh}})) - \log(C^B(P)). \end{cases} \tag{10.26}$$

The logarithm is introduced so that the ratios can be efficiently computed via convolution by a derivative filter. They also note that the invariance of the ratios can also be enhanced by "sharpening" [26] the sensors. Subsequently, Chong et al. [47] proposed these ratios in a sharpened sensor space as a good "perception-based" space for invariants.

The same assumptions as above, but with illumination assumed constant over larger neigborhoods, have also been used in other contexts. For example, Land [66] suggested normalizing the color channels by their local mean values. Others normalize by the local maximum value [17, 65]. Finlayson et al. proposed a generalization of these normalization approaches [24] and showed that normalization by the mean and max can be considered as special cases of normalization by the Minkowski norm over the local region.

Starting with the assumption that the light is uniform across a local region of N_{pix} pixels, Finlayson vectorizes the pixels into three N_{pix}-dimensional vectors $\mathbf{k}$, $k = \mathrm{R,G,B}$. The coordinates $\mathbf{k_1}, \mathbf{k_2}, \ldots, \mathbf{k_{N_{pix}}}$ of vector $\mathbf{k}$ are the component levels $C^k(P)$ of the pixels P in the region. Finlayson observes that under the given assumptions, the angles $angle^{kk'}$, $k, k' = \mathrm{R,G,B}$ between the vectors $\mathbf{k}$ and $\mathbf{k}'$ are invariant to the illumination [27]. If a change in the illumination is modeled as a multiplication of the pixel values by a constant factor for each color channel then this multiplication modifies the norm of each vector $\mathbf{k}$ associated with each component k, but does not modify its direction. Hence, the three values representing the angles between each pair of distinct vectors constitute a color invariant.

- *The m_1, m_2, m_3 color invariant of Gevers et al. [43].*

Gevers et al. propose a color invariant denoted $\{m_1, m_2, m_3\}$ based on the following four assumptions [43] :

- Dichromatic reflection ((10.4))
- Matte surface reflectance ((10.8))
- Narrowband sensors ((10.10))
- Locally constant light color ((10.15))

Under these assumptions, the color components $C^k(P)$, $k = \mathrm{R,G,B}$, become:

$$C^k(P) = m_{\mathrm{bod}}(\theta)\, \beta(\mathbf{x}, \lambda_k)\, e(\mathbf{x})\, E(\lambda_k)\, k(\lambda_k). \tag{10.27}$$

where θ is the light direction with respect to the surface normal at $\mathbf{x}$, assuming a distant point source. Likewise, the color $C^k(P_{\mathrm{neigh}})$, $k = \mathrm{R,G,B}$, of pixel P_{neigh} within the 3×3 neighborhood of P corresponding to scene location $\mathbf{x}_{neigh}$ can be expressed as:

$$C^k(P_{\mathrm{neigh}}) = m_{bod}(\theta_{\mathrm{neigh}})\, \beta(\mathbf{x}_{\mathrm{neigh}} P, \lambda_k)\, e(\mathbf{x}_{\mathrm{neigh}})\, E(\lambda_k)\, k(\lambda_k), \tag{10.28}$$

where θ_{neigh} is the light direction with respect to the surface normal at $\mathbf{x}_{\mathrm{neigh}}$. This direction may differ from θ. The introduction of the light direction parameter facilitates modeling the effects created by variations in surface orientation.

Taking a different color channel $k' \neq k$, the following ratio depends only on the spectral reflectances of the surface patches and on the sensitivity wavelength of the sensors, but not on the illumination:

$$\frac{C^k(P)C^{k'}(P_{\text{neigh}})}{C^k(P_{\text{neigh}})C^{k'}(P)} = \frac{\beta(\mathbf{x},\lambda_k)\beta(\mathbf{x}(P_{\text{neigh}},\lambda_k')}{\beta(\mathbf{x}(P_{\text{neigh}},\lambda_k)\beta(\mathbf{x},\lambda_k')}. \tag{10.29}$$

Thus, for the case of a matte surface, narrowband sensors and illumination of locally constant color, Gevers et al. show that the ratio between two different color channels from two neighboring pixels is invariant to the illumination's color, intensity, and direction, as well as to view direction.

Based on this analysis, Gevers et al. [43] define the color invariant $(X^1(P), X^2(P), X^3(P))^T$ for pixel P as:

$$\begin{cases} X^1(P) = \dfrac{C^R(P)C^G(P_{\text{neigh}})}{C^R(P_{\text{neigh}})C^G(P)}, \\ X^2(P) = \dfrac{C^R(P)C^B(P_{\text{neigh}})}{C^R(P_{\text{neigh}})C^B(P)}, \\ X^3(P) = \dfrac{C^G(P)C^B(P_{\text{neigh}})}{C^G(P_{\text{neigh}})C^B(P)}. \end{cases} \tag{10.30}$$

Ratio of Color Components at a Single Pixel

- *The color invariant l_1, l_2, l_3 of Gevers et al. [43]*

Gevers et al. also propose a color invariant denoted $\{l_1, l_2, l_3\}$ based on the following assumptions [43]:

- Dichromatic reflection ((10.4))
- Neutral interface reflection ((10.7))
- Sensor sensitivities balanced such that their integrals are equal ((10.9))
- Ideal white illumination ((10.17))

Under these assumptions, the kth color component is:

$$C^k(P) = m_{\text{bod}}(\theta)\, E(\mathbf{x}) \int_{\text{vis}} \beta(\mathbf{x},\lambda)\, k(\lambda)\text{d}\lambda + m_{\text{int}}(\theta,\alpha)\, F_{\text{int}}(\mathbf{x})\, E(\mathbf{x}) i_{\text{RGB}}. \tag{10.31}$$

From this, it follows that the ratio of the differences of the three color channels k, k', and k'' depends only on the surface reflectance and the sensor sensitivities:

$$\frac{C^k(P) - C^{k'}(P)}{C^k(P) - C^{k''}(P)} = \frac{\int_{\text{vis}} \beta(\mathbf{x},\lambda)\, k(\lambda)\text{d}\lambda - \int_{\text{vis}} \beta(\mathbf{x},\lambda)\, k'(\lambda)\text{d}\lambda}{\int_{\text{vis}} \beta(\mathbf{x},\lambda)\, k(\lambda)\text{d}\lambda - \int_{\text{vis}} \beta(\mathbf{x},\lambda)\, k''(\lambda)\text{d}\lambda} \tag{10.32}$$

In other words, for the case of neutral interface reflection, balanced sensors, and ideal white illumination, the ratio between the differences of color components at a single pixel is invariant to the light's intensity and direction, as well as to view direction and specularities.

Gevers et al. [43] combine the three possible ratios into the "color"-invariant feature $(X^1(P), X^2(P), X^3(P))^T$:

$$\begin{cases} X^1(P) = \dfrac{(C^R(P) - C^G(P))^2}{(C^R(P) - C^G(P))^2 + (C^R(P) - C^B(P))^2 + (C^G(P) - C^B(P))^2}, \\ X^2(P) = \dfrac{(C^R(P) - C^B(P))^2}{(C^R(P) - C^G(P))^2 + (C^R(P) - C^B(P))^2 + (C^G(P) - C^B(P))^2}, \\ X^3(P) = \dfrac{(C^G(P) - C^B(P))^2}{(C^R(P) - C^G(P))^2 + (C^R(P) - C^B(P))^2 + (C^G(P) - C^B(P))^2}. \end{cases} \tag{10.33}$$

Although this in an invariant, it is hardly a color invariant since the illumination is by assumption unchanging and ideal white. Nonetheless, we will continue to refer to all the invariants in this chapter as color invariants for the sake of consistency.

There are also some other color invariants that follow from these assumptions that are based on the ratios of channel-wise differences. For example, hue H expressed as

$$H(P) = \arctan\left(\frac{\sqrt{3}(C^G(P) - C^B(P))}{(C^R(P) - C^G(P)) + (C^R(P) - C^B(P))}\right) \tag{10.34}$$

has the same invariance properties as $(X^1(P), X^2(P), X^3(P))^T$ [43].

- *The color invariant c_1, c_2, c_3 of Gevers et al. [43]*

Gevers et al. propose a third color invariant denoted $\{c_1, c_2, c_3\}$ based on the following three assumptions [43]:

- Dichromatic reflection ((10.4))
- Matte surface ((10.8))
- Ideal white illumination ((10.17))

Under these assumptions, the kth color component becomes:

$$\begin{aligned} C^k(P) &= \int_{\text{vis}} m_{\text{bod}}(\theta)\, \beta(\mathbf{x},\lambda)\, E(\mathbf{x},\lambda)\, k(\lambda) \mathrm{d}\lambda \\ &= m_{\text{bod}}(\theta)\, E(\mathbf{x}) \int_{\text{vis}} \beta(\mathbf{x},\lambda)\, k(\lambda) \mathrm{d}\lambda. \end{aligned} \tag{10.35}$$

For two channels C^k and $C^{k'}$ ($k' \neq k$), their ratio depends only on the surface reflectance and on the sensors:

$$\frac{C^k(P)}{C^{k'}(P)} = \frac{\int_{\text{vis}} \beta(\mathbf{x},\lambda)\, k(\lambda) \mathrm{d}\lambda}{\int_{\text{vis}} \beta(\mathbf{x},\lambda)\, k'(\lambda) \mathrm{d}\lambda}. \tag{10.36}$$

In other words, for a matte surface under ideal white illumination, the ratio of color component pairs is invariant to the light's intensity and direction, and to the view direction. Gevers et al. [43] define the color invariant $(X^1(P), X^2(P), X^3(P))^T$:

$$\begin{cases} X^1(P) = \arctan\left(\dfrac{C^R(P)}{\max(\mathrm{C^G(P), C^B(P)})}\right), \\ X^2(P) = \arctan\left(\dfrac{C^G(P)}{\max(\mathrm{C^R(P), C^B(P)})}\right), \\ X^3(P) = \arctan\left(\dfrac{C^B(P)}{\max(\mathrm{C^R(P), C^G(P)})}\right). \end{cases} \tag{10.37}$$

The function $max(x,y)$ returns the maximum of its two arguments.

Standard chromaticity space is also an invariant feature of a similar sort. It is defined by:

$$\begin{cases} C^r(P) = \dfrac{C^R(P)}{C^R(P)+C^G(P)+C^B(P)}, \\ C^g(P) = \dfrac{C^G(P)}{C^R(P)+C^G(P)+C^B(P)}, \\ C^b(P) = \dfrac{C^B(P)}{C^R(P)+C^G(P)+C^B(P)}. \end{cases} \tag{10.38}$$

Finlayson et al. observed that computing invariants can be broken into two parts—one for the illuminant color and a second for its direction [28]—and then handled iteratively. Noting that the features proposed by Funt et al. [35] remove the dependence on the color of the light, while the features $\{C^r(P), C^g(P), C^b(P)\}$ remove the dependence on its direction, Finlayson et al. proposed an iterative normalization that applies the color normalization proposed by Funt followed by the $\{C^r(P), C^g(P), C^b(P)\}$ normalization. They report that it converges quickly on color features that are invariant to both the color and the direction of the light [28].

- *Finlayson et al. approach [23]*

Finlayson et al. propose color invariants based on the following three assumptions [23]:

- Lambertian model ((10.3))
- Narrowband sensors ((10.10))
- Blackbody illuminant ((10.14))

Under these assumptions, the color components are:

$$C^k(P) = \beta(\lambda_k)\,\frac{e\,c_1}{\lambda_k^5 \exp(\frac{c_2}{T\,\lambda_k})}\,k(\lambda_k). \tag{10.39}$$

Applying the natural logarithm yields:

$$\ln(C^k(P)) = \ln(e) + \ln\left(\frac{\beta(\lambda_k)c_1 k(\lambda_k)}{\lambda_k^5}\right) - \frac{c_2}{T\lambda_k}. \tag{10.40}$$

This equation can be rewritten as:

$$\ln(C^k(P)) = \text{Int} + \text{Ref}_k + T^{-1}\, L_k, \tag{10.41}$$

where :

- $\text{Int} = \ln(e)$ relates to the light intensity.
- $\text{Ref}_k = \ln\left(\frac{\beta(\lambda_k)\, c_1\, k(\lambda_k)}{\lambda_k^5}\right)$ depends on the surface reflectance properties and on the sensor sensitivity wavelength λ_k.
- $T^{-1}\, L_k = -\frac{c_2}{T\lambda_k}$ depends on the temperature (hence, color) and on the sensor sensitivity.

Given a second channel $k' \neq k$, the logarithm of their ratio is then independent of the intensity:

$$\begin{aligned}\ln\left(\frac{C^k(P)}{C^{k'}(P)}\right) &= \ln(C^k(P)) - \ln(C^{k'}(P)) \\ &= \text{Ref}_k - \text{Ref}_{k'} + T^{-1}(L_k - L_{k'}).\end{aligned} \tag{10.42}$$

Similarly, for a second pair of channels $\{k'', k'\}$:

$$\ln\left(\frac{C^{k''}(P)}{C^{k'}(P)}\right) = \text{Ref}_{k''} - \text{Ref}_{k'} + T^{-1}(L_{k''} - L_{k'}). \tag{10.43}$$

The points $(\ln(\frac{C^k(P)}{C^{k'}(P)}), \ln(\frac{C^{k''}(P)}{C^{k'}(P)}))$ define a line as a function of T^{-1}. The orientation of the line depends on the sensitivity peaks of the sensors. Finlayson et al. project the points from $(\ln(\frac{C^G(P)}{C^R(P)}), \ln(\frac{C^B(P)}{C^R(P)}))$ space onto the orthogonal direction and show that the resulting coordinates do not depend on either the intensity or the color of the light. With this approach, they reduce the dimension of the representational space from 3 to 1 since they have only one invariant feature for each pixel. They also show how shadows can be removed from images using this projection. Subsequently, Finlayson et al. [30] also showed that the projection direction can be found as that which minimizes the entropy of the resulting invariant image, thereby eliminating the need to know the sensor sensitivities in advance.

10.2.2.2 Normalizations Based on an Analysis of Color Distributions

Rank Measure Invariance

Finlayson et al. propose a model of illumination change based on the assumption that its effect is represented by a strictly increasing function on each color channel ((10.22)) [31]. Consider two pixels P_1 and P_2 from the local region around a detected keypoint. If their kth color components are such that $C^k(P_1) > C^k(P_2)$, then after an illumination change $f^k(C^k(P_1)) > f^k(C^k(P_2))$ because the illumination change is assumed to be represented by a strictly increasing function f^k for each component. In other words, if the assumption holds then the rank ordering of the pixel values is preserved across a change in illumination. Based on this observation, they propose an invariant based on ranking the pixels from a local region into increasing order of their value for each channel and normalizing by the number of pixels in the region. The color rank measure $\mathrm{Rc}^k(P)$ for the pixel P in terms of channel k is given by:

$$\mathrm{Rc}^k(P) = \frac{\mathrm{Card}\{P_i / c^k(P_i) \leq c^k(P)\}}{\mathrm{Card}\{P_j\}}, \qquad (10.44)$$

where Card is set cardinality.

The color rank measure ranges from 0 for the darkest value to 1 for the brightest. The pixels in a region are then characterized not by the color component levels $C^k(P)$ themselves, but rather by their ranks $\mathrm{Rc}^k(P)$, $k = \mathrm{R,G,B}$. This turns out to be equivalent to applying histogram equalization on each component independently. Under a similar assumption for the sensors, rank-based color features are invariant to changes in both the sensors and the illumination.

Normalization by Color Distribution Transformation

Some authors propose first normalizing the color distributions of an image (or local region) in RGB color space and then basing color invariants on the normalized values. For example, Lenz et al. assume that illumination variation can be modeled by a 3×3 matrix (cf. (10.22)) [67] and propose normalizing RGB color space such that the result is invariant to such a 3×3 transformation. The general idea of this normalization is to make the matrix of the second-order moments of the RGB values equal to the unity matrix. Similarly, Healey et al. [49] normalize the color distribution such that the eigenvalues of the moment matrices are invariant to the color of the illumination. The transform is based on a transform matrix obtained by computing the Cholesky decomposition of the covariance matrix of the color distribution. Both these normalizations [49, 67] are insensitive to the application of a 3×3 matrix in color space, and, hence, are invariant across variations in both the intensity and color of the incident light (cf. (10.22)).

Finlayson et al. have shown that a diagonal transformation in conjunction with a translation component models the effects of a change in the light quite well (cf. (10.21)) [31]. Normalizing the color distribution so as to make its mean 0 and its standard deviation 1 leads to color values that are illumination invariant:

$$\begin{cases} X^1(P) = \dfrac{C^R(P) - \mu(C^R(P_i))}{\sigma(C^R(P_i))}, \\ X^2(P) = \dfrac{C^G(P) - \mu(C^G(P_i))}{\sigma(C^G(P_i))}, \\ X^3(P) = \dfrac{C^B(P) - \mu(C^B(P_i))}{\sigma(C^B(P_i))}, \end{cases} \tag{10.45}$$

where $\mu(C^k(P_i))$ and $\sigma(C^k(P_i))$ represent the mean and standard deviation, respectively, of the pixels P_i in the given region. Subtraction of the mean removes the translational term, while division by the standard deviation removes the diagonal terms. These color features are invariant to the intensity and color of the illumination as well as to changes in the ambient illumination.

The Invariant Color Moments Of Mindru et al. [80]

The moments of local color distributions do not take into account the spatial information provided by the image. To overcome this limitation, Mindru et al. proposed generalized color moments defined as [80]:

$$M_{pq}^{abc} = \int_y \int_x x^p y^q C^R(P_{xy})^a C^G(P_{xy})^b C^B(P_{xy})^c \mathrm{d}x\mathrm{d}y, \tag{10.46}$$

where (x, y) is the position of pixel P_{xy} in the image. M_{pq}^{abc} is called the generalized color moment of order $p+q$ and degree $a+b+c$. Mindru et al. use only the generalized color moments of order less than or equal to 1 and degree less than or equal to 2, and consider different models of illumination change. In particular, they are able to find some combination of the generalized moments that is invariant to a diagonal transform (cf. (10.18)), to a diagonal transform with translation (cf. (10.19)), and to an affine transform (cf. (10.21)) [80]. Since spatial information is included in these generalized color moments, they show that there are combinations that are invariant to various types of geometric transformations as well.

10.2.2.3 The Invariant Derivatives

The Spectral Derivatives of Geusebroek et al. [42]

Geusebroek et al. propose several color invariants based on spectral derivatives of the surface reflectance [42]. They introduce several assumptions, three of which are common to all of the invariants they propose:

– Kubelka–Munk model ((10.5)).
– The transform from linearized *RGB* camera space to CIE 1964 *XYZ* space can be modeled by a 3×3 matrix ((10.11)).
– Neutral interface reflection ((10.7)).

Geusebroek et al. base their color invariants on the Gaussian color model. The Gaussian color model considers three sensors $\mathbb{G}$, $\mathbb{G}_\lambda$, and $\mathbb{G}_{\lambda\lambda}$ whose spectral sensitivities are, respectively, the 0, 1st- and 2nd-order derivatives of the Gaussian function $\mathbb{G}(\lambda)$ having central wavelength $\lambda_0 = 520$ nm and standard deviation $\sigma_\lambda = 55$ nm [41]. For the color signal $S(\mathbf{x},\lambda)$ reflected from surface location $\mathbf{x}$, Geusebroek et al. show that the color components expressed in this Gaussian space represent the successive coefficients of the Taylor expansion of the color signal $S(\mathbf{x},\lambda)$ weighted by the Gaussian $\mathbb{G}(\lambda)$ [41]. In other words, these components represent the successive spectral derivatives of the color signal spectrum. They show that the color components $\{C^{\mathbb{G}}(P), C^{\mathbb{G}_\lambda}(P), C^{\mathbb{G}_{\lambda\lambda}}(P)\}$ of the pixel P in this Gaussian space can be approximately obtained from its color expressed in the CIE 1964 *XYZ* space using the following transform [41] :

$$\begin{bmatrix} C^{\mathbb{G}}(P) \\ C^{\mathbb{G}_\lambda}(P) \\ C^{\mathbb{G}_{\lambda\lambda}}(P) \end{bmatrix} = \begin{pmatrix} -0.48 & 1.2 & 0.28 \\ 0.48 & 0 & -0.4 \\ 1.18 & -1.3 & 0 \end{pmatrix} \begin{bmatrix} C^X(P) \\ C^Y(P) \\ C^Z(P) \end{bmatrix}. \tag{10.47}$$

Combining this with Assumption 10.2.1.3 ((10.11)), the global transformation from the camera RGB components to the Gaussian components becomes [41]:

$$\begin{bmatrix} C^{\mathbb{G}}(P) \\ C^{\mathbb{G}_\lambda}(P) \\ C^{\mathbb{G}_{\lambda\lambda}}(P) \end{bmatrix} = \begin{pmatrix} 0.06 & 0.63 & 0.27 \\ 0.3 & 0.04 & -0.35 \\ 0.34 & -0.6 & 0.17 \end{pmatrix} \begin{bmatrix} C^R(P) \\ C^G(P) \\ C^B(P) \end{bmatrix}. \tag{10.48}$$

Geusebroek et al. then go on to define a series of new color invariants (described below) based on the spectral derivatives of color signals $S_{\mathrm{KM}}(\mathbf{x},\lambda)$ satisfying the Kubelka–Munk model.

- *The color-invariant feature H*

Adding the additional assumption of ideal white illumination (10.17) the color signal $S_{\mathrm{KM}}(\mathbf{x},\lambda)$ reflected from $\mathbf{x}$ is:

$$S_{\mathrm{KM}}(\mathbf{x},\lambda) = (1 - F_{\mathrm{int}}(\mathbf{x}))^2\, \beta(\mathbf{x},\lambda)\, E(\mathbf{x}) + F_{\mathrm{int}}(\mathbf{x})\, E(\mathbf{x}). \tag{10.49}$$

The first and second spectral derivatives of this are:

$$S_{\mathrm{KM}_\lambda}(\mathbf{x},\lambda) = (1 - F_{\mathrm{int}}(\mathbf{x}))^2\, \frac{\partial \beta(\mathbf{x},\lambda)}{\partial \lambda}\, E(\mathbf{x}), \tag{10.50}$$

and

$$S_{KM_{\lambda\lambda}}(\mathbf{x},\lambda) = (1 - F_{\text{int}}(\mathbf{x}))^2 \frac{\partial^2 \beta(\mathbf{x},\lambda)}{\partial \lambda^2} E(\mathbf{x}). \tag{10.51}$$

Observing that their ratio depends only on the surface reflectance and recalling that the Gaussian color model provides these spectral derivatives leads to the following color invariant [42]:

$$X^H(P) = \frac{C^{\mathbb{G}_\lambda}(P)}{C^{\mathbb{G}_{\lambda\lambda}}(P)}. \tag{10.52}$$

In other words, for a surface with neutral interface reflection under ideal white illumination, the ratio at pixel P of the Gaussian color components $C^{\mathbb{G}_\lambda}(P)$ and $C^{\mathbb{G}_{\lambda\lambda}}(P)$ is independent of the intensity and incident angle of the light as well as of the view direction and the presence of specular highlights.

- *Color invariant C [42]*

Adding the further assumption that the surface is matte ((10.8)) leads Geusebroek et al. to another invariant they call C [42]. For a matte surface, there is no specular component so the color signal $S_{KM}(\mathbf{x},\lambda)$ becomes:

$$S_{KM}(\mathbf{x},\lambda) = \beta(\mathbf{x},\lambda)\, E(\mathbf{x}). \tag{10.53}$$

The first derivative is:

$$S_{KM_\lambda}(\mathbf{x},\lambda) = \frac{\partial \beta(\mathbf{x},\lambda)}{\partial \lambda} E(\mathbf{x}). \tag{10.54}$$

The ratio of these depends only on the surface reflectance, so Geusebroek et al. propose the color invariant [42]:

$$X^C(P) = \frac{C^{\mathbb{G}_\lambda}(P)}{C^{\mathbb{G}}(P)}. \tag{10.55}$$

Thus, in case of matte surface under ideal white illumination, the ratio between the component levels $C^{\mathbb{G}_\lambda}(P)$ and $C^{\mathbb{G}}(P)$ depends neither on the intensity or direction of the incident light nor on the view direction.

- *Color invariant W of Geusebroek et al.*

The W color invariant differs from the previous ones in that it is based on the spatial derivative instead of the spectral derivative. For this invariant, Geusebroek et al. make three assumptions in addition to their standard 3:

- Lambertian surface reflectance ((10.3))
- Ideal white illumination ((10.17))
- Spatially uniform illumination ((10.16)). Together these last two assumptions imply $E(\mathbf{x},\lambda) = E$

Under these assumptions the reflected color signal $S_{\mathrm{KM}}(\mathbf{x},\lambda)$ becomes:

$$S_{\mathrm{KM}}(\mathbf{x},\lambda) = \beta(\mathbf{x},\lambda)\, E. \tag{10.56}$$

Its spatial derivative is:

$$S_{\mathrm{KM}_{\mathbf{x}}}(\mathbf{x},\lambda) = \frac{\partial \beta(\mathbf{x},\lambda)}{\partial \mathbf{x}}\, E. \tag{10.57}$$

The ratio of the previous two quantities depends only on the spectral reflectance of the surface and not the intensity of the light. This is also the case in color space instead of spectral space. Based on this analysis, Geusebroek et al. propose the color invariant [42]:

$$X^W(P) = \frac{C^{\mathbb{G}_{\mathbf{x}}}(P)}{C^{\mathbb{G}}(P)}, \tag{10.58}$$

where $C^{\mathbb{G}_{\mathbf{x}}}(P) = \frac{\partial C^{\mathbb{G}}(P)}{\partial \mathbf{x}}$ is the spatial derivative of $C^{\mathbb{G}}(P)$ image. This links the spatial derivative in the image space with the corresponding spatial derivative in the scene space.

Thus, for a Lambertian surface under locally constant white illumination, the ratio of the spatial derivative of the Gaussian color component $C^{\mathbb{G}}(P)$ over the Gaussian color component itself $C^{\mathbb{G}}(P)$ is independent of the light intensity.

- *Color invariant N of Guesebreok et al.*

  For color invariant N, Geusebroek et al. supplement their three standard assumptions with two additional ones:

- Lambertian reflection ((10.3)),
- Illumination with spatially uniform relative SPD ((10.15)).

Under these assumptions, the reflected color signal $S_{\mathrm{KM}}(\mathbf{x},\lambda)$ is:

$$S_{\mathrm{KM}}(\mathbf{x},\lambda) = \beta(\mathbf{x},\lambda)\, e(\mathbf{x})\, E(\lambda). \tag{10.59}$$

The first spectral derivative is:

$$S_{\mathrm{KM}_{\lambda}}(\mathbf{x},\lambda) = e(\mathbf{x}) \left[\frac{\partial \beta(\mathbf{x},\lambda)}{\partial \lambda} E(\lambda) + \frac{\partial E(\lambda)}{\partial \lambda} \beta(\mathbf{x},\lambda) \right]. \tag{10.60}$$

The ratio of these is:

$$\frac{S_{\mathrm{KM}_{\lambda}}(\mathbf{x},\lambda)}{S_{\mathrm{KM}}(\mathbf{x},\lambda)} = \frac{1}{\beta(\mathbf{x},\lambda)} \frac{\partial \beta(\mathbf{x},\lambda)}{\partial \lambda} + \frac{1}{E(\lambda)} \frac{\partial E(\lambda)}{\partial \lambda}. \tag{10.61}$$

Differentiating this removes the light dependency yielding a result depending only on the surface reflectance:

$$\frac{\partial}{\partial \mathbf{x}} \left\{ \frac{S_{\mathrm{KM}_{\lambda}}(\mathbf{x},\lambda)}{S_{\mathrm{KM}}(\mathbf{x},\lambda)} \right\} = \frac{\partial}{\partial \mathbf{x}} \left\{ \frac{1}{\beta(\mathbf{x},\lambda)} \frac{\partial \beta(\mathbf{x},\lambda)}{\partial \lambda} \right\}. \tag{10.62}$$

Applying the Gaussian color model to this yields the color invariant [42]:

$$X^N(P) = \frac{C^{\mathbb{G}_{\lambda\mathbf{x}}}(P)C^{\mathbb{G}}(P) - C^{\mathbb{G}_{\lambda}}(P)C^{\mathbb{G}_{\mathbf{x}}}(P)}{(C^{\mathbb{G}}(P))^2}. \tag{10.63}$$

The Spatial Derivatives of van de Weijer et al.

van de Weijer et al. propose color invariants based on the four following assumptions [118]:

- Dichromatic reflectance ((10.4))
- Neutral interface reflection ((10.7))
- Illumination with spatially uniform relative SPD ((10.15))
- Known illumination chromaticity (Assumption 13 page 336). If the illumination chromaticity is unknown, they assume it to be white ((10.17))

Under these assumptions, the color components $C^k(P)$, $k = \mathrm{R,G,B}$ are:

$$\begin{aligned}
C^k(P) &= \int_{\text{vis}} m_{\text{bod}}(\theta)\, \beta(\mathbf{x},\lambda)\, e(\mathbf{x})\, E(\lambda) \\
&\quad + m_{\text{int}}(\theta,\alpha)\, F_{\text{int}}(\mathbf{x})\, e(\mathbf{x})\, E(\lambda)\mathrm{d}\lambda \\
&= m_{\text{bod}}(\theta)e(\mathbf{x}) \int_{\text{vis}} \beta(\mathbf{x},\lambda)\, E(\lambda)\mathrm{d}\lambda \\
&\quad + m_{\text{int}}(\theta,\alpha)\, F_{\text{int}}(\mathbf{x})\, e(\mathbf{x}) \int_{\text{vis}} E(\lambda)\mathrm{d}\lambda \\
&= m_{\text{bod}}(\theta)e(\mathbf{x})C^k_{\text{diff}}(P) + n_{\text{int}}(\theta,\alpha,\mathbf{x})\, e(\mathbf{x})\, C^k_{\text{int}} \\
&= e(\mathbf{x})(m_{\text{bod}}(\theta)C^k_{\text{diff}}(P) + n_{\text{int}}(\theta,\alpha,\mathbf{x})C^k_{\text{int}})
\end{aligned} \tag{10.64}$$

where

$$\begin{cases}
C^k_{\text{diff}}(P) = \int_{\text{vis}} \beta(\mathbf{x},\lambda)\, E(\lambda)\, k(\lambda)\mathrm{d}\lambda, \\
C^k_{\text{int}} = \int_{\text{vis}} E(\lambda)\, k(\lambda)\mathrm{d}\lambda, \\
n_{\text{int}}(\theta,\alpha,\mathbf{x}) = m_{\text{int}}(\theta,\alpha)\, F_{\text{int}}(\mathbf{x}).
\end{cases} \tag{10.65}$$

If we take the derivative of this equation with respect to $\mathbf{x}$, we obtain:

$$\begin{aligned}
C^k_{\mathbf{x}}(P) &= e_{\mathbf{x}}(\mathbf{x})(m_{\text{bod}}(\theta)C^k_{\text{diff}}(P) + n_{\text{int}}(\theta,\alpha,\mathbf{x})C^k_{\text{int}}) \\
&\quad + e(\mathbf{x})(m_{\text{diff}_{\mathbf{x}}}(\theta)C^k_{\text{diff}}(P) + m_{\text{bod}}(\theta)C^k_{\text{diff}_{\mathbf{x}}}(P) \\
&\quad + n_{\text{int}_{\mathbf{x}}}(\theta,\alpha,\mathbf{x})C^k_{\text{int}}), \\
&= (e(\mathbf{x})m_{\text{bod}}(\theta))C^k_{\text{diff}_{\mathbf{x}}}(P) \\
&\quad + (e_{\mathbf{x}}(\mathbf{x})m_{\text{bod}}(\theta) + e(\mathbf{x})m_{\text{diff}_{\mathbf{x}}}(\theta))C^k_{\text{diff}}(P) \\
&\quad + (e_{\mathbf{x}}(\mathbf{x})n_{\text{int}}(\theta,\alpha,\mathbf{x}) + e(\mathbf{x})n_{\text{int}_{\mathbf{x}}}(\theta,\alpha,\mathbf{x}))C^k_{\text{int}}.
\end{aligned} \tag{10.66}$$

Consequently, the spatial derivative $\mathbf{C_x}(P) = (C^{\mathrm{R}}_{\mathbf{x}}(P), C^{\mathrm{G}}_{\mathbf{x}}(P), C^{\mathrm{B}}_{\mathbf{x}}(P))^T$ of the color vector $\mathbf{C}(P) = (C^{\mathrm{R}}(P), C^{\mathrm{G}}(P), C^{\mathrm{B}}(P))^T$ is:

$$\begin{aligned}\mathbf{C_x}(P) &= (e(\mathbf{x})m_{bod}(\theta))\mathbf{C}_{diff_{\mathbf{x}}}(P)\\ &+ (e_{\mathbf{x}}(\mathbf{x})m_{bod}(\theta) + e(\mathbf{x})m_{\mathrm{diff}_{\mathbf{x}}}(\theta))\mathbf{C}_{\mathrm{diff}}(P)\\ &+ (e_{\mathbf{x}}(\mathbf{x})n_{int}(\theta,\alpha,\mathbf{x}) + e(\mathbf{x})n_{int_{\mathbf{x}}}(\theta,\alpha,\mathbf{x}))\mathbf{C}_{int}.\end{aligned} \tag{10.67}$$

The spatial derivative of the image color, therefore, can be interpreted as the sum of three vectors. van de Weijer et al. associate a specific underlying physical cause to each:

- $(e(\mathbf{x})m_{\mathrm{bod}}(\theta))\mathbf{C}_{\mathrm{diff}_{\mathbf{x}}}(P)$ correlates with the spatial variation in the surface's body reflection component.
- $(e_{\mathbf{x}}(\mathbf{x})m_{\mathrm{bod}}(\theta) + e(\mathbf{x})m_{\mathrm{diff}_{\mathbf{x}}}(\theta))\mathbf{C}_{\mathrm{diff}}(P)$ correlates with spatial changes in shadowing and shading. The shading term is $e_{\mathbf{x}}(\mathbf{x})m_{\mathrm{bod}}(\theta)$, while the shadowing term is $e(\mathbf{x})m_{\mathrm{diff}_{\mathbf{x}}}(\theta)$. In the absence of specular reflection, the observed color is $\mathbf{C}(P) = e(\mathbf{x})(m_{\mathrm{bod}}(\theta)\mathbf{C}_{\mathrm{diff}}(P)$. In this case, the shadowing/shading derivative $(e_{\mathbf{x}}(\mathbf{x})m_{\mathrm{bod}}(\theta) + e(\mathbf{x})m_{\mathrm{diff}_{\mathbf{x}}}(\theta))\mathbf{C}_{\mathrm{diff}}(P)$ shares the same direction.
- $(e_{\mathbf{x}}(\mathbf{x})n_{\mathrm{int}}(\theta,\alpha,\mathbf{x}) + e(\mathbf{x})n_{\mathrm{int}_{\mathbf{x}}}(\theta,\alpha,\mathbf{x}))\mathbf{C}_{\mathrm{int}}$ correlates with the spatial variation in the specular component of the reflected light arising from two different physical causes. The first term corresponds to a shadow edge superimposed on a specular reflection, and the second to variations in the lighting direction, viewpoint, or surface orientation. By the neutral interface assumption, C_{int} is the same as the color of the illumination; therefore, the authors conclude that this vector is also in the same direction as the illumination color.

Through this analysis, van de Weijer et al. show that given the illumination color $(E^{\mathrm{R}}, E^{\mathrm{G}}, E^{\mathrm{B}})^T = (\int_{\mathrm{vis}} E(\lambda)\, R(\lambda)\mathrm{d}\lambda,\ \int_{\mathrm{vis}} E(\lambda)\, G(\lambda)\mathrm{d}\lambda,\ \int_{\mathrm{vis}} E(\lambda)\, B(\lambda)\mathrm{d}\lambda)^T$, and the color $\mathbf{C}(P)$ of a pixel occurring near an edge, it is possible to determine the direction (in the color space) of two underlying causes of the edge:

- Shadow/shading direction of the Lambertian component: $\mathbf{O}(P) = \frac{(C^R(P), C^G(P), C^B(P))^T}{\sqrt{C^R(P)^2 + C^G(P)^2 + C^B(P)^2}}$,
- Specular direction: $\mathbf{Sp} = \frac{(E^R, E^G, E^B)^T}{\sqrt{(E^R)^2 + (E^G)^2 + (E^B)^2}}$.

van de Weijer et al. call the vector crossproduct of these two vectors $\mathbf{T} = \frac{\mathbf{O} \times \mathbf{Sp}}{|\mathbf{O} \times \mathbf{Sp}|}$ the hue direction, and argue that this direction is not inevitably equal to the direction of body reflectance change but at least the variations along this direction are due only to body reflectance. Klinker et al. [61] have previously defined these directions for use in image segmentation.

- *Shadow/shading invariance and quasi-invariance [116]*

Since image edges created by shadow/shading have the same direction in color space as $\mathbf{O}$, projecting the derivative $\mathbf{C_x}(P)$ on $\mathbf{O}(P)$, namely, $(\mathbf{C_x}(P).\mathbf{O}(P))\mathbf{O}(P)$

provides the component due to shadow/shading. If we subtract this component from the derivative $\mathbf{C_x}(P)$ itself, we obtain a vector that is independent of shadow/shading. Hence, van de Weijer et al. define the invariant

$$\mathbf{X}^{oq}(P) = \mathbf{C_x}(P) - (\mathbf{C_x}(P).\mathbf{O}(P))\mathbf{O}(P), \tag{10.68}$$

where the symbol "." is the dot product. While being insensitive to shadow/shading, this color feature contains information about reflectance variations across the surface and about specular reflection. van de Weijer et al. show that this feature can be obtained by applying spatial derivatives of color components after transformation to a spherical color space $r\theta\varphi$ in which the direction r is aligned with the shadow/shading direction $\mathbf{O}$ [116].

To provide invariance to intensity and direction as well, they divide the previous color invariant by the norm of the color vector [118]:

$$\mathbf{X}^{of}(P) = \frac{\mathbf{X}^{oq}(P)}{|\mathbf{C}(P)|}. \tag{10.69}$$

The color invariant $\mathbf{X}^{of}(P)$ is called the "shadow/shading full invariant" while $\mathbf{X}^{oq}(P)$ is called the "shadow/shading quasi-invariant" [118]. They show that the "quasi-invariants" are more stable than the "full invariants" in the presence of noise. In general, they are more applicable to keypoint detection than to local keypoint description.

- *Highlight invariance [116]*

A similar approach leads to a highlight invariant. van de Weijer et al. project the derivative of the color vector onto the specular direction **Sp** and then subtract the result from this derivative to obtain the highlight-invariant color feature $\mathbf{X}^{sq}(P)$ [116]:

$$\mathbf{X}^{sq}(P) = \mathbf{C_x}(P) - (\mathbf{C_x}(P).\mathbf{Sp}(P))\mathbf{Sp}(P). \tag{10.70}$$

This feature preserves information related to body reflectance and shadow/shading, while being insensitive to specular reflection. They show that it can be obtained by applying spatial derivatives in a color opponent space $o_1o_2o_3$ [116]. For transformation to the opponent space, the illuminant color is required.

- *Highlight and shadow/shading invariance [116]*

To create a color feature $\mathbf{X}^{soq}(P)$ that is invariant to highlights and shadow/shading, van de Weijer et al. project the derivative of the color vector onto the hue direction $\mathbf{T}(P)$ [116]:

$$\mathbf{X}^{soq}(P) = (\mathbf{C_x}(P).\mathbf{T}(P))\mathbf{T}(P). \tag{10.71}$$

This color invariant contains information about the body reflectance, while being invariant to highlights and shadow/shading effects. van de Weijer et al. show that it can be obtained by applying spatial derivatives in the *HSI* color space [116].

Dividing this feature by the saturation of the color provides further invariance to the illumination direction and intensity as well as view direction. In this context, saturation is defined as the norm of the color vector $\mathbf{C}(P)$ after projection on the plane perpendicular to the specular direction $\mathbf{Sp}$:

$$\mathbf{X}^{sof}(P) = \frac{\mathbf{X}^{soq}(P)}{|\mathbf{C}(P) - (\mathbf{C}(P).\mathbf{Sp})\mathbf{Sp}|}. \qquad (10.72)$$

The color invariant $\mathbf{X}^{sof}(P)$ is called the "shadow-shading-specular full invariant" while $\mathbf{X}^{soq}(P)$ is called the "shadow-shading-specular quasi-invariant" [118].

10.2.3 Summary of Invariance Properties of the Classical Color Features

We have presented most of the color invariants used in the context of color-based object recognition and justified the invariance of each. The color invariants, their invariance properties, and their underlying assumptions are tabulated in Tables 10.3 and 10.4. The labels used in these two tables are defined in Tables 10.1 and 10.2. Tables 10.3 and 10.4 can help when choosing a color invariant for use in a particular application. We should emphasize that Table 10.4 cannot be used without Table 10.3, because the invariance properties hold only when the assumptions are satisfied.

The tables do not list invariance with respect to surface orientation because it is equivalent to invariance across viewpoint and lighting direction, since if a color feature is invariant to both viewpoint and light direction, it is necessarily invariant to surface orientation.

10.2.4 Invariance Summary

As we have seen, there are many ways color can be normalized in order to obtain invariant features across various illumination conditions. These normalizations are based on assumptions about color formation in digital images, surface reflection properties, camera sensor sensitivities, and the scene illumination. Color is a very interesting cue for object recognition since it increases the discriminative power of the system and at the same time makes it possible to create local descriptors that are more robust to changes in the lighting and imaging conditions. In the next sections, we describe how color invariants can be exploited in keypoint detection and the keypoint description—two main steps in object recognition.

Table 10.1 Definition of labels for the color-invariant features (CIF) used in Tables 10.3 and 10.4

Label	Description of the color invariant	Equation and/or page number
CIF_1	Component levels ratio proposed by Funt et al. [35]	(10.26) page 339
CIF_2	Color invariants m_1, m_2, m_3 of Gevers et al. [43]	(10.30) page 341
CIF_3	Color invariants l_1, l_2, l_3 of Gevers et al. [43]	(10.33) page 342
CIF_4	Color invariants c_1, c_2, c_3 of Gevers et al. [43]	(10.37) page 343
CIF_5	Entropy minimization by Finlayson et al. [23, 30]	Page 344
CIF_6	Histogram equalization by Finlayson et al. [31]	Page 345
CIF_7	Moment normalization by Lenz et al. [67]	Page 345
CIF_8	Eigenvalue normalization by Healey et al. [49]	Page 345
CIF_9	Mean and standard deviation normalization	(10.45) page 346
CIF_{10}	Color moments invariant to diagonal transform of Mindru et al. [80]	Page 346
CIF_{11}	Color moments invariant to diagonal transform and translation of Mindru et al. [80]	Page 346
CIF_{12}	Color moments invariant to affine transform of Mindru et al. [80]	Page 346
CIF_{13}	Color-invariant feature H of Geusebroek et al. [42]	(10.52) page 348
CIF_{14}	Color-invariant feature C of Geusebroek et al. [42]	(10.55) page 348
CIF_{15}	Color-invariant feature W of Geusebroek et al. [42]	(10.58) page 349
CIF_{16}	Color-invariant feature N of Geusebroek et al. [42]	(10.63) page 350
CIF_{17}	Color feature "quasi-invariant" to shadow/shading by van de Weijer et al. [116]	(10.68) page 352
CIF_{18}	Color feature "full invariant" to shadow/shading by van de Weijer et al. [116]	(10.69) page 352
CIF_{19}	Color feature "quasi-invariant" to highlights by van de Weijer et al. [116]	(10.70) page 352
CIF_{20}	Color feature "quasi-invariant" to shadow/shading and highlights by van de Weijer et al. [116]	(10.71) page 352
CIF_{21}	Color feature "full invariant" to shadow/shading and highlights by van de Weijer et al. [116]	(10.72) page 353

10.3 Color-based Keypoint Detection

The goal of keypoint detection is to identify locations on an object that are likely to be stable across different images of it. Some keypoint detectors find homogeneous regions, some detect corners, some detect sharp edges, but all have the goal that they remain stable from one image to the next. If a keypoint is stable, then the keypoint descriptor describing the local region surrounding the keypoint will also be reliable. However, detected keypoints should not only be stable, they should also identify salient regions so that their keypoint descriptors will be discriminative, and, hence, useful for object recognition. The majority of keypoint detectors have been defined for grayscale images; however, several studies [81, 83, 99] illustrate the advantages of including color in keypoint detection, showing how it can increase both the position robustness of keypoints and the discriminative power of their associated descriptors.

Table 10.2 Definition of assumption labels used in Tables 10.3 and 10.4

Label	Assumption and page
A_1	Kubelka–Munk model page 333
A_2	Shafer's dichromatic model page 333
A_3	Lambertian reflectance page 332
A_4	Neutral interface reflection page 333
A_5	Matte reflectance page 334
A_6	Normalized sensor sensitivities (integrals equal) page 334
A_7	Narrowband sensors page 334
A_8	Transformed from *RGB* space to CIE *XYZ* space page 334
A_9	Planck equation page 335
A_{10}	Locally constant illumination relative SPD page 335
A_{11}	Locally uniform illumination page 335
A_{12}	White illumination page 336
A_{13}	Known illumination color page 336
A_{14}	Diagonal model page 336
A_{15}	Diagonal model and translation page 337
A_{16}	Linear transformation page 337
A_{17}	Affine transformation page 337
A_{18}	Increasing functions page 338

Table 10.3 Assumptions upon which each color-invariant feature depends

Invariant	A_1	A_2	A_3	A_4	A_5	A_6	A_7	A_8	A_9	A_{10}	A_{11}	A_{12}	A_{13}	A_{14}	A_{15}	A_{16}	A_{17}	A_{18}
CIF_1			⊕				⊕				⊕							
CIF_2		⊕			⊕		⊕			⊕								
CIF_3		⊕		⊕		⊕						⊕						
CIF_4		⊕			⊕							⊕						
CIF_5			⊕				⊕		⊕									
CIF_6																		⊕
CIF_7																⊕		
CIF_8																⊕		
CIF_9															⊕			
CIF_{10}														⊕				
CIF_{11}															⊕			
CIF_{12}																	⊕	
CIF_{13}	⊕			⊕				⊕				⊕						
CIF_{14}	⊕				⊕			⊕				⊕						
CIF_{15}	⊕				⊕						⊕	⊕						
CIF_{16}	⊕				⊕			⊕		⊕								
CIF_{17}		⊕		⊕						⊕			⊕					
CIF_{18}		⊕		⊕						⊕			⊕					
CIF_{19}		⊕		⊕						⊕			⊕					
CIF_{20}		⊕		⊕						⊕			⊕					
CIF_{21}		⊕		⊕						⊕			⊕					

Table 10.4 Variations to which each color-invariant feature is invariant

Feature	Light int.	Light color	Light dir.	Viewpoint	Amb. light	Shadow/shading	Highlights
CIF_1	$\oplus$	$\oplus$		$\oplus$			
CIF_2	$\oplus$	$\oplus$	$\oplus$	$\oplus$			
CIF_3	$\oplus$		$\oplus$	$\oplus$			$\oplus$
CIF_4	$\oplus$		$\oplus$	$\oplus$			
CIF_5	$\oplus$	$\oplus$				$\oplus$	
CIF_6	$\oplus$	$\oplus$					
CIF_7	$\oplus$	$\oplus$					
CIF_8	$\oplus$	$\oplus$					
CIF_9	$\oplus$	$\oplus$			$\oplus$		
CIF_{10}	$\oplus$	$\oplus$					
CIF_{11}	$\oplus$	$\oplus$			$\oplus$		
CIF_{12}	$\oplus$	$\oplus$			$\oplus$		
CIF_{13}	$\oplus$		$\oplus$	$\oplus$			$\oplus$
CIF_{14}	$\oplus$		$\oplus$	$\oplus$			
CIF_{15}	$\oplus$						
CIF_{16}	$\oplus$	$\oplus$	$\oplus$	$\oplus$			
CIF_{17}						$\oplus$	
CIF_{18}	$\oplus$		$\oplus$	$\oplus$		$\oplus$	
CIF_{19}							$\oplus$
CIF_{20}						$\oplus$	$\oplus$
CIF_{21}	$\oplus$		$\oplus$	$\oplus$		$\oplus$	$\oplus$

10.3.1 Quality Criteria of Keypoint Detectors

The literature contains many keypoint detectors, and although some are more widely used than others, none provides perfect results in all the cases. The choice of the detector depends on the application, so it is necessary to know the advantages and disadvantages of the various detectors, and have a method for evaluating them. There are many ways to evaluate the quality of a detector [107], of which the three most widely used criteria are: repeatability, discriminative power, and complexity.

A detector is considered to be repeatable if it detects the same points (or regions) in two images of the same scene but acquired under different conditions (e.g., changes in illumination, viewpoint, sensors). Thus, the repeatability measures the robustness of the detector across variations in the acquisition conditions.

The discriminative power of a detector relates to the usefulness of the information found in the neighborhood of the points it detects. Various methods are used in evaluating that information. For example, one is to compare the recognition rates [105] of a given system on a particular database—keeping all the other elements of the system (local descriptor, comparison measure, etc.) fixed—while varying the choice of keypoint detector. The higher the recognition rate, the greater the discriminative power is presumed to be. A second method measures the average entropy (in the information-theoretic sense) of the detected keypoints' neighborhoods, and concludes that the greater the entropy, the more discriminative the detector [99].

A detector's complexity relates to the processing time required per keypoint.

10.3.2 Color Keypoint Detection

Color information can be introduced into classical grayscale detectors such as the Harris detector or the Hessian-based detector. Part of the advantage is that color can provide information about the size of the local region around a detected keypoint.

Color Harris

The color Harris detector generalizes the grayscale version of Harris and Stephens [48], which itself generalizes the Moravec detector [86]. The Moravec detector measures the similarity between a local image patch and neighboring (possibly overlapping) local patches based on their sum-of-squares difference. High similarity to the patches in all directions indicates a homogeneous region. High similarity in a single direction (only the eight cardinal and intercardinal are considered) indicates an edge, while low similarity in all directions indicates a corner. The Moravec detector first computes the maximal similarity at each pixel and then searches for local minima of these maxima throughout the image.

Based on a similar intuition, Harris and Stephens rewrite the similarity between the local patches in terms of the local partial derivatives combined into the following matrix [48]:

$$M = \begin{pmatrix} I_x^2 & I_x I_y \\ I_x I_y & I_y^2 \end{pmatrix}, \tag{10.73}$$

where I_x and I_y are the partial derivatives evaluated with respect to x and y directions in image space. This is a simplified version of the original Harris matrix in that it does not include a scale parameter for either the derivative or the smoothing operators.

The eigenvalues of this matrix represent the strength of the variation in the local neighborhood along the primary and orthogonal directions (the eigenvectors). Both eigenvalues being small indicates a homogeneous region; one small and the other large indicates an edge; both large indicates a corner. For efficiency, Harris and Stephens propose a related measure that can be efficiently evaluated from the determinant and the trace of the matrix, which is simpler than the eigenvalue decomposition. Local maxima represent edges in the image. This detector is widely used, with one of its main advantages being that it is based on first derivatives, thereby making it is less sensitive to noise than detectors based on higher order derivatives.

Montesinos et al. extend the Harris matrix to include color based on simply summing the Harris matrices of the three color channels taken separately [83]:

$$M_{\mathrm{Mont}} = \begin{pmatrix} R_x^2 + G_x^2 + B_x^2 & R_xR_y + G_xG_y + B_xB_y \\ R_xR_y + G_xG_y + B_xB_y & R_y^2 + G_y^2 + B_y^2 \end{pmatrix}, \tag{10.74}$$

where R_x, G_x, B_x and R_y, G_y, B_y are, respectively, the red, green, and blue partial derivatives. The subsequent steps for corner detection are then exactly as for the grayscale Harris detector. This color corner detection has been widely used [37,46,83,84,98,99,105,116]. Gouet et al. compared the color Harris detector to the grayscale version on synthetic images [46] and showed that color improved its repeatability under 2-D rotations in the image plane, as well as under lighting and viewpoint variations.

Montesinos et al. based the color Harris detector on *RGB* space. Sebe et al. [99] tried it in opponent color space $o_1o_2o_3$ and an invariant color space based on the color ratios between neighboring pixels [43] (see (10.30)) along with "color boosting" (described later in this chapter). The results for real images [76] were:

- Repeatability: Under changes in illumination or the effects of JPEG compression, the grayscale Harris detector outperformed the color versions. For the other variations such as blurring, rotation, scaling, or change of viewpoint the performance was more or less similar.
- Discriminative power: Color significantly improved the discriminative power of keypoint detection, however, as mentioned above, the evaluation of the discriminative power of the keypoint detector depends on the keypoint descriptor involved. Sebe et al. tested two descriptors, one based on grayscale information and the other on color. From these tests, they show that color detection improves the discriminative power. However, these tests also show that, on average, the more repeatable a detector, the less discriminating it is likely to be. There is a trade-off between repeatability and discriminative power.
- Complexity: Color increases the computation required for keypoint detection. However, Sebe et al. show that, with color, fewer keypoints are needed to get similar results in terms of discriminative power. Thus, the increased complexity of keypoint detection may be counterbalanced by the corresponding reduction in the number of keypoints to be considered in subsequent steps.

Color Hessian

One alternative to the Harris detector is the Hessian-based detector [9]. The Hessian matrix H comes from the Taylor expansion of the image function I:

$$H = \begin{pmatrix} I_{xx} & I_{xy} \\ I_{xy} & I_{yy} \end{pmatrix}, \tag{10.75}$$

where I_{xx}, I_{xy}, and I_{yy} are the second-order partial derivatives and encode local shape information. The trace of the Hessian matrix is the Laplacian. Once again, this version does not include the derivative and smoothing scales of the original. Beaudet showed that keypoints based on the locations of the local maxima of the determinant of the Hessian matrix (i.e., $I_{xx} \times I_{yy} - I_{xy}^2$) are rotation invariant [9].

There have been several interesting color extensions to the Hessian approach. For example, Ming et al. [81], represents the second derivatives of a color image as chromaticity-weighted sums of the second derivatives of the color channels:

$$H_{\text{Ming}} = \begin{pmatrix} C^r R_{xx} + C^g G_{xx} + C^b B_{xx} & C^r R_{xy} + C^g G_{xy} + C^b B_{xy} \\ C^r R_{xy} + C^g G_{xy} + C^b B_{xy} & C^r R_{yy} + C^g G_{yy} + C^b B_{yy} \end{pmatrix}, \quad (10.76)$$

where R_{xx}, R_{xy}, etc. are the second-order partial derivatives of the respective color channels, and C^r, C^g, and C^b are the corresponding chromaticities ((10.38)).

The second color extension of the Hessian matrix involves the use of quaternions to represent colors [101]. Quaternion algebra is described in Chap. 6 of this book. Shi's goal was not keypoint detection but rather to find vessel-like structures (e.g., blood vessels) in color images. In earlier work, Frangi et al. [34] showed that vessels could be located via an analysis of the eigenvalues of the Hessian matrix. Shi et al. extend that approach to color images using quaternions. The eigenvalues of the quaternion Hessian matrix are found via quaternion singular value decomposition (QSVD). The results show that using color in this way improves the overall accuracy of vessel detection.

The third color extension is to work directly with a Hessian matrix containing vectors, not scalars. In particular, Vigo et al. [110] replace the scalars I_{xx}, I_{yy}, and I_{xy} by vectors $(R_{xx}G_{xx}B_{xx})^T$, $(R_{yy}G_{yy}B_{yy})^T$, and $(R_{xy}G_{xy}B_{xy})^T$, respectively, and generalize Beaudet's [9] use of the determinant to the norm of the vectorial determinant $\text{Det}_{\text{coul}} = ||(R_{xx}R_{yy} \;\; G_{xx}G_{yy} \;\; B_{xx}B_{yy})^T - (R_{xx}^2 G_{xx}^2 B_{xx}^2)^T||$ as the criterion for keypoint detection.

Color Harris-Laplace and Hessian-Laplace

The Harris and Hessian detectors extract keypoints in the images but do not provide any information about the size of the surrounding neighborhood on which to base the descriptors. Since the detectors are quite sensitive to the scale of the derivatives involved, Mikolajczyk et al. propose grayscale detectors called Harris–Laplace and Hessian–Laplace, which automatically evaluate the scale of the detected keypoints [76]. Around each (Harris or Hessian) keypoint, the image is convolved with Laplacian of Gaussian filters (LoG) [69] of increasing scale (i.e., standard deviation of the Gaussian). The local maximum of the resulting function of scale is taken as the scale of the detected keypoint.

Stoettinger et al. extend this approach [105] to color to some extent by using PCA (principal components analysis) to transform the color information to grayscale. PCA is applied to the color-image data and the resulting first component is used as the grayscale image. The authors show that this 1-D projection of the color leads to more stable and discriminative keypoints than regular grayscale.

Ming et al. propose a different color extension of the Laplace methods (i.e., trace of Hessian) in which they replace the grayscale Laplacian by the trace of the color Hessian matrix (cf. (10.76)):

$$\text{Laplace}_{\text{Ming}} = C^r R_{xx} + C^g G_{xx} + C^b B_{xx} + C^r R_{yy} + C^g G_{yy} + C^b B_{yy}. \tag{10.77}$$

The repeatability and discriminative power of this color scale selection are not assessed in their paper.

10.3.3 Color Key-region Detection

Key-region detection is like keypoint detection except the idea is to locate sets of connected pixels that share the same properties (color and texture). For object recognition, the two main requirements of such key regions are the stability of their borders across different images and the discriminative information that can be extracted from the region. As with keypoint detection, numerous grayscale key-region detectors have been extended in various ways to include color.

Most Stable Color Regions

The maximally stable extremal region (MSER) method [75] is one method for finding key regions in grayscale images. MSER is based on extracting sets of connected pixels (regions) characterized by gray values greater than a threshold t. The regions will vary with t, and the MSER method finds the regions that are the most stable across variations in t.

Forssén [32] extended this approach to color images using color differences in place of grayscale differences. First, the Chi^2 color distance is calculated between all the neighboring pairs of pixels (P_i, P_j):

$$\text{dist}_{Chi^2}(P_i, P_j) = \frac{(C^R(P_i) - C^R(P_j))^2}{C^R(P_i) + C^R(P_j)} + \frac{(C^G(P_i) - C^G(P_j))^2}{C^G(P_i) + C^G(P_j)} + \frac{(C^B(P_i) - C^B(P_j))^2}{C^B(P_i) + C^B(P_j)}. \tag{10.78}$$

Second, regions of connected pixels are established such that no pair of neighboring pixels has a color distance greater than t. Finally, as in grayscale MSER, the value of

t is varied in order to find the regions that are the most stable relative to changing t. The authors show that the borders of the regions detected by this approach are stable across the imaging conditions.

Hierarchical Color Segmentation

Vázquez-Martin et al. use a color segmentation algorithm for region detection [109] involving two main steps. The first step involves a hierarchical (pyramid) algorithm that extracts homogeneously colored regions. This step generates a rough, over-segmented image. The second step merges the initial regions based on three factors:

- The Euclidean distance between the mean colors of the two regions calculated in L*a*b* color space [122]. If the distance is small, the regions are candidates for merging.
- The number of edge pixels found by the Canny edge detector [13] along the boundary between two regions. If few edge pixels are found, the regions are candidates for merging.
- The stereo disparity between the two regions. Vázquez-Martin et al. used stereo imagery and tended to merge regions whenever the disparity was small, but if only monocular data is available then merging is based on the first two criteria alone.

They define a region dissimilarity/similarity measure based on these three factors and show that the merged regions extracted are very stable with respect to the imaging conditions.

Forssén et al. also use a hierarchical approach to detect regions in color images [33]. They create a pyramid from the image by at each level regrouping the sets of connected pixels that have similar mean values at the next lower level. Each extracted region is then characterized by an ellipse whose parameters (major axis, minor axis, and orientation) are determined from the inertia matrix of the region.

10.3.4 HVS-based Detection

The human visual system (HVS) extracts a large amount of information from an image within a very short time in part because attention appears to be attracted first to the most informative regions. Various attempts have been made to try and model this behavior by automatically detecting visually salient regions in images [19, 57] with the goal of improving the performance of object-recognition systems [94, 112]. Most visual saliency maps have been based on color [59], among other cues. Visual saliency can be integrated into a system in various ways in order to aid in keypoint detection.

Gao et al. and Walther et al. [38, 112] start from the point of view that if the HVS focusses on visually salient regions, keypoints should be extracted primarily from them too. Consequently, they apply Itti's model [57] to locate the salient regions and then apply SIFT [72] keypoint detection, retaining only the keypoints found within the salient regions. The standard SIFT detector relies only grayscale data, but Itti's visual saliency map includes color information. Retaining as few as 8% of the initial SIFT keypoints (i.e., just those in salient regions) has been found to be as effective for recognition as using all SIFT keypoints [38]. Likewise, Walther et al. [112] compared using a random selection of SIFT keypoints to using the same number obtained from visually salient regions, and found that the performance increased from 12% to 49% for the salient set over the random set. In a similar vein, Marques et al. [74] determine saliency by combining two visual saliency models. In their approach, the centers of salient regions are found by Itti's model followed by a growing-region step that starts from those centers and then relies on Stentiford's model [104] to define the region borders.

Wurz et al. [121] model the HVS for color corner detection. They use an opponent color space [116] with achromatic, red–green, and blue–yellow axes to which they apply Gabor filters at different scales and orientations in order to simulate the simple and complex cells of the primary visual cortex. The outputs of these filters are differentiated perpendicular to the filter's orientation. The first derivative is used for modeling the simple end-stopped cells, and the second derivative for the double end-stopped ones. Since the outputs of end-stopped cells are known to be high at corners [52], Wurz et al. detect corners based on the local maxima of these derivatives.

In a different saliency approach, Heidemann [51] uses color symmetry centers as keypoints based on the results of Locher et al. [70] showing that symmetry catches the eye. He extends the grayscale symmetry center detector of Reisfeld et al. [92] adding color to detect not only the intra-component symmetries but also the inter-component ones as well. The symmetry around a pixel is then computed as the sum of the symmetry values obtained for all the component combinations at that pixel. Heidemann shows that the keypoints detected by this method are highly robust to changes in the illumination.

10.3.5 Learning for Detection

The saliency methods in the previous section are based on bottom-up processing mechanisms that mimic unconscious visual attention in the HVS. At other times, attention is conscious and lead by the task being done. For example, when driving a car, attention is directed more toward the road signs than when walking, even though the scene may be identical. In the context of object detection, it is helpful to detect quickly the regions that might help identify the desired object. In order to simulate the top-down aspects of conscious visual attention, several authors add a learning step to improve keypoint detection.

For example, van de Weijer et al. [108, 117] find that applying some information-theoretic, color-based preprocessing increases the discriminative power of the points extracted by classical detectors. Information theory says that the more seldom an event, the higher its information content. To find the more distinctive points, they evaluate the gradients for each component (C^{R}, C^{G}, and C^{B}) independently over a large database of images and plot the corresponding points in 3-D. They notice that the iso-probability (iso-frequency) surfaces constitute ellipsoids and not spheres. This means that the magnitude of the gradient at a pixel does not accurately reflect the probability of its occurrence and, hence, its discriminative power as a keypoint either. Its true discriminative power relates to the shape of the ellipsoid. In order to directly relate the gradient magnitude to its discriminative power, van de Weijer et al. apply a linear transformation to the C^{R}, C^{G}, and C^{B} components transforming the ellipsoids into spheres centered on $(0,0,0)$. After this transformation, the gradient magnitude at a pixel will be directly related to its frequency of occurrence and therefore to its discriminative power. For example, this preprocessing benefits the color Harris detector since it extracts keypoints characterized by high gradient magnitude. With the preprocessing it finds corners with higher discriminative power. Overall, testing shows that this preprocessing called "color boosting" significantly increases recognition rates.

Other learning approaches try to classify pixels as "object" versus "background." Given such a classification, the image content is then characterized based on only the object pixels. Excluding the background pixels can improve the speed of object recognition without decreasing its accuracy. For the learning phase, Moosmann et al. [85] use hue saturation luminance (HSL) color space and for each HSL component compute the wavelet transform of subwindows of random position and size across all the images in the training set of annotated (object versus background) images. Each subwindow is rescaled to 16×16 so that its descriptor always has dimension 768 ($16 \times 16 \times 3$). After training, a decision tree is used to efficiently classify each subwindow in a given query image as to whether it belongs to an object or the background.

A second object-background classification approach is that of Alexe et al. [2] who base their method on learning the relative importance of 4 cues. The first cue relates to visual saliency and involves detecting subwindows that are likely to appear less frequently. They analyze the spectral residual (a measure of visual saliency [53]) of the Fourier transform along each color component. The second cue is a measure of color contrast based on comparing the distance between the color histogram of a subwindow to that of its neighboring subwindows. The third cue is the density of edges found by the Canny edge detector along the subwindow's border. The final cue measures the degree to which superpixels straddle the subwindow boundary. Superpixels are regions of relatively uniform color or texture [22]. If there are a lot of superpixel regions at the subwindow boundary then it is likely that the subwindow covers only a portion of the object. The relative importance of these four cues is learned in a Bayesian framework via training on an annotated database of images. Alexe et al. [2] report this method to be very effective in determining subwindows belonging to objects.

10.4 Local Color Descriptors

Once the keypoints (or key regions) have been identified, local descriptors need to be extracted from their surrounding neighborhoods. The SIFT descriptor is perhaps the most widely used [72] local descriptor. It is based on local gradient orientation histograms at a given scale. As such, SIFT mainly encodes shape information. A recent study [96] compared the performance of SIFT descriptors to that of local color descriptors such as color histograms or color moments and found the recognition rates provided by the SIFT to be much higher than those for the color descriptors. This result is a bit surprising since the test images contained many objects for which color would appear to be discriminative. Clearly, this means that if color is going to be of use, it will be best to combine both color and shape information in the same descriptor. We distinguish four different approaches to combining them. The first concatenates the results from two descriptors, one for shape and the other for color, via simple weighted linear combination. The second sequentially evaluates a shape descriptor followed by a color descriptor. The third extracts in parallel, and generally independently, both a shape descriptor and a color descriptor and the two are nonlinearly fused in order to get a color-shape similarity. The final approach extracts spatio-chromatic descriptors that represent both shape and color information in a single descriptor.

10.4.1 Descriptor Concatenation

In terms of the concatenation of shape and color descriptors, Quelhas et al. [90] use the SIFT descriptor to represent the shape information. The SIFT descriptor they use is initially 128-dimensional, which they reduce to 44-dimensional via PCA. The color information is summarized by 6 values comprising the mean and standard deviation of each of the L*u*v* [122] color components. Based on tests on the database of Vogel et al. [111], the relative weights given to the shape and color cues are 0.8 and 0.2, respectively. They assess the effectiveness of the shape–color combination for scene classification rather than for object recognition. For scene classification, color information will be less discriminative for some classes than others. Indeed, their classification results show that including color increases the classification rate from 67.6% to 76.5% for the class sky/clouds but leads to no increase on some other classes. Overall, the scene classification rate increases from 63.2% to 66.7% when color accompanies the grayscale SIFT.

van de Weijer et al. [114] propose concatenating the SIFT descriptor with a local color histogram based on a choice of one of the following color spaces:

- Normalized components C^r, C^g (2-D) (see page 343)
- Hue (1-D)
- Opponent angle (1-D) [116] based on the ratio of derivatives in opponent color space $o_1o_2o_3$ (see page 352)

- Spherical angle (1-D) [116] based on the ratio of derivatives in the spherical color space $r\theta\varphi$ (see page 352)
- Finlayson's [28] normalized components (see page 343)

Following [44], they propose weighting each occurrence of these components according to their certainty in order to get robust histograms. The certainty is estimated by the sum-of-square-roots method that is related to the component standard deviation derived from homogeneously colored surface patches in an image under controlled imaging conditions. They keep the 128 dimensions of SIFT and quantize the color histograms into 37 bins for the $1-D$ cases and 121 bins for the $2-D$ ones. The SIFT descriptor is combined with each of the histogram types in term, which leads to five possible combinations to test. Based on weightings of 1 for shape and 0.6 for color, performance is evaluated under (i) varying light but constant viewpoint; (ii) constant light but varying viewpoint, and (iii) varying light and viewpoint. For case (i), shape alone provides better results than color alone, and adding color to shape does not help; and vice versa in case (ii) where color alone provides better results than shape alone, and adding shape to color does not help. For case (iii), the best performance is obtained when color and shape are combined.

Dahl et al. [18] summarize a region's color information simply by its mean RGB after equalizing the $1-D$ histograms to obtain invariant components [31] (see page 345 for justification). Their local descriptor is then the equally weighted combination of the RGB mean and the SIFT descriptor reduced via PCA to 12 dimensions. Testing on the database of Stewnius and Nistr [91] shows that adding this simple bit of color information raises the recognition rate to 89.9% from 83.9% for SIFT alone.

Like van de Weijer et al., Zhang et al. combine SIFT descriptors with local color histograms [124]. They work with histograms of the following types:

- Normalized components C^r, C^g quantized to 9 levels each
- Hue quantized to 36 levels
- Components l_1, l_2 [43] quantized to 9 levels each (see page 341)

To add further spatial information to the color description, they divide the region around a keypoint into three concentric rings and compute each of the three types of histograms separately on each of the three rings. The color descriptor is then a concatenation of three color histograms with a resulting dimension of 594 ($3 \times (9 \times 9 + 36 + 9 \times 9)$). PCA reduces this dimension to 60. The final shape–color descriptor becomes 188-dimensional when this color descriptor is combined with the 128-dimensional SIFT descriptor. The relative weightings are 1 for shape and 0.2 for color. Tests on the INRIA database [56], showed that the addition of color increased the number of correct matches between two images from 79 with SIFT alone to 115 with the shape–color descriptor.

10.4.2 Sequential Combination

The concatenation methods discussed above use color and shape information extracted separately from each local image region and combine the results. An alternative approach is to use color to locate regions having similar color distributions and then, as a second step, compare the shape information within these regions to confirm a match. We will examine three of these sequential methods.

The first, proposed by Khan et al. [60, 110] characterizes each region by either a histogram of hues or a histogram of "color names," where the color names are based on ranges of color space that are previously learned [115]. Object recognition is then based on a bag-of-words approach using two dictionaries, one for color and one for shape. The idea is to weight the probability that each shape word belongs to the target object by a value determined by each color word. An interesting aspect of this approach is that if the color is not discriminative for a particular target object, the "color probabilities" will be uniformly spread over all the shape words, with the result being that only shape will be used in recognizing the object. As such, this approach addresses the issue of how to determine the relative importance of color versus shape.

In a similar vein, Elsayad et al. [20] and Chen et al. [15] both use bags of visual shape words that are weighted by values coming from the color information. Although there are some differences in the details, they both use a 5-D space (3 color components and 2 spatial coordinates) and determine the parameters of the Gaussian mixture model that best fits the distribution of the points in this space. The probabilities deduced from these Gaussians are the weights used in the bag of visual shape words. Indeed, the authors claim that accounting the number of occurrences of each visual shape words is not enough to discriminate some classes and rather propose to weight the visual words according to the spatial distribution of the colors in the images.

Another sequential combination method was developed by Farag et al. [21] for use in the context of recognizing road signs. It involves a Bayesian classifier based on hue to find pixels that are likely to belong to a specific road sign, followed by a SIFT-based shape test. In other words, the candidate locations are found based on color and then only these locations are considered further in the second step, which compares SIFT descriptors of the query road sign with those from the candidate location in order to make the final decision. Similarly, Wu et al. [119] use color to direct attention to regions for shape matching. The color descriptors are the mean and standard deviation of the CIE a*b* components of subwindows, while SIFT descriptors are once again used for shape.

The sequential methods discussed above used color then shape; however, the reverse order can also be effective. Ancuti et al. [3] observed that some matches are missed when working with shape alone because sometimes the matching criteria for grayscale SIFT are too strict and can be relaxed when color is considered. They weaken the criteria on SIFT matches and then filter out potentially incorrect matches based on a match of color co-occurrence histograms. The color co-occurrence histogram (Chang [14]) counts the number of times each possible color pair occurs. In contrast to Ancuti et al., Goedeme et al. [45] compare color moments [79] of the SIFT-matching regions to reduce the number of false matches.

10.4.3 Parallel Comparison

The parallel comparison methods simultaneously exploit shape and color information in a manner that differs somewhat from the concatenation methods described earlier. For example, Hegazy et al. [50], like van de Weijer et al. [114], combine the standard SIFT shape descriptor with a color descriptor based on the histogram of opponent angles [116]; however, rather than combining the two descriptors linearly, they evaluate the probability of an object of a given query class being in the image by considering the probability based on the SIFT descriptor and on the corresponding probability based on the color descriptor. The AdaBoost classifier is used to combine the probabilities. Tests on the Caltech [6] and Graz02 [7] databases show that this combination always provides better classification results than when either descriptor is used alone.

For object tracking, Wu et al. [120] combine the SIFT descriptor with the 216-dimensional (6^3) HSV color histograms. A particle filter [4] is used in which the particle weights are iteratively updated by alternatively considering color and shape. The test results show that the parallel use of color and shape improves the performance of the system with respect to the case when only color is used. Unfortunately, Wu does not include results using shape alone.

Hu et al. [54] use auto-correlograms as color descriptors. An auto-correlogram is an array with axes of color C and distance d [55] in which the cell at coordinates (i, j) represents the number of times that a pair of pixels of identical color C_i are found at a distance d_j from one another. In other words, this descriptor represents the distribution of the pairs of pixels with the same color with respect to the distance in the image. Hu et al. evaluate the similarity between two objects as the ratio between a SIFT-based similarity measure and an auto-correlogram-based distance (i.e., dissimilarity) measure.

Schugerl et al. [97] combine the SIFT descriptor with a color descriptor based on the MPEG-7 [102] compression standard for object re-detection in video, which involves identifying the occurrence of specific objects in video. Each region detected initially by the SIFT detector is divided into 64 blocks organized on an 8×8 grid, and each block is then characterized by its mean value in YCrCb color space [122]. A discrete cosine transform (DCT) is applied to this 8×8 image of mean values, separately for each color component. The low frequency terms are then used to describe the region. In other words, the color descriptor is a set of DCT coefficients representing each color component Y, Cb, and Cr. Given a target (or targets) to be found in the video, the SIFT and color descriptors of each detected region are independently compared to those of the target(s). Each match votes for a target object, and a given video frame is then associated with the target objects that have obtained a significant number of votes.

Finally, Nilsback et al. [87] combine color and shape to recognize flowers. The shape descriptors are SIFT ones along with a histogram of gradients, while the color descriptor is the HSV histogram. They use a bag-of-words approach to characterize the image content using shape-based and color-based dictionaries. For the classification, they apply a multiple kernel SVM (support vector machine) classifier, where two of the kernels correspond to the two dictionaries. The final kernel is a linear combination of these kernels, with the relative weighting determined experimentally. In the context of flower classification, Nilsback et al. show that this combination of shape and color cues significantly improves the results provided by either cue in isolation.

10.4.4 Spatio-chromatic Descriptors

A spatio-chromatic descriptor inextricably encodes the information concerning both color and shape. The most widely used spatio-chromatic descriptor is color SIFT, which is simply grayscale SIFT applied to each color component independently. There have been several versions of color SIFT differing mainly in the choice of the color space or invariant features forming the color components. For example, Bosch et al. [11] apply SIFT to the components of HSV color space and show that the results are better than for grayscale SIFT. Abdel-Hakim et al. [1] and Burghouts et al. [12] use the Gaussian color space (see page 346) invariants of Geusebroek et al. [42]. The main potential drawback of color SIFT is that its already high dimension of 128 becomes multiplied by the number of color components. Since many of the classification methods use the bag-of-words approach, the extra dimensionality matters since it can significantly increase the time required to create the dictionary of words.

Based on tests on a variety of grayscale, color, shape, and spatio-chromatic descriptors, Burghouts et al. conclude that SIFT applied to Geusebroek's Gaussian-based color invariant C (see page 348 (10.55)) provides the best results. Van de Sande et al. [96] also tested several descriptors and found that SIFT in opponent color space (see page 352) and SIFT applied to color invariant C provide the best results on average. Chu et al. [16] show that grayscale SURF [8] can also be applied independently to the color components to good effect. In their case, they show that SURF applied to the color-invariant feature C (see page 348) provides slightly better results than applied to the opponent color space.

There are other non-SIFT approaches to spatio-chromatic descriptors as well. For example, Luke et al. [73] use a SIFT-like method to encode the color information instead of geometric information. SIFT involves a concatenation of gradient orientation histograms weighted by gradient magnitudes, so for color they use a concatenation of local hue histograms weighted by saturation, the intuition being that the hue of a pixel is all the more significant when its saturation is high. These spatio-chromatic descriptors provide better results than the grayscale SIFT on the tested database.

Geusebroek [40] proposes spatio-chromatic descriptors based on spatial derivatives in the Gaussian color space [41]. The derivatives are histogrammed and then the histograms are characterized by 12 Weibull parameters. The advantage of this approach is that it is not as sensitive as the SIFT descriptors to 3-D rotations of the object since the resulting descriptors have no dependence on the gradient orientations in image space.

Qiu [89] combines color and shape information by vectorizing the image data in a special way, starting with color being represented in YCrCb color space. To build the descriptor for a detected image region, the region is initially downsampled (with averaging) to 4×4, and from this the Y-component is extracted into a 16-element vector. The 4×4 representation is then further downsampled to 2×2 from which the chromatic components Cr and Cb are formed into two 4-dimensional vectors. Finally, clustering is applied to the chromatic and achromatic descriptor spaces to build a dictionary (one for each) so that each region can then be described using a chromatic and achromatic bag of words representation. Qiu shows that the resulting descriptors provide better recognition results than auto-correlograms.

As a final example of a spatio-chromatic descriptor, Song et al. [103] propose a local descriptor that characterizes how the colors are organized spatially. For a given pixel, its 2 spatial coordinates and 2 of its 3 RGB components (the three pairs RG, RB, and GB are independently considered) are taken together so that each pixel is characterized by 2 points, one in a 2-D spatial space and one in a 2-D color space. The idea is to evaluate the affine transform that best projects the points in the spatial domain to the corresponding points in the color domain and to apply this affine transform to the region's corners (see Fig. 10.2). The authors show that the resulting coordinates in the color space are related both to the region's colors and to their spatial organization. Color invariance is obtained by considering the rank measures $\mathrm{Rc}^k(P)$ (see page 345) of the pixels instead of their original color components. The final descriptor encodes only the positions of the three defining corners of the local keypoint region that result after the affine transformation. The tests show that this compact descriptor is less sensitive than the SIFT descriptor to 3-D rotations of the query object.

10.5 Conclusion

Color can be a very important cue for object recognition. The *RGB* values in a color image, however, are only indirectly related to the surface "color" of an object since they depend not only on the object's surface reflectance but also on such factors as the spectrum of the incident illumination, shininess of the surface, and view direction. As a result, there has been a great deal of research into color invariants—features that encode color information that are invariant to these confounding factors. Different object-recognition situations call for different classes of invariants depending on the particular surface reflectance and lighting conditions likely to be encountered. It is important to choose the optimal color

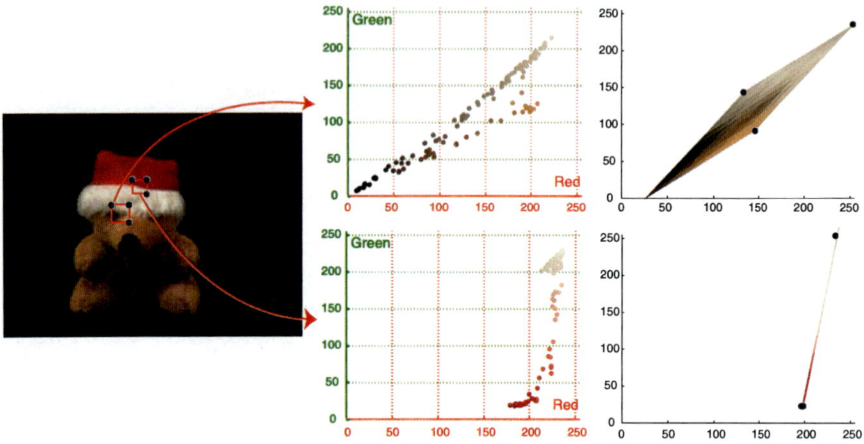

Fig. 10.2 Transformation from image space to *RG* color space [103]. *Left*: Example of two rectangular regions around two detected keypoints. For each region, the best affine transform from the points (pixels) within it (*left*) to their corresponding points (pixel colors) in the *RG* space (*center*) is found. Applying this transform to the corners of the rectangles provides discriminative positions (*right*) that depend both on the colors present in the considered region and their relative spatial positions in this region. The positions of just three of the corners (*black points*) are sufficient to summarize the result

invariant for each particular situation because there is a trade-off between invariance and discriminative power. All unnecessary invariance is likely to decrease the discriminative power of the system. Section 10.2.1 describes the assumptions that underly the various color invariants, followed by the invariants themselves and a description of their invariance properties.

Object recognition involves image comparison based on two important prior steps: (i) Keypoint detection, and (ii) description of the local regions centered at these keypoints. Using color invariants, color information can be introduced into both these steps, and generally color improves the recognition rate of most object-recognition systems. In particular, including color in keypoint detection increases the likelihood that the surrounding region contains useful information, so descriptors built around these keypoints tend to be more discriminative. Similarly, color keypoint detection is more robust to the illumination when color-invariant features are used in place of standard grayscale features. Also local region descriptors based on color invariants more richly characterize regions, and are more stable relative to the imaging conditions than their grayscale counterparts. Color information encoded in the form of color invariants has proven very valuable for object recognition.

References

1. Abdel-Hakim A, Farag A (2006) Csift: A sift descriptor with color invariant characteristics. In: 2006 IEEE computer society conference on computer vision and pattern recognition, New York, USA, vol 2, pp 1978–1983
2. Alexe B, Deselaers T, Ferrari V (2010) What is an object? IEEE computer society conference on computer vision and pattern recognition 4:73–80
3. Ancuti C, Bekaert P (2007) Sift-cch: Increasing the sift distinctness by color co-occurrence histograms. In: Proceedings of the 5th international symposium on image and signal processing and analysis, Istambul, Turkey, pp 130–135
4. Arulampalam M, Maskell S, Gordon N, Clapp T (2002) A tutorial on particle filters for online nonlinear/non-gaussian bayesian tracking. IEEE Trans Signal Process 50(2):174–188
5. Barnard K, Martin L, Coath A, Funt B (2002) A comparison of computational color constancy algorithms. II. Experiments with image data. IEEE Trans Image Process 11(9):985–996
6. Base caltech. URL http://www.vision.caltech.edu/html-files/archive.html
7. Base graz02. URL http://www.emt.tugraz.at/~pinz/data/GRAZ_02/
8. Bay H, Ess A, Tuytelaars T, Gool LV (2008) Surf: Speeded up robust features. Comput Vis Image Understand 110:346–359
9. Beaudet PR (1978) Rotationally invariant image operators. In: Proceedings of the International Conference on Pattern Recognition, Kyoto, Japan, pp 579–583
10. Beckmann P, Spizzichino A (1987) The scattering of electromagnetic waves from rough surfaces, 2nd edn. Artech House Inc, Norwood, USA
11. Bosch A, Zisserman A, Munoz X (2006) Scene classification via plsa. In: Proceedings of the European conference on computer vision, Graz, Austria, pp 517–530
12. Burghouts G, Geusebroek JM (2009) Performance evaluation of local colour invariants. Comput Vis Image Understand 113(1):48–62
13. Canny J (1986) Computational approach to edge detection. IEEE Trans Pattern Anal Mach Intell 8(6):679–698
14. Chang P, Krumm J (1999) Object recognition with color cooccurrence histograms. In: In IEEE conference on computer vision and pattern recognition (CVPR), vol 2, p 504
15. Chen X, Hu X, Shen X (2009) Spatial weighting for bag-of-visual-words and its application in content-based image retrieval. In: Advances in knowledge discovery and data mining, lecture notes in computer science, vol 5476, pp 867–874
16. Chu DM, Smeulders AWM (2010) Color invariant surf in discriminative object tracking. In: ECCV workshop on color and reflectance in imaging and computer vision, Heraklion, Crete, Greece
17. Ciocca G, Marini D, Rizzi A, Schettini R, Zuffi S (2001) On pre-filtering with retinex in color image retrieval. In: Proceedings of the SPIE Conference on Internet Imaging II, San Jos, California, USA, vol 4311, pp 140–147
18. Dahl A, Aanaes H (2008) Effective image database search via dimensionality reduction. IEEE Computer Society Conference on Computer Vision and Pattern Recognition Workshop, Anchorage, Alaska, pp 1–6
19. Dinet E, Kubicki E (2008) A selective attention model for predicting visual attractors. In: Proceedings of IEEE international conference on acoustics, speech, and signal processing, États-Unis, pp 697–700
20. Elsayad I, Martinet J, Urruty T, Djeraba C (2010) A new spatial weighting scheme for bag-of-visual-words. In: Proceedings of the international workshop on content-based multimedia indexing (CBMI 2010), Grenoble, France, pp 1 –6
21. Farag A, Abdel-Hakim A (2004) Detection, categorization and recognition of road signs for autonomous navigation. In: Proceedings of Advanced Concepts in Intelligent Vision Systems, Brussel, Belgium, pp 125–130
22. Felzenszwalb PF, Huttenlocher DP (2004) Efficient graph-based image segmentation. Int J Comput Vis 59:167–181

23. Finlayson G, Hordley S (2001) Colour constancy at a pixel. J Opt Soc Am 18(2):253–264
24. Finlayson GD, Trezzi E (2004) Shades of gray and colour constancy. In: Proceeding color imaging conference, Scottsdale, Arizona, pp 37–41
25. Finlayson G, Drew M, Funt B (1994) Color constancy : generalized diagonal transforms suffice. J Opt Soc Am 11(A):3011–3020
26. Finlayson GD, Drew MS, Funt BV (1994) Spectral sharpening : sensor transformations for improved color constancy. J Opt Soc Am 11(A):1553–1563
27. Finlayson G, Chatterjee S, Funt B (1995) Color angle invariants for object recognition. In: Proceedings of the 3rd IS&T/SID color imaging conference, Scottsdale, Arizona, pp 44–47
28. Finlayson G, Schiele B, Crowley J (1998) Comprehensive colour image normalization. Lecture notes in computer science 1406:475–490. URL citeseer.nj.nec.com/finlayson98comprehensive.html
29. Finlayson G, Hordley S, Hubel P (2001) Color by correlation: a simple, unifying framework for color constancy. IEEE Trans Pattern Anal Mach Intell 23(11):1209–1221
30. Finlayson G, Drew M, Lu C (2004) Intrinsic images by entropy minimization. In: Proceedings of the European conference on computer vision, Prague, Czech Republic, pp 582–595
31. Finlayson G, Hordley S, Schaefer G, Tian GY (2005) Illuminant and device invariant colour using histogram equalisation. Pattern Recogn 38:179–190
32. Forssén PE (2007) Maximally stable colour regions for recognition and matching. In: IEEE conference on computer vision and pattern recognition, IEEE computer society, IEEE, Minneapolis, USA
33. Forssén P, Moe A (2009) View matching with blob features. Image Vis Comput 27(1–2): 99–107
34. Frangi A, Niessen W, Vincken K, Viergever M (1998) Multiscale vessel enhancement filtering. In: Proceeding of the MICCAI98 lecture notes in computer science, Berlin, vol 1496, pp 130–137
35. Funt B, Finlayson G (1995) Color constant color indexing. IEEE Trans Pattern Anal Mach Intell 17(5):522–529
36. Funt B, Cardei VC, Barnard K (1999) Method of estimating chromaticity of illumination using neural networks. In: United States Patent, USA, vol 5,907,629
37. Gabriel P, Hayet JB, Piater J, Verly J (2005) Object tracking using color interest points. In: IEEE conference on advanced video and signal based surveillance, IEEE computer society, Los Alamitos, CA, USA, vol 0, pp 159–164
38. Gao K, Lin S, Zhang Y, Tang S, Ren H (2008) Attention model based sift keypoints filtration for image retrieval. In: Proceedings of seventh IEEE/ACIS international conference on computer and information science, Washington, DC, USA, pp 191–196
39. Geusebroek J (2000) Color and geometrical structure in images. PhD thesis, University of Amsterdam
40. Geusebroek J (2006) Compact object descriptors from local colour invariant histograms. In: British machine vision conference, vol 3, pp 1029–1038
41. Geusebroek JM, van den Boomgaard R, Smeulders AWM, Dev A (2000) Color and scale: the spatial structure of color images. In: Proceedings of the European conference on computer vision, Dublin, Ireland, pp 331–341
42. Geusebroek JM, van den Boomgaard R, Smeulders AWM, Geerts H (2001) Color invariance. IEEE Trans Pattern Anal Machine Intell 23(12):1338–1350
43. Gevers T, Smeulders A (1999) Color-based object recognition. Pattern Recogn 32:453–464
44. Gevers T, Stokman H (2004) Robust histogram construction from color invariants for object recognition. IEEE Trans Pattern Anal Mach Intell 23(11):113–118
45. Goedem T, Tuytelaars T, Gool LV (2005) Omnidirectional sparse visual path following with occlusion-robust feature tracking. In: 6th workshop on omnidirectional vision, camera networks and non-classical cameras, OMNIVIS05, in Conjunction with ICCV 2005, Beijing, China

46. Gouet V, Montesinos P, Deriche R, Pel D (2000) Evaluation de dtecteurs de points d'intrt pour la couleur. In: Proceeding congrs Francophone AFRIF-AFIA, Reconnaissance des Formes et Intelligence Artificielle, Paris, vol 2, pp 257–266
47. Hamilton Y, Gortler S, Zickler T (2008) A perception-based color space for illumination invariant image processing. In: Proceeding of the special interest group in GRAPHics (SIGGRAPH), Los Angeles, California, USA, vol 27, pp 1–7
48. Harris C, Stephens M (1988) A combined corner and edge detector. In: Proceedings of the 4th Alvey vision conference, Manchester, pp 147–151
49. Healey G, Slater D (1995) Global color contancy:recognition of objects by use of illumination invariant properties of color distributions. J Opt Soc Am 11(11):3003–3010
50. Hegazy D, Denzler J (2008) Boosting colored local features for generic object recognition. Pattern Recogn Image Anal 18(2):323–327
51. Heidemann G (2004) Focus-of-attention from local color symmetries. PAMI 26(7):817–830
52. Heitger F, Rosenthaler L, von der Heydt R, Peterhans E, Kubler O (1992) Simulation of neural contour mechanisms: from simple to end-stopped cells. Vis Res 32(5):963–981
53. Hou X, Zhang L (2007) Saliency detection: a spectral residual approach. IEEE computer society conference on computer vision and pattern recognition 0:1–8
54. Hu L, Jiang S, Huang Q, Gao W (2008) People re-detection using adaboost with sift and color correlogram. In: Proceedings of the IEEE international conference on image processing, San Diego, California, USA, pp 1348–1351
55. Huang J, Kumar SR, Mitra M, Zhu W, Zabih R (1997) Image indexing using color correlogram. IEEE conference on computer vision and pattern recognition pp 762–768
56. Inria database. URL http://lear.inrialpes.fr/data
57. Itti L, Koch C, Niebur E (1998) A model of saliency-based visual attention for rapid scene analysis. IEEE Trans Pattern Anal Mach Intell 20(11):1254–1259
58. (ITU) IRCC (1990) Basic parameter values for the hdtv standard for the studio and for international programme exchange. Tech. Rep. 709-2, CCIR Recommendation
59. Jost T, Ouerhani N, von Wartburg R, Muri R, Hugli H (2005) Assessing the contribution of color in visual attention. Comput Vis Image Understand 100:107–123
60. Khan F, van de Weijer J, Vanrell M (2009) Top-down color attention for object recognition. In: Proceedings of the international conference on computer vision, Japan, pp 979–986
61. Klinker G, Shafer S, Kanade T (1991) A physical approach to color image understanding. Int J Comput Vis 4(1):7–38
62. von Kries J (1970) Influence of adaptation on the effects produced by luminous stimuli. In: MacAdam, D.L. (ed) Sources of color vision. MIT Press, Cambridge
63. Kubelka P (1948) New contribution to the optics of intensity light-scattering materials, part i. J Opt Soc Am A 38(5):448–457
64. Lambert JH (1760) Photometria sive de mensure de gratibus luminis, colorum umbrae. Eberhard Klett
65. Land E (1977) The retinex theory of color vision. Sci Am 237:108–129
66. Land E (1986) An alternative technique for the computation of the designator in the retinex theory of color vision. In: Proceedings of the national academy science of the United State of America, vol 83, pp 3078–3080
67. Lenz R, Tran L, Meer P (1999) Moment based normalization of color images. In: IEEE workshop on multimedia signal processing, Copenhagen, Denmark, pp 129–132
68. Li J, Allinson NM (2008) A comprehensive review of current local features for computer vision. Neurocomput 71(10-12):1771–1787. DOI http://dx.doi.org/10.1016/j.neucom.2007.11.032
69. Lindeberg T (1994) Scale-space theory in computer vision. Springer, London, UK
70. Locher P, Nodine C (1987) Symmetry catches the eye. Eye Movements: from physiology to cognition, North-Holland Press, Amsterdam
71. Logvinenko AD (2009) An object-color space. J Vis 9:1–23
72. Lowe D (2004) Distinctive image features from scale-invariant keypoints. Int J Comput Vis 60(2):91–110

73. Luke RH, Keller JM, Chamorro-Martinez J (2008) Extending the scale invariant feature transform descriptor into the color domain. Proc ICGST Int J Graph Vis Image Process, GVIP 08:35–43
74. Marques O, Mayron L, Borba G, Gamba H (2006) Using visual attention to extract regions of interest in the context of image retrieval. In: Proceedings of the 44th annual Southeast regional conference, ACM, ACM-SE 44, pp 638–643
75. Matas J, Chum O, Martin U, Pajdla T (2002) Robust wide baseline stereo from maximally stable extremal regions. In: Proceeding of the British machine vision conference, pp 384–393
76. Mikolajczyk K, Schmid C (2004) Scale & affine invariant interest point detectors. Int J Comput Vision 60:63–86
77. Mikolajczyk K, Schmid C (2005) A performance evaluation of local descriptors. IEEE Trans Pattern Anal Mach Intell 27:1615–1630
78. Mikolajczyk K, Tuytelaars T, Schmid C, Zisserman A, Matas J, Schaffalitzky F, Kadir T, Gool LV (2005) A comparison of affine region detectors. Int J Comput Vis 65(1/2):43–72. URL http://lear.inrialpes.fr/pubs/2005/MTSZMSKG05
79. Mindru F, Moons T, van Gool L (1999) Recognizing color patterns irrespective of viewpoints and illuminations. In: IEEE conference on computer vision and pattern recognition (CVPR), pp 368–373
80. Mindru F, Tuytelaars T, Gool LV, Moons T (2004) Moment invariants for recognition under changing viewpoint and illumination. Comput Vis Image Understand 1(3):3–27
81. Ming A, Ma H (2007) A blob detector in color images. In: Proceedings of the 6th ACM international conference on image and video retrieval, ACM, New York, NY, USA, CIVR '07, pp 364–370
82. Mollon J (2006) Monge: The verriest lecture, lyon, july 2005. Visual Neurosci 23:297–309
83. Montesinos P, Gouet V, Deriche R (1998) Differential invariants for color images. In: Proceedings of the international conference on pattern recognition, Brisbane (Australie), vol 1, pp 838–840
84. Montesinos P, Gouet V, Deriche R, Pel D (2000) Matching color uncalibrated images using differential invariants. Image Vis Comput 18(9):659–671
85. Moosmann F, Larlus D, Jurie F (2006) Learning Saliency Maps for Object Categorization. In: ECCV international workshop on the representation and use of prior knowledge in vision
86. Moravec H (1977) Towards automatic visual obstacle avoidance. In: Proceedings of the 5th international joint conference on artificial intelligence, p 584
87. Nilsback ME, Zisserman A (2008) Automated flower classification over a large number of classes. In: Proceedings of the indian conference on computer vision, graphics image processing, pp 722 –729
88. Poynton's web page. URL http://www.poynton.com/notes/colour_and_gamma/GammaFAQ.html
89. Qiu G (2002) Indexing chromatic and achromatic patterns for content-based colour image retrieval. Pattern Recogn 35(8):1675–1686
90. Quelhas P, Odobez J (2006) Natural scene image modeling using color and texture visterms. In: Proceedings of conference on image and video retrieval, Phoenix, USA, pp 411–421
91. Recognition benchmark images. URL http://www.vis.uky.edu/stewe/ukbench/
92. Reisfeld D, Wolfson H, Yeshurun Y (1995) Context-free attentional operators: the generalized symmetry transform. Int J Comput Vis 14:119–130
93. Rosenberg C, Hebert M, Thrun S (2001) Color constancy using kl-divergence. In: IEEE international conference on computer vision, pp 239–246
94. Rutishauser U, Walther D, Koch C, Perona P (2004) Is bottom-up attention useful for object recognition? In: IEEE conference on computer vision and pattern recognition (CVPR), pp 37–44
95. van de Sande K, Gevers T, Snoek C (2010) Evaluating color descriptors for object and scene recognition. IEEE Trans Pattern Anal Mach Intell 32:1582–1596
96. van de Sande KE, Gevers T, Snoek CG (2010) Evaluating color descriptors for object and scene recognition. IEEE Trans Pattern Anal Mach Intell 32:1582–1596

97. Schugerl P, Sorschag R, Bailer W, Thallinger G (2007) Object re-detection using sift and mpeg-7 color descriptors. In: Proceedings of the international workshop on multimedia content analysis and mining, pp 305–314
98. Sebe N, Gevers T, Dijkstra S, van de Weije J (2006) Evaluation of intensity and color corner detectors for affine invariant salient regions. In: Proceedings of the 2006 conference on computer vision and pattern recognition workshop, IEEE computer society, Washington, DC, USA, CVPRW '06, pp 18–25
99. Sebe N, Gevers T, van de Weijer J, Dijkstra S (2006) Corners detectors for affine invariant salient regions: is color important? In: Proceedings of conference on image and video retrieval, Phoenix, USA, pp 61–71
100. Shafer SA (1985) Using color to separate reflection components. Color Res Appl 10(4):210–218
101. Shi L, Funt B, Hamarneh G (2008) Quaternion color curvature. In: Proceeding IS&T sixteenth color imaging conference, Portland, pp 338–341
102. Sikora T (2001) The mpeg-7 visual standard for content description - an overview. IEEE Trans Circ Syst Video Technol 11:696–702
103. Song X, Muselet D, Tremeau A (2009) Local color descriptor for object recognition across illumination changes. In: Proceedings of the conference on advanced concepts for intelligent vision systems (ACIVS'09), Bordeaux (France), pp 598–605
104. Stentiford FWM (2003) An attention based similarity measure with application to content-based information retrieval. In: Proceedings of the storage and retrieval for media databases conference, SPIE electronic imaging
105. Stoettinger J, Hanbury A, Sebe N, Gevers T (2007) Do colour interest points improve image retrieval? In: Proceedings of the IEEE international conference on image processing, San Antonio (USA), vol 1, pp 169–172
106. Stokes M, Anderson M, Chandrasekar S, Motta R (1996) A standard default color space for the internet-srgb, Available from http://www.w3.org/Graphics/Color/sRGB.html
107. Tuytelaars T, Mikolajczyk K (2008) Local invariant feature detectors: a survey. Found Trends Comput Graph Vis 3(3):177–280
108. Vazquez E, Gevers T, Lucassen M, van de Weijer J, Baldrich R (2010) Saliency of color image derivatives: a comparison between computational models and human perception. J Opt Soc Am A 27(3):613–621
109. Vázquez-Martína R, Marfila R, nez PN, Bandera A, Sandoval F (2009) A novel approach for salient image regions detection and description. Pattern Recogn Lett 30:1464–1476
110. Vigo DAR, Khan FS, van de Weijer J, Gevers T (2010) The impact of color on bag-of-words based object recognition. In: International conference on pattern recognition, pp 1549–1553
111. Vogel J, Schiele B (2004) A semantic typicality measure for natural scene categorization. In: Rasmussen CE, Blthoff HH, Schlkopf B, Giese MA (eds) Pattern recognition, lecture notes in computer science, vol 3175, Springer Berlin/Heidelberg, pp 195–203
112. Walther D, Rutishauser U, Koch C, Perona P (2005) Selective visual attention enables learning and recognition of multiple objects in cluttered scenes. Comput Vis Image Understand 100:41–63
113. Wandell B (1987) The synthesis and analysis of color images. IEEE Trans Pattern Anal Mach Intell 9:2–13
114. van de Weijer J, Schmid C (2006) Coloring local feature extraction. In: Proceedings of the ninth European conference on computer vision, Graz, Austria, vol 3954, pp 334–348
115. van de Weijer J, Schmid C (2007) Applying color names to image description. In: Proceedings of the IEEE international conference on image processing, San Antonio (USA), vol 3, pp 493–496
116. van de Weijer J, Gevers T, Geusebroek JM (2005) Edge and corner detection by photometric quasi-invariants. IEEE Trans Pattern Anal Mach Intell 27(4):625–630
117. van de Weijer J, Gevers T, Bagdanov A (2006) Boosting color saliency in image feature detection. IEEE Trans Pattern Anal Mach Intell 28(1):150–156
118. van de Weijer J, Gevers T, Smeulders A (2006) Robust photometric invariant features from the colour tensor. IEEE Trans Image Process 15(1):118–127

119. Wu P, Kong L, Li X, Fu K (2008) A hybrid algorithm combined color feature and keypoints for object detection. In: Proceedings of the 3rd IEEE conference on industrial electronics and applications, Singapore, pp 1408–1412
120. Wu P, Kong L, Zhao F, Li X (2008) Particle filter tracking based on color and sift features. In: Proceedings of the international conference on audio, language and image processing, Shanghai
121. Wurtz R, Lourens T (2000) Corner detection in color images through a multiscale combination of end-stopped cortical cells. Image Vis Comput 18(6-7):531–541
122. Wyszecki G, Stiles WS (1982) Color science: concepts and methods, quantitative data and formulas, 2nd ed. Wiley, New York
123. Xiong W, Funt B (2006) Color constancy for multiple-illuminant scenes using retinex and svr. In: Proceeding of imaging science and technology fourteenth color imaging conference, pp 304–308
124. Zhang D, Wang W, Gao W, Jiang S (2007) An effective local invariant descriptor combining luminance and color information. In: Proceedings of IEEE international conference on multimedia and expo, Beijing (China), pp 1507–1510

Chapter 11
Motion Estimation in Colour Image Sequences

Jenny Benois-Pineau, Brian C. Lovell, and Robert J. Andrews

Mere color, unspoiled by meaning, and unallied with definite form, can speak to the soul in a thousand different ways

Oscar Wilde

Abstract Greyscale methods have long been the focus of algorithms for recovering optical flow. Yet optical flow recovery from colour images can be implemented using direct methods, i.e., without using computationally costly iterations or search strategies. The quality of recovered optical flow can be assessed and tailored after processing, providing an effective, efficient tool for motion estimation.

In this chapter, a brief introduction to optical flow is presented along with the optical flow constraint equation and proposed extensions to colour images. Methods for solving these extended equations are given for dense optical flows and the results of applying these methods on two synthetic image sequences are presented. The growing need for the estimation of large-magnitude optical flows and the filtering of singularities require more sophisticated approaches such as sparse optical flow. These sparse methods are described with a sample application in the analysis of High definition video in the compressed domain.

Keywords Colour • Optical flow • Motion estimation

J. Benois-Pineau (✉)
LaBRI, UMR CNRS 5800, Bordeaux University, France
e-mail: jenny.benois@labri.fr

B.C. Lovell
The University of Queensland (UQ), Brisbane, Australia
e-mail: lovell@itee.uq.edu.au

R.J. Andrews
The University of Queensland, Brisbane, Australia

C. Fernandez-Maloigne (ed.), *Advanced Color Image Processing and Analysis*,
DOI 10.1007/978-1-4419-6190-7_11,

11.1 Introduction

Since the inception of optical flow in the late 1970s, generally attributed to Fennema [10], many methods have been proposed to recover the flow field of a sequence of images. The methods can be categorized as either gradient, frequency, or correlation-based methods. The main focus of this chapter is to discuss methods suitable for colour images, including simple extensions of current gradient-based methods to colour space. We further choose to concentrate on differential methods since they generally perform well and have reasonable computational efficiency.

The direct extension of optical flow methods to colour image sequences has the same drawbacks as the original grey-scale approach—difficulties in the estimation of large-magnitude displacements resulting in noisy, non-regularized fields. Recently approaches based on sparse optical flow estimation [18] and the use of sparse fields to estimate dense optical flow or global motion model parameters have been quite successful—especially for the problem of the segmentation of motion scenes. Hence, the extension to colour information in estimation of sparse flow is major advance. We devote this chapter to presentation of both (1) the extension of classical optical flow methods, and (2) the extension of sparse optical flow methods to colour spaces.

Optical flow has been applied to problems of motion segmentation [31], time-to-contact [8, 23, 30], and three-dimensional reconstruction (structure from motion) [15] among many other applications in computer vision. Traditionally, most researchers in this field have focused their efforts on extending Horn and Shunck [12] or Lucas and Kanade's [20] methods, all working with greyscale intensity images. Colour image sequences have been largely ignored, despite the immense value of three planes of information being available rather than just the one.

Psychological and biological evidence suggests that primates use at least a combination of feature and optical flow-based methods in early vision (initial, unintelligent visual processing) [9, 22]. As many visual systems which occur in nature consist of both rods and cones (rods being stimulated purely by intensity, cones stimulated by light over specific wavelength ranges), it is natural to want to extend current optical flow techniques, mostly based on greyscale intensity, to incorporate the extra information available in colour images.

Golland proposed and discussed two simple methods which incorporate colour information [11]. She investigated RGB, normalized RGB, and HSV colour models. Her results indicated that colour methods provide a good estimate of the flow in image regions of non-constant colour. This chapter compares traditional greyscale with Golland's methods and two colour methods proposed in [2]. It also describes the logical extension of greyscale methods to colour.

11.2 Optical Flow

We start this chapter by comparing each image in a sequence to the reference image (the next or previous one) to obtain a set of vector fields called the optical flow. Each vector field represents the apparent displacement of each pixel from image to image. If we assume the pixels conserve their intensity, we arrive at the "brightness conservation equation",

$$I(x,y,t) = I(x+\mathrm{d}x, y+\mathrm{d}y, t+\mathrm{d}t), \tag{11.1}$$

where I is an image sequence, $(\mathrm{d}x,\mathrm{d}y)$ is the displacement vector for the pixel at coordinate (x,y), and t and dt are the frame and temporal displacement of the image sequence. The ideas of brightness conservation and optical flow were first proposed by Fennema [10]. The obvious solution to (11.1) is to use template-based search strategies. A template of a certain size around each pixel is created and the best match is searched for in the next image. Best match is usually found using correlation, sum of absolute difference or sum of squared difference metrics. This process is often referred to as block-matching and is commonly used in the majority of video codecs.

Such a search strategy is computationally costly and generally does not estimate sub-pixel displacements. Most methods for optical flow presented in the last thirty years have been gradient based. Such methods are reasonably efficient and can determine sub-pixel displacements. They solve the differential form of (11.1) derived by a Taylor expansion. Discarding higher order terms, (11.1) becomes

$$\frac{\partial I}{\partial x}u + \frac{\partial I}{\partial y}v + \frac{\partial I}{\partial t} = 0, \tag{11.2}$$

where u and v are the coordinates of the velocity vector function. This equation is known as the Optical Flow Equation (OFE).

Here we have two unknowns in one equation, the problem is ill posed and extra constraints must be imposed in order to arrive at a solution. The two most commonly used and earliest optical flow recovery methods in this category are briefly outlined below: the Horn and Shunck [12] and Lucas and Kanade [20] optical flow methods. These and other traditional methods are described and quantitatively compared in Barron et al. [5, 6].

11.2.1 Horn and Shunck

In 1981, Horn and Shunck [12] were the first to impose a global smoothness constraint which simply assumes the flow to be smooth across the image. Their minimization function,

$$E(u,v) = \int\int (I_x u + I_y v + I_t)^2 + \alpha^2(\|\nabla u\|_2^2 + \|\nabla v\|_2^2)\,(dxdy) \tag{11.3}$$

can be expressed as a pair of Gauss–Siedel iterative equations,

$$\bar{u}_{n+1} = \bar{u}_n - \frac{I_x[I_x\bar{u}_n + I_y\bar{v}_n + I_t]}{\alpha^2 + I_x^2 + I_y^2} \tag{11.4}$$

and

$$\bar{v}_{n+1} = \bar{v}_n - \frac{I_y[I_x\bar{u}_n + I_y\bar{v}_n + I_t]}{\alpha^2 + I_x^2 + I_y^2}. \tag{11.5}$$

where u bar and v bar are the weighted means in the neighbouhood of the current pixel.

11.2.2 Lucas and Kanade

Lucas and Kanade [20] proposed the assumption of constant flow in a local neighborhood. Their method is generally implemented with neighborhoods of size 5×5 pixels centered around the pixel whose displacement is being estimated. Measurements nearer the centre of the neighborhood are given greater weight in the weighted-least-squares formulation.

11.2.3 Other Methods

Later methods generally extended these two traditional methods. For example, researchers have been focusing on using concepts of robustness to modify Lucas and Kanade's method [1,4]. These methods choose a function other than the squared difference of the measurement to the line of fit (implicit in least squares calculation) to provide an estimate of the measurement's contribution to the best line. Functions are chosen so that outliers are ascribed less weight than those points which lie close to the line of best fit. This formulation results in a method which utilizes iterative numerical methods (e.g. gradient descent or successive over-relaxation).

11.3 Using Colour Images

Recovering optical flow from colour images seems to have been largely overlooked by researchers in the field of image processing and computer vision. Ohta [27] mentioned the idea, but presented no algorithms or methods. Golland proposed some methods in her thesis and a related paper [11]. She simply proposed using the three colour planes to infer three equations, then solving these using standard least squares techniques.

$$\frac{\partial I_R}{\partial x}u + \frac{\partial I_R}{\partial y}v + \frac{\partial I_R}{\partial t} = 0$$
$$\frac{\partial I_G}{\partial x}u + \frac{\partial I_G}{\partial y}v + \frac{\partial I_G}{\partial t} = 0$$
$$\frac{\partial I_B}{\partial x}u + \frac{\partial I_B}{\partial y}v + \frac{\partial I_B}{\partial t} = 0. \quad (11.6)$$

Another idea proposed by Golland was the concept of "colour conservation." By constructing a linear system to solve from only the colour components (e.g. hue and saturation from the HSV colour model) the illumination is allowed to change, the assumption is now that the colour, rather than the intensity, is conserved. This makes sense since colour is an intrinsic characteristic of an object, whereas intensity always depends on external lighting.

11.3.1 Colour Models

Four colour models are discussed in this chapter. These are RGB, HSV, normalized RGB, and YUV. The RGB (Red, Green, Blue) colour model decomposes colours into their respective red, green, and blue components. Normalized RGB is calculated by

$$N = R+G+B, \quad R_n = \frac{R}{N}, \quad G_n = \frac{G}{N}, \quad B_n = \frac{B}{N}, \quad (11.7)$$

where each colour is normalized by the sum of all colours at that point. If the colour value at that point is zero the normalized colour at that point is taken as zero.

The HSV (Hue, Saturation, Value) model expresses the intensity of the image (V) independently of the colour (H, S). Optical flow based purely on V is relying on brightness conservation. Conversely, methods which are based on H and S rely purely on colour conservation. Methods which combine the two incorporate both assumptions. Similar to HSV, the YUV model decomposes the colour as a brightness (Y) and a colour coordinate system (U,V). The difference between the two is the description of the colour plane. H and S describe a vector in polar form, representing the angular and magnitudinal components, respectively. Y, U, and V, however, form an orthogonal Euclidean space. An interesting alternative to these spaces is CIE perceptually linear colour space also known as UCS (Uniform Chromaticity Scale). This colour system has the advantage that Euclidean distances in colour space correspond linearly to perception of colour or intensity change.

11.4 Dense Optical Flow Methods

Two obvious methods for arriving at a solution to the extended brightness conservation (11.6) are apparent:

- Disregarding one plane so as to solve quickly and directly, using Gaussian Elimination.
- Solving the over-determined system as is, using either least squares or pseudo-inverse methods.

Disregarding one of the planes arbitrarily may throw away data that are more useful to the computation of optical flow than those kept. However, if speed of the algorithm is of the essence, disregarding one plane reduces memory requirements and computational cost. Another possibility is merging two planes and using this as the second equation in the system. Numerical stability of the solution should be considered when constructing each system. By using the simple method of pivoting, it is possible to ensure the best possible conditioning of the solution.

The methods of least squares and pseudo-inverse calculation are well known. A simple neighborhood least-squares algorithm, akin to Lucas and Kanade's [20], though not utilizing weighting, has also been implemented. Values in a $3 \times 3 \times 3$ neighborhood around the center pixel were incorporated into a large, overdetermined system.

Another option for the computation of optical flow from colour images is to estimate the optical flow of each plane using traditional greyscale techniques and then fuse these results to recover one vector field. This fusion has been implemented here by simply selecting the estimated vector with the smallest intrinsic error at each point. All of the methods mentioned above have been implemented and compared in this study.

11.4.1 Error Analysis

Image reconstruction is a standard technique for assessing the accuracy of optical flow methods, especially for sequences with unknown ground truth (see Barron and Lin [17]). The flow field recovered from an optical flow method is used to warp the first image into a reconstructed image, an approximation to the second image. If the optical flow is accurate, then the reconstructed image should be the same as the second image in the image sequence. Generally, the RMS error of the entire reconstructed image is taken as the image reconstruction error. However, it is advantageous to calculate the image reconstruction error at each point in the image. This enables a level of thresholding in addition to, or instead of culling estimates with high intrinsic error. The density of the flow field after thresholding at chosen image reconstruction errors can also be used to compare different methods. This is the chosen method for comparison in the next section.

Method	64×64	128×128	240×320
Color (Least Sq)	0.66	2.11	9.24
Color (G E, 3 rows)	0.05	0.24	1.58
Color (GE, pivot, 2 rows)	0.03	0.14	0.54
H & S (20 its)	0.23	0.45	2.18
Lucas & Kanade	0.73	3.52	15.19
Nagel	1.63	6.70	36.68
Uras, et. al.	2.06	8.19	37.07
NCC	3.00	9.08	40.20
Black & Anandan	0.60	2.94	19.08

Fig. 11.1 Time taken in seconds for computation on Pentium© III 700 MHz PC from [2]

11.4.2 Results and Discussion

Figure 11.1 compares the time taken for recovery of optical flow using Matlab©, excluding low-pass filtering and derivative calculation times. This highlights the drastic decrease in computational cost of direct colour methods. The two row partial pivoting Gaussian Elimination method is shown [2] to perform at approximately 20 fps on a Pentium© III 700 MHz PC with small images of 64×64 pixels, reducing to 2 fps when processing 240×320 images. Compared to Horn and Shunck's method [12], the best performer in the field of greyscale methods, this represents an approximately fourfold increase in speed.

Figure 11.2 compares three common greyscale optical flow methods; Horn and Shunck [12], Lucas and Kanade [20], and Nagel [26]. This figure illustrates the density of the computed flow field when thresholded at chosen image reconstruction errors. It is seen that Lucas and Kanade's method [20] slightly outperforms Horn and Shunck's [12] method, which itself performs better than Nagel's [26] method in terms of image reconstruction errors.

Figure 11.3 compares the performance of Lucas and Kanade's [20] with three colour methods. The first frame of this image sequence is shown in Fig. 11.4. This sequence was translating with velocity $[-1,-1]$ pixels per frame. The three colour methods shown here are (1) Gaussian elimination (with pivoting) of the saturation and value planes of HSV, (2) Gaussian elimination of RGB colour planes, and (3) neighborhood least squares. Neighborhood least squares is seen to perform the best out of the colour methods, closely approximating Lucas and Kanade at higher densities. Both Gaussian elimination versions performed poorly compared to the others.

An image sequence displaying a one degree anticlockwise rotation around the center of the image was used to assess three other colour optical flow methods. Pixel displacement ranges between zero and 1.5 pixels per frame. The methods compared were "Colour Constancy" [11], least squares solution to (11.6), and Combined-Horn and Shunck. Horn and Shunck's [12] (greyscale) algorithm was used as a yardstick

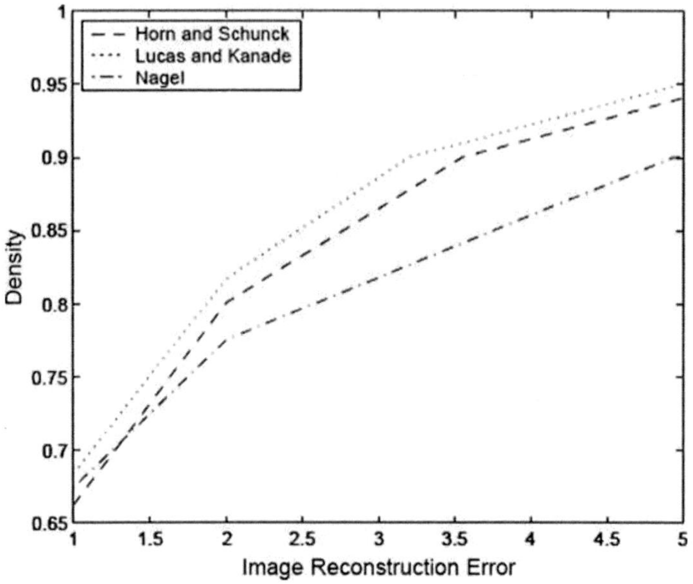

Fig. 11.2 Comparison of Greyscale methods applied to translating coloured clouds

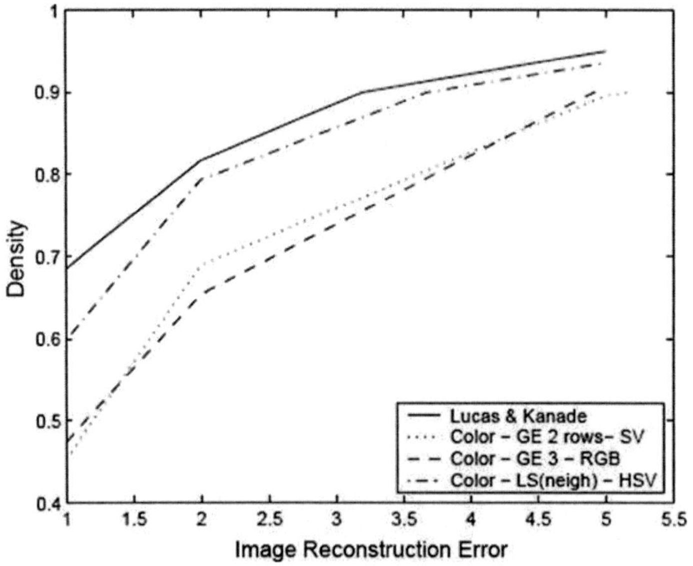

Fig. 11.3 Comparison of grey and colour methods applied to translating coloured clouds

Fig. 11.4 First frame of the translating RGB clouds sequence

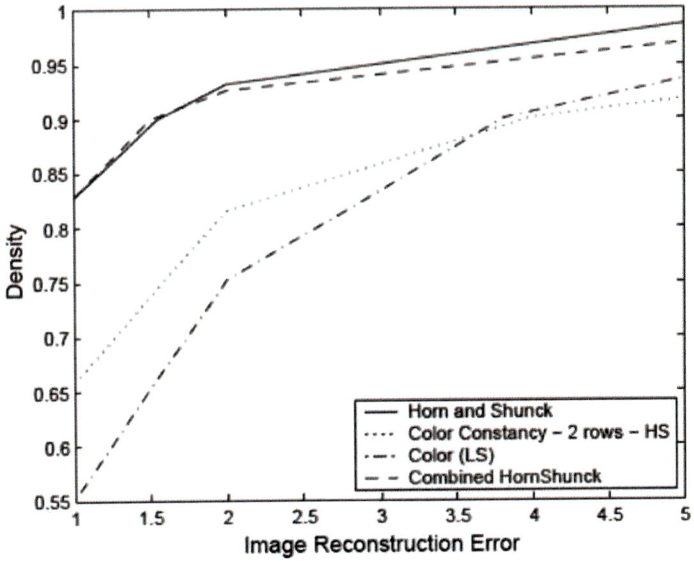

Fig. 11.5 Comparison of techniques applied to a rotating image sequence

for this comparison. The results are displayed in Fig. 11.5. Combined-Horn and Shunck applied Horn and Shunck optical flow recovery to each plane of the RGB image and fused them into one flow field utilizing a winner takes all strategy based on their associated error. It can be seen that the Combined-Horn and Shunck method performed similarly to Horn and Shunck [12]. The methods of least squares [11] and direct solution of the colour constancy equation [11] did not perform as well. Figure 11.6 gives an example of the optical flow recovered by the neighborhood least

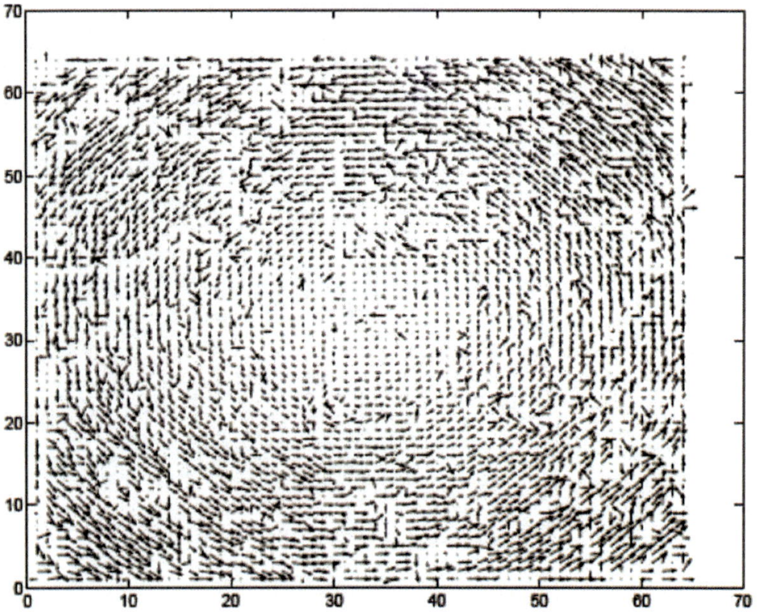

Fig. 11.6 Optical flow recovered by direct two-row optical flow and thresholding

squares algorithm. This corresponds to the rotating image sequence. Larger vectors (magnitude greater than 5) have been removed and replaced with zero vectors. This field has a density of 95%.

11.5 Sparse Optical Flow Methods: Patches and Blocks

Here we are interested in what is called sparse optical flow (OF) estimation in colour image sequences. First of all, a sparse optical flow can be used in a general OF estimation formulation as a supplementary regularization constraint. We will describe this in Sect. 11.5.1. Second, the sparse OF estimation has another very wide application area: efficient video coding or analysis of video content on partially decoded compressed streams.

11.5.1 Sparse of Constraints for Large Displacement Estimation in Colour Images

When calculated at full frame resolution, the above OF methods can handle only very small displacements. Indeed in the derivation of OFE (11.3), the fundamental assumption is the limited neigbourhood of (x,y,t) point in the R^3 space. To handle large displacements a reduction of resolution in a multi-scale/multi-resolution way with low pass filtering and subsampling allows us to solve the large displacement problem. This comes from a straightforward relation between the coordinates of the displacement vector at different scales. Indeed if r is a subsampling factor, a pixel with the coordinates $(x+u(x,y), y+v(x,y), t+\mathrm{d}t)$ at a full resolution will correspond to $((x+u(x,y))/r, (y+v(x,y))/r, t+\mathrm{d}t)$ in the subsampled image. Hence, the magnitude of displacement vector $(\mathrm{d}x, \mathrm{d}y)$ will be reduced by the factor of r.

Using an adapted multi-resolution estimation, the large displacements can be adequately estimated by differential techniques as presented in Sects. 11.2–11.4. Nevertheless, subsampling with a preliminary low pass-filtering not only smoothes the error functional, but may also eliminate details which are crucial for good matching of images.

To overcome this effect when seeking estimation of large-magnitude displacements in colour image sequences Brox et al. [7] proposed a new approach which uses colour information. The main idea consists of using sparse motion information to drive the variational estimation of OF in the manner of the Horn and Schunck estimator. Hence, in their work they first propose to extend the well-known SIFT optical flow estimator proposed by Liu et al. [18]. The principle of SIFT flow with regard to usual OF methods consists of comparing SIFT descriptors of characteristic points and not the original grey-level or colour values. A SIFT descriptor of dimensionality $N^2 \times m$ as introduced by Lowe [19] is a concatenation of N^2 histograms of m bins. The statistic here is the angle of gradient orientation of an $N \times N$ square grid of square regions surrounding a feature point. The histogram is weighted in each direction by the"strength" of direction expressed by the gradient magnitude. The descriptor is normalized and does not convey any colour information.

Instead of using feature points, Brox et al. [7] proposed to segment the frames into homogeneous arbitrary-shaped regions. Then, an ellipse is fitted to each region and the area around the centroid is normalized to a 32×32 patch. Then, they built two descriptors: one of them S being SIFT and another C the mean RGB colour of the same N^2 ($N^2 = 16$) subparts which served for SIFT computation. Both consecutive frames I_1 and I_2 are segmented and pairs of regions (i, j) are matched according to the balanced distance:

$$d^2(i,j) = \frac{1}{2}(d^2(C_i,C_j) + d^2(S_i,S_j)) \tag{11.8}$$

with $d^2(D_i,D_j) = \frac{\|D_i - D_j\|_2^2}{\sum_{k,l}\|D_k - D_l\|_2^2}$, where D is the descriptor-vector.

The results presented on the sequences with deformable human motion show that such a matching gives good ranking, but the method is not sufficiently discriminative between good and bad matches. Hence, in the sequences with deformable motion they propose to select small patches inside segmented regions and optimize an error functional per patch with the same regularization as proposed by Horn and Schunk:

$$E(u,v) = \int (P_2(x+u,y+v) - P_1(x,y))^2 \mathrm{d}x\mathrm{d}y + \alpha^2 \int (\|\nabla u\|_2^2 + \|\nabla v\|_2^2)\mathrm{d}x\mathrm{d}y, \quad (11.9)$$

where P denotes the patches, $u(x,y), v(x,y)$ denotes the deformation field to be estimated for the patch, and ∇ is the gradient operator. The first term in the equation is a nonlinearized version of Horn and Shunck energy.

The estimated optical flow $u(x,y), v(x,y)$ in patches is then used as a constraint in the extended Horn and Schunk formulation (11.10). To limit combinatorial complexity, they preselect patches to be matched by (11.9) accordingly to the (11.8). They retain 10 nearest neighbours to be matched and then limit the number of potential matches after the minimization of (11.9) to 5. Hence, each patch I will have 5 potential matches j, $j = 1,\ldots,5$ with an associated confidence $c_j(i)$ of this match on the basis of deviation of $d^2(i,j)$ from its mean value per patch $d^2(i)$. Finally Brox et al. formulate a global energy to minimize by

$$\begin{aligned} E(u,v) = & \int ((I_2(x+u,y+v) - I_1(x,y))^2)dxdy \\ & + \gamma^2 \int (|\nabla I_2(x+u,y+v) - \nabla I_1(x,y)|^2)dxdy \\ & + \beta^2 \sum_{j=1}^{5} \int \rho_j(x,y)((u(x,y) - u_j(x,y))^2 + (v(x,y) - v_j(x,y))^2)dxdy \\ & + \alpha^2 \int (|\nabla u(x,y)|^2 + |\nabla v(x,y)|^2 + g(x,y)^2)dxdy. \end{aligned} \quad (11.10)$$

Here I_1 and I_2 are the current and reference images, $s^2 = \sqrt{s^2 + \varepsilon}$ with ε small, is a robust function to limit the influence of outliers,$(u_j(x,y), v_j(x,y))$ is one of the motion vectors at the position (x,y) derived from patch matching: $\rho_j(x,y) = 0$ if there is no correspondence at this position. Otherwise, $\rho_j(x,y) = c_j$ where $g(x,y)$ is a boundary map value corresponding to the boundaries of initial segmentation into regions. The latter is introduced to avoid smoothing across edges. The regularization factors are chosen very strong with stronger value for the modified Horn and Shunck regularization term $\alpha^2 = 100$, the influence of patch matches is also stressed $\beta^2 = 25$. Finally, less importance is given to the smoothness of image gradient, which regulates "structural" correspondence of local dense matching.

Following the Horn and Schunck approach Brox et al. derive a Euler–Lagrange system and solve it by a fixed-point iterative scheme. The results of this estimation constrained by region matching are better than the optical flow obtained by a simple region matching. Unfortunately, Brox et al. do not compare their scheme with a multi-resolution optical flow method to assess the improvements obtained. In any case, the introduction of constraints allows for more accurate OF estimation, specifically in the case of occlusions [29]. The interest of this method for colour-based optical flow estimation resides in the intelligent use of colour as the information to build initial primitives—regions to match by constraining OF by to pre-estimated local OF on patches.

11.5.2 Block-based Optical Flow from Colour Wavelet Pyramids

Block-based optical flow estimation has been traditionally used for video coding. Since the very beginning of hybrid video coding up to the most recent standards H.264 AVC and SVC, block-based motion estimation has proved to be an efficient tool for decorrelation of image sequence to encode. In coding applications, both fixed-size blocks and variable-sized blocks have been used.

From the point of view of analysis of image sequences, block-based motion is a particular case of sparse optical flow obtained from patches P which are (1) squared and (2) the set of patches represents a partition of image plane at each time instant t. The optical flow is sparse as only one displacement vector $(dx, dy)_i$ is considered per patch P_i. Considering a normalized temporal distance between images in a moving sequence, this is equivalent to searching for piece-wise constant optical flow

$$\forall (x,y) \in D(P_i) \rightarrow u(x,y) = c_{1,i}, v(x,y) = c_{2,i} \tag{11.11}$$

with $c_{1,i}, c_{2,i}$ constants.

This OF is locally optimal: for each patch it minimizes an error criterion $F(u,v)$ between value of a patch at a current moment t and its value at a reference moment t_{ref}. Note that in most cases, this criterion is expressed as

$$F(u^*, v^*) = \min \int\int |P_i(x+u, y+v, t_{\mathrm{ref}}) - P_i(x,y,t)| dxdy \tag{11.12}$$

with constraints $-u_{\max} \leqslant u \leqslant u_{\max}, -v_{\max} \leqslant u \leqslant v_{\max}$. In its discrete version, the criterion is called MAD or minimal absolute difference and expressed by

$$\mathrm{MAD}_i(u^*, v^*) = \\ min_{-u_{\max} \leqslant u \leqslant u_{\max}, -v_{\max} \leqslant u \leqslant v_{\max}} \sum_{(x,y) \in D(P_i)} |P_i(x+u, y+v, t_{\mathrm{ref}}) - P_i(x,y,t)|. \tag{11.13}$$

In video coding applications the function $P(x,y)$ is a scalar and represents the Y-component in the YUV colour system. In contrast to RGB systems, YUV is not homogeneous. The Y component carries the most prominent information in images, such as local contrasts and texture. The U and V components are relatively "flat." The general consensus with regard to these components in the block-based motion estimation community is that due to the flatness of these components, they do not bring any enhancement in terms of the minimization of MAD or the mean squared motion estimation error,

$$MSE_i(u^*,v^*) = \frac{1}{\mathrm{Card}(D(P_i))} \sum_{(x,y)\in D(P_i)} (P_i(x+u,y+v,t_{\mathrm{ref}}) - P_i(x,y,t))^2 \quad (11.14)$$

but only increase computational cost.

Nevertheless, when block-based motion estimation is not required for video coding, but for comparison and matching of objects in video displaced in time as in [13], the three components of a colour system can be used.

In this chapter, we present a block-based motion estimation in the YUV system decomposed on pyramids of Daubechies wavelets [3] used for the high-quality high-definition compression standard for visual content: JPEG2000 [14]. The rationale for estimating OF in the wavelet domain originates from the "Rough Indexing Paradigm" defining the possibility of analysis and mining of video content directly in compressed form [21]. The wavelet analysis filtering and sub-sampling decomposes the original frames in video sequences into pyramids. At each level of the pyramid except its basis which is an original frame, the signal is represented by four subbands (LL, LH, HL, and HH). As illustrated in Fig. 11.7, the LL subband in upper left corner is a low-pass version of a colour video frame which may contain some aliasing artifacts.

In a similar way, the block-based sparse OF estimation has been realized on Gaussian pyramids from pairs of consecutive images in video sequences in a hierarchical manner layer by layer [16].

The block-based motion estimation on wavelet pyramids served as a basis for efficient segmentation of JPEG2000 compressed HD video sequences to detect outliers in the homogeneous sparse motion field $\{(u,v)_i\}, i = 1,\ldots,N$, with N being the number of rectangular patches in the partition of image plane. These outliers correspond to the patches belonging to the moving objects and also some flat areas [25]. In [24], we proposed a solution for detecting such areas with local motion on the basis of motion vectors estimated on the Low Frequency (LL) subbands of the Y component. To do this, the three steps of motion estimation are fulfilled.

First of all, the block-based motion estimation is realized on the LL component at the top of the wavelet pyramid optimizing the MAD criterion (11.13) per block. Here, the patch contains the values of coefficients of the LL subband and the estimation of the displacement vector is realized by a full search within a square domain surrounding the initial point $(u_i,v_i)^T$ of Z^2.

Next, these motion vectors are supposed to follow a complete first-order affine motion model:

$$\begin{aligned} u(x,y) &= a_1 + a_2(x-x_0) + a_3(y-y_0) \\ v(x,y) &= a_4 + a_5(x-x_0) + a_6(y-y_0). \end{aligned} \quad (11.15)$$

With (x_0,y_0) being a reference point in the image plane, usually the centre of the image, or in our case the LL subband at a given level of resolution in the

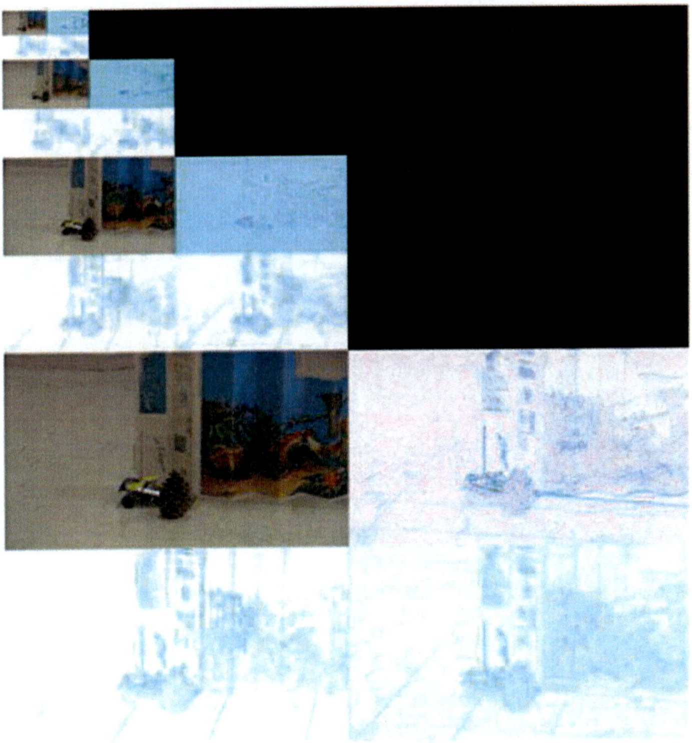

Fig. 11.7 Four levels of decomposition of a video frame into Dabechies Wavelet Pyramid. Sequence "VoiturePanZoom" from OpenVideo.org ICOS-HD corpus. Author(s): LaBRI, University of Bordeaux 1, Bordeaux, France

wavelet pyramid. The global model (11.15) is estimated by robust least squares using as measures the initially estimated block-based motion vectors. The outliers are then filtered as masks of the foreground objects [25].

Obviously the quality of the estimated model with regard to the scene content depends not only on the complexity of global camera motion, or "flatness" of the scene, but also on the quality of the initial block-based motion estimation.

In this case, we can consider the colour information of the LL subband of a given level of the wavelet pyramid and analyze the enhancement we can get from the use of complementary colour components. In the case of block-based motion estimation in the colour space, the quality criteria have to be redefined. Namely, the MSE metric (11.14) becomes

$$\mathrm{MSE}_i(u^*,v^*) = \frac{1}{\mathrm{Card}(D(P_i))} \sum_{(x,y)\in D(P_i)} \|\bar{P}_i(x+u,y+v,t_{\mathrm{ref}} - \bar{P}_i(x,y,t)\|_2^2 \quad (11.16)$$

with $\| \quad \|_2$ being the L2 or Euclidean norm of the colour vector function.

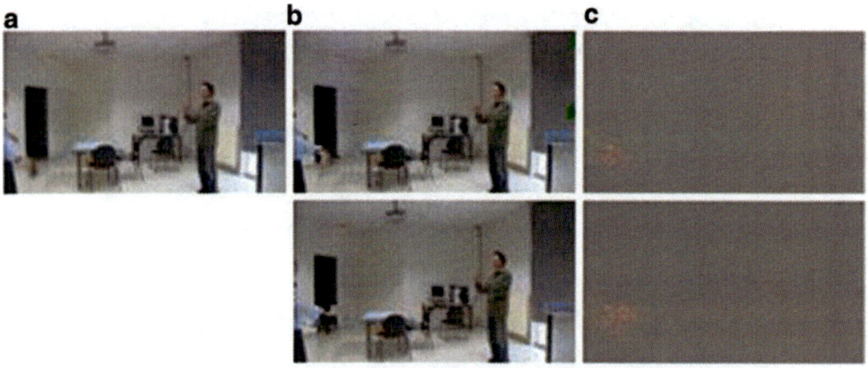

Fig. 11.8 Block-based colour motion compensation on wavelet pyramid. Sequence "Lancer Trousse" from OpenVideo.org ICOS-HD corpus. Author(s): LaBRI, University of Bordeaux 1, Bordeaux, France, 4th level of decomposition: (**a**) original frame, (**b**) motion-compensated frame, (**c**) error frame after motion compensation. *Upper row*: compensation with colour MV. *Lower row*: compensation with MV estimated on Y component

Furthermore, as in coding applications instead of using MSE quality measure, the Peak Signal to Noise Ratio is used to assess the quality of resulting sparse OF. It is derived from MSE (11.16) as

$$\text{PSNR} = 10\log_{10}\frac{\|P_{\max}\|_2^2}{\text{MSE}}. \tag{11.17}$$

Here, $\|P_{\max}\|^2$ is the squared Euclidean norm of a saturated colour vector $(255,255,255)^T$ in the case of 8-bit quantization of colour components.

We present the results for the sequence "Lancer Trousse" in Figs. 11.8 and 11.9. In the following, PSNRCinit means the PSNR computed on the LL colour wavelet frame without motion compensation. PSNRY-C means the PSNR computed with motion vectors estimated only on the Y component and applied to all three colour components. PSNRC-C denotes PSNR computed with motion vectors estimated by block matching on three colour components. As it can be seen from Fig. 11.9, colour block-based motion compensation is more efficient in areas with strong motion (e.g. the moving hand of the person on the left).

As it can be seen from Fig. 11.8, the PSNRC-C is in general higher than the PSNR on the Y component. The low resolution of the frames makes it difficult to match flat areas and the colour information enhances the results. With increasing resolution, the difference in PSNRs becomes more visible as the block-based motion estimator would better fit not only the local Y-contrast but also the U and V details. The examples on the sequence "Lancer Trousse" are given here as a critical case, as this video scene is not strongly "coloured." Much better behavior is observed on complex high-definition colour sequences, such as TrainTracking or Voitures PanZoom [28].

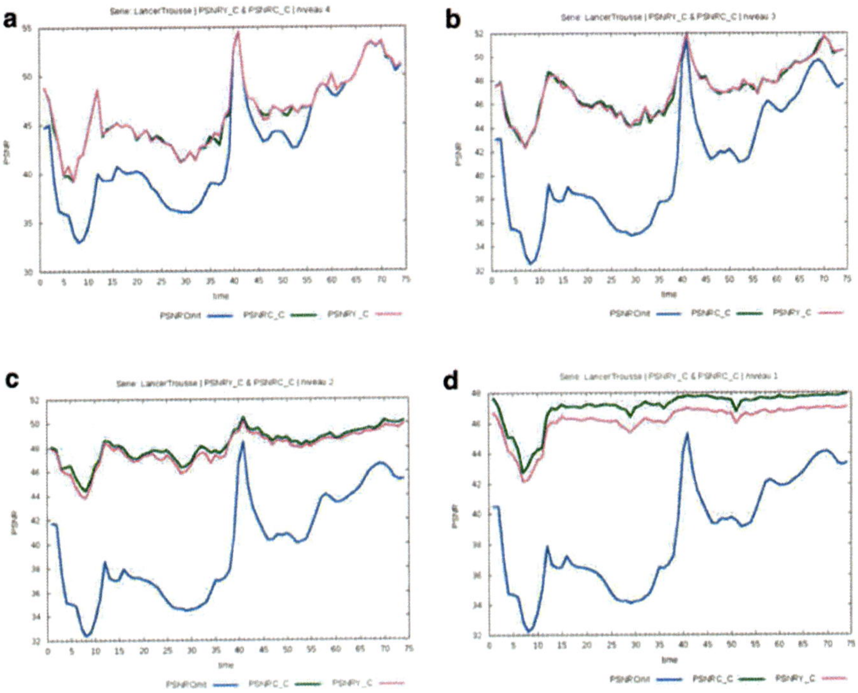

Fig. 11.9 PSNR of block-based motion compensation at the levels 4-1 of the pyramid on colour LL frames. Sequence "Lancer Trousse" from OpenVideo.org ICOS-HD corpus. Author(s): LaBRI, University of Bordeaux 1, Bordeaux, France: (**a**)–(**d**) PSNR at the 4th through 1st levels of the pyramid respectively

11.6 Conclusions

In this chapter, we presented two approaches to the estimation of optical flow in colour image sequences. The first trend is the extension of the classical Horn and Shunck and Lucas and Kanade methods to colour spaces. The second study described a sparse optical flow estimation via patches. The question of colour optical flow is very far from being exhausted, but nevertheless some conclusions can be made for both families of methods.

Dense colour optical flow has been shown to be quite simple to compute and to have a level of accuracy similar to traditional greyscale methods. The speed of these algorithms is a significant benefit; the linear optical flow methods presented run substantially faster than greyscale, nonlinear methods.

Accuracy of the neighborhood least squares approach can be improved in a number of ways. Using robust methods, e.g. least-median of squares [4], could provide a much better estimate of the correct flow. Applying the weighted least squares approach of Lucas and Kanade [20] could likewise improve the results.

A better data-fusion algorithm could be used to improve the Combined-Horn and Schunk method. The three flows being combined could be calculated using any greyscale method. Methods that iterate towards a solution usually perform better with a good initial starting estimate. Colour-optical flow could be used to provide this estimate, speeding the computation of some of the slower, well-known greyscale methods.

Also we discussed sparse OF estimation on colour image sequences. We have seen that "colour" information *per se* can be used either in an indirect or direct way. An example of the former is segmenting video sequence and using this as a constraint for general optical flow estimation of the Y component. As an example of the latter, the direct use of colour in sparse OF estimation per block improves the quality of optical flow in the pixel domain as well as in wavelet transform domain. This can help in the further application of sparse optical flow for fine video sequence analysis and segmentation.

References

1. Anandan P, Black MJ (1996) The robust estimation of multiple motions: parametric and piecewisesmooth flow fields. Comput Vis Image Underst 63(1):75–104
2. Andrews RJ, Lovell BC (2003) Color optical flow. In: Lovell, Brian C. and Maeder, Anthony J, Proceedings of the 2003 APRS Workshop on digital image computing: 135–139
3. Antonini M, Barlaud M, Mathieu P, Daubechies I (1992) Image coding using wavelet transform. IEEE Trans Image Process 1(2):205–220
4. Bab-Hadiashar A, Suter D (1996) Robust optic flow estimation using least median of squares. In: Proc IEEE ICIP, Lausanne, Switzerland: 513–516
5. Barron JL, Fleet D, Beauchemin SS, Burkitt T (1993) Performance of optical flow techniques. Technical report, Queens University RPL-TR-9107
6. Barron JL, Fleet DJ, Beauchemin SS (1994) Systems and experiment performance of optical flow techniques. Int J Computer Vis 12:43–77
7. Brox T, Malik J (2011) Large displacement optical flow. IEEE Trans Pattern Anal Mach Intell 33(3):500–13
8. Camus T (1994) Real-time optical flow. Ph.D. thesis, Brown University
9. Chhabra A, Grogan T (1990) Early vision and the minimum norm constraint. In: Proceedings of IEEE Int. Conf. on Systems, Man, and Cybernetics;Los Angeles, CA:547–550
10. Fennema CL, Thompson WB (1979) Velocity determination in scenes containing several moving objects. Computer Graph Image Process 9(4):301–315. DOI:10.1016/0146--664X(79) 90097--2
11. Golland P, Bruckstein AM (1997) Motion from color. Comput Vis Image Underst 68(3):346–362. DOI:10.1006/cviu.1997.0553
12. Horn B, Shunck B (1981) Determining optical flow. Artiff Intell 17:185–203
13. Huart J, Bertolino P (2007) Extraction d'objets-clés pour l'analyse de vidéos. In: Proc GRETSI(France), hal-00177260: 1–3
14. ISO/IEC 15444–1: 2004 Information technology—JPEG 2000 image coding system: core coding system (2004): 1–194
15. Kanatani K, Shimizu Y, Ohta N, Brooks MJ, Chojnacki W, van den Hengel (2000) A Fundamental matrix from optical flow: optimal computation and reliability estimation. J Electron Imaging 9(2):194–202

16. Lallauret F, Barba D (1991) Motion compensation by block matching and vector postprocessing in subband coding of tv signals at 15 mbit/s. In: Proceedings of SPIE, vol 1605: 26–36
17. Lin T, Barron J (1994) Image reconstruction error for optical flow. In: In vision interface, Scientific Publishing Co, pp 73–80
18. Liu C, Yuen J, Torralba A, Sivic J, Freeman W (2008) Sift flow: dense correspondence across different scenes. In: Proc ECCV. Lecture notes in computer science, Springer, Berlin, pp 28–42
19. Lowe DG (1999) Object recognition from local scale-invariant features. In: Proceedings of the international conference on computer vision, vol 2: 1150–1157
20. Lucas B, Kanade T (1981) An iterative image restoration technique with an application to stereo vision. In: Proc DARPA IU workshop: 121–130
21. Manerba F, Benois-Pineau J, Leonardi R (2004) Extraction of foreground objects from a MPEG2 video stream in "rough-indexing" framework. In: Proceedings of SPIE, vol 5307: 50–60
22. Mathur BP, Wang HT (1989) A model of primates. IEEE Trans Neural Network 2:79–86
23. Micheli ED, Torre V, Uras S (1993) The accuracy of the computation of optical flow and of the recovery of motion parameters. IEEE Trans Pattern Anal Mach Intell 15(15):434–447
24. Morand C, Benois-Pineau J, Domenger JP (2008) HD motion estimation in a wavelet pyramid in JPEG2000 context. In: Proceddings of IEEE ICIP: 61–64
25. Morand C, Benois-Pineau J, Domenger JP, Zepeda J, Kijak E (2010) Guillemot C: Scalable object-based video retrieval in hd video databases. Signal Process Image Commun 25(6):450–465
26. Nagel HH (1983) Displacement vectors derived from second-order intensity variations in image sequences. Comput Vis Graph Image Process 21(1):85–117. DOI:10.1016/S0734--189X(83)80030--9
27. Ohta N (1989) Optical flow detection by color images. In: Proc IEEE ICIP: 801–805
28. Open video: The open video project (2011) http://www.open-video.org, [Last Visited: 25-May-2011]
29. Roujol S, Benois-Pineau J, Denis de Senneville BD, Quesson B, Ries M, Moonen C (2010) Real-time constrained motion estimation for ecg-gated cardiac mri. In: Proc IEEE ICIP: 757–760
30. Tistarelli M, Sandini G (1993) On the advantages of polar and log-polar mapping for direct estimation of time-to-impact from optical flow. IEEE Trans Pattern Anal Mach Intell 14(4):401410
31. Verri A, Poggio T (1989) Motion field and optical flow: qualitative properties. IEEE Trans Pattern Anal Mach Intell 11(5):490–498

Chapter 12
Protection of Colour Images by Selective Encryption

W. Puech, A.G. Bors, and J.M. Rodrigues

The courage to imagine the otherwise is our greatest resource, adding color and suspense to all our life

Daniel J. Boorstin

Abstract This chapter presents methods for the protection of privacy associated with specific regions from colour images or colour image sequences. In the proposed approaches, regions of interest (ROI) are detected during the JPEG compression of the colour images and encrypted. The methodology presented in this book chapter performs simultaneously selective encryption (SE) and image compression. The SE is performed in a ROI and by using the Advanced Encryption Standard (AES) algorithm. The AES algorithm is used with the Cipher Feedback (CFB) mode and applied on a subset of the Huffman coefficients corresponding to the AC frequencies chosen according to the level of required security. In this study, we consider the encryption of colour images and image sequences compressed by JPEG and of image sequences compressed by motion JPEG. Our approach is performed without affecting the compression rate and by keeping the JPEG bitstream compliance. In the proposed method, the SE is performed in the Huffman coding stage of the JPEG algorithm without affecting the size of the compressed image. The most significant characteristic of the proposed method is the utilization of a single procedure to

W. Puech (✉)
Laboratory LIRMM, UMR CNRS 5506, University of Montpellier II, France
e-mail: william.puech@lirmm.fr

A.G. Bors
Department of Computer Science, University of York, UK
e-mail: adrian.bors@cs.york.ac.uk

J.M. Rodrigues
Department of Computer Science, Federal University of Ceara, Fortaleza, Brazil
e-mail: marconi@ufc.fr

C. Fernandez-Maloigne (ed.), *Advanced Color Image Processing and Analysis*,
DOI 10.1007/978-1-4419-6190-7_12,

simultaneously perform the compression and the selective encryption rather than using two separate procedures. Our approach reduces the required computational complexity. We provide an experimental evaluation of the proposed method when applied on still colour images as well as on sequences of JPEG compressed images acquired with surveillance video cameras.

Keywords Selective encryption • Colour image protection • JPEG compression • AES • Huffman coding

12.1 Introduction

Digital rights management (DRM) systems enforce the rights of the multimedia property owners while ensuring the efficient rightful usage of such property. A specific concern to the public has been lately the protection of the privacy in the context of video-camera surveillance. The encryption of colour images in the context of DRM systems has been attempted in various approaches. A secure coding concept for pairwise images using the fractal mating coding scheme was applied on colour images in [2]. A selective image encryption algorithm based on the spatiotemporal chaotic system is proposed to encrypt colour images in [32]. Invisible colour image hiding schemes based on spread vector quantization and encryption was proposed in [12]. Self-adaptive wave transmission was extended for colour images in [11] where half of image data was encrypted using the other half of the image data. A colour image encryption method based on permutation and replacement of the image pixels using the synchronous stream cipher was proposed in [3]. In [9, 14], visual cryptography was applied on colour images aiming to hide information. In this approach, the image is split into colour halftone images which are shared among n participants. Any $k < n$ participants can visually reveal the secret image by superimposing their shares together but it cannot be decoded by any fewer participants, even if infinite computational power is available to them. The security of a visual cryptography scheme for colour images was studied in [10]. A selective partial image encryption scheme of secure JPEG2000 (JPSEC) for digital cinema was proposed in [25]. While these approaches address various aspects of DRM systems, none of them provides a practical application to the problem of privacy protection of images which is efficient. The technical challenges are immense and previous approaches have not entirely succeeded in tackling them [13]. In this chapter, we propose a simultaneous partial encryption, selective encryption and JPEG compression methodology which has low computational requirements in the same time.

Multimedia data requires either full encryption or selective encryption depending on the application requirements. For example, military and law enforcement applications require full encryption. Nevertheless, there is a large spectrum of applications that demands security on a lower level, as, for example, that ensured by selective encryption (SE). Such approaches reduce the computational requirements in networks with diverse client device capabilities [4]. In this chapter, the first goal of SE of an image is to encrypt only regions of interest (ROI) which are defined within specific areas of the image. The second goal of SE is to encrypt a well-defined range of parameters or coefficients, as, for example, would be the higher spectrum of frequencies. SE can be used to process and transmit colour images acquired by a surveillance video camera. Indeed, in order to visualize these images in real time, they must be quickly transmitted and the full encryption is not really necessary. The security level of SE is always lower when compared with the full encryption. On the other hand, SE decreases the data size to be encrypted and consequently requires lower computational time which is crucial for wireless and portable multimedia systems. In this case, we have a trade-off between the amount of data that we encrypt and the required computational resources.

JPEG is a commonly used image compression algorithm which is used in both security and industrial applications [20]. JPEG image compression standard is employed in a large category of systems such as: digital cameras, portable telephones, scanners and various other portable devices. This study shows that SE can be embedded in a standard coding algorithm such as JPEG, JPEG 2000, MJPEG or MPEG, while maintaining the bitstream compliance. In fact, using a standard decoder it should be possible to visualize the SE data in the low-resolution image as well. On the other hand, with a specific decoding algorithm and a secret key it should be possible to correctly decrypt the SE data and get the high resolution whenever is desired.

In this chapter, we present new approaches of SE for JPEG compressed colour image sequences by using variable length coding (VLC). The proposed method is an improvement of the proposed methods from [21, 22]. We propose to encrypt selected bits in the Huffman coding stage of JPEG algorithm. By using a skin detection procedure, we choose image blocks that definitely contain the faces of people. In our approach, we use the Advanced Encryption Standard (AES) [5] in the Cipher Feedback (CFB) mode which is a stream cipher algorithm. This method is then applied to protect the privacy of people passing in front of a surveillance video camera. Only the authorized persons, possessing the decrypting code are able to see the full video sequences. In Sect. 12.2, we provide a short description of JPEG and AES algorithms as well as an overview of previous research results in the area of colour image encryption. The proposed method is described in Sect. 12.3. Section 12.4 provides a set of experimental results, while Sect. 12.5 draws the conclusion of this study.

12.2 Description of the JPEG Compressing Image Encryption System

Confidentiality is very important for low-powered systems such as, for example, wireless devices. Always, when considering image processing applications on such devices we should use minimal resources. However, the classical ciphers are usually

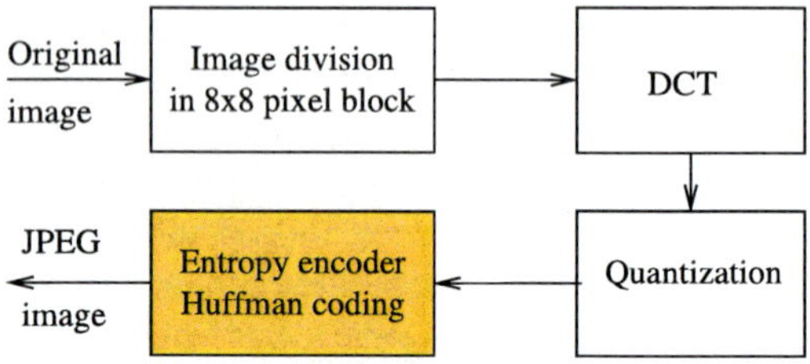

Fig. 12.1 Processing stages of the JPEG algorithm

too slow to be used for image and video processing in commercial low-powered systems. The selective encryption (SE) can fulfill the application requirements without the extra computational effort required by the full encryption. In the case of SE, only the minimal necessary data are ciphered. However, the security of SE is always lower when compared to that of the full encryption. The only reason to accept this drawback is the substantial computational reduction. We review the basic steps of the JPEG algorithm in Sect. 12.2.1, the AES algorithm in Sect. 12.2.2, while in Sect. 12.2.3 we present an overview of the previous work.

12.2.1 The JPEG Algorithm

The standard JPEG algorithm decomposes initially the image in blocks of 8×8 pixels. These pixel blocks are transformed from the spatial to the frequency domain using the Discrete Cosine Transform (DCT). The DC coefficient corresponds to zero frequency and depends on the average greylevel value in each 8×8 pixel block, while the AC coefficients correspond to the frequency information. Then, each DCT coefficient is divided by its corresponding parameter from a quantization table, corresponding to the chosen quality factor and rounded afterwards to the nearest integer. The quantized DCT coefficients are mapped according to a predefined zigzag order into an array according to their increasing spatial frequency. Then, this sequence of quantized coefficients is used in the entropy-encoding (Huffman coding) stage. The processing stages of the JPEG algorithm are shown in Fig. 12.1.

In the Huffman coding block, the quantized coefficients are coded by pairs $\{H,A\}$ where H is the head and A is the amplitude. The head H contains the control information provided by the Huffman tables. The amplitude A is a signed integer representing the amplitude of the nonzero AC, or in the case of DC is the difference between the DC coefficients of two neighboring blocks. Because the DC coefficients are highly predictable, they are treated separately in the Huffman coding. For the AC coefficients, H is composed of a pair $\{R,S\}$, where R is the runlength and S is the size of H, while for the DC coefficients, H is made up only by size S. The SE approach proposed in this chapter is essentially based on encrypting only certain AC coefficients.

For the AC coding, JPEG uses a method based on combining runlength and amplitude information. The runlength R is the number of consecutive zero-valued

AC coefficients which precede a nonzero value from the zigzag sequence. The size S is the amount of necessary bits to represent the amplitude A. Two extra codes that correspond to $\{R,S\} = \{0,0\}$ and $\{R,S\} = \{15,0\}$ are used to mark the end of block (EOB) and a zero run length (ZRL), respectively. The EOB is transmitted after the last nonzero coefficient in a quantized block. The ZRL symbol is transmitted whenever R is greater than 15 and represents a run of 16 zeros. One of the objectives of our method is to encrypt the image while preserving the JPEG bitstream compliance in order to provide a constant bit rate.

12.2.2 The AES Encryption Algorithm

The AES algorithm consists of a set of processing steps repeated for a number of iterations called rounds [5]. The number of rounds depends on the size of the key and of that of the data block. The number of rounds is 9, for example, if both the block and the key are 128 bits long. Given a sequence $\{X_1,\ldots,X_n\}$ of bit plaintext blocks, each X_i is encrypted with the same secret key k producing the ciphertext blocks $\{Y_1,\ldots,Y_n\}$. To encipher a data block X_i in AES, you first perform an AddRoundKey step by XORing a subkey with the block. The incoming data and the key are added together in the first AddRoundKey step. Afterwards, it follows the round operation. Each regular round operation involves four steps which are: SubBytes, ShiftRows, MixColumns and AddRoundKey. Before producing the final ciphered data Y_i, the AES performs an extra final routine that is composed of the steps: SubBytes, ShiftRows and AddRoundKey.

The AES algorithm can support several cipher modes: ECB (Electronic Code Book), CBC (Cipher Block Chaining), OFB (Output Feedback), CFB (Cipher Feedback), and CTR (Counter) [28]. The ECB mode is actually the basic AES algorithm. With the ECB mode, each plaintext block X_i is encrypted with the same secret key k producing the ciphertext block Y_i:

$$Y_i = E_k(X_i). \tag{12.1}$$

The CBC mode adds a feedback mechanism to a block cipher. Each ciphertext block Y_i is XORed with the incoming plaintext block X_{i+1} before being encrypted with the key k. An initialization vector (IV) is used for the first iteration. In fact, all modes (except the ECB mode) require the use of an IV. In the CFB mode, Y_0 is substituted by the IV as shown in Fig. 12.2. The keystream element Z_i is then generated and the ciphertext block Y_i is produced as:

$$\begin{cases} Z_i = E_k(Y_{i-1}), & \text{for } i \geq 1 \\ Y_i = X_i \oplus Z_i \end{cases}, \tag{12.2}$$

where $\oplus$ is the XOR operator.

In the OFB mode, Z_0 is substituted by the IV and the input data is encrypted by XORing it with the output Z_i. The CTR mode has very similar characteristics

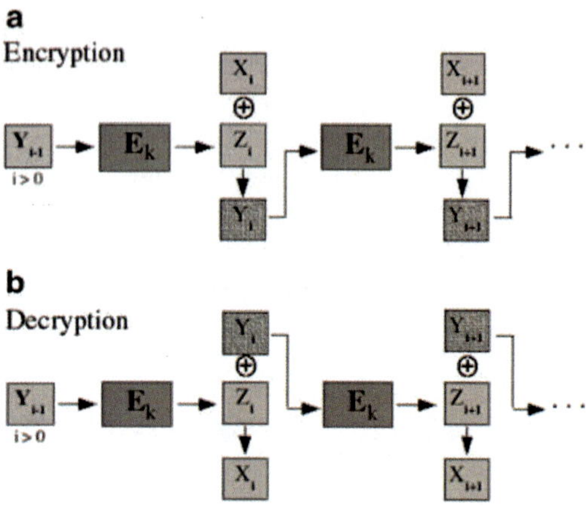

Fig. 12.2 The CFB stream cipher scheme: (**a**) Encryption, (**b**) Decryption

to OFB, but in addition it allows pseudo-random access for decryption. It generates the next keystream block by encrypting successive values of a counter.

Although AES is a block cipher, in the OFB, CFB and CTR modes it operates as a stream cipher. These modes do not require any specific procedures for handling messages whose lengths are not multiples of the block size because they all work by XORing the plaintext with the output of the block cipher. Each mode has its advantages and disadvantages. For example in the ECB and OFB modes, any modification in the plaintext block X_i causes the corresponding ciphered block Y_i to be altered, while other ciphered blocks are not affected. On the other hand, if a plaintext block X_i is changed in the CBC and CFB modes, then Y_i and all subsequent ciphered blocks will be affected. These properties mean that CBC and CFB modes are useful for the purpose of authentication while ECB and OFB modes treat separately each block. Therefore, we can notice that OFB mode does not spread noise, while the CFB mode does exactly that.

12.2.3 Previous Work

Selective encryption (SE) is a technique aiming to reduce the required computational time and to enable new system functionalities by encrypting only a portion of the compressed bitstream while still achieving adequate security [16]. SE as well as the partial encryption (PE) is applied only on certain parts of the bit stream corresponding to the image. In the decoding stage, both the encrypted and the non-encrypted information should be appropriately identified and displayed [4, 18, 22]. The protection of the privacy in the context of video-camera surveillance is a requirement in many systems. The technical challenges posed by such systems are high and previous approaches have not entirely succeeded in tackling them [13].

In [29] was proposed a technique called zigzag permutation applicable to DCT-based videos and images. On one hand this method provides a certain level of confidentiality, while on the other hand it increases the overall bit rate. Combining SE and image/video compression using the set partitioning in hierarchical trees was used in [4]. Nevertheless, this approach requires a significant computational complexity. A method that does not require significant processing time and which operates directly on the bit planes of the image was proposed in [17]. SE of video while seeking the compliance with the MPEG-4 video compression standard was studied in [30]. An approach that turns entropy coders into encryption ciphers using statistical models was proposed in [31]. In [6], it was suggested a technique that encrypts a selected number of AC coefficients. The DC coefficients are not ciphered since they carry important visual information and they are highly predictable. In spite of the constancy in the bit rate while preserving the bitstream compliance, this method produces codes which are not scalable. Moreover, the compression and the encryption process are separated and consequently the computational complexity is increased. Fisch et al. [7] proposed a method whereby the data are organized in a scalable bitstream form. These bitstreams are constructed with the DC and some AC coefficients of each block which are then arranged in layers according to their visual importance. The SE process is applied over these layers. Some encryption methods have been applied in the DCT coefficient representations of image sequences [4, 30, 36].

The AES [5] was applied on the Haar discrete wavelet transform compressed images in [19]. The encryption of colour images in the wavelet transform has been addressed in [18]. In this approach the encryption takes place on the resulting wavelet code bits. In [21], SE was performed on colour JPEG images by selectively encrypting only the luminance component Y. The encryption of JPEG 2000 codestreams has been reported in [8, 15]. SE using a mapping function has been performed in [15]. It should be noticed that wavelet-based compression employed by JPEG 2000 image-coding algorithm increases the computational demands and is not used by portable devices.

The robustness of selectively encrypted images to attacks which exploit the information from non-encrypted bits together with the availability of side information was studied in [23]. The protection rights of individuals and the privacy of certain moving objects in the context of security surveillance systems using viewer generated masking and the AES encryption standard has been addressed in [33]. In the following, we describe our proposed approach to apply simultaneously SE and JPEG compression in images.

12.3 The Proposed Selective Encryption Method

The SE procedure is embedded within the JPEG compression of the colour image. Our approach consists of three steps: JPEG compression, selective encryption according to the ROI detection performed during the Huffman coding stage of JPEG.

In Sect. 12.3.1, we present an overview of the proposed method. The colour range based ROI detection used for SE is described in Sect. 12.3.2 and the SE during the Huffman coding stage of JPEG is presented in Sect. 12.3.3. In Sect. 12.3.4, we explain the decryption of the protected image.

12.3.1 *Overview of the Method*

In the case of image sequences, each frame is treated individually. For each colour frame, we apply the colour transformation used by the JPEG algorithm, converting from the RGB to the YC_rC_b colour space. The two chrominance components C_r and C_b are afterwards subsampled. The DCT and the quantization steps of the JPEG algorithm are performed on the three components Y, C_r, and C_b. SE is applied only on particular blocks corresponding to the Y component, during the Huffman coding stage because the luminance carries the most significant information [21]. In order to detect the particular blocks that we have to encrypt, we use the quantized DC coefficients of the two chrominance components C_r and C_b. These quantized DC coefficients are not encrypted and could be used during the decryption stage. Using the quantized DC coefficients, we detect the ROI as it will be described in Sect. 12.3.2. As part of the SE process, after the ROI detection, selected AC coefficients corresponding to a chosen block are encrypted in the Y component during the Huffman coding stage of JPEG. The detected blocks are selectively encrypted by using the AES algorithm with the CFB mode as it will be described in Sect. 12.3.3. Afterwards, we encrypt in the area defined as ROI, and we combine with SE by encrypting only the AC coefficients corresponding to the chosen higher range of frequencies. The overview of the method is presented in the scheme from Fig. 12.3.

12.3.2 *Detection of the ROI Using the Chrominance Components*

The ROI's, representing skin information in our application, are selected using the average colour in a 8×8 pixel block, as indicated by the zero frequency (DC coefficients) from the DCT coefficients. We use the DC coefficients of the C_r and C_b components, denoted as DC_{C_r} and DC_{C_b}, respectively, to detect the human skin according to:

$$\sqrt{\left(\frac{DC_{C_r}}{8} - C_{r_s}\right)^2 + \left(\frac{DC_{C_b}}{8} - C_{b_s}\right)^2} < T, \tag{12.3}$$

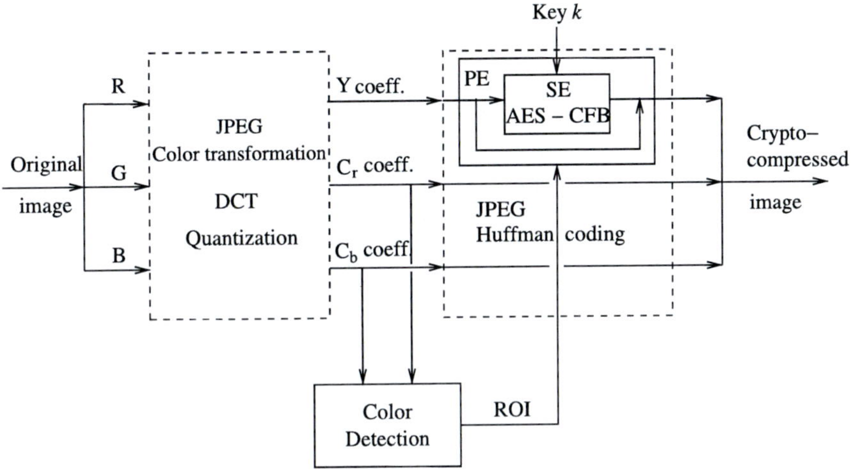

Fig. 12.3 Schematic of the proposed methodology for simultaneous PE and compression in images

where C_{b_s} and C_{r_s} are the reference skin colour in YC_rC_b space and T is a threshold [1, 35]. These parameters are chosen such that the entire range of human skin is detected.

The DC coefficients that fulfill the condition (12.3) are marked indicating the ROI. However, we segment areas which are not always contiguous due to the noise and the uncertainty when choosing a value for the threshold T. Consequently, we have to smooth the chosen image areas in order to ensure contiguity. For enforcing smoothness and contiguity of the ROI, we apply morphological opening (erosion followed by dilatation) [26] onto the mapping formed by the marked and non-marked DC coefficients. Smoothed regions of marked DC coefficients indicate the image areas that must be encrypted from the original image. Each marked DC coefficient corresponds to a block of 8×8 pixels. In the following, we describe the SE method which is applied to the Huffman vector corresponding to the Y component.

12.3.3 Selective Encryption of Quantified Blocks During the Huffman Coding Stage of JPEG

Let us consider $Y_i = X_i \oplus E_k(Y_{i-1})$ as the notation for the encryption of a n bit block X_i, using the secret key k with the AES cipher in the CFB mode as given by equation (12.2), and performed as described in the scheme from Fig. 12.2. We have chosen to use this mode in order to keep the original compression rate. Indeed, with

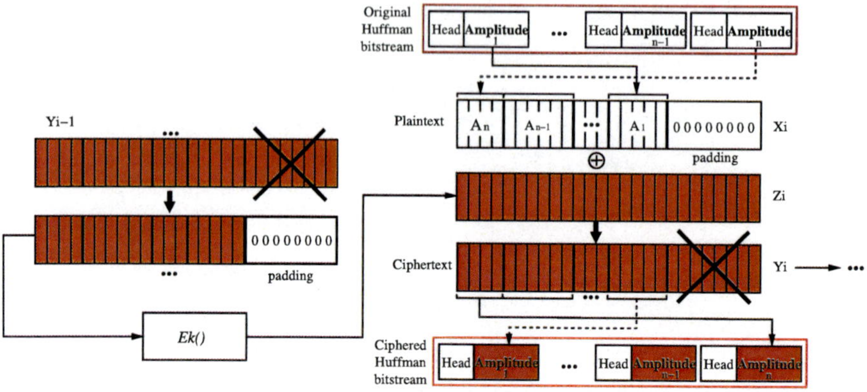

Fig. 12.4 Global overview of the proposed SE method

the CFB mode for each block, the size of the encrypted data Y_i can be exactly the same one as the size of the plaintext X_i. Let $D_k(Y_i)$ be the decryption of a ciphered text Y_i using the secret key k. In the CFB mode, the code from the previously encrypted block is used to encrypt the current one as shown in Fig. 12.2.

The proposed SE is applied in the entropy-encoding stage during the creation of the Huffman vector. The three stages of the proposed algorithm are: the construction of the plaintext X_i, described in Sect. 12.3.3.1, the encryption of X_i to create Y_i which is provided in Sect. 12.3.3.2 and the substitution of the original Huffman vector with the encrypted information, which is explained in Sect. 12.3.3.3. These operations are performed separately in each selected quantified DCT block. Consequently, the blocks that contain many details and texture will be strongly encrypted. On the other hand, the homogeneous blocks, i.e., blocks that contain series of identical pixels, are less ciphered because they contain a lot of null coefficients which are represented by special codes in the Huffman coding stage. The overview of the proposed SE method is provided in Fig. 12.4.

12.3.3.1 The Construction of Plaintext

For constructing the plaintext X_i, we take the non-zero AC coefficients of the current block i by accessing the Huffman vector in reverse order of its bits in order to create {H, A} pairs. The reason for ordering the Huffman code bits from those corresponding to the highest to those of the lowest frequencies (the reverse order of the zigzag DCT coefficient conversion from matrix to array as used in JPEG), is because the most important visual characteristics of the image are placed in the lower frequencies, while the details are located in the higher frequencies. The human visual system is more sensitive to the lower frequencies when compared to the higher range of frequencies. Therefore, by using the Huffman bits corresponding to the decreasing frequency ordering, we can calibrate the visual appearance of the resulting image. This means that we can achieve a progressive or scalable encryption with respect to the visual effect. The resulting image will have a higher level of encryption as we increasingly use the lower range of frequencies.

A constraint C is used in order to select the quantity of bits to encrypt from the plaintext X_i. The constraint C graduates the level of ciphering and the visual quality of the resulting image. For each block, the plaintext length $L(X_i)$ to be encrypted depends on both the homogeneity of the block and the given constraint C:

$$0 \leq L(X_i) \leq C, \tag{12.4}$$

where $C \in \{4, 8, 16, 32, 64, 128\}$ bits. When $C = 128$, AES will fully use the available block of Huffman bits while for the other values several blocks are grouped in order to sum up to 128 bits which is the standard size of AES as explained in Sect. 12.2.2. The constraint C specifies the maximum quantity of bits that must be considered for encryption in each block as in VLC. On the other hand, the homogeneity depends on the content of the image and limits the maximum quantity of bits that can be used for encryption from each Huffman block. This means that a block with great homogeneity will produce a small $L(X_i)$. The Huffman vector is encrypted as long as $L(X_i) \leq C$ and the sequence of selected bits does not include those corresponding to the DC coefficient. Then, we apply a padding function $p(j) = 0$, where $j \in \{L(X_i)+1, \ldots, C\}$, to fill in the vector X_i with zeros up to C bits. In cryptography, padding is the practice of adding values of varying length to the plaintext. This operation is done because the cipher works with units of fixed size, but messages to be encrypted can vary in length. Several padding schemes exist, but we will use the simplest one, which consists of appending null bits to the plaintext in order to bring its length up to the block size. Historically, padding was used to increase the security of the encryption, but in here it is used for rather technical reasons with block ciphers, cryptographic hashing and public key cryptography [24].

The length of amplitude A in bits is extracted using H. These values are computed and tested according to (12.4). In the proposed method, only the values of the amplitudes $(A_n, \ldots A_1)$ are considered to build the vector X_i. The Huffman vector is composed of a set of pairs {H, A} and of marker codes such as ZRL and EOB. If the smallest AC coefficients are zero, the Huffman bitstream for this block must contain the mark EOB. In turn, the ZRL control mark is found every time that sixteen successive AC coefficients which are zero are followed by at least one non-zero AC coefficient. In our method, we do not make any change in the head H or in the mentioned control marks. To guarantee the compatibility with any JPEG decoder, the bitstream should only be altered at places where it does not compromise the compliance with the original format.

The homogeneity in the image leads to a series of DCT coefficients of value almost zero in the higher range of frequencies. The DCT coefficients can be used to separate the image into spectral sub-bands. After quantization, these coefficients become exactly zero [34]. The plaintext construction is illustrated in Fig. 12.4.

12.3.3.2 Encryption of the Plaintext with AES in the CFB Mode

According to (12.2), in the encryption step with AES in the CFB mode, the previous encrypted block Y_{i-1} is used as the input of the AES algorithm in order to create Z_i. Then, the current plaintext X_i is XORed with Z_i in order to generate the encrypted text Y_i.

For the initialization, the IV is created from the secret key k according to the following strategy. The secret key k is used as the seed of the pseudo-random number generator (PRNG). Firstly, the secret key k is divided into 8 bits (byte) sequences. The PRNG produces a random number for each byte component of the key, that defines the order of IV formation. Then, we substitute Y_0 with the IV, and Y_0 is used in AES to produce Z_1.

As illustrated in Fig. 12.4, with the CFB mode of the AES algorithm, the generation of the keystream Z_i depends of the previous encrypted block Y_{i-1}. Consequently, if two plaintexts are identical $X_i = X_j$ in the CFB mode, then always the two corresponding encrypted blocks are different, $Y_i \neq Y_j$.

12.3.3.3 Substitution of the Original Huffman Bitstream

The third step is the substitution of the original information in the Huffman vector by the encrypted text Y_i. As in the first step (construction of the plaintext X_i), the Huffman vector is accessed in the sequential order, while the encrypted vector Y_i is accessed in the reversed order. Given the length in bits of each amplitude $(A_n, \ldots, A_1)$, we start substituting the original amplitude in the Huffman vector by the corresponding parts of Y_i as shown in Fig. 12.4. The total quantity of replaced bits is $L(X_i)$ and consequently we do not necessarily use all the bits of Y_i.

12.3.4 Image Decryption

In this section, we describe the decryption of the protected image. During the first step, we apply the Huffman decoding on the C_r and C_b components. After the Huffman decoding of the two chrominance components, we apply the colour detection in order to retrieve an identical ROI with the one that had been encrypted. By knowing the ROI, it is possible to know which blocks of the Y component should be decrypted during the Huffman decoding stage and which blocks should be only decoded.

The decryption process in the CFB mode works as follows. The previous block Y_{i-1} is used as the input to the AES algorithm in order to generate Z_i. By knowing the secret key k, we apply the same function $E_k(\cdot)$ as that used in the encryption stage. The difference is that the input of the encrypting process is now the ciphered Huffman vector. This ciphered vector is accessed in the reverse order of its bits in order to construct the plaintext Y_{i-1}. Then, it will be used in the AES to generate the

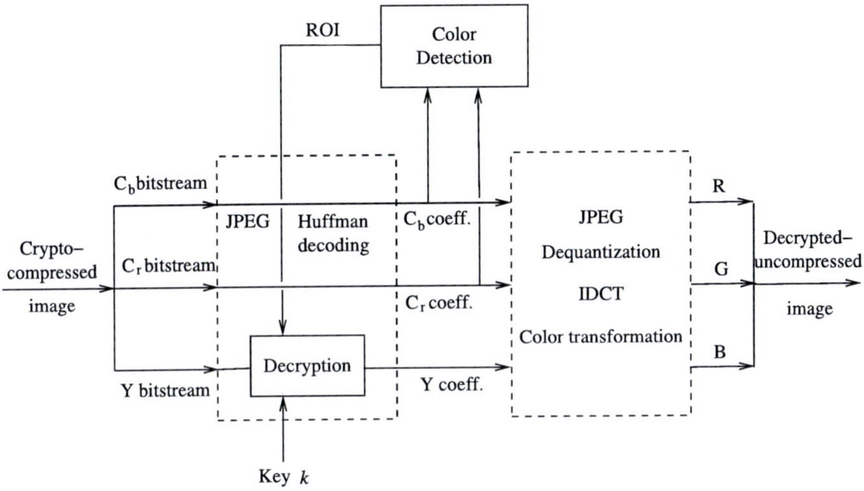

Fig. 12.5 Global overview of the decryption

keystream Z_i. The keystream Z_i is then XORed with the current block Y_i to generate X_i, as shown in Fig. 12.2b. The resulting plaintext vector is split into segments in order to substitute the amplitudes $(A_n, \ldots, A_1)$ in the ciphered Huffman code and to generate the original Huffman vector. Afterwards, we apply the Huffman decoding and retrieve the quantized DCT coefficients. After the dequantization and the inverse DCT, we transform the image from YC_rC_b colour space to RGB colour space. The overview of the decryption is shown in Fig. 12.5.

In order to decrypt the image, the user needs the secret key. Nevertheless, without the secret key it is still possible to decompress and visualize the image in low-resolution format because our approach fulfills the JPEG bitstream compliance and the Huffman bits corresponding to the DC coefficients of the DCT are not encrypted.

12.4 Experimental Results

In this section, we analyze the results when applying SE onto the Huffman coding of the high-frequency DCT coefficients of in the ROI from JPEG compressed colour images and image sequences.

12.4.1 Analysis of Joint Selective Encryption and JPEG Compression

We have applied simultaneously our selective encryption and JPEG compression as described in Sect. 12.3, on several images. In this section, we show the results of

Fig. 12.6 Original Lena image

SE when applied in the entire JPEG compressed image. The original Lena image 512×512 pixels is shown in Fig. 12.6. The compressed JPEG Lena image with a quality factor (QF) of 100% is shown in Fig. 12.7a and the compressed JPEG with a QF of 10% is shown in Fig. 12.7d.

In a first set of experiments, we have analyzed the available space for encryption in JPEG compressed images. In Table 12.1, we provide the number of bits available for selective encryption for Lena of 512×512 pixels corresponding to various JPEG quality factors. In the same table, for each QF we provide the distortion calculated as the PSNR (Peak to Signal Noise Ratio) as well as the average number of available bits for SE per block of quantized DCT coefficients. We can observe that when QF is lower and implicitly the image compression is higher, we are able to embed fewer bits in the compressed image. This is due to the fact that JPEG compression creates flat regions in the image blocks resulting in the increase of the number of AC coefficients equal to zero. Consequently, the Huffman coding creates special blocks for such regions which our method does not encrypt. Not all the available bits provided in the third column of Table 12.1 are actually used for SE because of the limit imposed by the constraint C. For optimizing the time complexity, C should be smaller than the ratio between the average number of bits and the block size.

In Fig. 12.8 we provide the graphical representation of the last column from Table 12.1, displaying the variance of the ratio between the number of available bits for SE and the total number of block bits. We can observe that this variance decreases together with the QF as the number of flat regions in the compressed image increases. For improving the time requirements of the proposed encryption method, a smaller constraint C should be used.

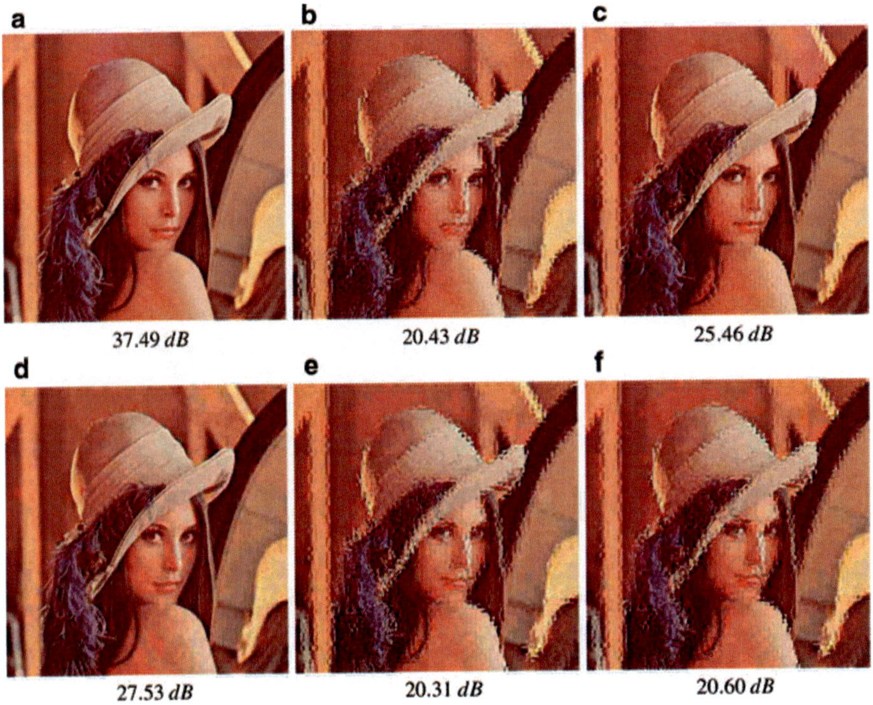

Fig. 12.7 (**a**) JPEG compressed image with QF=100%, (**b**) Image (a) with $C = 128$ bits/block, (**c**) Image (a) with $C = 8$ bits/block, (**d**) JPEG compressed image with QF = 10%, (**e**) Image (d) with $C = 128$ bits/block, (**f**) Image (d) with $S = 8$ bits/block

Table 12.1 Results for various JPEG quality factors

		Bits available for SE		
Quality factor	PSNR (dB)	Total in the *Y* component	Percentage from *Y* component	Average bits/block
100	37.49	537,936	25.65	131
90	34.77	153,806	7.33	38
80	33.61	90,708	4.33	22
70	32.91	65,916	3.14	16
60	32.41	50,818	2.42	12
50	32.02	42,521	2.03	10
40	31.54	34,397	1.64	8
30	30.91	26,570	1.27	6
20	29.83	17,889	0.85	4
10	27.53	8,459	0.40	2

In Fig. 12.9, we show the evaluation of the PSNR between the crypto-compressed Lena image and the original, for several QF and for various constraints C. In the same figure, for comparison purposes we provide the PSNR between the

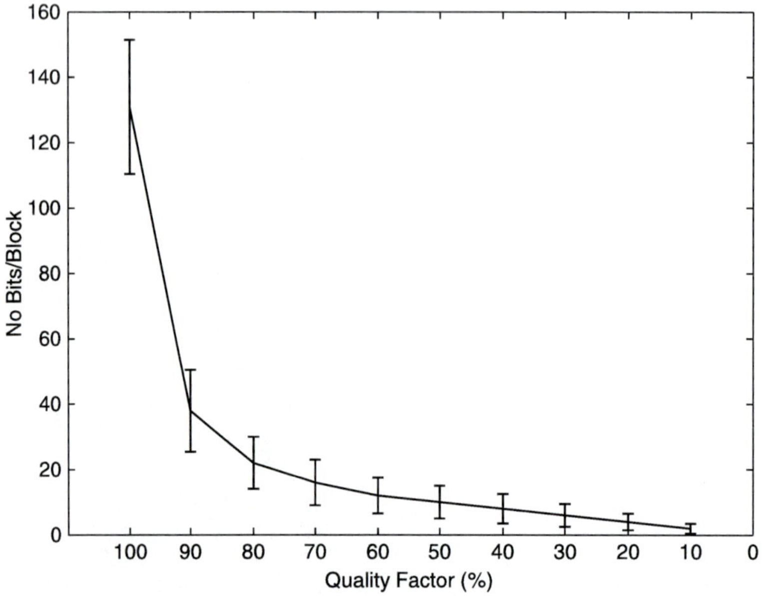

Fig. 12.8 The ratio between the average number of bits available for SE and the block size. The variance is indicated as a confidence interval

compressed image with different QF and the original image. From this figure, we can observe that for a higher C we encrypt a larger number of bits and consequently the image is more distorted with respect to the original. It can be observed that when $C \in \{32, 64, 128\}$, the difference in the PSNR distortion is similar and varies slowly when decreasing the QF.

In Fig. 12.7b we show the original Lena image encrypted using a constraint $C = 128$ bits per block of quantized DCT coefficients, while in Fig. 12.7c the same image is encrypted using a constraint of $C = 8$ bits/block. In Fig. 12.7e we show Lena image with QF of 10%, encrypted using a constraint $C = 128$ bits/block, while in Fig. 12.7f the same image is encrypted using a constraint $C = 8$ bits/block. We can see that the degradation introduced by the encryption in the image with $\mathrm{QF} = 100\%$, from Fig. 12.7b, is higher than the degradation in the image from Fig. 12.7c because in the latter we encrypt more bits per block. When combining a high JPEG compression level ($\mathrm{QF} = 10\%$) with selective encryption, as shown in the images from Figs. 12.7e and 12.7f, we can observe a high visual degradation with respect to the images from Figs. 12.7b and 12.7c, respectively. The higher distortion is caused by the increase in the number of block artifacts. The distortion is more evident when observing some image features as, for example, the eyes.

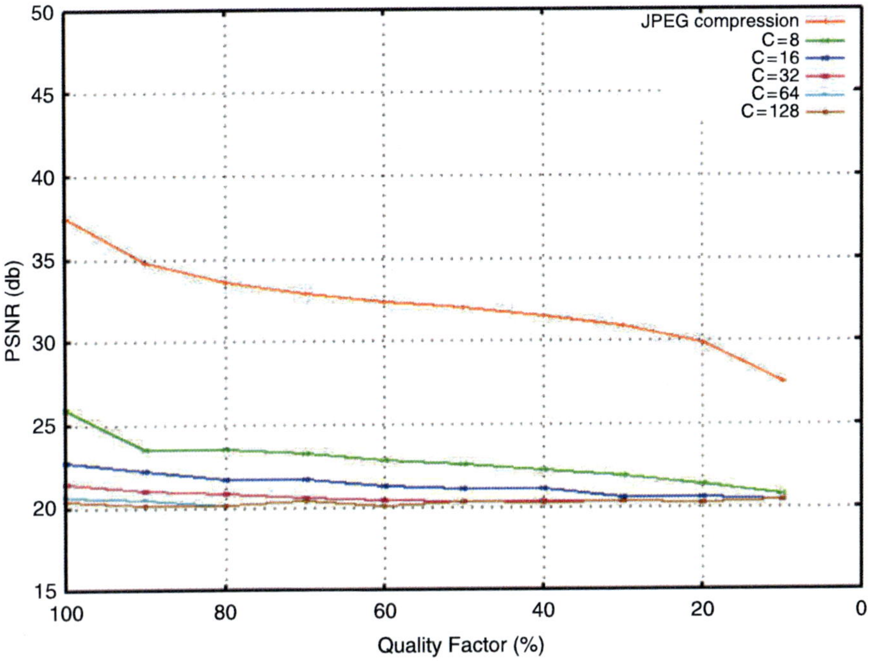

Fig. 12.9 PSNR of crypto-compressed Lena image for various quality factors and constraints

12.4.2 *Selective Encryption of the Region of Interest in Colour Images*

In this section, we have applied our encryption method to the colour image illustrated in Fig. 12.10a[1] and on a colour image sequence shown in Fig. 12.11a. We use the DC components of the chrominance in order to select the ROI which in this case corresponds to the skin. Based on several experimental tests, for the initial colour image in the RGB space displayed in Fig. 12.10a, we consider the following values in (12.3): $T = 15$, $C_{r_s} = 140$ and $C_{b_s} = 100$. The resulting ROIs are shown in Fig. 12.10b. We can observe that all the skin regions, including the faces are correctly detected. Each selected DC coefficient corresponds to a pixel block marked with white in Fig. 12.10b. Only these blocks are selectively encrypted. We can observe that a diversity of skin colours has been appropriately detected by our skin selection approach defined by equation (12.3). We have then selectively encrypted the original image from Fig. 12.10a by using the proposed skin detection procedure. We encrypt 3,597 blocks from a total of 11,136 blocks in the full image

[1] In order to display the image artifacts produced by our crypto-compression algorithm, we have cropped a sub-image of 416 × 200 pixels.

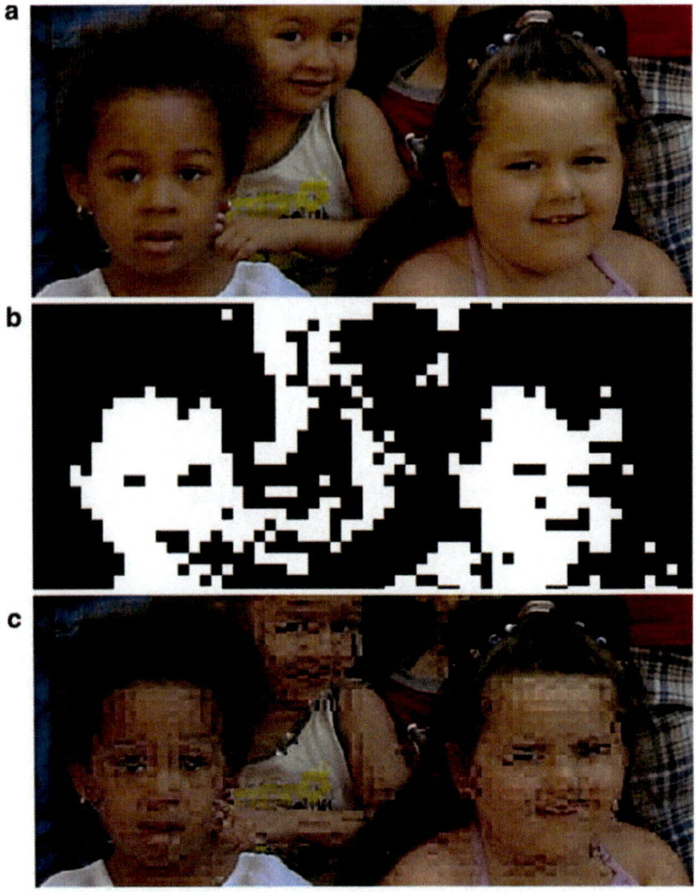

Fig. 12.10 Selective encryption of the ROI corresponding to the skin: (**a**) Original image 416 × 200 pixels, (**b**) ROI detection, (**c**) Protected image

of 1,024 × 696 pixels, resulting in the encryption of only 7.32% from the image. The resulting SE image is shown in Fig. 12.10c.

For our experiments on the colour image sequence illustrated in Fig. 12.11a, we have extracted four images (#083, #123, #135, #147) from a sequence of 186 images acquired with a surveillance video camera. Each one of them is in JPEG format with a QF of 100%. For the encryption, we have used the AES cipher in the CFB stream cipher mode with a key of 128 bits long.

Each RGB original image, 640 × 480 pixels, of the extracted sequence, shown in Fig. 12.11a was converted to YC_bC_r. An example of the image components Y, C_b and C_r for the frame #83 is shown in Fig. 12.12. For the skin selection, we have used the DC of chrominance components C_b and C_r. The binary images were filtered using a morphological opening operation (erosion followed by dilatation) [26] to obtain

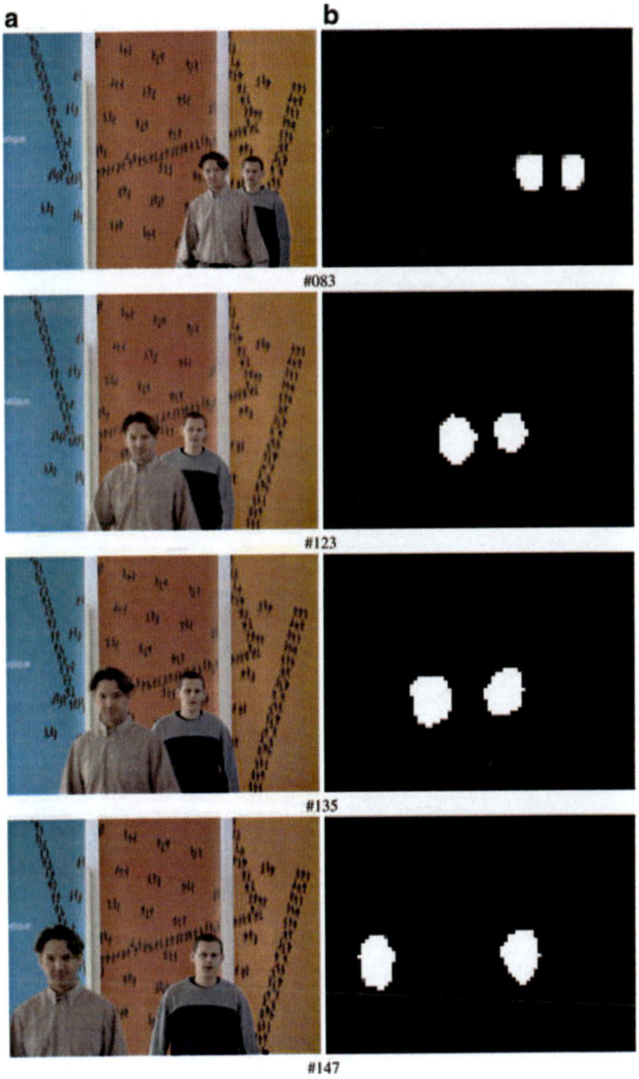

Fig. 12.11 (**a**) Sequence of original images, (**b**) Detection of the ROI representing the skin

the neat binary images illustrated in Fig. 12.11b. The detection of the human skin region, in this case mostly of human faces, is represented by the white pixels. We have mapped a white pixel in the binary image as corresponding to a block of 8×8 pixels from the original image. Finally, we have applied the method described in this study to generate the selectively encrypted images.

Table 12.2 shows the cryptography characteristics for each image. For the frame #083, we have detected 79 blocks representing people faces. This means that 2,547

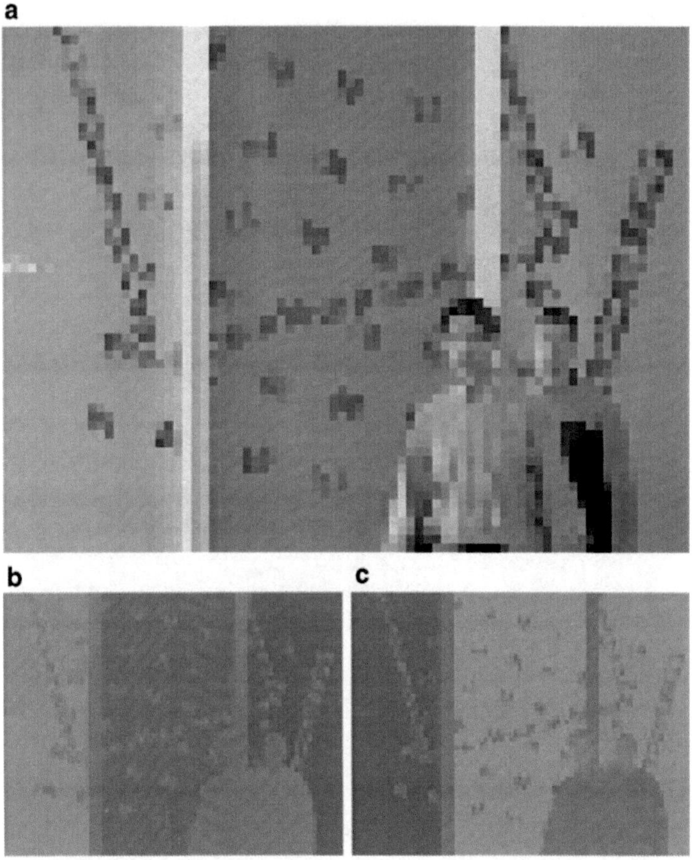

Fig. 12.12 DC coefficients of frame #083 for the three components YC_bC_r: (**a**) Y component, (**b**) C_r component, (**c**) C_b component

Table 12.2 Results of SE in a sequence of images acquired with a surveillance video-camera

	Total ciphered			
Image	Quant. blocks	Coeff.	Bits	Blocks %
083	79	2,547	10,112	1.65
123	113	3,042	14,464	2.35
135	159	4,478	20,352	3.31
147	196	5,396	25,088	4.08

AC coefficients are encrypted, corresponding to 10,112 bits in the Huffman code. The number of encrypted blocks corresponds to 1.6% of the total number of blocks from the original image. For the frame #123, we have 113 blocks. In this frame, we have encrypted 3,042 AC coefficients which represent 14,464 bits corresponding to 2.35% from the total number of blocks in the image. The quantity of blocks for

Fig. 12.13 Sequence of selectively encrypted images

encryption increases because the two persons are getting closer to the video camera. After analyzing Table 12.2, we can conclude that the amount of bits encrypted is very small relatively to the size of the whole image. This makes our method suitable for low-powered systems such as surveillance video camera. Fig. 12.13 shows the final results of face detection, tracking and the selective encryption of the chosen frames. In order to clearly show our results, we have cropped from frame #123 a detail of 216×152 pixels which is shown enlarged in Fig. 12.14.

12.4.3 Cryptanalysis and Computation Time of the SE Method

It should be noted that security is linked to the ability to guess the values of the encrypted data. For example, from a security point of view, it is preferable to encrypt the bits that look the most random. However, in practice this trade-off is challenging because the most relevant information, such as the DC coefficients in a JPEG encoded image are usually highly predictable [6].

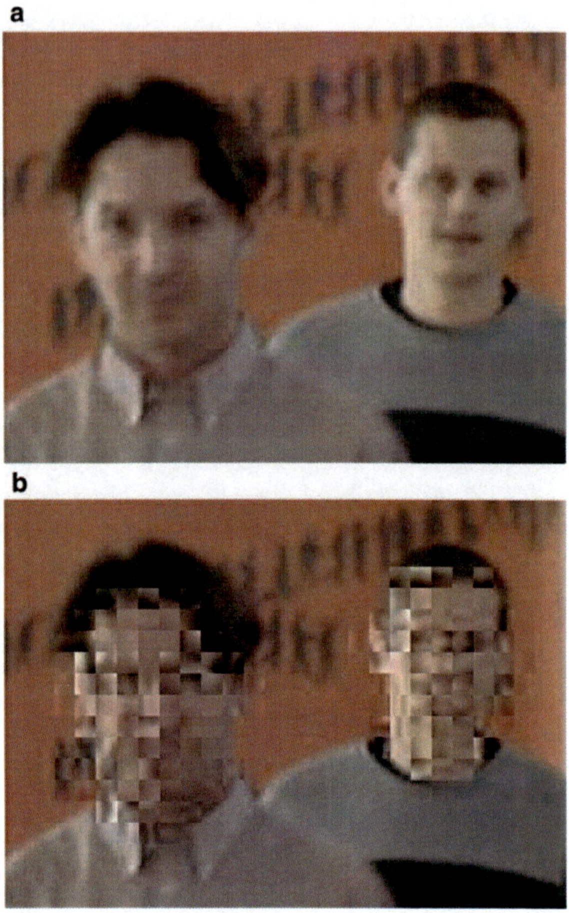

Fig. 12.14 Region of 216×152 pixels from frame #123: (**a**) Original image, (**b**) Protected image

In another experiment, we have replaced the encrypted AC coefficients with constant values. For example, if we set the encrypted AC coefficients of all blocks from Fig. 12.7b, which shows Lena with QF $= 100\%$, $C = 128$ having $PSNR =$ 20.43 dB, to zero, we get the image illustrated in Fig. 12.15. Its PSNR with respect to the original image is 23.44 dB. We can observe that in SE, since we do not encode the Huffman coefficients corresponding to the DC component, the rough visual information can be simply recovered by replacing the ciphered AC coefficients with constant values. This action will result in an accurate but low-resolution image.

Because of the SE, we concede that our method is slower when compared to a single standard JPEG compression processing. Nevertheless, it must be noted that when considering both compression and selective encryption of the image, our method is faster than when applying the two standard methods separately. Consequently, the proposed methodology provides a significant processing time reduction and can provide more than 15 image/s which is a good result in the context of video surveillance camera systems.

Fig. 12.15 Attack in the selectively encrypted image (Fig. 12.7b) by removing the encrypted data (23.44 dB)

12.5 Conclusion

In this chapter, selective encryption systems have been presented for colour images. For JPEG compression colour images, we have developed an approach where the encryption is performed in the Huffman coding stage of the JPEG algorithm using the AES encryption algorithm in the CFB mode. In this way, the proposed encryption method does not affect the compression rate and the JPEG bitstream compliance. The selective encryption (SE) is performed only on the Huffman vector bits that correspond to the AC coefficients as provided by the DCT block of JPEG. The SE is progressively performed according to a constraint onto the Huffman vector bits ordered in the reverse order of their corresponding frequencies. This procedure determines the desired level of selectivity for the encryption of the image content. The DC coefficient provided by the DCT is used as a marker for selecting the ROI for selective encryption. Due to the fact that the Huffman code corresponding to the DC component is not encrypted, a low-resolution version of the image can be visualized without the knowledge of the secret key. This facility can be very useful in various applications. In the decoding stage, we can use the DC coefficient value in order to identify the encrypted regions. The proposed methodology is applied for ensuring the personal privacy in the context of video surveillance camera systems. The colour range of skin is used to detect faces of people as ROI in video streams and afterwards to SE them. Only authorized users that possess the key can decrypt the entire encrypted image sequence. The proposed method has the advantage of being suitable for mobile devices, which currently use the JPEG image compression algorithm, due to its lower computational requirements. The experiments have shown that we can achieve the desired level of encryption in selected areas of the image, while maintaining the full JPEG image compression compliance, under a minimal set of computational requirements. Motion estimation and tracking can be used to increase the robustness and to speed up the detection of ROI. The proposed system can be extended to standard video-coding systems such as those using MPEG [27].

References

1. Chai D, Ngan KN (1999) Face segmentation using skin-color map in videophone applications. IEEE Trans Circ Syst Video Tech 9(4):551–564
2. Chang HT, Lin CC (2007) Intersecured joint image compression with encryption purpose based on fractal mating coding. Opt Eng 46(3):article no. 037002
3. Chen RJ, Horng SJ (2010) Novel SCAN-CA-based image security system using SCAN and 2-D Von Neumann cellular automata. Signal Process Image Comm 25(6):413–426
4. Cheng H, Li X (2000) Partial encryption of compressed images and videos. IEEE Trans Signal Process 48(8):2439–2445
5. Daemen J, Rijmen V (2002) AES proposal: the Rijndael block cipher. Technical report, Proton World International, Katholieke Universiteit Leuven, ESAT-COSIC, Belgium
6. Van Droogenbroeck M, Benedett R (2002) Techniques for a selective encryption of uncompressed and compressed images. In: Proceedings of advanced concepts for intelligent vision systems (ACIVS) 2002, Ghent, Belgium, pp 90–97
7. Fisch MM, Stgner H, Uhl A (2004) Layered encryption techniques for DCT-coded visual data. In: Proceedings of the European signal processing conference (EUSIPCO) 2004, Vienna, Austria, pp 821–824
8. Imaizumi S, Watanabe O, Fujiyoshi M, Kiya H (2006) Generalized hierarchical encryption of JPEG2000 codestreams for access control. In: Proceedings of IEEE internatinoal conference on image processing, Atlanta, USA, pp 1094–1097
9. Kang I, Arce GR, Lee H-K (2011) Color extended visual cryptography using error diffusion. IEEE Trans Image Process 20(1):132–145
10. Leung BW, Ng FW, Wong DS (2009) On the security of a visual cryptography scheme for color images. Pattern Recognit 42(5):929–940
11. Liao XF, Lay SY, Zhou Q (2010) A novel image encryption algorithm based on self-adaptive wave transmission. Signal Process 90(9):2714–2722
12. Lin C-Y, Chen C-H (2007) An invisible hybrid color image system using spread vector quantization neural networks with penalized FCM. Pattern Recognit 40(6):1685–1694
13. Lin ET, Eskicioglu AM, Lagendijk RL, Delp EJ. Advances in digital video content protection. Proc IEEE 93(1):171–183
14. Liu F, Wu CK, Lin XJ (2008) Colour visual cryptography schemes. IET Information Security 2(4):151–165
15. Liu JL (2006) Efficient selective encryption for JPEG2000 images using private initial table. Pattern Recognit 39(8):1509–1517
16. Lookabaugh T, Sicker DC (2004) Selective encryption for consumer applications. IEEE Comm Mag 42(5):124–129
17. Lukac R, Plataniotis KN (2005) Bit-level based secret sharing for image encryption. Pattern Recognit 38(5):767–772
18. Martin K, Lukac R, Plataniotis KN (2005) Efficient encryption of wavelet-based coded color images. Pattern Recognit 38(7):1111–1115
19. Ou SC, Chung HY, Sung WT (2006) Improving the compression and encryption of images using FPGA-based cryptosystems. Multimed Tool Appl 28(1):5–22

20. Pennebaker WB, Mitchell JL (1993) JPEG: still image data compression standard. Van Nostrand Reinhold, San Jose, USA
21. Rodrigues J-M, Puech W, Bors AG (2006) A selective encryption for heterogenous color JPEG images based on VLC and AES stream cipher. In: Proceedings of the European conference on colour in graphics, imaging and vision (CGIV'06), Leeds, UK, pp 34–39
22. Rodrigues J-M, Puech W, Bors AG (2006) Selective encryption of human skin in JPEG images. In: Proceedings of IEEE international conference on image processing, Atlanta, USA, pp 1981–1984
23. Said A (2005) Measuring the strength of partial encryption scheme. In: Proceedings of the IEEE international conference on image processing, Genova, Italy, vol 2, pp 1126–1129
24. Schneier B (1995) Applied cryptography. Wiley, New York, USA
25. Seo YH, Choi HJ, Yoo JS, Kim DW (2010) Selective and adaptive signal hiding technique for security of JPEG2000. Int J Imag Syst Tech 20(3):277–284
26. Serra J (1988) Image analysis and mathematical morphology. Academic Press, London
27. Shahid Z, Chaumont M, Puech W. Fast protection of H.264/AVC by selective encryption of CAVLC and CABAC for I and P frames. IEEE Trans Circ Syst Video Tech 21(5):565–576
28. Stinson DR (2005) Cryptography: theory and practice, (discrete mathematics and its applications). Chapman & Hall/CRC Press, New York
29. Tang L (1999) Methods for encrypting and decrypting MPEG video data efficiently. In: Proceedings of ACM Multimedia, vol 3, pp 219–229
30. Wen JT, Severa M, Zeng WJ, Luttrell MH, Jin WY (2002) A format-compliant configurable encryption framework for access control of video. IEEE Trans Circ Syst Video Tech 12(6):545–557
31. Wu CP, Kuo CCJ (2005) Design of integrated multimedia compression and encryption systems IEEE Trans Multimed 7(5):828–839
32. Xiang T, Wong K, Liao X (2006) Selective image encryption using a spatiotemporal chaotic system. Chaos 17(3):article no. 023115
33. Yabuta K, Kitazawa H, Tanaka T (2005) A new concept of security camera monitoring with privacy protection by masking moving objects. In: Proceedings of Advances in Multimedia Information Processing, vol 1, pp 831–842
34. Yang JH, Choi H, Kim T (2000) Noise estimation for blocking artifacts reduction in DCT coded images. IEEE Trans Circ Syst Video Tech 10(7):1116–1120
35. Yeasin M, Polat E, Sharma R (2004) A multiobject tracking framework for interactive multimedia applications. IEEE Trans Multimed 6(3):398–405
36. Zeng W, Lei S (1999) Efficient frequency domain video scrambling for content access control. In: Proceedings of ACM Multimedia, Orlando, FL, USA, pp 285–293

Chapter 13
Quality Assessment of Still Images

Mohamed-Chaker Larabi, Christophe Charrier, and Abdelhakim Saadane

Blueness doth express trueness

Ben Jonson

Abstract In this chapter, a description of evaluation methods to quantify the quality of impaired still images is proposed. The presentation starts with an overview of the mainly subjective methods recommended by both the *International Telecommunication Union* (ITU) and *International Organization for Standardization* (ISO) and widely used by *Video Quality Experts Group* (VQEG). Then, the algorithmic measures are investigated. In this context, low-complexity metrics such as *Peak Signal to Noise Ratio* (PSNR) and *Mean Squared Error* (MSE) are first presented to finally reach perceptual metrics. The general scheme of these latter is based on the *Human Visual System* (HVS) and exploits many properties such as the luminance adaptation, the spatial frequency sensitivity, the contrast and the masking effects. The performance evaluation of the objective quality metrics follows a methodology that is described.

Keywords Image quality assessment • Evaluation methods • Human visual system (HVS) • International telecommunication union (ITU) • International organization for standardization (ISO) • Video quality experts group (VQEG) • Low complexity metrics Peak signal to noise ratio (PSNR) • Mean squared error (MSE) • Perceptual metrics • Contrast sensitivity functions • Masking effects

---

M.-C. Larabi (✉) • A. Saadane
Laboratory XLIM-SIC, UMR CNRS 7252, University of Poitiers, France
e-mail: chaker.larabi@univ-poitiers.fr; hakim.saadane@univ-poitiers.fr

C. Charrier
GREyC Laboratory, UMR CNRS 6072, Image team 6 Bd. Maréchal Juin, 14050 Caen, France
e-mail: christophe.charrier@unicaen.fr

C. Fernandez-Maloigne (ed.), *Advanced Color Image Processing and Analysis*,
DOI 10.1007/978-1-4419-6190-7_13,

13.1 Introduction

With the first glance on an object, the human observer is able to say if its sight is pleasant for him or not. He makes then neither more nor less than one classification of the perception of this object according to the feeling gotten and felt in two categories: "I like" or "I don't like."

Such an aptitude to classify the visual feelings is indisputably to put in relation with the inherent conscience of each human being. The conscience is related so that *Freud* calls "the perception-conscience system." It concerns a peripheral function of the psychic apparatus which receives information of the external world and those coming from the memories and the internal feelings of pleasure or displeasure. The immediate character of this perceptive function involves an impossibility for the conscience of keeping a durable trace of this information. It communicates them to preconscious, place of a first setting in memory. The conscience perceives and transmits significant qualities. *Freud* employs formulas like "index of perception, of quality, of reality" to describe the content of the operations of the perception-conscience system.

Thus, perception is to be regarded as one of the internal scales of a process driving to an overall quality assessment of an object or an image.

We must, however, notice that by language abuse, it is often made an amalgam between the terms quality and fidelity [40]. The concept of quality could be to the Artist what the concept of fidelity would be to the forger. The Artist generally works starting from concepts, impressions related to its social environment and/or professional and places himself in an existing artistic current(relation Master–student) or in a new current that he creates. The works carried out are thus regarded as originals, and the experts speak about the quality of the works. Behind this approach, one realizes that for the word quality the concept of originality is associated. Who has never found himself faced with work which left him perplexed while his neighbor was filled with wonder? It is enough to saunter in the museums to see this phenomenon. Thus, one qualifies the quality of a work according to his conscience and his personal sensitivity preset from his economic and social environment.

The forger generally works starting from a model and tries to reproduce it with the greatest possible fidelity. In this case, the forger must provide an irreproachable piece and it is not rare that he uses the same techniques employed by the author several centuries before (combination of several pigments to carry out the color, use of a fabric of the same time, etc.). In this case, the copy must be faithful to the original. No one could claim that, in certain cases, the copy can exceed the quality of the original.

From a more pragmatic point of view, the quality of an image is one of the concepts on which research in image processing takes a dominating part. All the problem consists in characterizing the quality of an image, in the same way done by a human observer. Consequently, we should dissociate the two types of measurements: (1) fidelity measurement and (2) Quality measurement.

The fidelity measurement mainly allows to know if the reproduction of the image is faithful or not to the original one. In this case, the measurement set up calculates the distance between the two images. This distance numerically symbolizes the variation existing between the two reproductions of the image.

The quality measurement is close from what does naturally and instinctively the human observer in front of any new work: it gives him an appreciation according to its conscience. Consequently, the human observer cannot be dissociated from the measurement of quality. Thus, the study of the mechanisms allows to apprehend the internal scales used for the quality evaluation by a human observer became a imposing research field. Thus in 1860, *Gustav Theodor Fechner* proposed to measure physical events started intentionally by the experimenter, and of the express answers observers, answers obtained according to specified models.

Within a very general framework, the psychophysics study the quantitative relations shown between identified and measurable physical events, and of the answers given according to a proved experimental rule. These various relations are then interpreted according to models, what contributes to the deepening of our knowledge on the functions of the organism with regards to the environment.

The psychophysical methods allow, in general, to approach the situations studies in which the stimulus is not definable *a priori* but where its structure can be deduced starting from the structure of the observers judgements. The development of operation models of the human visual system is often the required goal at the time of the psychophysical experiments.

At the time of these experiments, the answers distribution integrates a part due to the sensory and perceptive processes, and a part relating to the development processes of the answers. This idea to separate these two components from the answers reflects the influence of the theory of signal detection, and of the design of the organism subjected to an experiment, as a system of data processing.

These experiments are usually used in the field of color-image compression since one wishes to quantify, using a human observer, the quality of a compressed image. In this case, one uses the expression of "subjective tests of quality" which answer a number of constraints.

In this chapter, the proposed tutorial does not concern areas such as medical imaging or control quality. Furthermore, the presentation of evaluation methods of the quality concerns only degraded natural images, and not segmented or classified images. In addition, the complexity associated to all evaluation techniques does not allow us to describe all existing methods for both color and gray-level images. Since color data can be transformed within other color representation containing one achromatic axis and two chromatic axis, only evaluation methods concerned by achromatic place, and so gray-level images, are presented.

The chapter is organized as follows: in Sect. 13.2, subjective measurements as well as experimental environment are described. A description of the aim and of the algorithmic measures is given in Sect. 13.3. Once such measures are designed, we have to evaluate the performances with regard to human judgments measured following requirements indicated in the previous section. The used criteria for performance evaluation are listed in Sect. 13.4. Section 13.5 concludes this tutorial.

13.2 Subjective Measurements

Images and their associated processing (compression, halftoning, ...) are produced for the enjoyment or education of human viewers so their opinion of the quality is very important. Subjective measurements have always been, and will continue to be, used to evaluate system performance from the design lab to the operational environment [25,42]. Even with all the excellent objective testing methods available today, it is important to have human observation of the pictures. There are impairments which are not easily measured yet but which are obvious to a human observer. This situation will certainly go worst with the addition of modern digital compression. Therefore, casual or informal subjective testing by a reasonably expert viewer remains an important part of system evaluation or monitoring. Formal subjective testing has been used for many years with a relatively stable set of standard methods until the advent of digital compression subjective testing described in the ITU recommendation [24] and ISO standards [23]. In the framework of this section, we will only focus on double stimuli methods which means that all the techniques will be with reference.

13.2.1 Specifications of the Experimental Conditions

13.2.1.1 Observer's Characteristics

Observers shall be free from any personal involvement with the design of the psychophysical experiment or the generation of, or subject matter depicted by, the test stimuli. Observers shall be checked for normal vision characteristics insofar as they affect their ability to carry out the assessment task. In most cases, observers should be confirmed to have normal color vision and should be tested for visual acuity at approximately the viewing distance employed in the psychophysical experiment. The number of observers participating in an experiment shall be significant (15 are recommended).

13.2.1.2 Stimulus Properties

The number of distinct scenes represented in the test stimuli shall be reported and shall be equal to or exceed three scenes (and preferably should be equal to or exceed six scenes). If less than six scenes are used, each shall be preferably depicted or alternatively briefly described, particularly with regard to properties that might influence the importance or obviousness of the stimulus differences. The nature of the variation (other than scene contents) among the test stimuli shall be described in both subjective terms (image quality attributes) and objective terms (stimulus treatment or generation).

13.2.1.3 Instructions to the Observer

The instructions shall state what is to be evaluated by the observer and shall describe the mechanics of the experimental procedure. If the test stimuli vary only in the degree of a single artifactual attribute, and there are no calibrated reference stimuli presented to the observer, then the instructions shall direct the observer to evaluate the attribute varied, rather than to evaluate overall quality. A small set of preview images showing the range of stimulus variations should be shown to observers before they begin their evaluations, and the differences between the preview images should be explained.

13.2.1.4 Viewing Conditions

For monitor viewing, if the white point u',v' chromaticities are closer to D50 than D65, the white point luminance shall exceed 60 cd/m^2; otherwise, it shall exceed 75 cd/m^2.

The viewing conditions at the physical locations assumed by multiple stimuli that are compared simultaneously shall be matched in such a degree as critical observers see no consistent differences in quality between identical stimuli presented simultaneously at each of the physical locations. The observer should be able to view each stimulus merely by changing his glance, without having to move his head.

13.2.1.5 Experimental Duration

To avoid fatigue, the median duration (over observer) of an experimental session, including review of the instructions, should not exceed 45 min.

13.2.2 *Paradigms*

13.2.2.1 Comparative Tests

From these kinds of tests, one can distinguish forced-choice experiments and rank-ordering tasks.

The use of a forced-choice experiment [9, 31] lets us determine the sensitivity of an observer. During this test, the observer asks the following question: "Which one of the two displayed images is the best in terms of quality?" Depending on the final application, the original image can be displayed (or not) between the two images (*see* Fig. 13.1).

Another way for judging the image quality is the use of the rank-ordering tests. In that case, the observer has to rank the image quality from the best to the worst

Fig. 13.1 Example of the forced-choice experiment

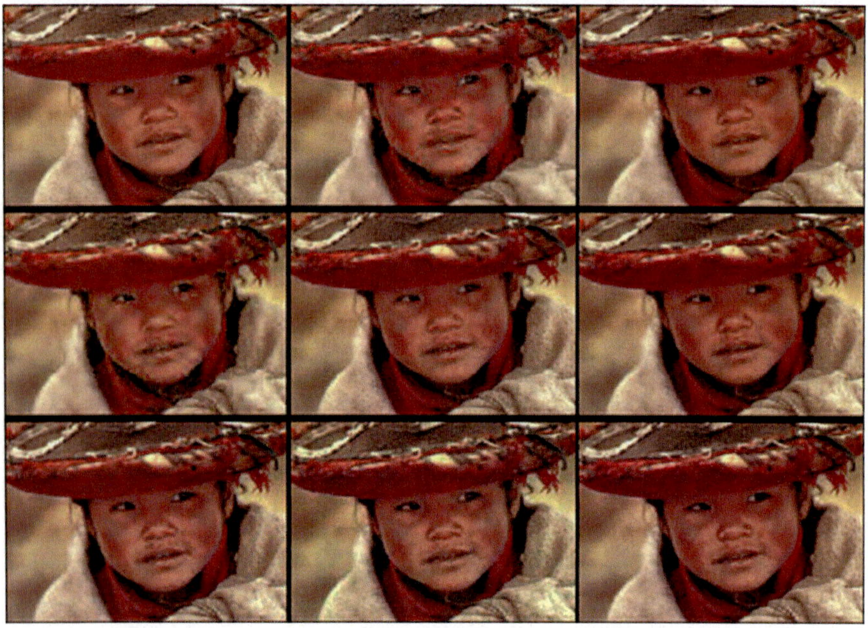

Fig. 13.2 Example of the rank-ordering experiment based on a categorical ordering task

(*see* Fig. 13.2). Nevertheless, this task is not really obvious for the observers. This test can be performed in two ways:

- *An individual ordering*: the observer ranks the images from the best to the worst (or vice versa).
- *A categorical ordering*: the observer classifies the image by the same levels of quality.

These two kinds of tests can be complementary. Indeed, the individual ordering test can be validated by the categorical ordering one.

Table 13.1 Used scores to quantify the image quality

Quality	Score	Signification
Excellent	5	Imperceptible defaults
Good	4	Perceptible defaults but not annoying
Quite good	3	Slightly annoying perceptible defaults
Mediocre	3	Annoying perceptible defaults
Bas	1	Very annoying perceptible defaults

13.2.2.2 Absolute Measure Tests

For such tests, the observer is asked to score the quality of an image. The observer is asked to score the quality of an image. This process is widely used to evaluate the performance of a quality metric. Indeed, we are able to compute the mean opinion score (described below) and compare this score with the one obtained from a particular metric. Table 13.1 shows the widely used scores [24]:

13.2.3 MOS Calculation and Statistical Analysis

13.2.3.1 MOS Calculation

The interpretation of the obtained judgements from many psychophysical tests is not really appropriate, due to their variation with the domain.

The MOS $\bar{u}_{jkr}$ is computed for each presentation:

$$\bar{u}_{jkr} = \frac{1}{N}\sum_{i=1}^{N} u_{ijkr}, \tag{13.1}$$

where u_{ijkr} is the score of the observer i for the degradation j of the image k and the rth iteration. N represents the observers number. In a similar way, we can calculate the global average scores, $\bar{u}_j$ and $\bar{u}_k$, respectively, for each test condition (degradation) and each test image.

13.2.3.2 Calculation of Confidence Interval

In order to evaluate as well as possible the reliability of the results, a confidence interval is associated to the MOS. It is commonly adopted that the 95% confidence interval is enough. This interval is designed as:

$$\lfloor \bar{u}_{jkr} - \delta_{jkr}, \bar{u}_{jkr} + \delta_{jkr} \rfloor, \tag{13.2}$$

where:

$$\delta_{jkr} = 1.95\,\frac{\sigma_{jkr}}{\sqrt{N}}, \tag{13.3}$$

σ_{jkr} represents the standard deviation defined as:

$$\sigma_{jkr} = \sqrt{\sum_{i=1}^{N} \frac{(\bar{u}_{jkr} - u_{ijkr})^2}{N-1}}. \tag{13.4}$$

13.2.3.3 Outliers Rejection

One of the objectives of results analysis is also to be able to eliminate from the final calculation either a particular score, or an observer. This rejection allows to correct influences induced by the observer's behavior, or bad choice of test images. The most obstructing effect is incoherence of the answers provided by an observer, which characterizes the non-reproducibility of a measurement. In the ITU-R 500–10 standard [24], a method to eliminate the incoherent results is recommended.

To that aim, it is necessary to calculate the MOS and the standard deviations associated with each presentation. These average values are function of two variables the presentations and the observers. Then, check if this distribution is normal by using the β_2 test. β_2 is the kurtosis coefficient (i.e., the ratio between the fourth-order moment and the square of the second-order moment). Therefore, the β_{2jkr} to be tested is given by:

$$\beta_{2jkr} = \frac{\frac{1}{N}\sum_{i=1}^{N}(\bar{u}_{jkr} - u_{ijkr})^4}{\left(\frac{1}{N}\sum_{i=1}^{N}(\bar{u}_{jkr} - u_{ijkr})^2\right)^2}. \tag{13.5}$$

If β_{2jkr} is between 2 and 4, we can consider that the distribution is normal. In order to compute P_i and Q_i values that will allow to take the final decision regarding the outliers, the observations u_{ijkr} for each observer i, each degradation j, each image k, and each iteration r, is compared thanks to a combination of the MOS and the associated standard deviation. The different steps of the algorithm are summarized below:

13.2.4 Conclusion

Image and video processing engineers often use subjective viewing tests in order to obtain reliable quality ratings. Such tests have been standardized in ITU-R Recommendation 500 [24] and have been used for many years. While they undoubtedly represent the benchmark for visual quality measurements, these tests are complex and time consuming, hence, expensive and highly impractical or not feasible at all. It assumes that a room with the recommended characteristics has been constructed for this kind of tests. Consequently, researchers often turn to very basic error measures such as root mean squared error (RMSE) or peak signal-to-noise ratio (PSNR) as alternatives, suggesting that they would be equally valid. However,

Algorithm 1: Steps for outliers rejection

if $(2 \leq \beta_{2jkr} \leq 4)$ /* *(normal distribution)* */ **then**
  if $(u_{ijkr} \geq \bar{u}_{ijkr} + 2\sigma_{jkr})$ **then**
    $P_i = P_i + 1;$
  end if
  if $(u_{ijkr} \leq \bar{u}_{ijkr} - 2\sigma_{jkr})$ **then**
    $Q_i = Q_i + 1;$
  end if
end if
else
  if $(u_{ijkr} \geq \bar{u}_{ijkr} + \sqrt{20}\sigma_{jkr})$ **then**
    $P_i = P_i + 1;$
  end if
  if $(u_{ijkr} \leq \bar{u}_{ijkr} - \sqrt{20}\sigma_{jkr})$ **then**
    $Q_i = Q_i + 1;$
  end if
end if
/* *Finally, we can carry out the following eliminatory test:* */
if $\left(\frac{P_i+Q_i}{J.K.R} > 0.05\right)$ *and* $\left(\left|\frac{P_i-Q_i}{P_i+Q_i}\right| < 0.3\right)$ **then**
  Eliminate scores of observer i;
end if
/* *Where J is the total number of degradations, K is the total number of images and R is the total number of iterations.* */

these simple error measures operate solely on a pixel-by-pixel basis and neglect the much more complex behavior of the human visual system. These aspects will be discussed in the following section.

13.3 Objective Measures

As mentioned above, the use of psychophysical tests to evaluate the image quality is time consuming and heavy to start off. This is one of the main reasons justifying the persistence in using algorithmic measures. From these measures, one finds PSNR (*Peak Signal to Noise Ratio*) and MSE (*Mean Squared Error*) [19] measures that directly result from signal processing. Yet, those measures with low complexity are not truly indicators of visual quality. Thus, many investigations have been provided to increase their visual performance.

13.3.1 Low-Complexity Measures

Some criteria based on a distance measure between an input image I and a degraded one $\tilde{I}$ are presented. All these measures are based on a L_p norm. Starting from the various values of p, we obtain:

The Average Difference (AD) between I and $\tilde{I}$ defined by

$$L_1\{I,\tilde{I}\} = \mathrm{AD} = \frac{1}{MN}\sum_{j=1}^{M}\sum_{i=1}^{N}|I(j,i)-\tilde{I}(j,i)|, \quad p=1. \tag{13.6}$$

The root mean square error (RMSE) defined as

$$L_2\{I,\tilde{I}\} = \mathrm{RMSE} = \left\{\frac{1}{MN}\sum_{j=1}^{M}\sum_{i=1}^{N}|I(j,i)-\tilde{I}(j,i)|^2\right\}^{\frac{1}{2}}, \quad p=2. \tag{13.7}$$

In [28], it has been shown that for $p=2$, a good correlation with the human observer for homogeneous distortions (noise) is obtained. Practically, instead of L_2, L_2^2 is often used, and represents the MSE:

$$\left(L_2\{I,\tilde{I}\}\right)^2 = \mathrm{MSE} = \frac{1}{MN}\sum_{j=1}^{M}\sum_{i=1}^{N}|I(j,i)-\tilde{I}(j,i)|^2, \quad p=2. \tag{13.8}$$

These two measurements (13.7 and 13.8) have the same properties in terms of minima and maxima. However, the MSE is more sensitive to the important differences than the RMSE. The MSE can be balanced by the reference image, as given in the following equation:

$$\mathrm{NMSE} = \left\{\frac{1}{MN}\sum_{j=1}^{M}\sum_{i=1}^{N}|I(j,i)-\tilde{I}(j,i)|^2\right\} \Big/ \left\{\frac{1}{MN}\sum_{j=1}^{M}\sum_{i=1}^{N}(I(j,i))^2\right\}. \tag{13.9}$$

Using this balance, labeled to as Normalized Mean Square Error (NMSE), the distance values are less dependent on the reference image. In addition to the previously described measurements, the most frequently used criterion in the literature in order to quantify the quality of a processing on an image, is the *PSNR* (*peak signal noise ratio*), described by the following equation:

$$\mathrm{PSNR} = 10\mathrm{Log}_{10}\frac{(\text{Signal max. value})^2}{\mathrm{MSE}} = 10\mathrm{Log}_{10}\frac{255^2}{\mathrm{MSE}}\ (\mathrm{dB}). \tag{13.10}$$

Typically, each pixel of a monochromatic image is coded on 8 bits, i.e., 255 gray-levels.

Figure 13.3 shows an example of the upper limits of the interpretation of the PSNR. Both images have the same PSNR = 11.06 dB. Nevertheless, the visual perception is obviously not the same. Although this measure is not a true reliable indicator of visual perception, it is, even today, the most popular method used to evaluate the image quality, due to its low complexity. This lack of correlation is mainly due to the fact that this measure does not allow to take into account the

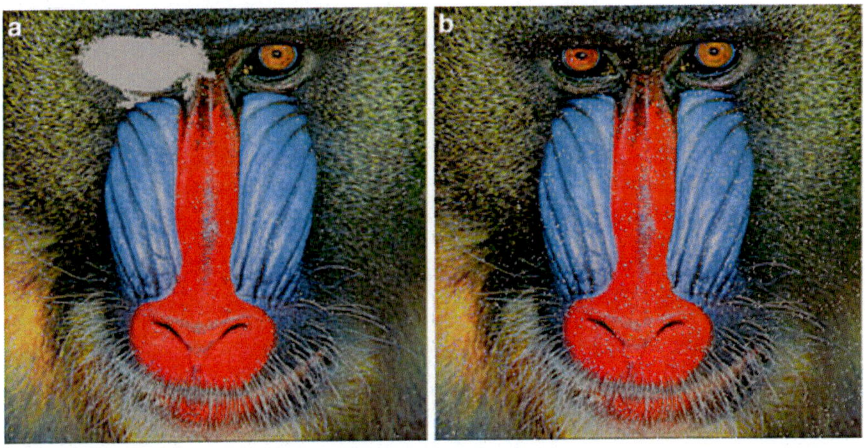

Fig. 13.3 Example of the upper limit of the PSNR (**a**) Addition of a 1,700 pixels region and a Gaussian noise on 300 pixels, (**b**) Addition of a Gaussian noise on 2,000 pixels

correlations between the components, nor the neighborhood of a pixel [12],[45]. In order to increase the correlation between PSNR and visual quality, many metrics have been developed based on *Linfoot* criteria.

Two families of these criteria can be distinguished:

1. The first family of these criteria is based on the study of the properties of the Spectral Power Density (*SPD*) of the reference image, of the degraded image and the error image (differences of the two images); these criteria allow to take into account the spectral properties of the images. One can find the image fidelity criterion (IFC) computed as a ratio of the SPD of the error image and the reference image. The fidelity is equal to one when the output image is equal to the input image.
2. The second family corresponds to correlation measurements on the *SPD* and on the images. *Linfoot* [29] introduces two other quality measurements, the structural content SC and the correlation Q. The structural content is the ratio of the *SPD* of the two images and the *Q* criterion represents the correlation between the different spectra. The structural content is connected to the two other criteria by:

$$\mathrm{Q} = \frac{1}{2}(\mathrm{IFC} + \mathrm{SC}). \tag{13.11}$$

These criteria were used to evaluate the quality of infrared image systems, such as FLIR (*forward-looking infrared*) [38]. *Huck* and *Fales* [13] used the concept of fidelity to build their criterion of mutual information *H*, which consists in an entropy measurement of the ratio between the SPD of the reference image and the degraded one. Then in [21], they used this criterion to evaluate the limiting parameters of a vision system.

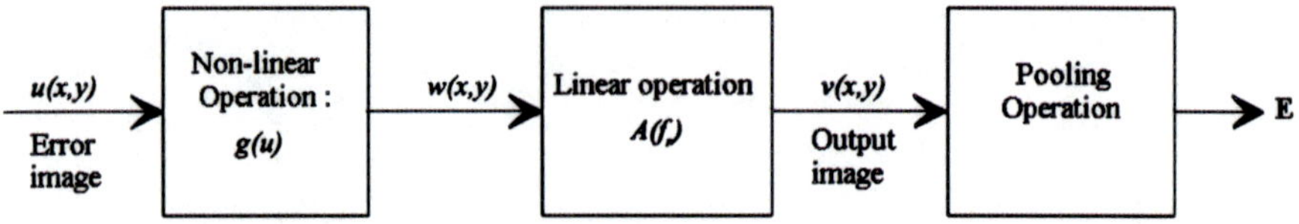

Fig. 13.4 *Mannos* and *Sakrison* model

Another tool based on the criterion of *Linfoot* such as the *Normalized Cross-Correlation* (NCC) is defined in [11]. The NCC represents the correlation between the reference image and the degraded one.

However, the interpretation of criteria like the *SC* and the *NCC* is more difficult than traditional measurement tools.

13.3.2 Measures Based on Error Visibility

In order to counterbalance the drawbacks of the low-complexity measures, many researches have been investigated to develop quality metrics based on one or several known properties of the HVS. The majority of the proposed quality assessment models have followed a weighting of the MSE measures by taking into account penalizing errors in accordance with their visibility [46].

The computations of the standard L_2 distance are carried out on both input and output images after a transformation by a single-channel model. This model consists in a succession of blocks representing the *HVS* low-level processes without taking into account the properties of the visual cortex. In general, the first block consists in a nonlinear transformation (logarithmic function) modeling the perception of light intensity. The second block corresponds to a frequential filtering of the image. This band-pass filter represents the sensitivity to contrast of the *HVS*. Finally, the last block corresponds to in taking into account the masking effect by measuring an activity function. This last is a measurement of the strong variations in the pixel neighborhood. A number of criteria are based on this model that mainly used the CSF.

The quality evaluation criterion of *Mannos* and *Sakrison* [32], was the first to use a vision model for the assessment of image-processing tools. This criterion consists in multiplying the error spectrum by the CSF and then to compute its energy by applying a nonlinear function (e.g., a logarithmic function). Figure 13.4 presents the diagram of this model.

Hall and *Hall* [20] have proposed a model that takes into account at the same time the nonlinearity and the frequential response of the *HVS*. Contrary to *Mannos* and *Sakrison* model, the band-pass filter (CSF) is replaced by a low-pass filter followed by high pass one. The model is represented by Fig. 13.5, a nonlinear function is placed between the two filters. This model represents as well as possible the

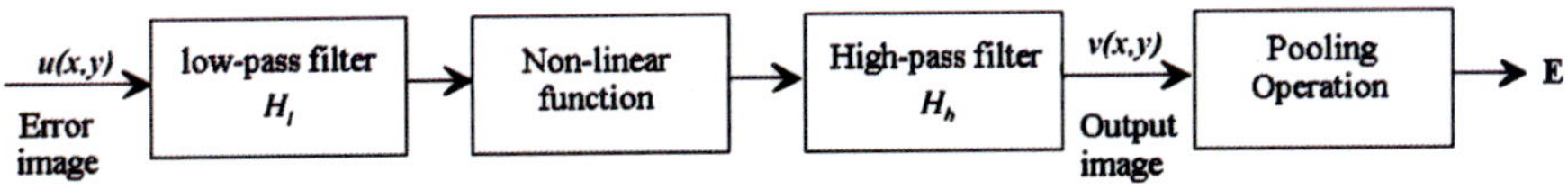

Fig. 13.5 Hall et Hall model

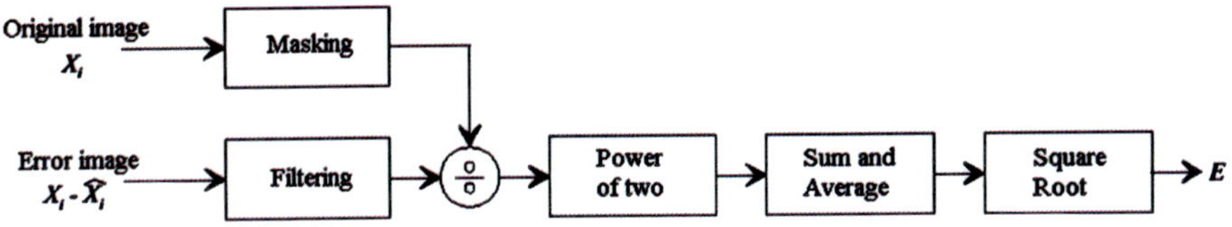

Fig. 13.6 Limb model

physiology of the HVS: the low-pass filter corresponds to the image construction on the retina, the nonlinearity represents the sensitivity of the retina cells and the high-pass filter corresponds to the construction of the neuronal image.

In his criterion, *Limb* [28] looked for a quality objective measurement which is as close as possible to the observer's judgement. To find this measurement, he asked a number of observers to evaluate five types of images having undergone 16 degradations (DPCM coding, noise, filtering, ...) of various intensities. For each family of images and degradations, *Limb* calculates a polynomial regression between the subjective scores and the values of objective measurement. The variance between the experimental points and the regression curves is taken as a performance measurement of the objective model (a weak variance implies a better performance). Finally, *Limb* proposed a complete model of human vision by including an error filtering by a low-pass filter and a weighting by a masking function (Fig. 13.6).

Limb was one of the first researches to take into account the masking effect in a quality evaluation and to study the correlation between objective and subjective measurements for various degradations. However, the simplified modeling of the masking effect and the filtering does not allow to obtain satisfactory results.

Miyahara and *Algazi* [33] proposed a new methodology of quality measurement of an image, called Picture Quality Scale (PQS). In fact, it is considered as a combination of a set of single-channel criteria. They have used the principle that the sensitivity of the HVS depends on the type of distortion introduced into the image. Thus, they use five objective criteria, each one of them address the detection of a particular type of distortion. These criteria could be correlated and a principal component analysis (PCA) allows to project them in an uncorrelated space. Then, a multivariable analysis between the principal components resulting from the PCA and the subjective measurements is carried out. The Fig. 13.7 represents a graph of construction of the different criteria of the PQS. For the computation of the last four criteria, *Miyahara* and *Algazi* used a simplified modeling of the HVS.

The first two criteria F_1 and F_2, are used for the measurement of the random distortions. The F_1 criterion corresponds to the weighting of the error image by a

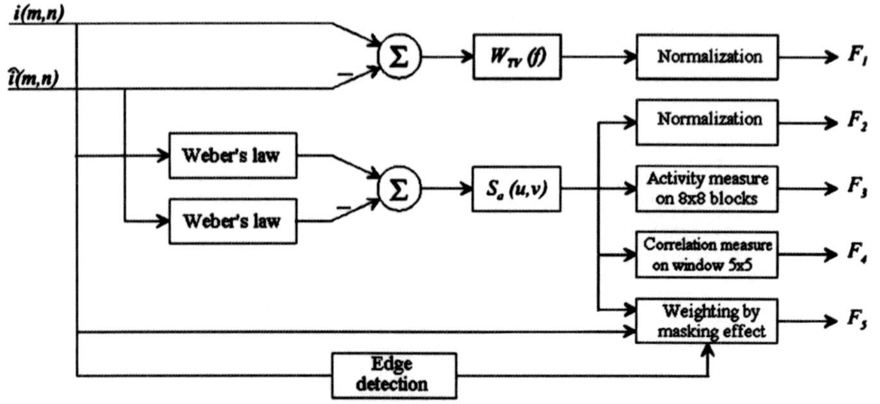

Fig. 13.7 Picture quality scale (PQS) criteria

low-pass filter and to a normalization by the energy of the input image. The F_2 criterion corresponds to the weighting of the error image by a nonlinear function and a band-pass filter, then to a normalization by the energy of the input image.

The three other criteria F_3, F_4, and F_5 are used for the measurement of geometrical and located distortions. The F_3 criterion is used more specifically for the measurement of the block effects. It carries out a sum of the errors between two adjacent blocks (8 pixels) in the vertical and horizontal directions. The F_4 criterion is a correlation measurement, carrying out a sum on the whole error image of the spatial correlation in a 5×5 window. This criterion takes into account degradations due to the perception of textured areas. Finally, the fifth criterion F_5 allows to measure the error on contours by using a weighting of these lasts by an exponential masking function. This criterion allows to take into account the sensitivity of the HVS to the masking effects.

In [44], *Wang* and *Bovik* proposed an objective metric based on a combination of three criteria: (1) a loss of correlation, (2) a luminance distortion, and (3) a contrast distortion. The proposed metric, labeled as a quality index (QI) by the authors, is defined as:

$$\mathrm{QI} = \frac{\sigma_{xy}}{\sigma_x \sigma_y} \cdot \frac{2\bar{x}\bar{y}}{\bar{x}^2 + \bar{y}^2} \cdot \frac{2\sigma_x \sigma_y}{\sigma_x^2 + \sigma_y^2}, \tag{13.12}$$

where the first component measures the degree of linear correlation existing between image x and image y. Its dynamic range is $[-1, 1]$. The second component measures how the mean luminance between x and y is similar. The third component measures how close the contrasts are. σ_x and σ_y can be interpreted as an estimate of the contrast of x and y. In their experiments, the authors claim that the obtained results from QI significantly outperformed the MSE. They suppose that it is due to the strong ability of the metric to measure structural distortions occurred during the image-degradation processes.

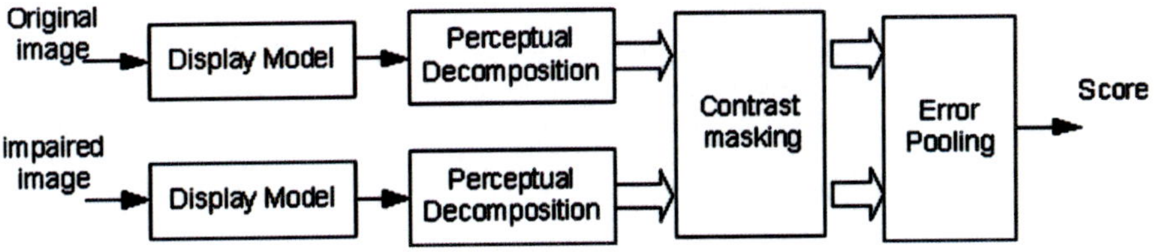

Fig. 13.8 General layout of perceptual quality metrics

Nevertheless, one can observe that in the metrics mentioned previously, there is no particular scheme to follow. Only one or several characteristics of the HVS are used. There is no doubt that more precise modeling of a particular scheme to integrate known characteristics of the HVS will be advantageous in the design of quality metrics.

13.3.3 Perceptual Quality Metrics

The previously described metrics do not always accurately predict the visual quality of images with varying content. To overcome this lack, the Human Visual System (HVS) models have been considered in the design of the perceptual quality metrics. The goal of the latter is to determine the difference between the original and impaired images that are visible to the HVS. For the last ten years, a large number of these metrics has been proposed in literature. Most of them are intended for a large applicability. In this case, the perceptual metrics are computationally intensive. They vary in complexity but use the same general layout [34] (*see* Fig. 13.8).

The common blocks are: display model, decomposition into perceptual channels, contrast sensitivity, contrast masking, and error pooling. The first two blocks vary slightly among authors and do not seem to be critical. The two last blocks vary significantly from one paper to another and metric performance highly depends on their choices. The best-known perceptual metrics will be presented by using this common structure.

13.3.3.1 Display Model

When the quality metrics are designed for a specific set of conditions, the gray levels of the input images are converted to physical luminance by considering the calibration, registration and display model. In most cases, this transformation is modeled by a cube root function. Obviously, the main disadvantage of such approach is that the model has to be adapted to each new set of conditions. At this level, another perceptual factor is considered by some authors [3, 8, 30, 36]. Called light adaptation, luminance masking or luminance adaptation, this factor represents the human visual system sensitivity to variations in luminance. It is admitted that this sensitivity depends on the local mean luminance and is well modeled by the *Weber-Fechner* law. The light adaptation is incorporated in the various metrics by including a nonlinear transformation, typically a logarithmic, cube root or square root function.

13.3.3.2 Perceptual Decomposition

This block models the selective sensitivity of the HVS. It is well known that the HVS analyzes the visual input by a set of channels, each of them selectively sensitive to a restricted range of spatial frequencies and orientations. Several psychophysical experiments have been conducted by different researchers to characterize these channels. Results show that the radial bandwidth is approximately about one octave while the angular selectivity varies between 20 and 60° depending on the spatial frequency. To model this selectivity, linear transforms are used. The requirements and properties of these transforms are well summarized in [10]. Most of well known linear transforms do not meet all the needed properties. The wavelet transform, for example, only has three orientation channels and is not shift invariant. *Gabor* transforms are not easily invertible and block transforms are not selective to diagonal orientations. Currently, two transforms are often used. The first one is the cortex transform and has been used by *Watson* [47], *Daly* [8]. Both authors use a radial frequency selectivity that is symmetric on a log frequency axis with bandwidths nearly constant at one octave. Their decompositions consist in one isotropic low pass and three band-pass channels. The angular selectivity is constant and is equal to 45° for Watson and 30° for Daly. Recently, the cortex transform has also been used by *Fontaine* et al [5] to compare different decompositions. The second transform is called the steerable pyramid [18]. It has the advantage of being rotation invariant, self-inverting, and computationally efficient. In the *Winkler* implementation [50], the basis filters have one octave bandwidth. Three levels plus one isotropic low-pass filter are used. The bands at each level are tuned to orientations of 0, 45, 90, and 135°.

13.3.3.3 Contrast Masking

This block expresses the variation of the visibility threshold of a stimulus induced by the presence of another signal called masker. Different models depending on stimulus/masker nature, orientation, and phase are used in the design of perceptual quality metrics. The best known are discussed in [16, 41]. After the perceptual decomposition, the masking model is applied to remove all errors which are below their visibility thresholds. Hence, only perceived errors are kept in each filtered channel. Two configurations of masking are generally considered in literature. The first one is the intra-channel masking and results from single neurons tuned to the frequency and orientation of the masker and the stimulus. The second one is the inter-channel masking and results from interaction between neurons tuned to the frequency and orientation of the masker and those tuned to the frequency and orientation of the stimulus.

In the case of intra-channel masking, [30] provides the most widely used model. For this model which sums excitation linearly over a receptive field, the masked visual subband (i, j) is computed as:

$$m_{i,j}(k,l) = \frac{a\left|c_{i,j}(k,l)\right|^{\alpha}}{b + \left|c_{i,j}(k,l)\right|^{\beta}}, \tag{13.13}$$

where α and β are constants.

The inter-channel masking has been outlined by several studies which showed that there is a broadband interaction in the vision process. The experiments conducted by *Foley* and *Boynton* [16] on simultaneous masking of *Gabor* patterns by sine wave gratings showed that the inter-channel masking can be significant both when the masker and the stimulus have the same orientation and when their orientations are different. Based on such results, more elaborate models have been presented [15, 41, 49]. *Teo* and *Heeger* model, which restrained the masking to channels having the same radial frequency as the stimulus, is given by:

$$m_{i,j}(k,l) = \gamma \frac{a\left|c_{i,j}(k,l)\right|^{2}}{\sigma^2 + \left|c_{i,j}(k,l)\right|^{2}}, \tag{13.14}$$

where γ is a scaling constant and σ is a saturation constant.

The masking induced by an additive broadband noise has also been examined [37]. The elevation of contrast discrimination has been shown to be proportional to the masker energy. This masking, much larger than the one observed with sinusoidal maskers, tends to decrease significantly according to the duration of the observation [48]. If the observer is given enough time to become familiar with the noise mask, the masking decreases to reach the same level as the one induced by a sinusoidal masker. All the models reported here are derived from experiments conducted with very basic test patches (sinewave, Gabor, noise). The complexity of the real images requires an adaptation of these models and constrains the quality metric designers to approximations. Two studies present a new way to do masking measures directly on natural images.

13.3.3.4 Error Pooling

The goal of this block is to combine the perceived errors as modeled by the contrast masking for each spatial frequency and orientation channel and each spatial location, into a single objective score for the image under test. Most metrics sum errors across frequency bands to get a visible error map and then sum across space. Even if all authors argue that this block requires more elaborate models, the

Minkowski summation is always used. For summation across frequency bands, the visible error map is given by

$$c^P(m,n) = \left[\frac{1}{IJ}\sum_{i=1}^{I}\sum_{j=1}^{J}\left(c_{i,j}^P(m,n)\right)^{\alpha}\right]^{\frac{1}{\alpha}}, \tag{13.15}$$

where $c^P(m,n)$ is the error contrast in frequency band (i,j) and at spatial position (m,n). IJ is the total number of subbands. The value of α varies from 2 to infinity (MAX operator). *Teo* and *Heeger* [41] use $\alpha = 2$. Lubin [30] uses $\alpha = 2.4$. *Daly* implements probability summation with $\alpha = 3.5$ and *Watson* uses the maximum value ($\alpha = \infty$). When a single number is needed, a *Minkowski* summation is also performed across space. *Lubin* and *Daly*, for example, use a MAX operator while *Watson* chooses $\alpha = 3.5$.

13.3.4 Conclusion

To improve the performance of perceptual metrics, higher level visual attention processes are modeled through the use of Importance maps. These maps, which determine the visual importance for each region in the image, are then used to weigh the visible errors before their pooling. A recent study [17] using this technique shows an improved prediction of subjective quality. An alternative way is described in another recent and interesting paper [46] where a new framework for quality assessment based on the degradation of structural information is proposed. The developed structural similarity index demonstrates its promise.

13.4 Performance Evaluation

This section lists a set of metrics to measure a set of attributes that characterizes the performance of an objective metric with regards to the subjective data [4,39]. These attributes are the following:

- Prediction accuracy
- Prediction monotonicity
- Prediction consistency

13.4.1 Prediction Accuracy of a Model: Root Mean Square Error

The RMSE indicates the accuracy and precision of the model and is expressed in the original units of measure. Accurate prediction capability is indicated by a small RMSE.

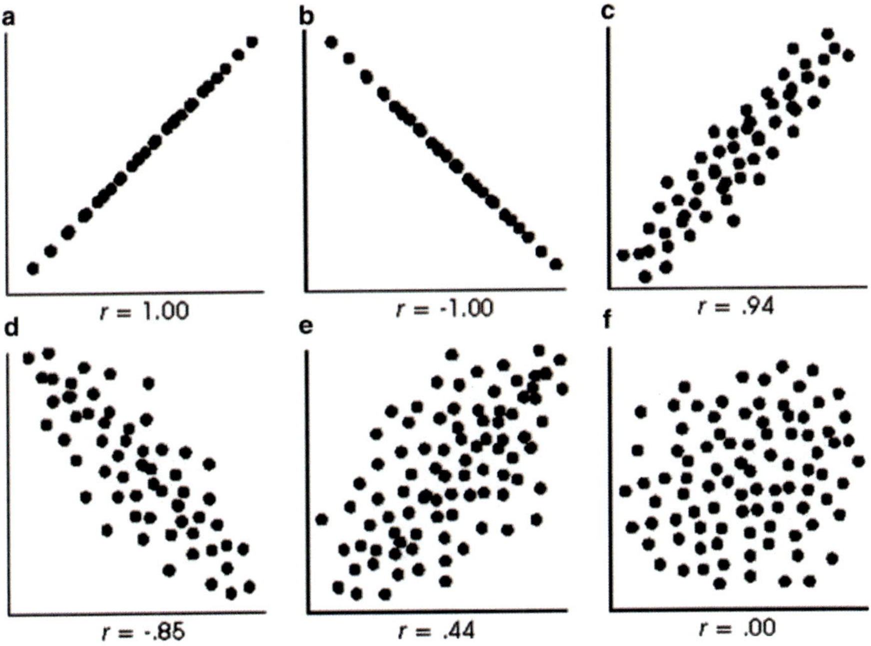

Fig. 13.9 Scatter plots indicating various degrees of linear correlation

$$\mathrm{RMSE} = \sqrt{\frac{1}{n}\sum_{N}(X-Y)^2}. \qquad (13.16)$$

13.4.2 *Prediction Monotonicity of a Model*

Correlation analysis [2, 26] tells us the degree by which the values of variable Y can be predicted, or be explained, by the values of variable X. A strong correlation means we can infer something about Y given X.

The strength and direction of the relationship between X and Y is given by the correlation coefficient. It is often easy to predict whether there is a correlation simply by examining the data using a scatter plot (see Fig. 13.9).

Although we quantitatively assess the relationship, we cannot interpret the relationship without a quantitative measure of effect and significance. The effect is the strength of the relationship—the correlation coefficient, r.

There are several tools on the correlation analysis in literature. In this section, we describe the two tools used in the framework of the VQEG group [43]: Pearson's correlation coefficient and Spearman rank order correlation.

Table 13.2 Critical values of the Pearson correlation coefficient

Level of significance				
df	0.10	0.05	0.02	0.01
1	0.988	0.997	0.9995	0.9999
2	0.900	0.950	0.980	0.990
3	0.805	0.878	0.934	0.959
4	0.729	0.811	0.882	0.917
5	0.669	0.754	0.833	0.874
6	0.622	0.707	0.789	0.834
⋮	⋮	⋮	⋮	⋮

13.4.2.1 Pearson's Correlation Coefficient

Pearson's correlation coefficient r [1, 12, 22, 35] is used for data on the interval or ratio scales, and is based on the concept of covariance. When an X, Y sample are correlated they can be said to covary; or they vary in similar patterns.

The product–moment r statistic is given by:

$$r = \frac{n\sum_{i=0}^{n} X_iY_i - (\sum_{i=0}^{n} X_i)(\sum_{i=0}^{n} Y_i)}{\sqrt{\left(\left[n\sum_{i=0}^{n} X_i^2 - (\sum_{i=0}^{n} X_i)^2\right]\left[n\sum_{i=0}^{n} Y_i^2 - (\sum_{i=0}^{n} Y_i)^2\right]\right)}}, \tag{13.17}$$

where n is the number of pairs of scores. The degree of freedom is $df = n - 2$.

Prior to collecting data, we have to predetermine an alpha level, which corresponds to the error we are going to tolerate when we state that there is a relationship between the two measured variables. A common alpha level for educational research is 0.05. It corresponds to 5% of the population of the experiment. Critical r values are shown in Table 13.2.

For example, If we collected data from seven pairs, the degrees of freedom would be 5. Then, we use the critical value table to find the intersection of alpha 0.05 and 5 degrees of freedom. The value found at the intersection (0.754) is the minimum correlation coefficient r that we would need to confidently state 95 times out of a hundred that the relationship we found with our seven subjects exists in the population from which they were drawn.

If the absolute value of the correlation coefficient is above 0.754, we reject our null hypothesis (there is no relationship) and accept the alternative hypothesis: There is a statistically significant relationship between the two studied properties.

If the absolute value of the correlation coefficient was less than 0.381, we would fail to reject our null hypothesis: There is no statistically significant relationship between the two properties.

13.4.2.2 Spearman Rank Order Correlation

The Spearman rank correlation coefficient [22], r_s (or Spearman's rho), is used with ordinal data and is based on ranked scores. Spearman's rho is the nonparametric analog to Pearson's r.

The process for Spearman's correlation first requires ranking the X and Y scores: the analysis is then performed on the ranks of the scores, and not the scores themselves. The paired ranks are then subtracted to get the values of d, which are then squared to eliminate the minus sign. If there is a strong relationship between X and Y, then paired values should have similar ranks. The test statistic is given by:

$$r_s = 1 - \frac{6\sum_{i=0}^{n} d_i^2}{n(n^2-1)}. \tag{13.18}$$

Significance testing is conducted as usual: the test statistic is r_s and for the critical values we use the ones given in Table 13.2.

13.4.3 *Prediction Consistency of a Model: Outliers Ratio*

Prediction consistency of a model prediction can be measured by number of outliers. An outlier is defined as a data point for which the prediction error is greater than a certain threshold. In general, the threshold is twice the standard deviation σ of the subjective rating differences for the data point.

$$|X_i - Y_i| > 2\sigma \tag{13.19}$$

$$r_o = \frac{N_o}{N}, \tag{13.20}$$

where N_o is the number of outliers.

13.4.4 *Metrics Relating to Agreement: Kappa Test*

The Kappa (K) nonparametric test [6, 7, 14], allows to quantify the agreement between two or several observers when the judgments are qualitative. Let us take the case of a quality assessment campaign where two observers give different judgments or the observers judgment is opposite to objective metric results (Table 13.3). When a reference judgment is missing (which is often the case), this multiplication of opinions does not bring a confidence on the results. A solution consists in carrying out an "agreement" measure by the Kappa coefficient. More generally, the statistical test Kappa is used in the studies of reproducibility which require to estimate agreement between two or several quotations when a discontinuous variable is studied.

In the case of an agreement study between two observers statistically independent, the Kappa coefficient is written:

$$K = \frac{P_o - P_e}{1 - P_e}, \tag{13.21}$$

Table 13.3 Joint proportions of two judgment with a scale of n categories

	Judgment A					
	Category	1	2	...	n	Total
Judgment B	1	p_{11}	p_{12}	...	p_{1n}	p_1
	2	p_{21}	p_{22}	...	p_{2n}	p_2
	.					
	.					
	.					
	n	p_{n1}	p_{n2}	...	p_{nn}	p_n
	Total	p_1	p_2	...	p_n	1

Table 13.4 Agreement degree according the Kappa value

Accord	Kappa
Excellent	0.81
Good	0.80–0.61
Moderate	0.60–0.41
Poor	0.40–0.21
Bad	0.20–0.0
Very bad	< 0.0

where P_o is the observed proportion of agreement and P_e the proportion of agreement by chance.

We call the observed agreement P_o, the proportion of the individuals classified in the diagonal boxes.

$$P_o = \sum_{i=1}^{n} p_{ii}. \tag{13.22}$$

And the agreement by chance P_e is given by:

$$P_e = \sum_{i=1}^{n} p_{i\cdot} p_{i\cdot} \tag{13.23}$$

The agreement will be all the higher as the value of Kappa is close to 1 and the maximum agreement is reached (K = 1) when $P_o = 1$ and $P_e = 0.5$.

Landis and *Koch*[27] have proposed a ranking of the agreement according the Kappa value (*see* Table 13.4).

13.4.5 Conclusion

There is a rich literature about statistical approaches. Most of it is dedicated to biology and medical researches. In this section, we only focused on the tools recommended by VQEG [43] and used in the framework of image quality assessment.

13.5 Final Conclusion

In this contribution, we have presented some approaches dedicated to still image quality assessment. We first presented the subjective measurements and the different protocols starting from the recommendations of the ITU and ISO. It was followed by the paradigms. Finally, we described the MOS calculation and the statistical analysis. As a conclusion, the subjective measurements are complex and time consuming. It assumes that a room with the recommended characteristics has been constructed for the tests.

In the second part, we described the objective measurements starting from the low-complexity metrics such as PSNR and MSE. Quality measures based on objectives methods can be greatly outperformed including low-level characteristics of the HVS (e.g., masking effects) and/or high-level characteristics (e.g., visual attention). Nevertheless, only low-level factors of vision are implemented, mainly due to the high complexity of high-level factors implementation. the most promising direction in this area will be the use of high-level characteristics of the HVS in the objective metrics modeling.

The third part of this contribution has addressed the performance evaluation of the objective metrics with regard to human judgement. In this part, we only mentioned the tools used by VQEG for prediction accuracy, prediction monotonicity, and prediction consistency.

References

1. Altman DG (1991) Practical statistics for medical research. Chapman & Hall, London
2. Ardito M, Visca M (1996) Correlation between objective and subjective measurements for video compressed systems. SMPTE J 105(12):768–773
3. Barten P (1990) Evaluation of subjective image quality with the square-root integral method. J Opt Soc Am 7(10):2024–2031
4. Bechhofer RE, Santner TJ, Goldsman DM (1995) Design and analysis of experiments for statistical selection, screening and multiple comparisons. Wiley, New York
5. Bekkat N, Saadane (2004) A coded image quality assessment based on a new contrast masking model. J Electron Imag 2:341–348
6. Cohen J (1960) A coefficient of agreement for nominal scales. Educ Psychol Meas 20:37–46
7. Cook RJ (1998) Kappa. In: Armitage TP, Colton T (eds.) The encyclopedia of biostatistics, Wiley, New York, pp 2160–2166
8. Daly S (1992) The visible difference predictor: an algorithm for the assessment of image fidelity. In: SPIE human vision, visual processing and digital display III, vol 1666. pp 2–15
9. David H (1988) The method of paired comparisons. Charles Griffin & Company, Ltd., London
10. Eckert MP, Bradley AP (1998) Perceptual quality metrics applied to still image compression. Signal Process 70:177–200
11. Eskicioglu AM, Fisher PS (1993) A survey of quality measures for gray scale image compression. In: AIAA, computing in aerospace 9, vol 939. pp 304–313
12. Eskicioglu M, Fisher PS (1995) Image quality measures and their performance. IEEE Trans Comm 43(12):2959–2965

13. Fales CL, Huck FO (1991) An information theory of image ghatering. Inform Sci 57–58: 245–285
14. Fleiss JL (1981) Statistical methods for rates and proportions. Wiley, New York
15. Foley JM (1994) Human luminance pattern mechanisms: masking experiments require a new model. J Opt Soc Am 11(6):1710–1719
16. Foley JM, Boynton GM (1994) A new model of human luminance pattern vision mechanisms: analysis of the effects of pattern orientation, spatial phase and temporal frequency. In: SPIE proceedings, vol 2054. San José, California, pp 32–42
17. Fontaine B, Saadane A (2004) Thomas a perceptual quality metrics: evaluation of individual components. In: International conference on image processing, Singapore, pp 24–27
18. Freeman WT, Adelson EH (1991) The design and use of steerable filters. IEEE Trans Pattern Anal Mach Intell 13(9):891–906
19. Girod B (1993) What's wrong with mean-squared error. In: Watson AB (ed.) Digital images and human vision, MIT Press, Cambridge, MA, pp 207-220
20. Hall CF, Hall F (1977) A nonlinear model for the spatial characteristics of the human visual system. IEEE Trans Syst Man Cybern 7(3):161–170
21. Huck FO, Fales CL, Alter-Gartenberg R, Rahman ZU, Reichenbach SE (1993) Visual communication: information and fidelity. J Visual Commun Image Represent 4(1):62–78
22. Huck S, Cormier WH (1996) Reading statistics and research. Harper Collins, London
23. ISO 3664:2000 (2000) Viewing conditions-graphic technology and photography. Technical report, ISO, Geneva, Switzerland
24. ITU-R Recommendation BT.500–10: Methodology for the subjective assessment of the quality of television pictures (2000) Technical report, ITU, Geneva, Switzerland
25. Keelan BW (2002) Handbook of image quality: characterization and prediction. Dekker, New York, NY
26. Kendall MG (1975) Rank correlation methods. Charles Griffin & Company, Ltd., London
27. Landis J, Koch G (1977) The measurement of observer agreement for categorical data. Biometrics 33:159–174
28. Limb JO (1979) Distorsion criteria of the human viewer. IEEE Trans Syst Man Cybern 9(12):778–793
29. Linfoot EH (1958) Quality evaluation of optical systems. Optica Acta 5(1–2):1–13
30. Lubin J (1993) The use of psychophysical data and models in the analysis of display system performance. In: Watson A (ed.) Digital images and human vision, MIT, Cambridge, MA, pp 163–178
31. Macmillan NA, Creelman CD (1990) Detection theory: a user's guide. Cambridge University Press, Cambridge
32. Mannos JL, Sakrison DJ (1974) The effects of visual fidelity criterion on the encoding of images. IEEE Trans Inform Theor 20(4):525–536
33. Miyahara M, Kotani K, Algazi VR (1998) Objective picture quality scale (PQS) for image 670 coding. IEEE Trans Comm 46(9):1215–1226
34. Pappas TN, Safranek RJ (2000) Perceptual criteria for image quality evaluation. In: Bovik, A (ed.) Handbook of image and video processing, Academic, pp 669–684
35. Pearson ES, Hartley HO (1966) Biometrika tables for statisticians, vol 1. Cambridge University Press, Cambridge
36. Peli E (1990) Contrast in complex images. J Opt Soc Am 7(10):2032–2040
37. Pelli DG (1990) The quantum efficiency of vision. In: Blakemore C (ed.) Vision: coding and efficiency, Cambridge University Press, Cambridge, pp 3–24
38. Reichenbach SE, Park SK, O'Brien GF, Howe JD (1992) Efficient high-resolution digital filtres for flir images. In: SPIE, Visual Information Processing, vol 1705. pp 165–176
39. Siegel S, Castellan NJ (1988) Nonparametric Statistics for the Behavioral Sciences. McGraw-Hill, Boston
40. Silverstein DA, Farrell JE (1996) The relationship between image fidelity and image quality. In: IEEE international conference image processing, pp 881–884

41. Teo PC, Heeger DJ (1994) Perceptual image distortion. In: International conference on image processing, pp 982–986
42. Thurstone LL (1927) Psychophysical analysis. Am J Psych 38:368–389
43. VQEG: Final report from the video quality experts group on the validation of objective models of video quality assessment. Technical report, ITU-R. http://www.vqeg.org/
44. Wang Z, Bovik AC (2002) A universal image quality index. IEEE Signal Process Lett 9(3): 81–84
45. Wang Z, Bovik AC, Lu L (2002) Why is image quality assessment so difficult. In: Proceedings of ICASSP, Vol 4. Orlando, FL, pp 3313–3316
46. Wang Z, Bovik AC, Sheikh HR, Simoncelli EP (2004) Image quality assessment: from error visibility to structural similarity. IEEE Trans Image Process 13(4):600–612
47. Watson AB (1987) The cortex transform: Rapid computation of simulated neural images. Comput Vis Graph Image Process 39:311–327
48. Watson AB, Borthwick R, Taylor M (1997) Image quality and entropy masking. In: SPIE proceedings, vol 3016. San José, California pp 358–371
49. Watson AB, Solomon JA (1997) Model of visual contrast gain control and pattern masking. J Opt Soc Am 14(9):2379–2391
50. Winkler S (1999) A perceptual distortion metric for digital color images. In: SPIE proceedings, vol 3644. San José, California

Chapter 14
Image Spectrometers, Color High Fidelity, and Fine-Art Paintings

Alejandro Ribés

Until justice is blind to color, until education is unaware of race, until opportunity is unconcerned with the color of men's skins, emancipation will be a proclamation but not a fact

Lyndon B. Johnson

Abstract This book chapter presents an introduction to image spectrometers with as example their application to the scanning of fine-art paintings. First of all, the technological aspects necessary to understand a camera as a measuring tool are presented. Thus, CFA-based cameras, Foveon-X, multi-sensors, sequential acquisition systems, and dispersing devices are introduced. Then, the simplest mathematical models of light measurement and light–matter interaction are described. Having presented these models, the so-called spectral reflectance reconstruction problem is presented. This problem is important because its resolution transforms a multi-wideband acquisition system into an image spectrometer. The first part of the chapter seeks to give the reader a grasp of how different technologies are used to generate a color image, and to which extent this image is expected to be high fidelity.

In a second part, a general view of the evolution of image spectrometers in the field of fine-art paintings scanning is presented. The description starts with some historical and important systems built during European Union projects, such as the pioneering VASARI or its successor CRISATEL. Both being sequential and filter-based systems, other sequential systems are presented, taking care to choose different technologies that show how a large variety of designs have been applied. Furthermore, a section about hyperspectral systems based on dispersing devices is included. Though not numerous and currently expensive, these systems are considered as the new high-end acquisition equipment for scanning art paintings.

A. Ribés (✉)
EDF Research & Development, 1 avenue du Général de Gaulle,
BP 408, 92141 Clamart Cedex, France
e-mail: alejandro.ribes@gmail.com

C. Fernandez-Maloigne (ed.), *Advanced Color Image Processing and Analysis*,
DOI 10.1007/978-1-4419-6190-7_14,

To finalize, some examples of applications such as the generation of underdrawings, virtual restoration of paintings or pigment identification are briefly described.

Keywords Spectral imaging • Color imaging • Image spectrometers • Multispectral imaging • Hyperspectral imaging • Art and technology • Color high fidelity • Spectral reflectance • Art paintings scanning

14.1 Introduction

We are all familiar with the term "High Fidelity" in the electronic audio context. This means that, what is actually reproduced in our audio system at home, is very similar to what we would hear if we were in the concert where the music was recorded. However, very few people have ever heard about color high fidelity. Is the color in our digital images so similar to the real color of the imaged objects that the question is irrelevant? In fact, the colors we see in our holiday digital photos are often far from high fidelity. Furthermore, they can be so low-fidelity that some people would be surprised to view the real scene and compare its in situ colors with its digital image. Fortunately, this is not a major issue for the general market. Is a tourist going to come back to the Fiji islands with his or her images to check if the colors are "high fidelity?" Probability not and, as long as the images are considered beautiful and make the tourist's friends jealous, the tourist will be satisfied of his/her camera and of his/her acquisition skills. However, there are some applications where high-fidelity color is crucial. One clear example is when capturing an image of a fine-art painting in a museum. In this case, the people that will look at the image (in printed form or in a screen) are curators, art-historians or other professionals that are used to perceiving small subtleties in color. Thus, a non-accurate acquisition would generate non-realistic color images and this will limit the range of uses of these images.

Coming back to the title of the chapter, the concept of a Spectral Image is fundamental to this text. Such an image contains a spectral reflectance per pixel (image element) instead of the traditional three values representing color. This is the reason why the first section of this chapter explains the relationship between color and spectral reflectance. Then, it can be understood that high fidelity is easily obtained through the acquisition of spectral images. Furthermore, acquiring a spectral image requires an image spectrometer, which can be considered as an advanced type of digital camera. This is the object of study of this chapter. Thus, the main technologies used to build cameras and image spectrometers are described in Sect. 14.3. Afterward, the basic mathematical models that explain most phenomena in spectral image acquisition are described in Sect. 14.4. Also, the so-called spectral reconstruction problem, that is used to convert a multiband system into an image spectrometer, is introduced in Sect. 14.5. Finally, Sect. 14.6 exemplifies the introduced technologies and methods by presenting existing spectral imaging systems for the capture of fine-art paintings. This choice is justified because accurate color reproduction is especially important in this domain.

In summary, this chapter intends to make the reader understand what is a spectral image-acquisition system for applications that require high–end color performance. Even if extensive bibliography is presented about existing systems for scanning fine-art paintings, it is not an aim to list all existing systems. The bibliography is necessarily incomplete in this sense; however, it should be enough to illustrate the theoretical and technological concepts presented in the chapter.

14.2 Color Or Spectral Reflectance?

It is fundamental before reading the rest of this chapter to understand what is the relationship between color and spectral reflectance. For the non-initiated reader, it can be surprising to learn that color is not a physical property of an object. Color is indeed psychophysical, meaning that is a sensation produced by our brain but induced by physics. On the other hand, the spectral reflectance is a physical property attached to a point of an object's surface. Moreover, color is tristimuli, meaning that is represented by three numbers, these usually are red, green, and blue (even if other color combinations are possible). By contrast, spectral reflectance is a continuous function that represents how much light an object reflects as a function of wavelength.

A color digital image is normally a matrix of tristimulus, typically containing three numbers by element (or pixel) that represents the color stimuli. The cameras based on the acquisition of three channels are strongly dependent on the characteristics of the imaging system, including the illuminant used for image acquisition. This is normal as we cannot precisely mimic an acquisition system that performs the same operations than our eyes and, even if we could, our eyes see different colors depending on external conditions of illumination, surrounding objects properties and other factors. Spectral reflectance, unlike color, is completely independent of the characteristics of the imaging system. Such information allows us to reproduce the image of the object under an arbitrary illuminant. This means that, in any illumination condition, appropriate color reproduction that includes the color appearance characteristics of the human visual system is possible.

In this section, the basic formal relationship between color and spectral reflectance is presented, and also the concept of metamerism introduced. It is not intended to substitute a complete course on colorimetry but to briefly present two important concepts for the understanding of the rest of this chapter.

14.2.1 From Spectral Reflectance to Color

The *Commission Internationale de l'Eclairage* (CIE) defines the CIE 1931 XYZ Standard Colorimetric Observer that is based on the so-called color-matching functions. These functions are designated $x(\lambda)$, $y(\lambda)$, and $z(\lambda)$, and are positively

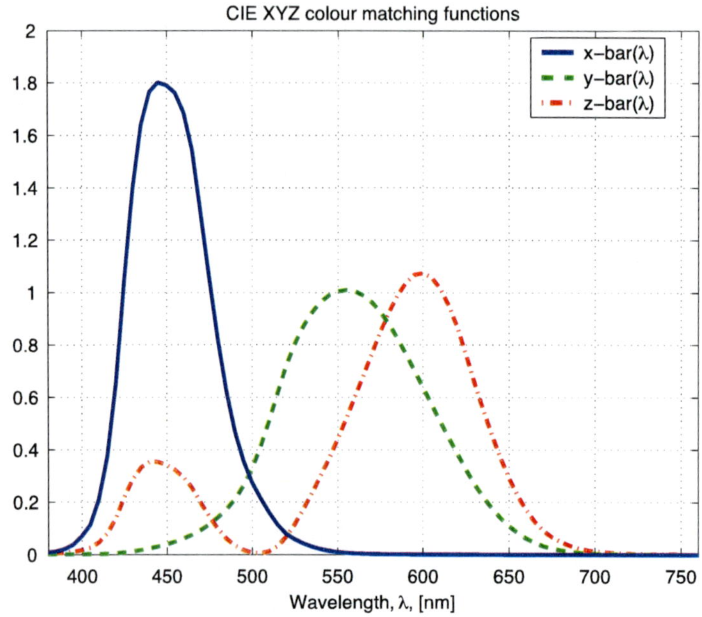

Fig. 14.1 CIE XYZ color-matching functions

valued (see Fig. 14.1). Thus, the X, Y, and Z tristimulus values of a surface point are calculated by integrating the product of its spectral reflectance $r(\lambda)$, the illuminant power distribution $l(\lambda)$ and the corresponding color-matching function as follows:

$$X = \int_{\lambda_{\min}}^{\lambda_{\max}} r(\lambda) l(\lambda) \bar{x}(\lambda)\, d\lambda$$

$$Y = \int_{\lambda_{\min}}^{\lambda_{\max}} r(\lambda) l(\lambda) \bar{y}(\lambda)\, d\lambda$$

$$Z = \int_{\lambda_{\min}}^{\lambda_{\max}} r(\lambda) l(\lambda) \bar{z}(\lambda)\, d\lambda \tag{14.1}$$

where usually $\lambda_{\min} = 380$ and $\lambda_{\max} = 760\,\text{nm}$. From the above equations it is simple to understand the relationship between spectral reflectance and color: color can be conceived as a projection of the spectral reflectance onto three numbers (for each color-matching function) that are modified by the illuminant of the scene. Moreover, the coordinates XYZ define a color space that is device independent. In any case, XYZ tristimulus values can be converted into device-dependent color spaces, such as RGB for monitors or CMYK for printers via a color profile, or alternatively into a psychometric color space such as CIE 1976 L*a*b* (CIELAB), [63].

14.2.2 Metamerism

Metameric color stimuli are color stimuli with the same tristimulus values but which correspond to different spectral power distributions. For color surfaces, metamers are different reflectance spectra that appear to have the same color (i.e., same tristimulus values) to the observer under a given illuminant, but may look different under other light sources.

The elimination of the metamerism phenomena is a fundamental reason for the use of spectral rather than trichomatric imaging when the highest-fidelity color reproduction is required. Color-imaging systems based on sensors with three color filters always exhibit metamerism. First, a metameric reproduction is always illuminant dependent. Therefore, a metameric match is not sufficient if the reproduction is viewed under a variety of illuminants. Imagine the repaired finish of a green car done under daylight to become a patchwork of green and brown under artificial illumination. Secondly, a metameric reproduction is observer dependent. The reproduced color and the original color only match as long as the standard observer is considered. A human observer, however, usually departs slightly from the standard observer, causing a mismatch between the original and the reproduced color.

14.3 Acquisition Systems and Color High Fidelity

In this section, technologies currently used to acquire color images are briefly described. The section aim is twofold: basic color and spectral reflectance capture technologies are presented, and their capacity to obtain color high-fidelity images discussed.

14.3.1 Color Filter Array

Most digital color cameras currently on the market are based on a single matrix sensor whose surface is covered by a color filter array (CFA) [18]. It is important to understand that each pixel receives only a specific range of wavelengths according to the spectral transmittance of the filter that is superposed to that specific pixel. Indeed, one pixel "sees" only one color channel.

There exist a wide variety of CFAs from the point of view of the transmittance of the filters used and the spatial arrangement of the color bands. The most popular CFA among digital cameras is the so-called Bayer CFA, based on red (R), green (G), and blue (B) filters [4]. Its configuration is shown in Fig. 14.2. In general, micro-lenses can be superimposed on the filters in order to increase the overall sensitivity of the sensor by reducing the loss of incident light; these micro-lenses

Fig. 14.2 Bayer color filter array (CFA)

focus the light rays on the sensitive area of each pixel. Also note that CFAs are used interchangeably with CCD (coupled charge device) or CMOS (complementary metal oxide semiconductors).

CFAs-based image acquisition necessitates an algorithm that converts the raw image containing one color channel per pixel into a three colors per pixel image. This operation is called demosaicing. Unfortunately, demosaicing is far from being simple and is an ill-posed problem: a problem that lacks the information to be properly solved. Only one color channel per pixel is acquired but three color channels per pixel should be reconstructed. Algorithms for demosaicing are numerous; there have been algorithms to solve this problem from the 1970s! This leads to different cameras using different demosaicing techniques, which are integrated in the hardware of the camera. In some cameras it is possible to obtain what are called "raw images", basically a copy of the signal responses of the camera sensor, then a non-demosaiced image. This allows the use of off-line advanced and time-consuming algorithms. In any case, a demosaicing algorithm is always applied, either integrated into the camera hardware or externally.

Concerning color high fidelity, the key problem of CFA-based cameras is that the demosaicing algorithms modify the acquired signals. The resultant image can be pleasant or seem visually coherent, but we do not know if it properly approximates the color of the scene. It could be argued that the high-fidelity capabilities of a demosaicing algorithm can be studied. In fact, the number of algorithms is enormous and the study far from simple. Moreover they all degrade the signals and the degradation ratio is, in general, pixel dependent. Furthermore, CFA-based systems are based in three filters and consequently present metamerism. It should be noted that, currently, there exist a small number of CFA-based cameras having four filters in the CFA, and that, these kind of systems also require demosaicing, [43].

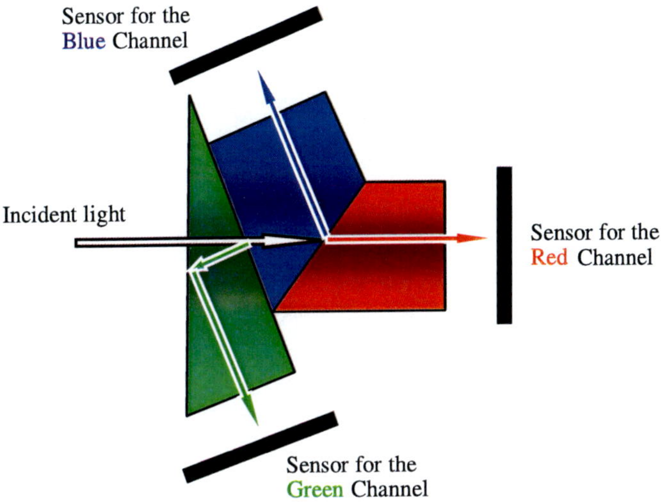

Fig. 14.3 Depiction of a tri-sensor acquisition system where three prisms separate the incoming polychromatic light into a *Green*, *Blue* and *Red* channels. These channels are simultaneously acquired

14.3.2 Multi-Sensor Acquisition

In a multi-sensor camera each color channel has its own sensor. An optical system, normally a dichroic prism is used to direct the incident light of each color band (depending on its wavelength range) to its appropriate sensor [60].

In practice, multi-sensor acquisition systems on the market have three channels whose spectral sensitivities match the "standard" colors R, G, B. See Fig. 14.3. In addition to its high cost, this approach faces a difficulty with alignment of optical elements with the sensors. The color information is obtained by directly combining the responses of different channels, and then it is essential that the convergence properties of light be perfectly controlled. In other words, the image of a point in space should be focused on a single pixel on each sensor with coordinates strictly identical on each one. Such control should be ensured for all points in the same image. This is complex because of the polychromatic nature of light refraction when it passes through the optical elements. In this context, two negative phenomena can appear: first, the image is formed slightly before or slightly rear of one or more of the sensors and, secondly, the pixels of the same coordinates do not receive the light reflected from the same portion of the scene. These phenomena are more or less marked according to the spectral composition of the incident light and depending on the area of the sensors. The impairments are generally most pronounced in the corners. It should be said, that the so-called *Chromatic Aberration* is also a problem in other kinds of cameras, but it is especially important in multi-sensor systems. In addition, problems can be largely enhanced if the angle of incidence of the light varies too strongly.

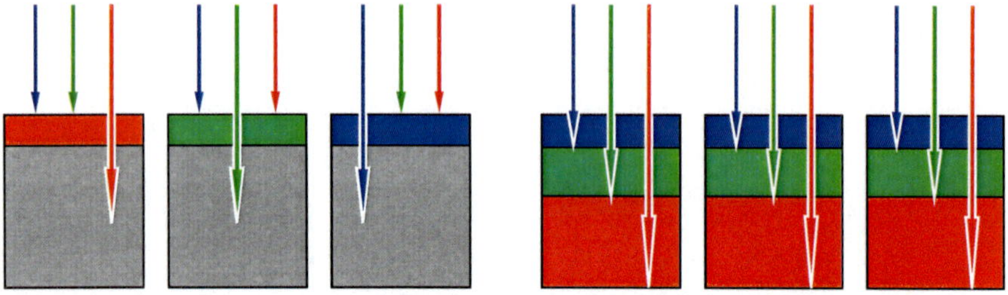

Fig. 14.4 Comparison between a CFA acquisition (*left*) and Foveon-X (*right*): (*left*) In CFA-based systems a per-pixel filter blocks all color channels but the one allowed to be sensed by the sensor, here represented in gray. (*right*) Depicted representation of three pixels of a CMOS Foveon X3 sensor where each color (*blue, green*, and *red*) response is generated in the same spatial location but at different depths in the sensor

Concerning color high fidelity, if properly built, a multi-sensor approach should be capable of high fidelity. However, current systems are mainly tri-sensors and consequently present metamerisms. Potentially more bands could be created but it appears the problem of how many bands and what price will be paid for them. Finally, one big advantage of such a system is that it can be also used for the generation of accurate color video sequences.

Some manufacturers offer color acquisition systems mixing multisensor and CFA. Such systems rely on a sensor covered with a mosaic of red and blue filters and one or two sensors covered with green filters [46]. The alignment problem of the channels is somehow simplified but in return a demosaicing algorithm is needed, that makes the system less adapted for the fidelity of color information.

14.3.3 Foveon X3

There exists a kind of light sensor based in the following property: the penetration depth of light in silicon directly depends on its wavelength. The sensor can then exploit this property by stacking photosensitive layers [25]. The light energy can be collected in silicon crystals where it is possible to differentiate, more or less finely, ranges of wavelengths, the shortest being absorbed at the surface and the longest in more depth. An implementation of this principle was proposed in 2002, the X3 sensor, by the Californian company Foveon, which in 2008 became Sigma Corporation [57]. The Foveon X3 technology is based on CMOS and contains three layers of photodiodes to collect and separate the energy received in three spectral bands corresponding to the usual colors red, green, and blue, see Fig. 14.4.

In theory, this approach has two advantages: (1) it relies only on a single sensor without requiring any system for filtering the incident light and thus, does not need to use a demosaicing algorithm; (2) the color information is acquired in a single exposure [31]. In practice, unfortunately things are not quite that simple. For example, it is not easy to ensure that a light beam, an image point of the scene, interacts only with a single sensor pixel. This is because the light will not propagate in a straight line in silicon and will also have a tendency to spread. In addition, all light rays do not arrive perpendicular to the surface of the sensor, which will reinforce the phenomenon. This directly translates into a degradation of the image resolution and the resulting color fidelity. Moreover, control transistors photodiodes disrupt the propagation of light and constitute a limiting factor in the sensitivity and dynamic range of the sensor. Finally, this approach has only been presented by using three color bands, consequently presents metamerism.

14.3.4 Sequential Acquisition

In this approach, the visual information is obtained by combining acquisitions, typically with a set of filters being inserted sequentially in front of a sensor. Such devices are based on a digital grayscale camera, normally using a non-masked sensor, normally CCD or CMOS. Several optical filters are interposed in the optical path, and several grayscale images using N filters are obtained. Consequently, an image becomes a compendium of N grayscale images that have been acquired using N different filters. In the case of $N = 3$, a sequential digital color camera is obtained. It should be noted that such a system is not equivalent to a usual color camera, among other differences, it does not necessitate a demosaicing algorithm. In general, when N is moderately bigger than three the system is called *multispectral.* If N becomes much bigger, for instance, 100, the system is called *hyperspectral.* However, it is not clear what is the boundary between multispectral and hyperspectral cameras, there is not a number of filters N accepted as a standard limit, some people call a camera with 30 bands multispectral, others hyperspectral.

Multiband sequential cameras can be used to build image spectrometers. If the number of channels is low, they necessitate a reconstruction system in order to generate the per-pixel spectral reflectance image. This step is called spectral reflectance reconstruction and will be formally discussed later in this chapter (Sect. 14.5). The historical trend has been to augment the number of filters to slowly move to hyperspectral systems, which have the advantage of having very simple or quasi-inexistent signal reconstruction. The justification of theses systems is that they are closer to the physical measurement of the reflectance than the multispectral ones. On the other side, narrow-band filters block most light in the spectrum, and so the use of numerous narrow-band filters implies long acquisition times. Thus, the sensors of hyperspectral cameras are often necessarily cooled to reduce noise. This makes the system potentially more expensive and slower but also potentially more precise.

A very common mechanical system found in multispectral and hyperspectral imaging is a grayscale camera with a barrel of filters that rotates to automatically change filters between acquisitions, see Fig. 14.9 (right) for an example. Although highly popular, mechanical scanning devices have some disadvantages: they hold only a limited number of filters with fixed transmittances and the mechanical rotation for the filter-changing process often causes vibrations or non-repeatable positioning that could result in image-registration problems, [11]. Moreover, the tuning speed (changing filters) is relatively slow. These are the reason that motivated the development of other sequential systems that do not need any mechanical displacement in order to change the filter transmittance. At the moment, two main technologies provide this ability:

(1) Liquid crystal tunable filters (LCTF). They are basically an accumulation of different layers, each layer containing linear parallel polarizers sandwiching a liquid crystal retarder element [24]. The reader willing to better understand this technology should be familiar with the concept of light polarization and understand what a retarder does. Indeed, polarization is associated with the plane of oscillation of the electric field of a light wave. Once light is polarized, the electric field oscillates in a single plane. On the opposite, unpolarized light presents no preferred oscillation plane. Furthermore, the plane of oscillation in polarized light can be rotated through a process called retardation. The corresponding optical device is called a *retarder*. The accumulation of the retarders is what generates an optical band-pass transmittance filter. The key idea is that an electric current can control the angle of polarization of the retarders. This is preformed by the application of an electric current in certain liquid crystals, which forces the molecules to align in a specific direction and thus determine the angle of polarization of the crystal.

(2) Acousto-optic tunable filters (AOTF). The operation of these devices is based on the interaction of electromagnetic and acoustic waves. The central component of an AOTF is an optically transparent crystal where both light and an acoustic wave propagate at the same time. The acoustic wave generates a refractive index wave within the crystal. Thus, the incident light beam, when passing through the crystal's refractive index wave, diffracts into its component wavelengths. Proper design allows the construction of the transmittance of a band-pass filter. See, for instance, [13] for more understanding of the basics of AOTF.

Both technologies (1) and (2) do not contain moving parts and their tuning speeds are fast: order of milliseconds for LCTF and microseconds for AOFT.

14.3.5 Dispersing Devices

Optical prisms are widely known to be able to spread a light beam into its component different wavelengths: this phenomenon is called *dispersion*. This is mostly due to the famous historical experiment of Isaac Newton to demonstrate the polychromatic nature of light. Dispersion occurs in a prism because the angle of refraction is dependent on the refractive index of a certain material, which in turn is slightly dependent on the wavelength of light that is traveling through it. Thus, if we attach a linear sensor (CCD or CMOS) to a glass prism, we could capture a sampled version of the spectral power distribution of the light incident to the prism. This basic device could be the base of a spectrometer.

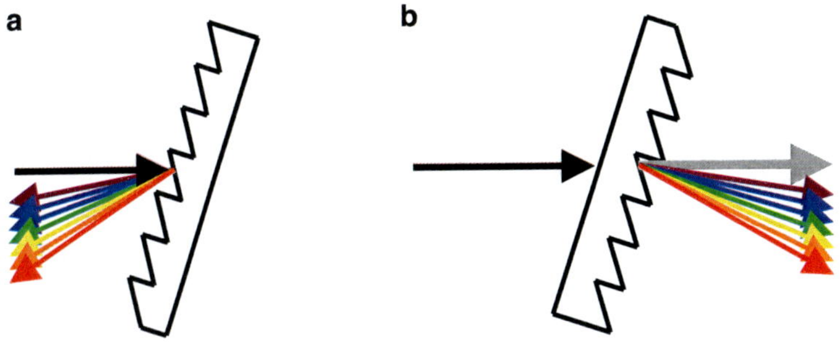

Fig. 14.5 Schematic representation of how dispersion of *white light* is performed

Already for long years, the prism is not the choice for building spectrometers but another optical technology called a *diffraction grating*. A *diffraction grating* is a collection of reflecting (or transmitting) elements periodically separated by a distance comparable to the wavelength of light under study, which produces diffraction. The aim of this section is not to give a course on the physics of diffraction gratings, but it suffices to know that their primary purpose is to disperse light spatially by wavelength. Even if this is the same objective as a dispersion prism, the underlying physical phenomena are of different nature. Instead of refracted, a beam of white light incident on a grating will be separated into its component wavelengths upon diffraction from the grating, with each wavelength diffracted along a different direction.

It is useful to know that there exist two types of diffraction gratings: A *reflection grating* consists of a grating superimposed on a reflective surface, whereas a *transmission grating* consists of a grating superimposed on a transparent surface. Figure 14.5 shows a schematic representation of both devices. Please note that a triangular groove profile is represented in the figures but gratings can also contain other grooves shapes for instance, a sinusoidal profile.

In the context of high-color fidelity acquisition, recent interest is being received by dispersing technologies for building image spectrometers, which have been traditionally used for point spectral reflectance measurements. In fact, displacement systems in combination with line spectrometers are a way of acquiring accurate spectral images. Such a system operates, in general, as follows: The camera fore optic images the scene onto a slit which only passes light from a narrow line in the scene. After collimation, a dispersive element (normally a transmission grating) separates the different wavelengths, and the light is then focused onto a detector array. This process is depicted in Fig. 14.6. Note that this depiction does not respect light angles, distances, and relative positions among components.

The effect of the presented system in Fig. 14.6 is that for each pixel interval along the line defined by the slit, a corresponding spectral reflectance spectrum is projected onto a column of detectors. Furthermore, the detector is not only a column but also a two-dimensional matrix of detectors. Thus, the acquired data contains

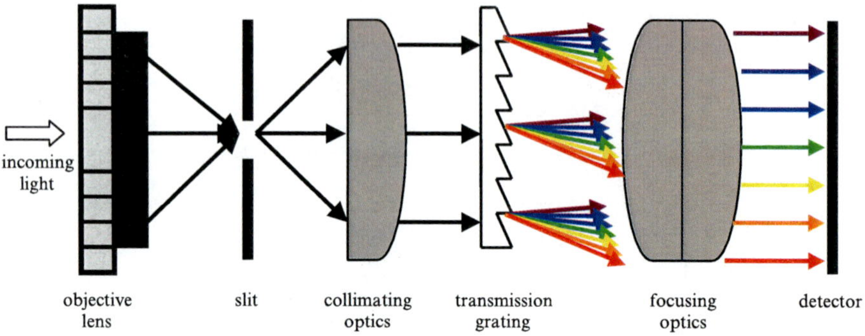

Fig. 14.6 Scheme of an image spectrometer based in a diffusion grating. For simplicity, the detector is depicted as a simple column but it is normally a two-dimensional matrix

a slice of a spectral image, with spectral information in one direction and spatial (image) information in the other. It is important to understand that scanning over the scene is to collect slices from adjacent lines, thus forming a spectral image with two spatial dimensions and one spectral dimension. Examples of the scan of art paintings by use of this technology will be given in Sect. 14.6.4. At the moment, it suffices to know that high-color fidelity is straightforward in this approach.

14.4 Image-Acquisition System Model

Although the general consumer does not conceive of a digital camera as a measurement device, digital acquisition devices are indeed measuring tools. Furthermore, the interaction of light with an object, and light transfer are fundamental for the understanding and design of an image-acquisition system. In this context, *radiometry* is the field of physics that studies the measurement of quantities associated with the transport of radiant energy. This science is well developed. Numerous mathematical models for light transfer, and how to perform measurements of light quantities exist for already long years. In this section, we present first (in Sect. 14.4.1) a very simple model of the measurement of radiant energy by a digital device. Even though simple, this model allows the understanding of the most important aspects of a digital camera or of an image spectrometer. This model has, of course, its limitations: first it does not consider a 3D model of the interaction of light and objects; second, it only considers reflectance on the object. In Sect. 14.4.2 it is shown how the equation becomes more complex when the 3D geometry is taken into account. In Sect. 14.4.3 light transport on the imaged object is briefly discussed.

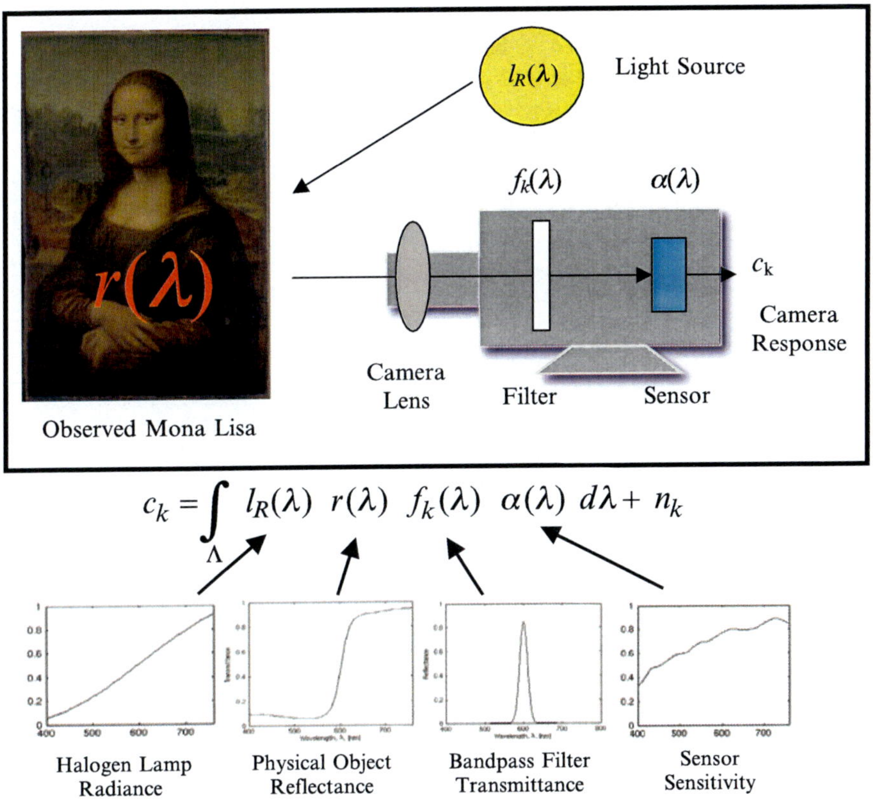

Fig. 14.7 Schematic view of the image acquisition process. The camera response depends on the spectral radiance of the light source, the spectral reflectance of the objects in the scene, the spectral transmittance of the color filter, and the spectral sensitivity of the sensor

14.4.1 A Basic Model

The main components involved in an image acquisition process are depicted in Fig. 14.7. We denote the spectral radiance of the illuminant by $l_R(\lambda)$, the spectral reflectance of the object surface imaged in a pixel by $r(\lambda)$, the spectral transmittance of the $k-$th optical color filter by $f_k(\lambda)$ and the spectral sensitivity of the CCD array by $\alpha(\lambda)$. Note that only one optical color filter is represented in Fig. 14.7 In a multichannel system, a set of filters is used. Furthermore, in a system using a dispersive device a set of delta-dirac shaped $f_k(\lambda)$ can model the acquisition.

Supposing a linear optoelectronic transfer function of the acquisition system, the camera response c_k for an image pixel is then equal to:

$$c_k = \int l_R(\lambda) r(\lambda) f_k(\lambda) \alpha(\lambda) \mathrm{d}\lambda + n_k = \int_\Lambda \phi_k(\lambda) r(\lambda) \mathrm{d}\lambda + n_k, \qquad (14.2)$$

where $\phi_k(\lambda) = l_R(\lambda) f_k(\lambda)\ \alpha(\lambda)$ denotes the spectral sensitivity of the k-th channel, n_k is the additive noise and Λ is the range of the spectrum where the camera is sensible. The assumption of system linearity comes from the fact that the CCD or CMOS sensor is inherently a linear device. However, for real acquisition systems this assumption may not hold, for example due to electronic amplification nonlinearities or stray light in the camera [22]. Stray light may be strongly reduced by appropriate black anodized walls inside the camera. Electronic nonlinearities may be corrected by an appropriate calibration of the amplifiers.

14.4.2 Taking Geometry into Account

The model introduced in Fig. 14.7 and (14.2) leads to an understanding of the light/object interaction as a simple product of the spectral reflectance and the spectral radiance. Even if this could be enough for most applications, it is not what is really happening. One of the main reasons is that object reflectance depends on the angle of viewing and illumination. Thus, the reflectance as presented before is just an approximation of the behavior of the imaged object. This approximation supposes that the light arriving to the object interacts in the same way no matter which direction is considered. Although some materials, called *Lambertians*, present an isotropic reflectance, in general, this is not the case.

In the following equation, the basic model is extended considering reflectance not only as a function of the incident wavelength but also of the viewing direction of the camera. This leads to:

$$c_{k,\psi} = \int_{\Lambda} \phi_k(\lambda) r(\psi,\lambda) \mathrm{d}\lambda + n_k, \tag{14.3}$$

where reflectance, $r(\psi,\lambda)$, is now also depending on ψ, and the camera responses, $c_{k,\psi}$, are different for different viewing directions. It is important to understand that the introduction of this parameter makes the measurement process much more complex. Indeed, if the new reflectance depending on two parameters is to be sampled, then the measuring equipment should move around the imaged object in order to regularly sample a half-sphere centered at each point of the imaged object. This process is, at the moment, only performed by the so-called *goniospectrometers*. A goniospectrometer is designed to measure the reflectance of an object, normally at a unique point of its surface, in numerous viewing directions. This process collects a large amount of data and, normally, requires a dedicated experimental set up. The problem of imaging a whole object by using the model presented in (14.3) seems, for this high complexity, not treated at the moment. In any case, serious practitioners are always aware of the dependence of the viewing angle on image acquisition. For instance, some well-known experimental set up exists, such us the so-called 0/45, where the acquisition device is positioned perpendicular to the imaged object and the light source is positioned at 45°.

In (14.3), the direction of the light source is intentionally not included. This is because, often, the acquisition is performed in a controlled environment, the light sources radiance distributions can be measured and their directions of illumination defined a priori. Even if this is not done, often, calibration data is collected before the acquisition, which helps eliminating the variability that light spectral distribution and direction introduces in the images, see, for instance, [54]. However, when the light source direction is taken into account too, the acquisition model deals with the *Bidirectional Reflectance Distribution Function* (BRDF), [44]. The BRDF at a point on an object, $r(\psi_L, \psi, \lambda)$, depends on three parameters: the wavelength, λ, the incident light direction, ψ_L, and viewing direction, ψ. An important property to the BDRF is its symmetry or reciprocity condition, which is based in the *Helmboltz reciprocity rule* [15]. This condition states that the BDRF for a particular point remains the same if the incident light direction and viewing direction are exchanged:

$$r(\psi_L, \psi, \lambda) = r(\psi, \psi_L, \lambda) \tag{14.4}$$

Although this property simplifies somehow the complexity of the model, the BDRF remains a difficult function to measure, store, and compute for a realistic rendering. This is due to its dependence on the already described three parameters but, also, this function is different for each point in the surface of an object. Now, the reader should have realized that a reflectance-oriented acquisition system presents non-trivial problems and a considerable explosion on acquired data if fidelity (spectral and directional) is required. The discussion about reflectance models will stop here; the interested reader can find details about realistic models of light/objects interactions in [44].

14.4.3 Not Only Reflectance in the Surface of the Imaged Object

The accurate modeling of interactions between light and matter requires considering not only reflectance on the surface of an object but also light transport in the object. In fact, light in a more or less degree "enters" the surface of objects. This kind of phenomenon requires the addition to the also-defined BDRF of the so-called *Bidirectional Transmittance Distribution Function* (BTDF), [44]. Furthermore, light transport in the object can be local or global. For instance, in a translucent object, light entering in a particular surface position can follow complicated paths under its surface emerging partially in other surface positions. In this general case of global transport inside the object, a complex model called *Bidirectional Scattering-Surface Distribution Function* (BSSDF) is required, [44]. In spectral image, sometimes it is necessary to consider light transport on the imaged object, but, when considered, this transport is always local.

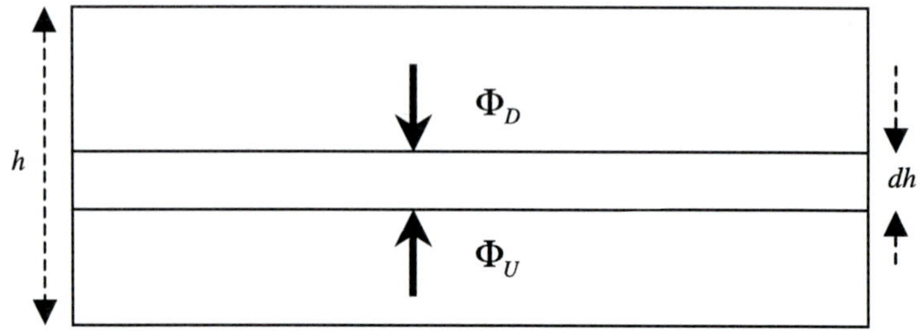

Fig. 14.8 Geometry used to formulate the Kubelka–Munk two-fluxes theory

In this section a local light-transfer model quite popular among spectral imaging practitioners is presented: the Kubelka–Munk model, [34]. Only this model is introduced due to its simplicity and popularity. Kubeka–Munk is a deterministic model, but the reader should be aware that other non-deterministic models exist, normally based in Monte-Carlo computations.

14.4.3.1 Kubelka–Munk

Early in the twentieth century, Kubelka and Munk [34] developed a simple relationship between the absorption coefficient (K) and the scattering coefficient (S) of paint and its overall reflectance: the so-called K–M theory. Originally it was developed to explain light propagation in parallel layers of paint (these layers are considered infinite). Currently K–M theory is widely used in the quantitative treatment of the spectral properties of a large variety of materials. Multiple extensions of this theory exist but they will not be treated here.

The original K-M theory applies two energy transport equations to describe the radiation transfer in diffuse scattering media using the K and S parameters. It is considered being a two-flux theory because only two directions are considered, namely a diffuse downward flux, Φ_{D}, and a diffusive upward flux, Φ_{U}. The relations between the fluxes are expressed by two simultaneous differential equations, [34]. Before presenting these equations, understanding of the passage of light thought an elementary layer is necessary. In Fig. 14.8, the basic elements of the equations are presented graphically. The thickness of the elementary layer is denoted dh while h is the thickness of the material. It is assumed that h is big compared to dh and that dh is larger than the diameter of colorant particles (pigments or dyes) embedded in the material. Due to absorption, the colorant particles reduce the diffuse radiant fluxes, formally expressed as $K\Phi_{\mathrm{D}}dh$. At the same time, scattering generates a flux $S\Phi_{\mathrm{D}}dh$ in the reversal direction of Φ_{D}. Symmetrically, the upward flux is also reduced by absorption by $K\Phi_{\mathrm{U}}dh$, and augmented by scattering by $S\Phi_{\mathrm{U}}dh$. Furthermore, the amount $S\Phi_{\mathrm{D}}dh$ is added to the upward flux and $S\Phi_{\mathrm{U}}dh$ to the downward flux.

Thus, the differential equations for the downward and upward fluxes are given by

$$-\,\mathrm{d}\Phi_{\mathrm{D}} = -(S+K)\Phi_{\mathrm{D}}dh + S\Phi_{\mathrm{U}}dh \tag{14.5}$$

and

$$\mathrm{d}\Phi_{\mathrm{U}} = -(S+K)\Phi_{\mathrm{U}}dh + S\Phi_{\mathrm{D}}dh \tag{14.6}$$

respectively. Kubelka obtained explicit hyperbolic solutions for these equations, [35]. Most of the time simplifications of the general solutions are used because they lead to very tractable expressions. No solution is presented here as it is out of the scope of this introduction.

Finally, it is important to know that the K and S parameters of the K–M theory present an interesting linear behavior, [20]. Indeed, if a mixture of colorants is present in a layer, the overall K and S coefficients of the layer are the result of a linear combination of the individual parameters of each of the n colorants:

$$K = C_1K_1 + C_2K_2 + \cdots + C_nK_n \tag{14.7}$$

and

$$S = C_1S_1 + C_2S_2 + \cdots + C_nS_n \tag{14.8}$$

where the C_i, $I = 1\ldots n$, are the *concentrations* of each colorant. This is, in fact, one of the main reasons of the popularity of this model. Easy predictions of the reflectance of a material can be formulated from its colorants.

14.5 Modeling Spectral Acquisition and Spectral Reconstruction

Before using the spectral reflectance for high-fidelity color reproduction or other uses, we stress that this property of the materials can be estimated for each pixel of a multispectral image. In order to understand how this estimation is performed, we first discretize the integral equation (14.2). Then, we should understand how to reconstruct a spectral reflectance function in each pixel of the image. If this is properly done, we should have created an image spectrometer giving the spectral signature of each image element. This section is dedicated to this issue.

14.5.1 Discretization of the Integral Equation

By uniformly sampling the spectra at N equal wavelength intervals, we can rewrite (14.2) as a scalar product in matrix notation:

$$c_k = \boldsymbol{\phi}_k^t\mathbf{r} + n_k \tag{14.9}$$

where $r = [r(\lambda_1)r(\lambda_2)\ldots r(\lambda_N)]^t$ and $\boldsymbol{\phi}_k = [\phi_k(\lambda_1)\phi_k(\lambda_2)\ldots\phi_k(\lambda_N)]^t$ are vectors containing the sampled spectral reflectance function, and the sampled spectral sensitivity of the k-th channel of the acquisition system, respectively. The vector $\mathbf{c}_K = [c_1 c_2 \ldots c_K]^t$ representing the responses of all K channels may then be described using matrix notation as:

$$\mathbf{c}_k = \Theta\mathbf{r} + \mathbf{n}, \tag{14.10}$$

where $\mathbf{n} = [n_1 n_2 \ldots n_K]^t$, and $\boldsymbol{\Theta}$ is the K-line, N-column matrix defined as $\boldsymbol{\Theta} = [\phi_k(\lambda_n)]$, where $\phi_k(\lambda_n)$ is the spectral sensitivity of the k-th channel at the n-th sampled wavelength.

14.5.2 A Classification of Spectral Reflectance Reconstruction Methods

We decided to base this section on a classification of the methods for spectral reflectance reconstruction. The methods are divided into three families: direct inversion, indirect inversion, and interpolation. For a more detailed description of reflectance reconstruction methods and inverse problems please refer to [56].

14.5.2.1 Direct Reconstruction

Direct reconstruction appears in the case where operator $\boldsymbol{\Theta}$ in (14.10) is known. Then, the problem consists in finding vector $\mathbf{r}$ when $\mathbf{c}_K$ is given. This should be reached by inversing matrix $\boldsymbol{\Theta}$, in the absence of noise: $\mathbf{r} = \text{inv}(\boldsymbol{\Theta})\ \mathbf{c}_K$. From this apparently simple linear system we remark that the matrix $\boldsymbol{\Theta}$ is in general not a square matrix, consequently the system itself is over or underdetermined by definition. This means that either the system has no solution or it has many. This is a so-called *ill-posed problem*

The notion of a *well-posed problem* goes back to a famous paper by Jacques Hadamard published in 1902, [26]. A well-posed problem in the sense of Hadamard is a problem that fulfils the following three conditions:

1. The solution exists
2. The solution is unique
3. The solution depends continuously on the problem data

Clearly, inversing matrix $\boldsymbol{\Theta}$ does not respect conditions 1 or 2 of Hadamard definition: the problem is ill posed. The third condition is not as straightforward to see as the others, but modern numerical linear algebra presents enough resources for the analysis of the stability of a matrix. If the matrix is singular its inverse will be unstable. The condition number, the rank of the matrix or the Picard condition, among others, are good analytical tools to determine if we are dealing with an ill-posed problem, see [27] for a valuable reference on this subject. In this context it is important to know the meaning of *regularization*. In fact, regularization means to make an ill-posed problem well posed. The reader should be aware that this simply defined regularization process can be the object of complex mathematics, especially when working on nonlinear systems. However, the spectral reflectance reconstruction problem is mainly linear, as can be seen from (14.2)

Some representative methods of this approach are based on a Wiener filter, which is a classical method for solving inverse problems in signal processing. From [52] this method, indeed, continues to be used, see [28] or [61]. Other approaches involving a priori knowledge over the imaged objects are also found, see for instance [29].

Difficulties Characterizing the Direct Problem

In direct reconstruction Θ is supposed to be known, but knowing Θ means that a physical characterization of the acquisition system has been performed. This characterization requires at least the measurement of the sensor sensitivity, filter transmittances and transmittance of the optics. This characterization involves the realization of physical experiments in which, typically, a monochromator is used for measuring the sensor sensitivity and a spectroradiometrer for measuring the spectral transmittances of the filters and of the other optical systems of the camera. For a sensor the noise model can be considered Gaussian, this assumption is justified by the physics of the problem. To study the noise a series of images is acquired with the camera lens being occluded with a lens cap or with the whole equipment placed in a dark room.

14.5.2.2 Indirect Reconstruction

Indirect reconstruction is possible when spectral reflectance curves of a set of P color patches are known and a multispectral camera acquires an image of these patches. From this data a set of corresponding pairs ($\mathbf{c}_p$, $\mathbf{r}_p$), for $p = 1,\ldots P$, is obtained; where $\mathbf{c}_p$ is a vector of dimension K containing the camera responses and $\mathbf{r}_p$ is a vector of dimension N representing the spectral reflectance of the p-th patch. Corresponding pairs $(\mathbf{c}_p, \mathbf{r}_p)$ are easy to obtain, professional calibrated color charts such as the GretagMacbeth™ DC are sold with the measurements of the reflectances of their patches. In addition, if a spectroradiometer is available, performing the measure is a fairly simple experiment. Obtaining the camera responses from the known spectral curves of the color chart is just a matter of taking a multispectral image.

A straightforward solution is given by

$$\Theta^{-}_{\text{Indirect}} = \mathbf{R} \cdot \mathbf{C}^{t} \cdot (\mathbf{C} \cdot \mathbf{C}^{t})^{-1}, \tag{14.11}$$

where $\mathbf{R}$ is a $N \times P$ matrix with columns containing all the $\mathbf{r}_p$'s and $\mathbf{C}$ is a $K \times P$ matrix with columns containing their corresponding $\mathbf{c}_p$'s. Most methods of this paradigm can be understood as a variation of the above formula.

Methods based on indirect reconstruction are numerous, see for instance [9,33] or [65]. Historically, they appeared later than the direct inversion methods and, due to their relatively easy to use approach, are currently quite spread. Even if we presented them in a linear perspective, nonlinear versions are also possible, see [53].

14.5.2.3 Solutions Based on Interpolation

A multispectral system can be considered as a tool that samples spectral reflectance curves. Instead of using delta Dirac functions for the sampling as in the classical framework, the spectral transmittance functions $f_k(\lambda)$ of the K filters are considered to be the sampling functions. This approach just requires the camera response itself, $\mathbf{c}$. The methods based on this paradigm interpolate the camera responses acquired by a multispectral camera by using a smooth curve. The smoothness properties of the interpolating curve introduce a natural constraint which regularizes the solutions. However, there are two underlying problems to take into account before representing the camera responses in the same space as spectral curves:

- Positioning the camera response samples in the spectral range. For instance, in the case of Gaussians-shaped filters, the camera responses can be positioned at the center of the filter. However, real filters are rarely Gaussian-shaped. In general, it is admitted that if a filter is narrow, positioning the camera responses can be done with low uncertainty. Unfortunately, when wide-band filters are used this uncertainty increases with the spectral width of the filter. This is the reason why interpolation methods should be used only with multispectral cameras using narrowband pass filters.
- The camera must be radiometrically calibrated. This means that camera responses must be normalized, i.e. to belong to the [0, 1] interval for all the camera channels. In high-end applications this normalization implies the use of a radiometric standard white patch. This reference patch is imaged for normalization as part of a calibration procedure.

Most practical applications of interpolation to spectral reconstruction are used in cases where the sensor is cooled and Gaussian like filters are available, see [30]. Such methods are reported not to be well adapted to filters having more complex wide-band responses, suffering from quite severe aliasing errors [9, 37]. Cubic splines were applied in this context by [36]. They are well adapted to the representation and reconstruction of spectral reflectance curves because they generate smooth curves, C^2 continuity being assured. Keusen [36] also introduced a technique called modified discrete sine transform (MDST) that is based upon Fourier interpolation.

14.6 Imaging Fine Art Paintings

This section was originally intended to present multi-domain applications of spectral imaging to high-end color performance environments. However, due to the difficult task of listing all, or at least a representative part of the existing systems, the discussion is reduced to the imaging of fine-art paintings. This choice is justified, as accurate color reproduction is especially important in this domain. Moreover, sophisticated spectral imaging equipment has always been applied to art paintings, which historically present some pioneering systems worth being known. Although extensive bibliography is presented, this bibliography is necessary incomplete due to the large amount of existing applications. In any case, what is intended is to provide enough examples to illustrate the preceding sections of this chapter.

The history of spectral imaging of fine art paintings starts with a pioneering system called VASARI. This system was developed inside a European Community ESPRIT II project that started in July 1989. The project name was indeed VASARI (visual art system for archiving and retrieval of images), [42]. Research about imaging art paintings continued to be funded by the European Community in successive projects, like MARC (methodology for art reproduction in colour), [17], the last project about this subject being called CRISATEL (conservation restoration innovation systems for image capture and digital archiving to enhance training, education, and lifelong learning) that finished in 2004. In this section, we will describe the VASARI and CRISATEL projects, but not MARC because it is not based in spectral technology. These systems being all sequential and filter based, other sequential systems are presented, choosing different technologies to show how a large variety of designs have been applied to the scanning of art paintings. Moreover, a section about hyperspectral systems based on dispersing devices is included. These kind of systems appeared recently in the scanning of art paintings and some researchers consider them as the new high-end acquisition equipment. To finalize this section, some examples of specific problems such as virtual restoration or pigment identification are briefly described. They aim to help the reader better understand the potential applications of the presented systems.

Other points of view about the digital scanning of paintings can be found in tutorials such as [6] or [23].

14.6.1 VASARI

A principal motivation behind the original VASARI project, [42], was to provide an accurate means of measuring the color across the entire surface of a fine art painting, to provide a definitive record of the state of an object at a given time. Against this image, future images recorded with equal precision could be compared to give an indication of any changes that had occurred. This would be extremely useful for

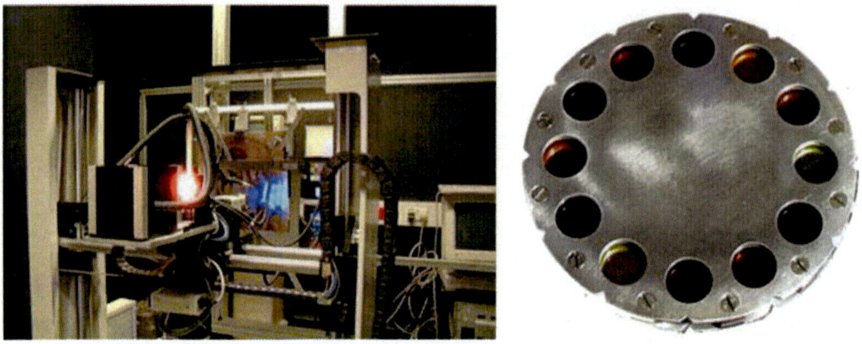

Fig. 14.9 (*left*) View of the VASARI system at the National Gallery of London in 2001. (*right*) Updated filters wheel of the VASARI system containing the 12 CRISATEL filters

studying the change of the paintings' colors over the years. In 1989, the only way to perform this study was by taking spot measures with a spectrometer. This technique was previously developed at the National Gallery in London [64].

At the time that the project started, the goals of VASARI were sufficiently difficult to inspire incredulity. It has to be remembered that the images produced by the VASARI system were very high resolution, despite dating from the early 1990s. The system aimed to use the capacity of one CD-ROM to store one image, with hard disks still at the 1-GB stage. The 2,000 paintings in the National Gallery would occupy around 1 TB and museums such as the Louvre have more than ten times that number. Despite this initial challenge a first VASARI system was built in the early 1990s. This system was designed to scan the paintings while vertical, since the shape of a canvas painting changes when it is laid flat.

The system was based on a $3{,}000 \times 2{,}300$ pixel monochrome camera, the ProgRes 3,000, designed at the technical university of Munich and marketed by Kontron, [38]. It used the microscanning principle, which is based on the idea of sensor displacement to improve the number of pixels. The camera employed a smaller CCD-sensor (typically containing 580×512 sensor elements) which was displaced by very small increments to get several different images. These single images were later interleaved to produce one $3{,}000 \times 2{,}300$ image. The camera was used in combination with a set of seven broadband filters (50 nm band) covering the 400–700 nm spectral range. The filters were attached to the lighting system: fiber-optic guides passed light through a single filter and then illuminate a small patch of the painting. These filters were exchanged using a wheel. This kind of approach is also modeled by (14.2) and presents the same theoretical properties. In this case, this had the additional important advantage of exposing the painting to less light during scanning. Finally, both camera and lighting unit were mounted on a computer-controlled positioning system, allowing the scanning of paintings of up to 1.5×1.5 m in size [58]. Two VASARI systems were originally operational: at the National Gallery of London and at the Doerner Institute in Munich. A photograph of the system at the National Gallery is shown in the left side of Fig. 14.9.

The VASARI system was basically mosaic based. The individual 3,000 × 2,300 pixels sub-images acquired at each displacement of the mechanical system were assembled to obtain image sizes up to 20,000 × 20,000 pixels. The camera's field of view covered an area on the painting of about 17 × 13 cm, giving a resolution of approximately 18 pixel/mm. Scanning took around three hours for a 1 × 1-m object. The project developed their own software and image format for the mosaicing, treatment, and storage of the images, it was called VIPS [16].

Even if initially developed for the monitoring of color changes in paintings over time, the VASARI system was very successfully used for documentation and archiving purposes [59]. However, the cumbersome mechanical system and the mosaicing approach lack of portability, and acquisition was a time-consuming procedure. Despite this, the VASARI system remained, for years, unrivalled in term of resolution, although in the late 1990s most of the digitizing work of paintings was carried out with the faster and even higher resolution MARC systems [10, 17]. However, and as already stated, the MARC system was not a spectral-based system.

14.6.2 CRISATEL

The CRISATEL European project started in 2001 and finished in 2004. Its goals were the spectral analysis and the virtual removal of varnish of art painting masterpieces. It is interesting to compare these goals with the ones of VASARI because the term "spectral analysis" was openly stated in comparison with "color" that was mainly used in VASARI. Indeed, the CRISATEL project was launched in order to treat paintings spectrally. In this context, the creation of a camera was an effort to build an image spectrometer that could finely acquire a spectral reflectance function at each pixel of the image.

A multispectral digital camera was built on a CCD, a 12,000 pixel linear array. This linear array was mounted vertically and mechanically displaced by a stepper motor. The system was able to scan up to 30,000 horizontal positions. This means that images up to 12,000 × 30,000 pixels could potentially be generated. For practical reasons concerning the format of the final images, the size was limited to 12,000 × 20,000 pixels. The camera was fitted with a system that automatically positions a set of 13 interference filters, ten filters covering the visible spectrum, and the other three covering the near infrared. The sensor being a linear array, the interference filters were cut in linear shape too. This is the reason for the shape of the filter-exchange mechanism presented in the left image of Fig. 14.10. In this mechanism, there also was an extra position without filter allowing panchromatic acquisitions. The system was built by *Lumière Technologie* (Paris, France) and its "skeleton" can be seen in the left image of Fig. 14.11. The signal processing methods and calibration were designed at the formerly called *École Nationale Supérieure des Télécommunications*, Paris France, [54].

The CRISATEL apparatus was intended to be used at the Louvre Museum (Paris, France) and included a camera and a dedicated lighting system. The lighting system

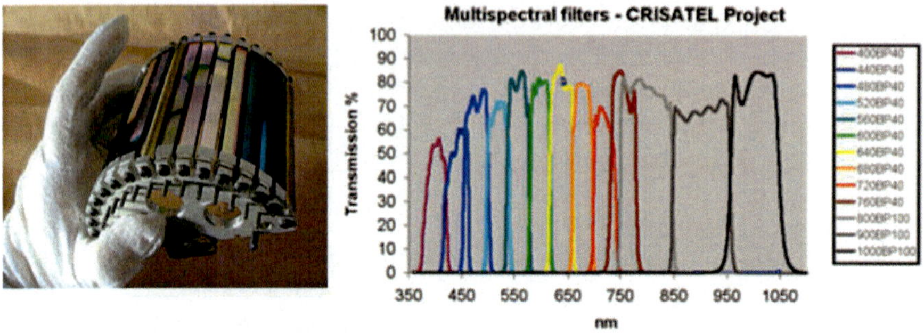

Fig. 14.10 (*left*) Image of the CRISATEL system mounted on the exchange system. (*right*) Spectral transmittances of the CRISTEL filters

Fig. 14.11 CRISATEL camera (*left*) and its experimental configuration (*right*)

was composed of two elliptical projectors. In the right side of Fig. 14.11, an image of a experimental setup using the CRISATEL system is shown where: (a) the optical axis of the camera should be perpendicular to the painting surface; (b) the two elliptical projectors of light are usually positioned at left and right sides of the camera and closer to the painting. Both projectors rotated synchronously with the CCD displacement and their projected light scanned the surface of the painting. This procedure intended to minimize the quantity of radiant energy received by the surface of the painting. This received energy is controlled because paintings in museums have a "per-year maximum" that must not be reached.

The CRISATEL scan time usually took about 3 min per filter (around 40 min per painting), if the exposure time for each line was set to 10 ms. A typical painting size was 1×1.5 m. Since each pixel delivers a 12-bit value the file size results in around 9.4 GB (uncompressed, coded on 16-bit words). A big advantage of this system was that could be transported to acquire the images, while keeping a high resolution both spatially and spectrally. This compared favorably with mosaic systems like

VASARI that are static. Moving a masterpiece painting can indeed be an extremely complicated bureaucratic issue when working in museological environments.

The National Gallery of London was also a member of the CRISATEL project. It was decided to update VASARI. This was done by a grayscale 12 bits cooled camera, and a filter wheel containing the same filters as the system built in Paris. This filter wheel can be seen in the right image of Fig. 14.9. Halogen lamps were used on the acquisition of the images, [39].

The CRISATEL system was finally used to acquire high-quality spectral images of paintings. As an example, the Mona Lisa by Leonardo da Vinci was digitized in October 2004, [55].

14.6.3 Filter-Based Sequential Systems

The equipment developed in the above-presented VASARI and CRISATEL projects are not the only sequential cameras that have been used for scanning paintings. In this section, some other examples are given to illustrate the variety of applied techniques. In fact, the technologies introduced in Sect. 14.3 all have been used for imaging paintings. This includes CFA-based cameras, which are still used in museums in daily digitizing routines.

Probably most quality art paintings spectral scanners found in museums use interferential filters mounted in some kind of mechanical device. VASARI or CRISATEL belong to this category of scanners but more modern designs are also based on these principles. For instance, [50] recently used a set of 15 interferometer filters for scanning paintings in the visible and near-infrared domain. They also performed ultraviolet visibly induced fluorescence. Of course, numerous examples of older systems exist, for instance, [5] that in 1998 obtained a twenty-nine-band image (visible and infrared) of the Holy Trinity Predellaby of Luca Signorelli, which was displayed at the *Uffizi Gallery* in Florence.

Before continuing the presentation of other multispectral systems, it is interesting to clarify a point concerning the number of channels. Normally, the more channels are used in an acquisition system, the smaller their bandwith. Typical broadband filters used for the scanning of paintings have 50 nm full width half maximum (FWHM) while 10 nm filters are considered narrow band. Broadband filters allow shorter sensor integration time and produce less noise. It is not to be forgotten that 50 or 10 nm are not just arbitrary numbers. Spectral reflectance functions are smooth and band limited; this fact has been extensively demonstrated by the study of sample reflectances of diverse materials, such as pigments in the case of paintings. See [48] for an example of this kind of study. It is generally accepted that a spectral reflectance sampled at 10 nm intervals is enough for most purposes.

When dealing with a filter-based system, the filters are not necessarily attached to the camera but can be, instead, attached to the lighting system. The VASARI system presented in Sect. 14.6.1 is an example. As we know, this kind of approach is also modeled by (14.2). In this context, the works of [3] go a step further by designing an optical monochromator in combination with a grayscale CCD. This approach was able to acquire bands of 5 nm FWHM, with 3 nm tuning step, in the spectral range 380–1,000 nm. Another interesting example of this approach is the digital acquisition, in August 2007, of the Archimedes palimpsest [21] (http://www.archimedespalimpsest.org), where a set of light-emitting diodes (LED) was used to create the spectral bands. These two examples present the advantage of eliminating the filters in the system, which straightforward reduces the costs associated to the filters that are often expensive. Moreover, it avoids several important problems in the calibration such as the misregistration of the channels (slightly bigger or smaller projections of the image on the sensor plane), which are caused by the optical filters. In any case, these approaches require a controlled narrow-band lighting system.

An example of LCTF-based (Liquid Crystal Tunable Filters already described in Sect. 14.3.4) scanning of paintings is found in [32], where 16 narrowband channels were used for the scan of van Gogh's Self-portrait at the National Gallery of Art, Washington DC, USA. This was the starting point of a series of collaborations between researchers at RIT (Rochester Institute of Technology) and several American museums (extensive information can be found in www.art-si.org.) One of their developed acquisition systems illustrate well the variety of deployed techniques for scanning paintings. It is a hybrid system based in a CFA camera and two filters that are mounted on top of the objective by using a wheel. In theory, it provides six bands and has spectral reflectance reconstruction capabilities. This system is currently in use at the Museum of Modern Art in New York City, USA, details can be found in [8].

Another example of a Tunable Filter-based system can be found in [45]. In this case, the authors search for an affordable multispectral system that could be used for documentation purposes, mainly for creating virtual museums. In fact, this discussion is quite pertinent as most existing multispectral systems based on interferential filters are seen as state-of-the-art expensive equipment. Tunable filters are not only comparatively cheaper but also smaller and faster than a traditional filter wheel, which normally facilitates the acquisition set up. However, as we know, they suffer from low absolute transmittance, which increases acquisition times and augments image noise.

The systems already introduced in this section do not take into account the BRDF and are simply based in the model presented in Sect 14.4.1. Recently, some interest has been shown in the measurements of some aspects of the BDRF. In general, the position of the light is moved and a spectral image is taken. This is the case of [62] which proposes a technique for viewpoint and illumination-independent digital archiving of art paintings in which the painting surface is regarded as a 2-D rough surface with gloss and shading. They acquire images of a painting using a multiband imaging system with six spectral channels at different illumination directions. They use that to finally combine all the estimates for rendering the painting under arbitrary illumination and viewing conditions on oil paintings.

This kind of "moving-the-light" study is not generalized but becoming popular in museological environments. An interesting example is the approach of the CHI (Cultural Heritage Imaging) corporation that uses flashlight domes or simply hand flashes to acquire polynomial texture maps (PTMs) [41], currently called reflectance transformation images (RTI). Basically, controlled varying lighting directions are used for acquiring stacks of images using a CFA color camera, thus these images are used in a dedicated viewer to explore the surface of the object by manually changing the light incident angle. In our knowledge, in this approach the CFA camera has not still been substituted by multiband systems, but several researchers already expressed this interest that could lead to a partial representation of the paintings surface BDRF.

14.6.4 Hyper-Spectral Scan of Pictorial Surfaces

Point spectrometers taking 1 nm sampled spot measures on the surface of art paintings have been used for many years, see, for instance, [64], but the interest of acquiring whole images of densely sampled spectral reflectances is a relatively new trend in museums. Although not very common, these systems currently exist. Some researchers consider them as candidates for being one of the scanning technologies of the future. In this section, this approach is described by the use of a few existing examples.

One straightforward approach is to use point-based spectrometers that are physically translated to scan the complete surface of the painting. This is actually performed by [14] who used what it is called *Fiber Optics Reflectance Spectroscopy*. In this case, the collecting optics gathers the radiation scattered from the scanned point on the painting and focuses it on the end of a multimode optical fiber that carries the light to the sensitive surface of the detector. The detection system is made of a 32-photomultiplier array, each element being filtered to select different 10 nm-wide (FWHM) bands in the 380–800 nm spectral range. This system is actually based on 32 filters but could also be, for instance, in a dispersive device and presents a potential high-spectral resolution. In general, the fact that the measuring device is physically translated makes acquisition times extremely long. Indeed, a megapixel scan of a painting with an instrument requiring a 250 ms dwell time per point, would take approximately 4,000 min. An imaging spectrometer provides a great improvement over such a point-based system. This is related to the number of simultaneously acquired points. For example, a scanning imaging spectrometer having 1,024 pixels across the slit reduces the scan time to approximately 4 min for the same dwell time.

In [12], an imaging spectrometer based on a transmission grating is applied in the case study on the Lansdowne version of the *Madonna dei fusi*. This system was realized by means of a hyper-spectral scanner assembled at the "Nello Carrara" Istituto di Fisica Applicata. The characteristics of the scanner were: 0.1 mm spatial sampling over a $1 \times 1\,\mathrm{m}^2$ surface and $\sim$1nm spectral sampling in the wavelength range from 400 to 900 nm. Antonioli et al. [1] is another example of systems based on a transmission grating (Imspector V8 manufactured by Specim, Finland) that disperses light from each line of the spectrum on the sensitive surface of the detector,

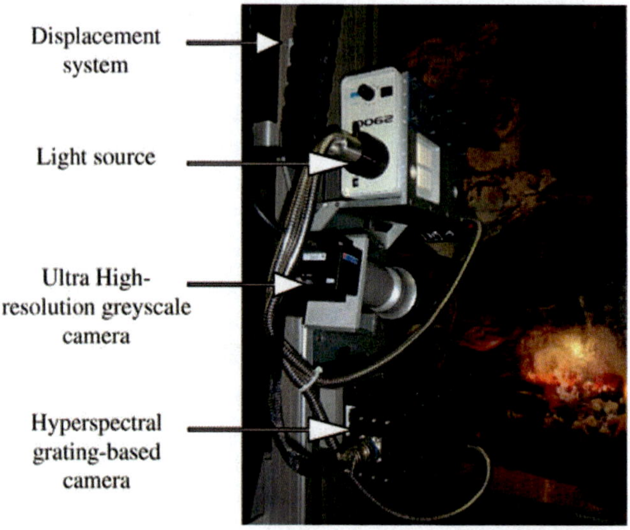

Fig. 14.12 Transmission grating-based image spectrometer of C2RMF scanning a degraded painting at the Louvre Museum (Paris, France)

in this case a CCD. The reflectance is scanned in the 400–780 nm spectral range with a spectral resolution of about 2 nm. A fundamental fact of such systems, already seen in Sect. 14.3.5, is that the presence of the slit converts a matrix camera in a line camera so, to capture the image of a painted surface, a displacement-motorized structure is necessary. Logically, the motor should displace the acquisition system in the spatial direction not covered by the matrix sensor. However, paintings can be much bigger than the spatial range covering a single translation and normally an XY displacement-motorized structure is used. This necessitates a post-processing mosaic reconstruction. Recently [49] proposed a grating-based system (Imspector V10 by Specim) for the scanning of large-sized pictorial surfaces such as frescoed halls or great paintings. Their approach is based in a rotating displacement system instead of the usual XY motorized structure.

Another example of densely sampled image spectrometry is the recent capture, at the *National Gallery of Washington*, of visible and infrared images of Picasso's *Harlequin Musician* having 260 bands (441–1680 nm) and processed using convex geometry algorithms, [19]. Finally, a transmission grating-based scanner was recently acquired by the C2RMF (*Centre de Restauration des Musees de France*) in January 2011 and deployed in the restoration department of the Louvre Museum, Paris. This equipment is based on a HySpex VNIR-1600 acquisition system combined with a large 1×1.5 m automated horizontal and vertical scanning stage and fiber-optic light sources. A photograph of this system is shown in Fig. 14.12. It scans at a resolution of 60 μm (15 pixels/mm) and at up to 200 spectral bands in the visible and near infrared region (400–1,000 nm). The system will be used to obtain high-resolution and high-fidelity spectral and imaging data on paintings for archiving and to provide help to restorers. Interested readers can refer to [51].

14.6.5 Other Uses of Spectral Reflectance

In this section, we briefly describe some important uses of spectral images of paintings that goes beyond color high fidelity.

14.6.5.1 Digital Archiving

As already shown, a principal motivation behind the original VASARI project, [42], was to provide a definitive record of the state of an object at a given time. The concept of creating digital archives of paintings for study and dissemination has been for a long time attached to spectral images.

14.6.5.2 Monitoring of Degradation

Once a digital archive of spectral images exists, future acquired images recorded with equal precision could be compared to give an indication of any changes that had occurred.

14.6.5.3 UnderDrawings

Underdrawings are drawings done on a painting ground before paint is applied. They are fundamental for the study of *pentimenti*: an alteration in a painting evidenced by traces of previous work, showing that the artist has changed his mind as to the composition during the process of painting. The underdrawings are usually hidden by covering pigment layers and therefore invisible to the observer in the visible light spectrum. Classically *infrared reflectograms* [2] are used to study pentimenti. They are obtained with infrared sensors sensitive from 1,000 to 2,200 nm. This range is often called the *fingerprint* region. This is the reason why numerous existing acquisition systems for art paintings present infrared channels. Normally they use the fact that CCD and CMOS sensors are sensitive in the near infrared and thus create near-infrared channels. In commercial cameras a cut-off filter is systematically added on top of the sensor to avoid its near-infrared response. As an example of *pentimenti*, Fig. 14.13 shows an image of the Mona Lisa taken by the first infrared channel of the CRISATEL project. We can compare this image with a color rendering, shown on thc left panel. Indeed, we see that under the hands of the Mona Lisa there is an older contour of the fingers in a different position relative to the hand.

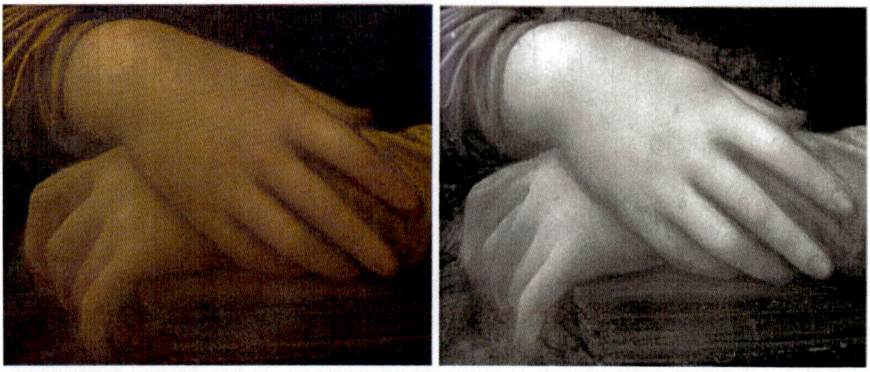

Fig. 14.13 Detail of the hands of the Mona Lisa. (*left*) color projection from the reconstructed spectral reflectance curves, (*right*) a near infrared band where we can observe on the *bottom-left* part of the image that the position of two fingers have been modified

14.6.5.4 Pigment Identification

In 1998, [5] applied principal component analysis (PCA) to spectral images of paintings. This aimed to reduce their dimensionality in more meaningful sets and to facilitate their interpretation. In the processed images, materials were identified and regions in the painting with similar spectral signatures were mapped. This application is currently very extended and some already cited references such as [12, 50] or [19] are mostly interested in pigment identification. Moreover, it can be said with no hesitation that all group working with hyperspectral acquisition systems are basically interesting in the identification of pigments. In general, dictionaries of reference pigment's reflectances are used to classify the spectral images. As a pixel can contain mixtures of several (normally few) pigments, some assumptions about the mixture models are necessary, at this point the theory of Kubelka–Munk presented in Sect. 14.4.3 is one of the popular choices.

14.6.5.5 Virtual Restoration

In most paintings, light is the main cause of deterioration. Exposure to light causes color changes due to photooxidation or photoreduction of the painted layer. Photo-damaging is cumulative and irreversible. There is no known way of restoring colors once they have been altered by the process. Although transparent UV-absorbing varnishes can be used to prevent or slow down photo-damaging in oil paintings, they also become photooxidized and turn yellow, thus requiring periodic restoration. Unfortunately, restoration is not only costly, but can also be harmful, since each time a painting is restored, there is the risk of removing some of pigment along with the unwanted deteriorated varnish. Virtual restoration is then an interesting application that can assist restorers to take decisions. This problem was already treated, using trichromatic images where color changes are calculated by comparing deteriorated with non-deteriorated areas [47]. In spectral imaging the same strategy has been used, see, for instance, [55] where a simple virtual restoration of the Mona Lisa colors is presented. See also as an example [7] where paintings of Vincent Van Gogh and Georges Seurat are rejuvenated by used of Kubelka–Munk turbid-media theory (see Sect. 14.4.3).

14.6.6 Conclusion

This book chapter presents an introduction to image spectrometers exemplified by the application of fine-art paintings scanning. Certainly, image spectrometers are used in high-end color-acquisition systems, first, because of their ability to obtain high-fidelity color images. This is a straightforward consequence of the capture of spectral reflectance instead of color, which eliminates metamerism and most problems associated with equipment dependence. Furthermore, the spectral images can be used for other purposes that go beyond color reproduction. Indeed, we have seen some examples of advanced image processing problems for fine-art paintings: performing physical simulations based in the spectral reflectance curves can solve problems such as virtual restoration or pigment identification.

This chapter was written to be self-contained, in the sense that the last part, which is a description of some existing systems, should be understandable without further reading. This implies the presentation of the basic technological and signal-processing aspects involved in the design of image spectrometers. For this, I decided to first present the technological aspects necessary to understand a camera as a measuring tool. Thus, CFA-based cameras, Foveon-X, multi-sensors, sequential acquisition systems, and dispersing devices were presented to give the reader a grasp of how these different technologies are used to generate a color image, and in which extent this image is expected to be high-fidelity. Once the basic technological aspects of color acquisition presented, I introduced the simplest mathematical models of light measurement and light-matter interaction. I hope these models will help the reader understand the difficulties associated with the acquisition of realistic color images. Knowing that the capture depends on the viewing and illumination directions, it is important to understand the explosion of data size performed if a realistic image is to be acquired. Moreover, it helps understanding limitations on the current technology only using one camera point of view and, sometimes, even unknown light sources. Furthermore, the interaction of light and matter does not only imply reflectance but also light propagation under the objects surface. This point is important in many applications willing to understand what is happening in the objects surface. Once models present, the so-called spectral reflectance reconstruction problem is introduced. This problem is important because its resolution transforms a multi-wideband acquisition system in an image spectrometer.

Concerning applications, I tried to present a general view of the evolution of image spectrometers in the field of fine-art paintings scanning. For this, I started by describing some historical and important systems built in European Union projects, such as the pioneering VASARI or its successor CRISATEL. Both being sequential and filter-based systems, other sequential systems were presented, taking care in choosing different technologies that show how a large variety of designs have been applied. Moreover, I included a section about hyperspectral systems based in dispersing devices. Even not numerous and currently expensive, these systems are worth to be known. They appeared recently in the scanning of art paintings and some researchers consider them as the new high-end acquisition equipment. To finalize the applications, some examples of specific problems such as underdrawings, virtual restoration or pigment identification are briefly described. I hope they will help the reader better understand the potential applications of the presented systems.

Finally, I hope this chapter will be useful for the reader interested in color acquisition systems viewed as physical measuring tools. This chapter was conceived as an introduction but extensive bibliography has been included in order to help the reader further navigate in this subject.

Acknowledgments I would like to thank Ruven Pillay for providing information on the VASARI project and on the C2RMF hyperspectral imaging system; as well as for having corrected and proof-read parts of the manuscript. Thanks also to Morwena Joly for the photograph of the transmission grating-based scanner recently acquired by the *Centre de Restauration des Musees de France*. I also extend my sincere thanks to: the *Département des peintures* of *Musée du Louvre*, for permission to use the images of the Mona Lisa; *Lumière Technologie* for the images of the CRISATEL camera and its filters; and Kirk Martinez for making available the images of the VASARI project.

References

1. Antonioli G, Fermi F, Oleari C, Riverberi R (2004) Spectrophotometric scanner for imaging of paintings and other works of art. In: Proceedings of CGIV, Aachen, Germany, 219–224
2. Asperen de Boer JRJVan (1968) Infrared reflectography: a method for the examination of paintings. Appl Optic 7:1711–1714
3. Balas C, Papadakis V, Papadakis N, Papadakis A, Vazgiouraki E, and Themelis G (2003) A novel hyper-spectral imaging apparatus for the non-destructive analysis of objects of artistic and historic value. J Cult Herit 4(1):330–337
4. Bayer BE (1976) Color imaging array. US Patent 3,971,065
5. Baronti S, Casini A, Lotti F and Porcinai S (1998) Multispectral imaging system for the mapping of pigments in works of art by use of principal component analysis. Appl Optic 37:1299–309
6. Berns RS (2001) The science of digitizing paintings for color-accurate image archives. J Imag Sci Tech 45(4):305–325
7. Berns RS (2005) Rejuvenating the appearance of cultural heritage using color and imaging science techniques. In: Proceedings of the 10th Congress of the International Colour Association, 10th Congress of the International Colour Association, AIC Colour 05, Granada, Spain, 369–375

8. Berns RS, Taplin LA, Urban P, Zhao Y (2008) Spectral color reproduction of paintings. In: Proceedings CGIV 2008/MCS, 484–488
9. Burns PD (1997) Analysis of image noise in multitraitement color acquisition. Ph.D. Dissertation, Center for Imaging Science, Rochester Institute of Technology, Rochester, NY
10. Burmester A, Raffelt L, Robinson G, and Wagini S (1996) The MARC project: from analogue to digital reproduction. In: Burmester A, Raffelt L, Renger K, Robinson G and Wagini S (eds) Flämische Barockmalerei: Meisterwerke der alten Pinakothek München. Flemish Baroque painting: masterpieces of the Alte Pinakothek München. Hirmer Verlag, Munich, pp 19–26
11. Brauers J, Schulte N, and Aach T (2008) Multispectral filter-wheel cameras: geometric distortion model and compensation algorithms, IEEE Trans Image Process 17(12):2368-2380
12. Casini A, Bacci M, Cucci C, Lotti F, Porcinai S, Picollo M, Radicati B, Poggesi M and Stefani L (2005) Fiber optic reflectance spectroscopy and hyper-spectral image spectroscopy: two integrated techniques for the study of the Madonna dei Fusi. In: Proceedings of SPIE 5857, 58570M doi:10.1117/12.611500. http://www.ifac.cnr.it/webcubview/WebScannerUK.htm
13. Chang IC (1981) Acousto-optic tunable filters. Opt Eng 20:824–829
14. Carcagni P, Patria AD, Fontana R, Grecob M, Mastroiannib M, Materazzib M, Pampalonib E and Pezzatib L (2007) Multispectral imaging of paintings by optical scanning. Optic Laser Eng 45(3):360–367
15. Clarke FJJ and Parry DJ (1985) Helmholtz reciprocity: its validity and application to reflectometry. Lighting Research and Technology 17(1):1–11
16. Cupitt J and Martinez K (1996) VIPS: an image processing system for large images. In: Proceedings of IS&T/SPIE Symp. Electronic imaging: science and technology, very high resolution and quality imaging, vol 2663, pp 19–28
17. Cupitt J, Martinez K, and Saunders D (1996) A methodology for art reproduction in colour: the MARC project. Comput Hist Art 6(2):1–19
18. Dillon PLP, Lewis DM, Kaspar FG (1978) Color imaging system using a single CCD area array. IEEE J Solid State Circ 13(1):28–33
19. Delaney JK, Zeibel JG, Thoury M, Littleton R, Palmer M, Morales KM, de la Rie ER, Hoenigswald A (2010) Visible and infrared imaging spectroscopy of picasso's harlequin musician: mapping and identification of artist materials in situ. Appl Spectros 64(6):158A-174A and 563–689
20. Duncan DR (1940) The color of pigment mixtures. Proc of the phys soc 52:390
21. Easton RL, Noel W (2010) Infinite possibilities: ten years of study of the archimedes palimpsest. Proc Am Phil Soc 154(1):50–76
22. Farrell JE, Wandell BA (1993) Scanner linearity. Journal of electronic imaging and color 3:147–161
23. Fischer C and Kakoulli I (2006) Multispectral and hyperspectral imaging technologies in conservation: current research and potential applications. Rev Conserv 7:3–12
24. Gat N (2000) Imaging spectroscopy using tunable filters: a review. In Proceedings of SPIE, 4056:50–64
25. Gilblom DL, Yoo SK, Ventura P (2003) Operation and performance of a color image sensor with layered photodiodes. Proc SPIE 5074:318–331
26. Hadamard J (1902) Sur les problèmes aux dérivées partielles et leur signification physique. Bulletin University, Princeton, pp 49–52
27. Hansen PC (1998) Rank-deficient and discrete ill-posed problems: numerical aspects of linear inversion. SIAM, Philadelphia
28. Haneishi H, Hasegawa T, Hosoi A, Yokoyama Y, Tsumura N, and Miyake Y (2000) System design for accurately estimating the spectral reflectance of art paintings. Appl Optic 39(35):6621–6632
29. Hardeberg JY, Schmitt F, Brettel H, Crettez J, and Maître H (1999) Multispectral image acquisition and simulation of illuminant changes. In: MacDonald LW and Luo MR (eds) Color imaging: vision and technology. Wiley, New York, pp 145–164
30. Herzog P and Hill B (2003) Multispectral imaging and its applications in the textile industry and related fields. In: Proceedings of PICS03: The Digital Photography Conf., pp 258–263

31. Hubel PM, Liu J, Guttosch RJ (2004) Spatial frequency response of color image sensors: Bayer color filters and Foveon X3. Proceedings SPIE 5301:402–407
32. Imai FH Rosen MR Berns RS (2001) Multi-spectral imaging of a van Gogh's self-portrait at the National Gallery of Art, Washington, D.C. In: Proceedings of IS&T Pics Conference, IS&T, PICS 2001: image processing, image quality, image capture systems conference, Rochester, NY, USA, pp 185–189
33. Imai FH, Taplin LA, and Day EA (2002) Comparison of the accuracy of various transformations from multi-band image to spectral reflectance. Tech Rep, Rochester Institute of Technology, Rochester, NY
34. Kubelka P and Munk F (1931) ."Ein beitrag zur optik der farbanstriche", Zurich Tech., Physik 12:pp. 543.
35. Kubelka P (1948) New contributions to the optics of intensely light-scattering materials, part I. J Opt Soc Am 38:448–460
36. Keusen T (1996) Multispectral color system with an encoding format compatible with the conventional tristimulus model. J Imag Sci Tech 40(6):510–515
37. König F and Praefcke W (1999) A multispectral scanner. Chapter in MacDonald and Luo. pp 129–144
38. Lenz R (1990) Calibration of a color CCD camera with 3000x2300 picture elements. In: Proceeding of Close Range Photogrammetry Meets Machine Vision, Zurich Switzerland, 3–7 Sept 1990. Proc SPIE, 1395:104–111 ISBN: 0–8194–0441–1
39. Liang H, Saunders D, and Cupitt J (2005) A new multispectral imaging system for examining paintings. J Imag Sci Tech 49(6):551–562
40. Maître H, Schmitt F, Crettez J-P, Wu Y and Hardeberg JY (1996) Spectrophotometric image analysis of fine art paintings. In: Proc. of the Fourth Color Imaging Conference, Scottsdale, Arizona, pp 50–53
41. Malzbender T, Gelb D, Wolters H (2001) Polynomial texture maps. In: SIGGRAPH: Proceedings of the 28th annual conference on Computer graphics and interactive techniques, ACM press, New york, NY, USA, pp 519–528
42. Martinez K, Cupitt J, Saunders D, and Pillay R (2002) Ten years of art imaging research. Proc IEEE 90:28–41
43. Miao L and Qi HR (2006) The design and evaluation of a generic method for generating mosaicked multispectral filter arrays. IEEE Trans Image Process 15(9):2780–2791
44. Nicodemus FE, Richmond JC, Hsia JJ, Ginsberg IW, Limperis T (1977) Geometrical considerations and nomenclature for reflectance. US Department of Commerce, National Bureau of Standards
45. Novati G, Pellegri P, and Schettini R (2005) An affordable multispectral imaging system for the digital museum. Int J Dig Lib 5(3): 167–178
46. Okano Y (1995) Electronic digital still camera using 3-CCD image sensors. In: Proceedings of IS&T's 48th Annual Conf., 428–432
47. Pappas M and Pitas I (2000) Digital color restoration of old paintings. Trans Image Process (2):291–294
48. Parkkinen JPS, Hallikainen J, and Jaaskelainen T (1989) Characteristic spectra of Munsell color. J Opt Soc Am 6:318–322
49. Paviotti A, Ratti F, Poletto L, and Cortelazzo GM (2009) Multispectral acquisition of large-sized pictorial surfaces. EURASIP Int J Image Video Process Article ID 793756, 17
50. Pelagotti A, Mastio AD, Rosa AD, and Piva A (2008) Multispectral imaging of paintings. IEEE Signal Processing Mag 25(4):27–36
51. Pillay R (2011) Hyperspectral imaging of paintings, web article accessed on October http://merovingio.c2rmf.cnrs.fr/technologies/?q=hyperspectral
52. Pratt WK and Mancill CE (1976) Spectral estimation techniques for the spectral calibration of a color image scanner. Appl Opt 15(1):73–75
53. Ribés A, and Schmitt F (2003) A fully automatic method for the reconstruction of spectral reflectance curves by using mixture density networks. Pattern Recogn Lett 24(11):1691–1701

54. Ribés A, Schmitt F, Pillay R, and Lahanier C (2005) Calibration, spectral reconstruction and illuminant simulation for CRISATEL: an art paint multispectral acquisition system. J Imaging Sci Tech 49(6):463–473
55. Ribés A, Pillay R, Schmitt F, and Lahanier C (2008) Studying that smile: a tutorial on multispectral imaging of paintings using the Mona Lisa as a case study. IEEE Signal Process Mag 25(4):14–26
56. Ribés A and Schmitt F (2008) Linear inverse problems in imaging. IEEE Signal Process Mag 25(4):84–99
57. Rush A, Hubel PM (2002) X3 sensor characteristics Technical Repport, Foveon, Santa Clara, CA
58. Saunders D, and Cupitt J (1993) Image processing at the National Gallery: the VASARI project. National Gallery Technical Bulletin 14:72–86
59. Saunders D (1998) High quality imaging at the National Gallery: origins, implementation and applications. Comput Humanit 31:153–167
60. Sharma G, Trussell HJ (1997) Digital color imaging. IEEE Trans on Image Process 6(7): 901–932
61. Shimano N (2006) Recovery of spectral reflectances of objects being imaged without prior knowledge. IEEE Trans Image Process 15:1848–1856
62. Tominaga S, Tanaka N (2008) Spectral image acquisition, analysis, and rendering for art paintings J Electron Imaging 17:043022
63. Wyszecki G and Stiles WS (2000) Color science: concepts and methods, quantitative data and formulae. John Wiley and Sons, 2nd edition
64. Wright W (1981) A mobile spectrophotometer for art conservation Color Res Appl 6:70–74
65. Zhao Y and Berns RS (2007) Image-based spectral reflectance reconstruction using the matrix R method. Color Res Appl 32:343–351. doi: 10.1002/col.20341

Chapter 15
Application of Spectral Imaging to Electronic Endoscopes

Yoichi Miyake

I have a dream, that my four little children will one day live in a nation where they will not be judged by the color of their skin but by the content of their character. I have a dream today

Martin Luther King, Jr

Abstract This chapter deals with image acquisition by CCD-based electronic endoscopes, developed from color film-based recording devices called gastrocameras. The quality of color images, particularly reproduced colors, is influenced significantly by the spectral characteristics of imaging devices, as well as illumination and visual environments. Thus, recording and reproduction of spectral information on the object rather than information on three primary colors (RGB) is required in electronic museums, digital archives, electronic commerce, telemedicine, and electronic endoscopy in which recording and reproduction of high-definition color images are necessary. We have been leading the world in developing five-band spectral cameras for digital archiving (Miyake, Analysis and evaluation of digital color images, 2000; Miyake, Manual of spectral image processing, 2006; Miyake and Yokoyama, Obtaining and reproduction of accurate color images based on human perception, pp 190–197, 1998). Spectral information includes information on all visible light from the object and may be used for new types of recording, measurement, and diagnosis that cannot be achieved with the three primary colors of RGB or CMY. This chapter outlines the principle of FICE (flexible spectral imaging color enhancement), a spectral endoscopic image processing that incorporates such spectral information for the first time.

Y. Miyake (✉)
Research Center for Frontier Medical Engineering, Chiba University, 133 Yayoi-cho, Inage-Ku 263–8522, Chiba, Japan
e-mail: miyake@faculty.chiba-u.jp

C. Fernandez-Maloigne (ed.), *Advanced Color Image Processing and Analysis*,
DOI 10.1007/978-1-4419-6190-7_15,

Keywords FICE Flexible spectral imaging color enhancement • Electronic endoscopy • Endoscope spectroscopy system • Spectral endoscope • Color reproduction theory • Spectrometer • Multiband camera • Wiener estimation • Spectral image enhancement

15.1 Introduction

Light we can perceive (visible light) consists of electromagnetic waves with a wavelength of 400–700 nm. The light may be separated with a prism or a diffraction grating into red to purple light as shown in Fig. 15.1.

When illuminated with visible light as described above, the object reflects some light, which is received by the L, M, or S cone in the retina that is sensitive to red (R), green (G), or blue (B), and then perceived as color in the cerebrum. Image input systems, such as CCD cameras and color films, use sensors or emulsions sensitive to RGB light to record the colors of the object.

Image reproduction is based on the trichromatic theory that is characterized by additive and subtractive color mixing of the primary colors of R, G, and B or cyan (C), magenta (M), and yellow (Y). Briefly, imaging systems, such as a television set, camera, printing machine, copier, and printer, produce color images by integrating spectra in terms of R, G, and B or C, M, and Y, each of which has a wide bandwidth, and mixing these elements. For example, the use of R, G, and B, each of which has eight bits or 256 levels of gray scale, will display 2^{24} colors ($256 \times 256 \times 256 = 16.7$ million). The theory for color display and measurement has been organized to establish the CIE-XYZ color system on the basis of the trichromatic theory. Uniform color spaces, such as $L^*a^*b^*$ and $L^*u^*v^*$, which have been developed from the color system, provide a basis for the development of a variety of imaging systems in the emerging era of multimedia. The development of endoscopes is not an exception, in which CCD-based electronic endoscopes have developed from color film-based recording devices called gastrocameras. However, the quality of color images, particularly reproduced colors, is influenced significantly by the spectral characteristics of imaging devices, as well as illumination and visual environments. Thus, recording and reproduction of spectral information on the object rather than

Fig. 15.1 Dispersion of visible light with a prism

information on three primary colors (RGB) is required in electronic museums, digital archives, electronic commerce, telemedicine, and electronic endoscopy in which recording and reproduction of high-definition color images are necessary. We have been leading the world in developing five-band spectral cameras for digital archiving [1,2,7]. Spectral information includes information on all visible light from the object and may be used for new types of recording, measurement, and diagnosis that cannot be achieved with the three primary colors of RGB or CMY. This chapter outlines the principle of FICE (flexible spectral imaging color enhancement), a spectral endoscopic image processing that incorporates such spectral information for the first time.

15.2 Color Reproduction Theory

Image recording and reproduction aim to accurately record and reproduce the three-dimensional structure and color of an object. Most commonly, however, an object with three-dimensional information is projected onto the two-dimensional plane for subsequent recording, transmission, display, and observation. For color information, three bands of R, G, and B have long been recorded as described above, rather than spectral reflectance. Specifically, colors are reproduced by additive color mixing of the three primary colors of R, G, and B or by subtractive color mixing of the three primary colors of C, M, and Y.

In general, the characteristics of an object can be expressed as the function $O(x, y, z, t, \lambda)$ of three-dimensional space (x, y, z), time (t), and wavelength (λ) of visible light (400–700 nm). More accurate description of object characteristics requires the measurement of the bidirectional reflectance distribution function (BRDF) of the object. For simplicity, however, this section disregards time, special coordinates, and angle of deviation and focuses on wavelength information of the

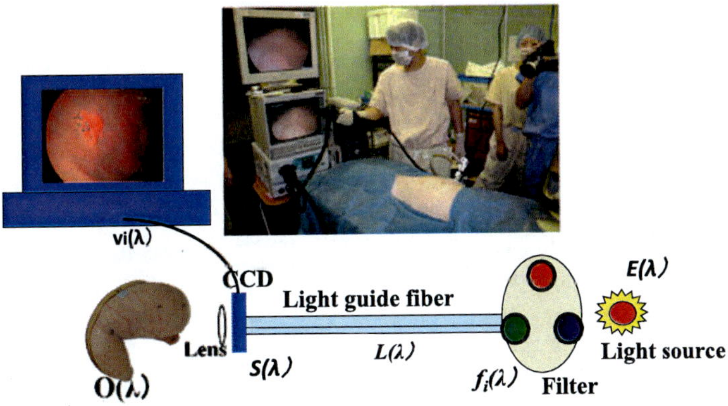

Fig. 15.2 Color reproduction process for electronic endoscopy

object $O(\lambda)$ to address color reproduction in the electronic endoscope as shown in Fig. 15.2. When an object (such as the gastric mucosa) with a spectral reflectance of $O(\lambda)$ is illuminated with a light source having a spectral emissivity of $E(\lambda)$ through a filter with a spectral transmittance of $f_i(\lambda)$ (i = R, G, B), and an image obtained through a lens and fiber with a spectral transmittance of $L(\lambda)$ is recorded with a CCD camera with a spectral sensitivity of $S(\lambda)$. Camera output V_i (i = R, G, B) can be expressed by (15.1) (for simplicity, noise is ignored).

$$V_i = \int_{400}^{700} O(\lambda)L(\lambda)f_i(\lambda)S(\lambda)E(\lambda)\mathrm{d}\lambda$$
$$i = R, G, B \tag{15.1}$$

Equation (15.1) can be expressed with a vector as follows:

$$\mathbf{v}_i = \mathbf{f}_i{}^t\mathbf{ELSO} = \mathbf{F}_i^t\mathbf{O} \tag{15.2}$$

$$\mathbf{F}_i^t = \mathbf{f}_i^t\mathbf{ELS} \tag{15.3}$$

Where F_i^t is the system's product, and t indicates transposition.

This means that the colors reproduced by the endoscope are determined after input of v into a display, such as a CRT and LCD, and the addition of the characteristics of the display and visual environment. When psychological factors, such as visual characteristics, are disregarded, the colors recorded and displayed with an electronic endoscope are determined by the spectral reflectance of the gastric mucosa (object) and the spectral characteristics of the light source for the illumination and imaging system. Thus, the spectral reflectance of the gastric mucosa allows the prediction of color reproduction by the endoscope.

In the 1980s, however, there was no report of direct measurement of the spectral reflectance of the gastric mucosa. Thus, we developed an endoscope spectroscopy system to quantitatively investigate color reproduction for endoscopy and measured, for the first time in the world, the spectral reflectance of the gastric mucosa at Toho University Ohashi Medical Center, Cancer Institute Hospital, and the National Kyoto Hospital [3, 4].

Figure 15.3 shows a block diagram and photograph of a spectral endoscope. This spectroscope consists of a light source, optical endoscope, spectroscope, and spectroscopic measurement system (optical multichannel analyzer, or OMA). The object is illuminated with light from the light source through the light guide. Through an image guide and half mirror, the reflected light is delivered partly to the camera and partly to the spectroscope. The luminous flux delivered to the spectroscope has a diameter of 0.24 mm and is presented as a round mark in the eyepiece field. When the distance between the endoscope tip and the object is 20 mm, the mark corresponds to a diameter of 4 mm on the object. A 1024-channel CCD line sensor is placed at the exit pupil of the spectroscope, and the output is transmitted to the PC for analysis. Wavelength calibration was performed with mercury spectrum and a standard white plate. The wavelength measured range from 400 to 700 nm when an infrared filter is removed from the endoscope.

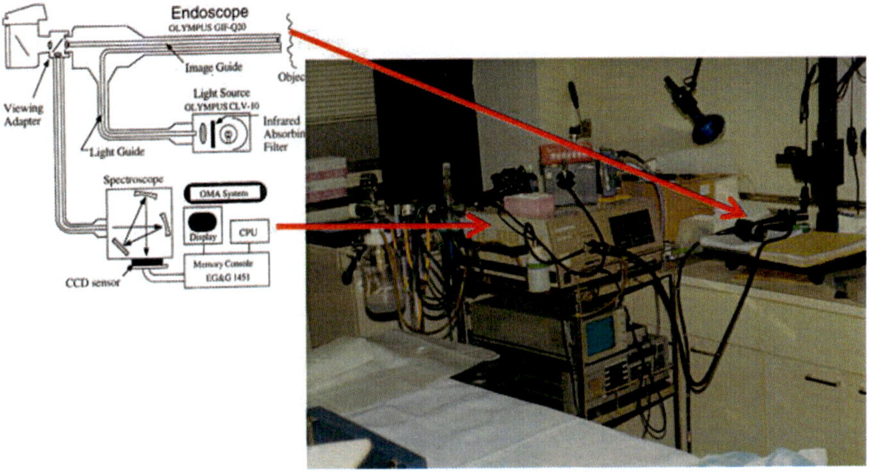

Fig. 15.3 Configuration and photograph of endoscope spectroscopy system

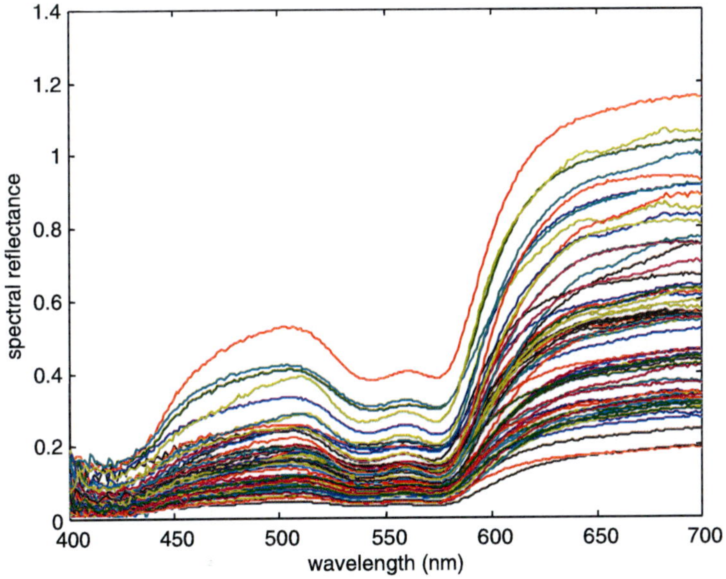

Fig. 15.4 Spectral reflectance of the colorectal mucosa (normal region)

Figure 15.4 shows an example of the spectral reflectance of normal, colorectal mucosa after denoising and other processing of measurements. As shown in (15.1), the measurement of $O(\lambda)$ allowed simulation of color reproduction. Initially, the spectral sensitivity of color films was optimized to be used for the endoscope.

However, the measurement of $O(\lambda)$ represented a single spot on the gastric mucosa. The measurement of the spectral reflectance at all coordinates of the object required huge amounts of time and costs and was not feasible with this spectroscope. Thus, an attempt was made to estimate the spectral reflectance of the gastric mucosa from the camera output.

15.3 Estimation of Spectral Reflectance

The spectral reflectance of an object may be estimated from the camera output by solving the integral equations (15.1) and (15.2). Compared with the camera output, however, the spectral reflectance generally has a greater number of dimensions. For example, the measurement of visible light with a wavelength of 400–700 nm at intervals of 5 nm is associated with 61 dimensions. Thus, it is necessary to solve an ill-posed equation in order to estimate 61 dimensions of spectral information from three-band data (RGB) in conventional endoscopy. This chapter does not detail the problem because a large body of literature is available, and I have also reported it elsewhere [1,2]. For example, an eigenvector obtained from the principal component analysis of spectral reflectance may be used for estimation as shown in (15.4),

$$\mathbf{o} = \sum_{i=1}^{n} \mathbf{a_i u_i} + \mathbf{m\bar{o}} \tag{15.4}$$

where $\mathbf{u}$ is an eigenvector obtained by the principal component analysis of the mucosal spectral reflectance, α is a coefficient calculated from the system spectral product, and $\mathbf{m}$ is the mean vector.

Figure 15.5 shows the eigenvectors of spectral reflectances of the colorectal mucosa and cumulative contribution. Figure 15.5 indicates that three principal component vectors allow good estimation of the spectral reflectance of the rectal mucosa. It was also found that the use of three principal components allowed estimation of the spectral reflectances of the gastric mucosa and skin [5, 6]. For example, when a comparison was made between 310 spectral reflectances estimated from three principal component vectors and those actually measured in the gastric mucosa, the maximum color difference was 9.14, the minimum color difference was 0.64, and the mean color difference was 2.66 as shown in Fig. 15.6.

These findings indicated that output of a three-channel camera allowed estimation of the spectral reflectance with satisfactory accuracy. When the system spectral product is not known, the Wiener estimation may be used to estimate the spectral reflectance of an object [7, 8]. This section briefly describes the estimation of the spectral reflectance by the Wiener estimation by (15.5)

$$\mathbf{O} = \mathbf{H}^{-1}\mathbf{V} \tag{15.5}$$

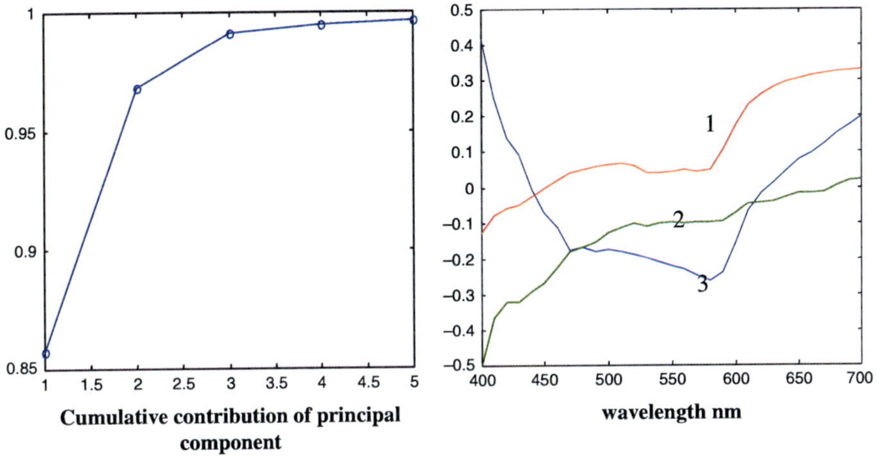

Fig. 15.5 Principal component analyses of the spectral reflectance of the colorectal mucosa

Fig. 15.6 Color difference between measured and estimated spectral reflectance of gastric mucous membrane

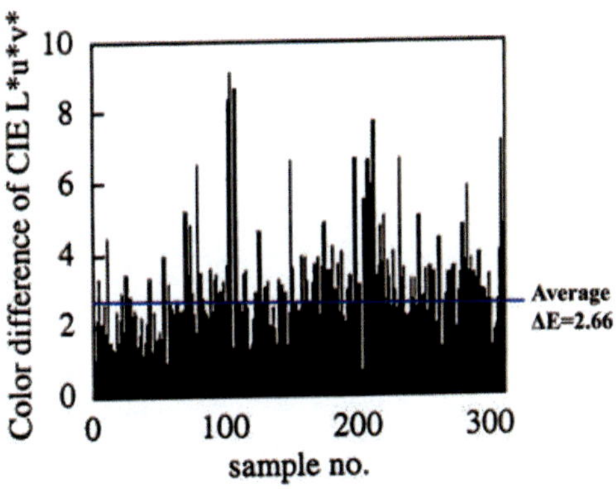

The pseudo-inverse matrix $\mathbf{H}^{-1}$ of the system matrix should be computed to obtain $\mathbf{o}$ from (15.2) For determination of the estimation matrix, an endoscope is used to capture sample color charts corresponding to spectral radiance $\mathbf{o}$ as shown in Fig. 15.7, and the camera output $\mathbf{v}$ should be measured. In this case, the estimate of spectral radiance of sample k can be expressed with the camera output as shown below.

According to the Wiener estimation method, the pseudo-inverse matrix that minimizes the error between the actual spectral radiance and the estimate for all sample data can be obtained.

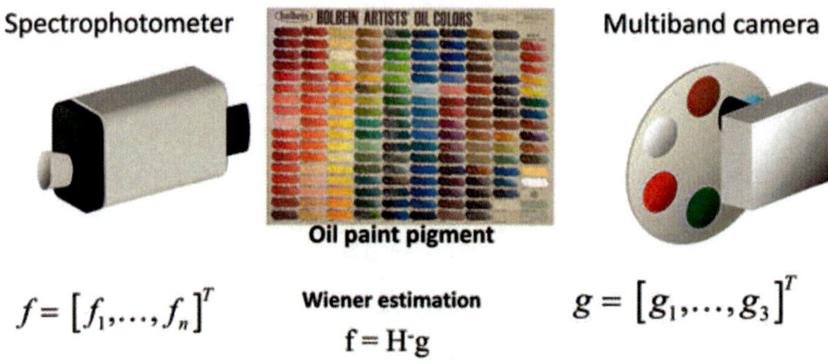

Fig. 15.7 Measurement of the spectral reflectance by the Wiener estimation

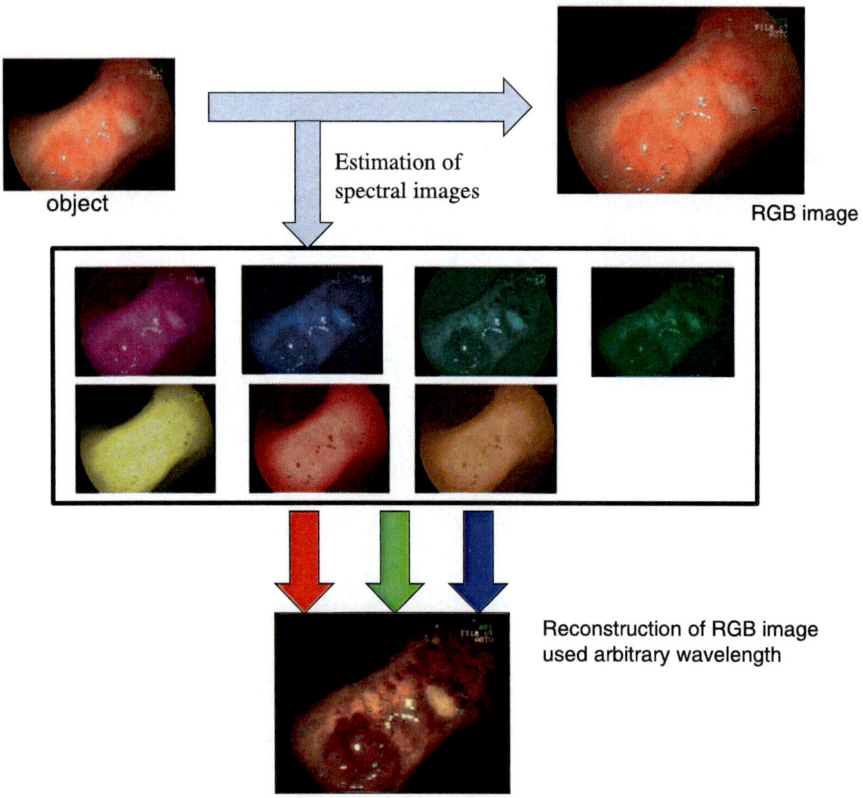

Fig. 15.8 Method for image construction using spectral estimation

15.4 Spectral Image

Figure 15.8 schematically shows the spectral estimation and image reconstruction based on the principle.

Figure 15.9 shows examples of spectral images at (a) 400 nm, (b) 450 nm, (c) 500 nm, (d) 550 nm, (e) 600 nm, (f) 650 nm, and (g) 700 nm estimated from and RGB image (h) of the gastric mucosa.

FICE [9, 10] has pre-calculated coefficients in a look-up table and estimates images at three wavelengths (λ_1, λ_2, λ_3), or spectral images, by using the following 3×3 matrix.

$$\begin{bmatrix} \lambda_1 \\ \lambda_2 \\ \lambda_3 \end{bmatrix} = \begin{bmatrix} k_{1r} & k_{1g} & k_{1b} \\ k_{2r} & k_{2g} & k_{2b} \\ k_{3r} & k_{3g} & k_{3b} \end{bmatrix} \begin{bmatrix} R \\ G \\ B \end{bmatrix} \tag{15.6}$$

Thus, FICE assigns estimated spectral images to RGB components in a display device and allows reproduction of color images at a given set of wavelengths in real time. Figure 15.10 shows a FICE block diagram.

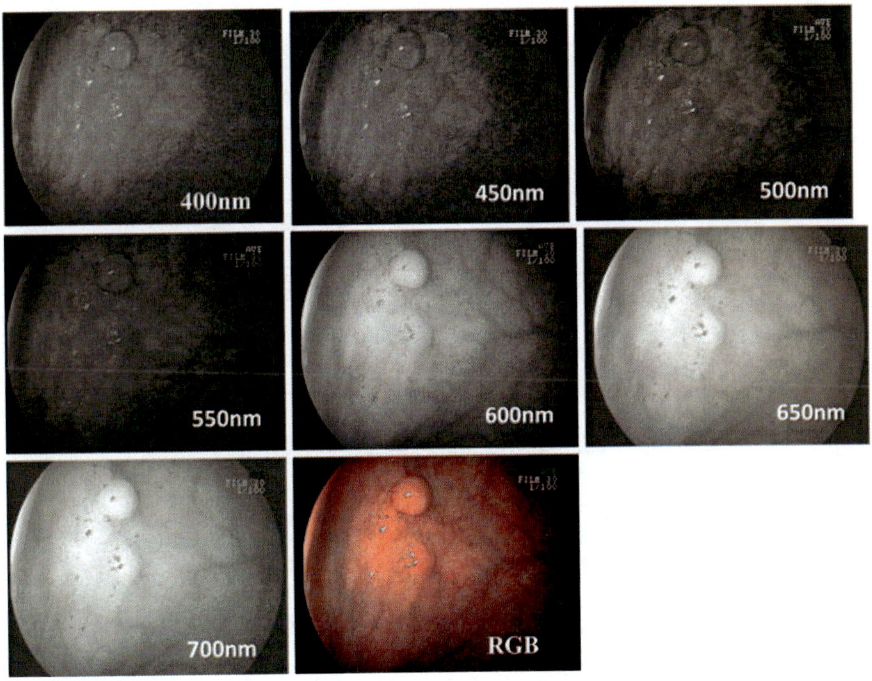

Fig. 15.9 Examples of spectral images at 400–700 nm estimated from an RGB image of the gastric mucosa

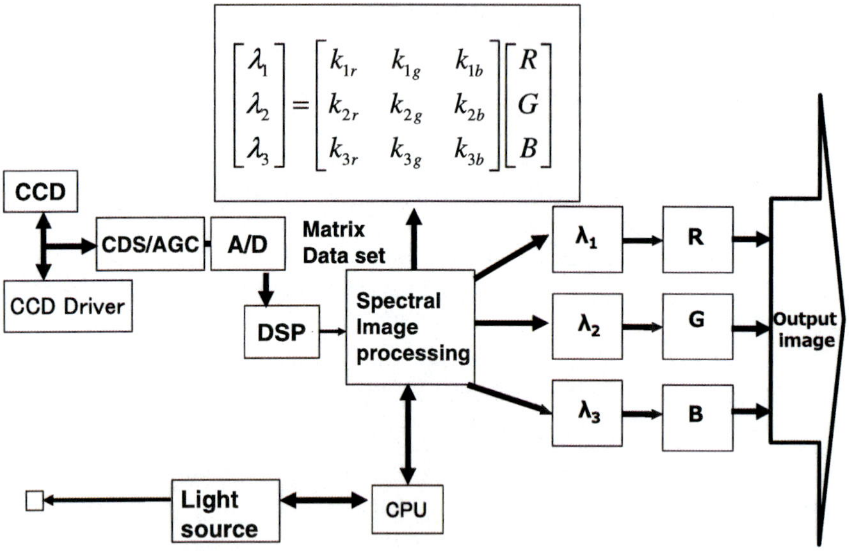

Fig. 15.10 Block diagram of the FICE

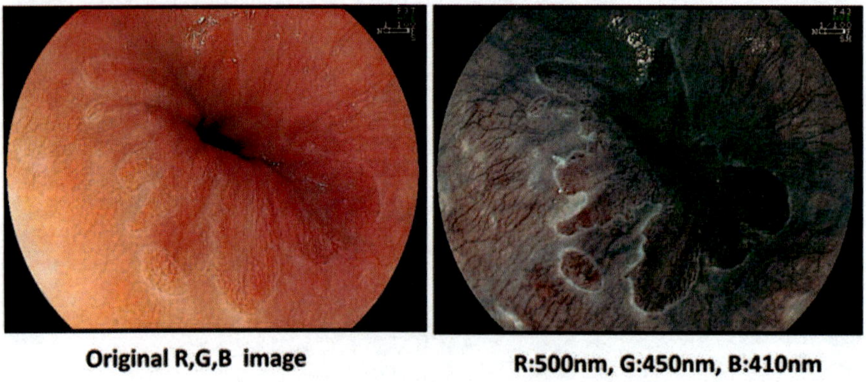

Fig. 15.11 Esophageal mucosa visualized by FICE (Images provided by Dr. T. Kouzu, Chiba University Hospital)

Figure 15.11 [10] shows an example of an endoscopic image of the esophagus taken with this endoscopy system. Figure 15.11a shows an image produced with conventional RGB data, and Fig. 15.11b shows an example of an image in which RGB components are replaced with spectral components (R, 500 nm; G, 450 nm; B, 410 nm). In Fig. 15.11b blood vessels and the contours of inflammatory tissue associated with reflux esophagitis are highlighted.

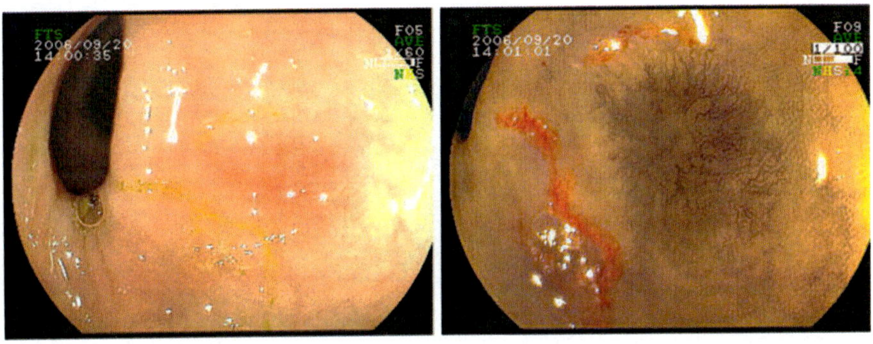

Original R:520nm, G:500nm, B:405nm

Fig. 15.12 Image of gullet (Images provided by Dr. H. Nakase, Kyoto University Hospital

Figure 15.12 shows images of the mucosa of the gullet. Figure 15.12a shows an image reproduced with conventional RGB data, and Fig. 15.12b shows an image reproduced with spectra (R, 520 nm; G, 500 nm; B, 405 nm).

Thus, the FICE endoscope produces images of an object with given wavelengths, thereby enhancing the appearance of mucosal tissue variations. Unlike processing with narrow-band optical filters, this system allows the combination of a huge number of observation wavelengths and rapid switching of the wavelengths using a keyboard. The system also allows switching between conventional and spectral images with the push of a button on the endoscope, providing the physician fingertip control, enabling simple and convenient enhancement of diagnostic procedures.

15.5 Summary

This chapter outlined the principle of FICE spectral image processing for endoscopy using the Fujifilm VP-4400 video processor (Fig. 15.13). FICE was commercialized by combining basic research, development of the endoscope spectroscopy system, measurement of the spectral reflectance of the gastrointestinal mucosa, principal component analysis of the spectral reflectance, and the Weiner estimation method. Ongoing development will realize even better systems in the future with more powerful capabilities and unleashing the full potential of spectral image enhancement.

Images contained in this atlas use the below convention to signify the application of FICE settings.

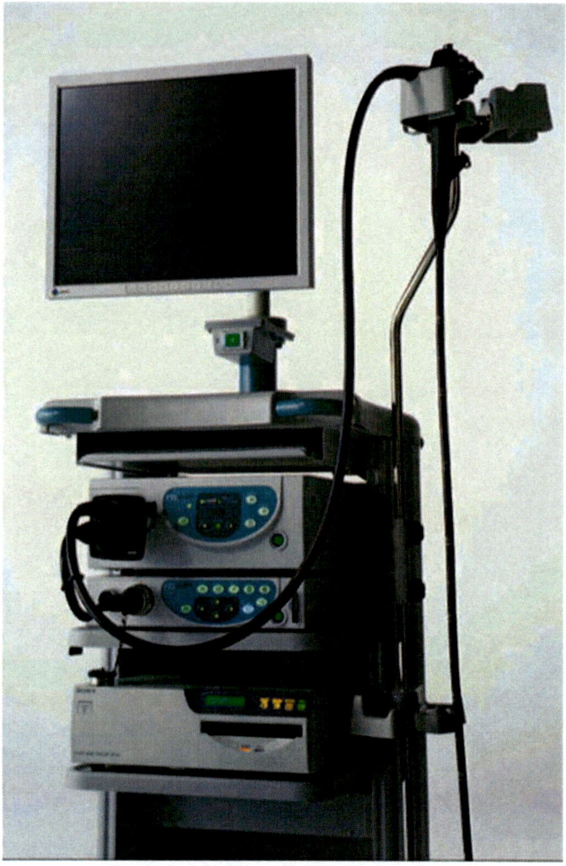

Fig. 15.13 Developed spectral endoscopes

References

1. Miyake Y (2000) Analysis and evaluation of digital color images. University of Tokyo Press, Tokyo
2. Miyake Y, editor (2006) Manual of spectral image processing. University of Tokyo Press, Tokyo
3. Miyake Y, Sekiya T, Kubo S, Hara T (1989) A new spectrophotometer for measuring the spectral reflectance of gastric mucous membrane. J Photogr Sci 37:134–138
4. Sekiya T, Miyake Y, Hara T (1990) Measurement of the spectral reflectance of gastric mucous membrane and color reproduction simulation for endoscopic images vol 736 Kyoto University Publications of the Research Institute for Mathematical Sciences, 101–130
5. Shiobara T, Haneishi H, Miyake Y (1995) Color correction for colorimetric color reproduction in an electronic endoscope. Optic Comm 114:57–63
6. Shiobara T, Zhou S, Haneishi H, Tsumura N, Miyake Y (1996) Improved color reproduction of electronic endoscopes. J Imag Sci Tech 40(6):494–501
7. Miyake Y, Yokoyama Y (1998) Obtaining and reproduction of accurate color images based on human perception. In: Proceedings of SPIE 3300, pp 190–197

8. Tsumura N, Tanaka T, Haneishi H, Miyake Y (1998) Optimal design of mosaic color electronic endoscopes. Optic Comm 145:27–32
9. Miyake Y, Kouzu T, Takeuchi S, Nakaguchi T, Tsumura N, Yamataka S (2005) Development of new electronic endoscopes using the spectral images of an internal organ. In: Proceedings of 13th CIC13 Scottsdale, pp 261–263
10. Miyake Y, Kouzu T, Yamataka S (2006) Development of a spectral endoscope. Image Lab 17:70–74

Index

C. Fernandez-Maloigne (ed.), *Advanced Color Image Processing and Analysis*,
DOI 10.1007/978-1-4419-6190-7,

J